Experimental Methods in Survey Research

WILEY SERIES IN SURVEY METHODOLOGY

A complete list of the titles in this series appears at the end of this volume.

Experimental Methods in Survey Research

Techniques that Combine Random Sampling with Random Assignment

Edited by

Paul J. Lavrakas
NORC, IL, US

Michael W. Traugott
University of Michigan, MI, US

Courtney Kennedy
Pew Foundation, DC, US

Allyson L. Holbrook
University of Illinois-Chicago, IL, US

Edith D. de Leeuw
University of Utrecht, NL, TC

Brady T. West
University of Michigan, MI, US

This edition first published 2019

Registered Office
John Wiley & Sons, Inc., 111 River Street, Hoboken, NJ 07030, USA

Editorial Office
111 River Street, Hoboken, NJ 07030, USA

For details of our global editorial offices, customer services, and more information about Wiley products visit us at www.wiley.com.

Wiley also publishes its books in a variety of electronic formats and by print-on-demand. Some content that appears in standard print versions of this book may not be available in other formats.

Library of Congress Cataloging-in-Publication Data

Names: Lavrakas, Paul J., editor.
Title: Experimental methods in survey research : techniques that combine random sampling with random assignment / edited by Paul J. Lavrakas, [and five others].
Description: New Jersey : Wiley, 2020. | Series: Wiley series in survey methodology | Includes bibliographical references and index. |
Identifiers: LCCN 2019015585 (print) | LCCN 2019021945 (ebook) | ISBN 9781119083757 (Adobe PDF) | ISBN 9781119083764 (ePub) | ISBN 9781119083740 (hardback)
Subjects: LCSH: Social surveys. | Social sciences–Research–Methodology. | Surveys–Evaluation. | BISAC: SOCIAL SCIENCE / Statistics.
Classification: LCC HM538 (ebook) | LCC HM538 .E97 2020 (print) | DDC 300.72/3–dc23
LC record available at https://lccn.loc.gov/2019015585

Cover design: Wiley
Cover image: © Angelina Babii/Shutterstock

Set in 10/12pt WarnockPro by SPi Global, Chennai, India

10 9 8 7 6 5 4 3 2 1

We dedicate this book with gratitude to the contributions made to the use of experimental design by Donald T. Campbell, Thomas D. Cook, Robert Boruch, Marilyn B. Brewer, Jeanne Converse, Stephen Fienberg, Stanley Presser, Howard Schuman, and Eleanor Singer, and to their influence on our own usage of survey-based experiments.

Contents

List of Contributors

Katrin Auspurg
Ludwig Maximilian University of Munich
Munich, Germany

Paul Beatty
Center for Behavioral Science Methods
U.S. Census Bureau
Washington, DC, USA

A. Bianchi
Department of Management, Economics and Quantitative Methods
University of Bergamo
via dei Caniana 2, 24127 Bergamo, Italy

S. Biffignandi
Department of Management, Economics and Quantitative Methods
University of Bergamo
via dei Caniana 2, 24127 Bergamo, Italy

Philip S. Brenner
Department of Sociology and Center for Survey Research
University of Massachusetts
100 Morrissey Blvd., Boston, MA 02125, USA

Alexandru Cernat
Social Statistics Department
University of Manchester
Manchester, M13 9PL, UK

David J. Ciuk
Department of Government
Franklin & Marshall College
Lancaster, PA, USA

Carol Cosenza
Center for Survey Research
University of Massachusetts
Boston, MA, USA

Mathew J. Creighton
School of Sociology, Geary Institute for Public Policy
University College Dublin
Stillorgan Road, Dublin 4, Ireland

Andrew W. Crosby
Department of Public Administration
Pace University
New York, NY 10038, USA

Richard Curtin
Institute for Social Research
University of Michigan
Ann Arbor, MI 48106, USA

Edith D. de Leeuw
Department of Methodology & Statistics
Utrecht University
Utrecht, the Netherlands

Stefanie Eifler
Department of Sociology
Catholic University of Eichstätt-Ingolstadt
85072 Eichstätt, Germany

Mahmoud Elkasabi
ICF, The Demographic and Health Surveys Program
Rockville, MD 20850, USA

Floyd J. Fowler Jr
Center for Survey Research
University of Massachusetts
Boston, MA, USA

Marek Fuchs
Darmstadt University of Technology
Institute of Sociology
Karolinenplatz 5, 64283 Darmstadt, Germany

Thomas Hinz
University of Konstanz
Konstanz, Germany

Allyson L. Holbrook
Departments of Public Administration and Psychology and the Survey Research Laboratory
University of Illinois at Chicago
Chicago, IL 60607, USA

Joop Hox
Department of Methodology and Statistics
University of Utrecht
Utrecht, the Netherlands

Ronaldo Iachan
ICF International
Fairfax, VA 22031, USA

Annette Jäckle
University of Essex
Institute for Social and Economic Research
Colchester CO4 3SQ, UK

Timothy P. Johnson
Department of Public Administration and the Survey Research Laboratory
University of Illinois at Chicago
Chicago, IL 60607, USA

Jenny Kelly
NORC, University of Chicago
55 East Monroe Street, Chicago, IL 60603, USA

Courtney Kennedy
Pew Research Center
Washington, DC, USA

Florian Keusch
Department of Sociology, School of Social Sciences
University of Mannheim
Mannheim, Germany

Samara Klar
School of Government & Public Policy
University of Arizona
Tucson, AZ 85721, USA

Maria Krysan
Department of Sociology and the Institute of Government & Public Affairs
University of Illinois at Chicago
Chicago, IL, USA

Tanja Kunz
Leibniz-Institute for the Social Sciences
P.O.Box 12 21 55, 68072, Mannheim, Germany

Paul J. Lavrakas
NORC, University of Chicago
55 East Monroe Street, Chicago, IL 60603, USA

Thomas J. Leeper
Department of Methodology
London School of Economics and Political Science
London WC2A 2AE, UK

James M. Lepkowski
Institute for Social Research
University of Michigan
Ann Arbor, MI 48106, USA

Mingnan Liu
Facebook, Inc.
Menlo Park, CA, USA

Peter Lynn
University of Essex
Institute for Social and Economic Research
Colchester CO4 3SQ, UK

Jaki S. McCarthy
United States Department of Agriculture
National Agricultural Statistics Service (USDA/NASS)
1400 Independence Avenue, SW,
Washington, DC 20250-2054, USA

Colleen McClain
Program in Survey Methodology at the University of Michigan, Institute for Social Research
426 Thompson Street, Ann Arbor, MI 48104, USA

Daniel L. Oberski
Department of Methodology & Statistics
Utrecht University
Utrecht, 3584 CH, the Netherlands

Kristen Olson
Department of Sociology
University of Nebraska-Lincoln
Lincoln, NE, USA

Colm O'Muircheartaigh
Harris School of Public Policy
University of Chicago and NORC at the University of Chicago
Chicago, IL, United States

Linda K. Owens
Stephens Family Clinical Research Institute
Carle Foundation Hospital, 611W. Park Street
Urbana, IL 61801, USA

Jennifer A. Parsons
Survey Research Laboratory
University of Illinois at Chicago
Chicago, IL, USA

Knut Petzold
Sociology Section
Ruhr-Universität Bochum
44801 Bochum, Germany

Annette Scherpenzeel
Chair for Economics of Aging, School of Management
Technical University of Munich
Munich, Germany

Peter Schmidt
Department of Political Science and Centre for Environment and Development (ZEU)
University of Giessen
Karl-Glöcknerstrasse 21 E, 35394 Giessen
Germany

Barbara Schneider
College of Education and Sociology Department
Michigan State University
East Lansing, MI 48824, USA

Stephen Smith
NORC at the University of Chicago
Chicago, IL, United States

Tom W. Smith
Center for the Study of Politics and Society
NORC at the University of Chicago
1155 East 60th Street Chicago, IL 60637, USA

Jolene D. Smyth
Department of Sociology
University of Nebraska-Lincoln
Lincoln, NE, USA

Jaesok Son
Center for the Study of Politics and Society
NORC at the University of Chicago
1155 East 60th Street Chicago, IL 60637, USA

Mathew Stange
Mathematica Policy Research
Ann Arbor, MI, USA

Marina Stavrakantonaki
Department of Public Administration
University of Illinois at Chicago
Chicago, IL 60607, USA

David Sterrett
NORC at the University of Chicago
Chicago, IL 60603, USA

Z. Tuba Suzer-Gurtekin
Institute for Social Research
University of Michigan
Ann Arbor, MI 48106, USA

Elizabeth Tipton
Human Development Department, Teachers College
Columbia University
New York, NY 10027, USA

and

Statistics Department
Northwestern University
Evanston, IL 60201, USA

Michael W. Traugott
Center for Political Studies
Institute for Social Research, University of Michigan
Ann Arbor, MI, USA

Jan A. van den Brakel
Department of Statistical Methods
Statistics Netherlands
Heerlen, the Netherlands

and

Department of Quantitative Economics
Maastricht University School of Business and Economics
Maastricht, the Netherlands

Susanne Vogl
Department of Education
Department of Sociology, University of Vienna
Vienna, Austria

Sandra Walzenbach
University of Konstanz,
Konstanz, Germany

and

ISER/University of Essex
Colchester, UK

Xiaoheng Wang
Department of Public Administration
University of Illinois at Chicago
Chicago, IL 60607, USA

Brady T. West
Survey Research Center
Institute for Social Research, University of Michigan
Ann Arbor, MI, USA

Diane K. Willimack
United States Department of Commerce
U.S. Census Bureau
4600 Silver Hill Road, Washington, DC 20233–0001, USA

Jaclyn S. Wong
Department of Sociology
University of South Carolina
Columbia, SC, United States

Ting Yan
Westat
Rockville, MD, USA

David S. Yeager
Psychology Department
University of Texas at Austin
Austin, TX 78712, USA

Berwood A. Yost
Franklin & Marshall College
Floyd Institute for Public Policy and Center for Opinion Research
Lancaster, PA, USA

Diana Zavala-Rojas
Research and Expertise Centre for Survey Methodology and European Social Survey ERIC, Department of Political and Social Sciences
Universitat Pompeu Fabra
C/de Ramon Trias Fargas, 25-27, 08005 Barcelona, Spain

Tianshu Zhao
Department of Public Administration
University of Illinois at Chicago
Chicago, IL 60607, USA

Preface by Dr. Judith Tanur

I am enormously flattered to be asked to supply a preface for this path-breaking volume of essays about experiments embedded in surveys. I had hoped to contribute a chapter to the volume with my friend and long-term collaborator, Stephen Fienberg, but Steve, active to the end, lost his long battle with cancer before he was able to make the time to work on the chapter we had planned to write. So, I would like to use this opportunity to write something about the work we had done, and had hoped to continue, on the parallels between experimental and survey methodology, on experiments embedded in surveys (and vice-versa), and some of the considerations for analysis occasioned by such embedding.

We had long noted (e.g. Fienberg and Tanur 1987, 1988, 1989, 1996) that there are a great many parallels between elements of survey design and experimental design. Although in fact surveys and experiments had developed very long and independent traditions by the start of the twentieth century, it was only with the rise of ideas associated with mathematical statistics in the 1920s that the tools for major progress in these areas became available. The key intellectual idea was the role of randomization or random selection, both in experimentation and in sampling, and both R.A. Fisher and Jerzy Neyman utilized that idea, although in different ways. The richness of the two separate literatures continues to offer new opportunities for cross-fertilization of theory and tools for survey practice in particular.

Although the ideas are parallel, often the purpose these ideas serve is different in the two domains. For example, experimental randomization is used in experimental design in order to justify the assumption that the experimental and control groups are equivalent a priori, but it finds its analog in probability sampling in surveys in order to assure the "representativeness" of the sample *vis a vis* the population to which generalizations are to be made. On the other hand, similar processes to create homogeneous groups occur in both experimental and sampling designs and serve similar purposes. In experimentation, blocking creates homogenous groups, exerting control of experimental error by segregating the effects of extraneous sources of variation. In sampling, stratification serves the same purpose by drawing sample members from each of the homogenous groups into which the population can be divided. The list could be lengthened (as indeed, Steve and I did in the papers cited above) to include such parallels in design as those between Latin and Graeco-Latin squares on the one hand and lattice sampling or "deep stratification" on the other and between split plot designs and cluster sampling. In the analysis stage, we pointed to the parallel between covariance adjustment in experiments and poststratification in surveys. We embarked on a project to find and describe more modern parallels and to encourage researchers in each of these fields to look in the literature of the other for ideas about design and analysis.

We wrote a series of papers and envisioned a book, *Reaching Conclusions: The Role of Randomized Experiments and Sample Surveys*, that will now, unfortunately, remain a draft, that explored the ramifications of these parallels, pointed out newer and less obvious parallels between experiments and surveys, and noted the frequent embedding of experiments in surveys and vice versa. In particular, we urged – sometimes by example – that when such embedding took place, the analyst should take advantage of the embedded structure in planning and carrying out the analysis.

In our 1989 *Science* paper, we addressed these issues of analysis of embedded experiments, reiterating the point that although there are formal parallels between the structures of surveys and experiments there are fundamental inferential differences as described above. We pointed out three inferential stances that could be used. In the context of this volume, it is perhaps worth quoting that discussion at some length (p. 243).

> (1) One can use the *standard experiment paradigm*, which relies largely on internal validity based on randomization and local control (for example, the device of blocking) and on the assumption that the unique effects of experimental units and the treatment effects can be expressed in a simple additive form, without interaction (Fisher 1935, Ch. 4; Kempthorne 1952, Ch. 9). Then inference focuses on within-experiment treatment differences.
>
> (2) One can use the *standard sampling paradigm*, which, for a two-treatment experiment embedded in a survey relies largely on external validity and generalizes the observations for each of the treatments to separate but paired populations of values. Each unit or individual in the original population from which the sample was drawn is conceived to have a pair of values, one for each treatment. But only one of these is observable, depending on which treatment is given. Then inferences focus on the mean difference or the difference in the means of the two populations.
>
> (3) One can conceptualize a *population of experiments*, of which the present embedded experiment is a unit or sample of units, and thus capitalize on the internal validity created by the design of the present embedded experiment as well as the external validity created by the generalization from the present experiment to the conceptual population of experiments. Then inferences focus on treatment differences in a broader context than simply the present embedded experiment.

Then we took advantage of the generosity of Roger Tourangeau and Kenneth Rasinski to reanalyze data they had collected in a probability sample survey on context effects in attitude surveys (1989). They had crossed four issues at differing levels of familiarity with four orders of presentation (balanced in the form of a Latin Square), two versions of the context questions, and two methods of structuring the context question. Thus, there were 16 versions of the questionnaire plus two extra versions with neutral context questions. These 18 versions of the questionnaire were crossed with four interviewers. We considered the interviewers as blocks, and so within each block, we had five replications of an 18-treatment experiment, where 16 of the treatments represent a $4 \times 2 \times 2$ factorial design. Focusing separately on two of the issues (abortion and welfare), we had four treatment combinations (positive vs. negative contexts by scattered vs. massed structure of the context-setting questions). The context-setting questions for the abortion question dealt with women's rights or traditional values, while those for the welfare question concerned fraud and waste in government programs or government responsibility to provide services. Using logit models to predict a favorable response on the target question and contrasting the four basic treatment combinations with the neutral context, we carried out detailed analyses and found very complicated results. I shall try to sketch only a fraction of results of those analyses – I urge the reader with a serious interest in these issues to refer to the original paper (Fienberg and Tanur, 1987). In short, we found that context matters – but its effect depends of how often the respondent agrees with the context-setting questions. And we found that all three of the inferential stances detailed above gave similar results for the abortion question, while the first and third gave similar results for the welfare question.

This volume concentrates on the embedding of experiments within surveys, often to answer questions about survey methodology as did the Tourangeau/Rasinski experiment discussed above. I would guess that such procedures are the most common uses of embedding. But perhaps it is worth bearing in mind that there are many good examples of the reverse – the embedding of surveys in experiments and such embedded surveys often serve a more substantive purpose. A set of particularly good examples were the massive social experiments of the mid-twentieth century such as the Negative Income Tax experiments and the Health Insurance Study. Many of the measurements of the outcome variables in those randomized experiments were necessarily carried out via surveys of the participants.

I find the contents of this volume fascinating, both for the broad sweep of the topics examined and for the variety of disciplines represented by the contributing authors and what I know of their expertise. I look forward to reading many of the articles.

I wish Steve's work could have been included in a more specific and current way than I have been able to achieve in this preface – he had plans for updating that I am not able to carry out. In that context, I am especially pleased to note the inclusion of a chapter by Jan Van den Brakel, whose work I know Steve admired and whose chapter I expect will contain much of the new material that would have informed our chapter had Steve lived to complete it.

Prolific as Steve was, he left so much good work unfinished and so much more not even yet contemplated. I miss him.

Montauk, New York
December 2017

Judith Tanur

References

Fienberg, S.E. and Tanur, J.M. (1987). Experimental and sampling structures: parallels diverging and meeting. *International Statistical Review* 55 (1): 75–96.

Fienberg, S.E. and Tanur, J.M. (1988). From the inside out and the outside in: combining experimental and sampling structures. *Canadian Journal of Statistics* 16: 135–151.

Fienberg, S.E. and Tanur, J.M. (1989). Experimental combining cognitive and statistical approaches to survey design. *Science* 243: 1017–1022.

Fienberg, S.E. and Tanur, J.M. (1996). Reconsidering the fundamental contributions of Fisher and Neyman on experimentation and sampling. *International Statistical Review* 64: 237–253.

Fisher, R.A. (1935). *The Design of Experiments*. Edinburgh: Oliver and Boyd.

Kempthorne, O. (1952). *The Design and Analysis of Experiments*. New York: Wiley.

Tourangeau, R. and Rasinski, K.A. (1989). Carryover effects in attitude surveys. *Public Opinion Quarterly* 53: 495–524.

About the Companion Website

This book is accompanied by a companion website:

www.wiley.com/go/Lavrakas/survey-research

The website includes:

- Letters
- Brochures
- Appendices

1

Probability Survey-Based Experimentation and the Balancing of Internal and External Validity Concerns[1]

Paul J. Lavrakas[1], Courtney Kennedy[2], Edith D. de Leeuw[3], Brady T. West[4], Allyson L. Holbrook[5], and Michael W. Traugott[6]

[1] *NORC, University of Chicago, 55 East Monroe Street, Chicago, IL 60603, USA*
[2] *Pew Research Center, Washington, DC, USA*
[3] *Department of Methodology & Statistics, Utrecht University, Utrecht, the Netherlands*
[4] *Survey Research Center, Institute for Social Research, University of Michigan, Ann Arbor, MI, USA*
[5] *Departments of Public Administration and Psychology and the Survey Research Laboratory, University of Illinois at Chicago, Chicago, IL, USA*
[6] *Center for Political Studies, Institute for Social Research, University of Michigan, Ann Arbor, MI, USA*

The use of experimental designs is an extremely powerful scientific methodology for directly testing casual relationships among variables. Survey researchers reading this book will find that they have much to gain from taking greater advantage of controlled experimentation with random assignment of sampled cases to different experimental conditions. Experiments that are embedded within probability-based survey samples make for a particularly valuable research method as they combine the ability to more confidently draw causal attributions based on a true experimental design that is used with the ability to generalize the results of an experiment with a known degree of confidence to the target population which the survey has sampled (cf. Fienberg and Tanur 1987, 1989, 1996). For example, starting in the late 1980s, the rapid acceptance of using technology to gather data via computer-assisted telephone interviewing (CATI), computer-assisted personal interviewing (CAPI), and computer-assisted web interviewing (CAWI) has made it operationally easy for researchers to embed experimental designs within their surveys.

Prior to the 1990s, the history of experimentation in the social sciences, especially within psychology, essentially reflected a primary concern for maximizing the ability to identify empirically based cause-and-effect relationships – ones with strong internal validity (Campbell and Stanley 1966) – with little regard for whether the study could be generalized with confidence – i.e. to what extent the study had strong external validity (Campbell and Stanley 1966) – beyond the particular group of subjects/respondents that participated in the experiment. This latter concern remains highly pertinent with the current "replication crisis" facing the health, social, and behavioral sciences and its focus on "reproducibility" as a contemporary criterion of good experimentation (Wikipedia 2018a,b).

This is not to say that prior to 1990 that all experimental social scientists were unconcerned about external validity (cf. Orwin and Boruch 1982), but rather that the research practices of most suggested that they were not. In contrast, practical experience within the field of survey

1 The authors very much appreciate the review and valuable input received from Professor Joop Hox on an earlier version of the chapter.

Experimental Methods in Survey Research: Techniques that Combine Random Sampling with Random Assignment, First Edition.
Edited by Paul J. Lavrakas, Michael W. Traugott, Courtney Kennedy, Allyson L. Holbrook, Edith D. de Leeuw, and Brady T. West.

Companion Website: www.wiley.com/go/Lavrakas/survey-research

research suggests that many survey researchers have focused on external validity, but failed to use experimental designs to enhance their research aims by increasing the internal validity of their studies. An example of this is in the realm of the political polling that is done by and for news media organizations. Far too often, political polls merely generate point estimates (e.g. 23% of the public approves of the job that the President is doing) without investigating what drives the approval and disapproval through the usage of question wording experiments (cf. Traugott and Lavrakas 2016).

A case in point: The results of a *nonexperimental* preelection study on the effects of political advertising were posted on the list-server of the American Association for Public Opinion Research (AAPOR) in 2000. This survey-based research was conducted via the Internet and reported finding that a certain type of advertising was more persuasive to potential voters than another type. By using the Internet as the data collection mode, this survey was able to display the ads – which were presented as digitized video segments – in real-time to respondents/subjects as part of the data collection process and, thereby, simulate the televised messages to which voters routinely are exposed in an election campaign. Respondents were shown all of the ads and then asked to provide answers to various questions concerning their reactions to each type of ad and its influence on their voting intentions. This was done in the individual respondent's own home in a room where the respondent normally would be watching television. Here, the Internet was used in a very effective and creative way to provide *mundane realism*[2] (and there by contribute to "ecological validity") to the research study by having survey respondents react to ads in a context quite similar to one in which they would be exposed to real political ads while they were enjoying a typical evening at home viewing television. Unlike the many social science research studies that are conducted under conditions far removed from "real life," this study went a long way toward eliminating the potential artificiality of the research environment as a serious threat to its overall validity.

Another laudable design feature of this study was that the Internet sample of respondents was chosen with a rigorous scientific sampling scheme so that it could reasonably be said to represent the population of potential American voters. The sample came from a large, randomly selected panel of U.S. households that had received Internet technology (*WebTV*) in their homes, allowing the researchers to survey subsets of the sample at various times. Unlike most social science research studies that have studied the effects of political advertising by showing the ads in a research laboratory setting (e.g. a centralized research facility on a university campus), the validity of this study was not threatened by the typical convenience sample (e.g. undergraduates "volunteering" to earn course credit) that researchers often rely upon to gather data. Thus, the results of this Internet research were based on a probability sample of U.S. households and, thereby, could reasonably be generalized to the potential U.S. electorate.

As impressive as these features of this research design were, the design had a serious, yet unnecessary, methodological limitation – one that caused it to miss a golden opportunity to add considerably to the overall validity of the conclusions that could have been drawn from its findings. The research design that was used displayed all the political ads to each respondent, one ad at a time. There were no features built into the design that controlled either for the possible effects of the order in which the respondent saw the ads or for having each respondent react to more than one ad within the same data collection session. As such, the cause-and-effect conclusions that could be drawn from this nonexperimental study design about which ads "caused" stronger respondent reactions rested on very weak footing. Since no design feature

2 Mundane realism (and ecological validity) refers to the extent to which the stimuli and procedures in a research study reflect everyday "real-life" experiences (e.g. Aronson and Carlsmith 1968; Crano and Brewer 1973; Wikipedia 2018b).

was used to control for the fact that respondents viewed multiple ads within the same data collection session, the validity of the conclusions drawn about the causality underlying the results remained little more than speculations on the part of the researchers, because such factors as the order of the ads and number of ads were not varied in a controlled manner by the researchers. Unfortunately, this missed opportunity is all too common in many survey-based research studies in the social sciences.

This study could have lent itself to the use of various experimental designs whereby a different political ad (i.e. the experimental stimuli) or different subsets of ads could have been *randomly assigned* to different subsamples of respondents. Or, the order of the presentation of the entire set of political ads could have been randomly assigned across respondents. In either case, an experimental design with random assignment would have provided the researchers with a far stronger basis (i.e. a study with greater internal validity) from which to draw causal inferences. Furthermore, such an experimental approach would have had little or no cost implications for the research budget, and under a design where one and only one ad was shown to any one respondent, would likely have saved data collection costs.

Using this example as our springboard, it is the goal of this book to explain how experimental designs *can and should* be deployed more often in survey research. It is most likely through the use of a true experimental design, with the random assignment of subjects/respondents to experimental conditions, that researchers gain the strong empirical basis from which they can make confident statements about causality. Furthermore, because of the widespread use of computer-assisted technologies in survey research, the use of a true experiment within a survey often adds little or no cost at all to the budget. In addition, embedding an experiment into a survey most often provides the advantage that the cell sizes of the different experimental groups can be much larger than in traditional experimental designs (e.g. in social psychology), which has the benefit that the statistical power of the analyses is much greater.

It is for these reasons and others that we believe that many survey researchers should utilize these powerful designs more often. And, toward that end, we believe that this consideration should be an explicit step in planning a survey, on par with traditional planning steps such as deciding upon one's sampling frame, sample size, recruitment methods, and data collection mode(s).

1.1 Validity Concerns in Survey Research

In their seminal book, *Experimental and Quasi-Experimental Designs for Research* (1966), Don Campbell and Julian Stanley identified four types of validity that can threaten the accuracy of any research study.

1. *Statistical conclusion validity* refers to the adequacy of the researcher's statistical methods to draw conclusions from the data in an accurate (unbiased) fashion. Without adequate statistical conclusion validity, the researchers cannot know whether they have data that demonstrate reliable relationships among their variables.
2. *Construct validity* refers to the adequacy of the measurement tools that are used for data collection to operationalize the constructs of interest (e.g. questionnaire items in a survey) in both a reliable and valid manner so that they actually are measuring what the researchers intended to measure. Without adequate construct validity, the researchers have no reliable basis from which to know what behaviors, knowledge, or attitudes actually have been measured. In survey methodology, lack of construct validity is related to *specification error and measurement errors* (cf. Biemer et al. 2004; Groves 1989).

3. *External validity* refers to the extent to which the findings of a particular study can be generalized beyond the sample from which the data were gathered, to other persons, places, or times. Without adequate external validity, researchers are hard pressed to know whether, for example, their findings are limited to merely those respondents/subjects who provided the data. In survey methodology, lack of external validity is related to *coverage error, sampling error, nonresponse error, and adjustment error* (cf. Groves 1989; Kalaian and Kasim 2008).
4. *Internal validity* refers to the extent to which the methodological design used by the researchers supports direct cause-and-effect reasoning. Without adequate internal validity, researchers may offer logical, reasoned arguments to speculate about the possible casual nature of any relationships they observe in their data, but they cannot use the internal strength of their research design itself to bolster such reasoning.

Groves' (1989) seminal work *Survey Errors and Survey Costs* provided a comprehensive and critical review of how different scientific disciplines have conceptualized validity concerns, and he correctly noted that the Campbell and Stanley framework is not embraced throughout all of social science. Groves adopted and built upon the growing tradition within survey research to utilize the *Total Survey Error* (TSE) framework to help survey researchers plan and interpret their research design. The TSE approach identifies major types of errors (bias and/or variance) inherent in survey designs including coverage error, sampling error, nonresponse error, and measurement error; see also Biemer et al. (2017) and Groves and Lyberg (2010).

1. *Coverage error* refers to error (primarily in the form of bias) that can result if one's sampling frame does not well represent one's target population. Coverage error can also result from within unit noncoverage associated with the respondent selection technique that is used (cf. Gaziano 2005).
2. *Sampling error* refers to the error, i.e. uncertainty (in the form of variance), in every sample survey that occurs merely by chance due to the fact that data are gathered from a sample of the population of interest rather than a full census being conducted. When a probability sample design is employed, the researcher can calculate the size of the sampling variance associated with the particular sample design.
3. *Nonresponse error* refers to the possible error (primarily in the form of bias) that can result if those sampled units from whom data are not gathered (e.g. not contacted or refusing households) are different in meaningful ways from those sampled units from which data are gathered (cf. Groves and Couper 1998). Nonresponse error can occur at both the unit level and the item level.
4. *Measurement error* refers to the possible error (in the form of bias and/or variance) associated with a survey's questionnaire, respondents, interviewers, and/or mode of data collection (cf. Biemer et al. 2004).

In creating this book, we embraced both the Campbell and Stanley validity framework and the TSE framework, which we view as compatible. However, many researchers who are familiar with traditional survey research methods come from disciplines that do not routinely use true random experimentation as part of their research methodologies and may be less familiar with the Campbell and Stanley framework. Thus, we expect that many using this book have not yet developed enough familiarity with the language of experimentation that is necessary to best understand what is presented in the chapters that follow. To that end, we proceed in this opening chapter with more explication about the interrelationships among the different types of validities and errors, followed by an even more detailed review of internal validity and external validity.

1.2 Survey Validity and Survey Error

Cook and Campbell (1979) discuss how the four forms of validity in the Campbell and Stanley framework are interrelated. Here, we also discuss how they interrelate to the phraseology and conceptualizations common to survey research.

Unless a research study utilizes statistical procedures that are adequate for the purposes to which they are deployed, analyses will be more prone to Type 1 (false positive, in finding a relationship when there actually is none) and Type 2 (false negative, in not finding a relationship when there actually is one) errors than would be the case with the use of more appropriate statistical procedures. This is especially of concern when the statistical procedures are used to test data that have been gathered in an experimental design. Type 1 errors refer to the possibility that any statistical relationship observed among two or more variables measured in a sample is a spurious one.[3] This concept is similar to sampling error: the recognition that bivariate and multivariate relationships observed between, and among, variables within a given sample of respondents may not hold up within another independent sample drawn from the same population.[4] Thus, the prudent survey researcher recognizes that the covariation that is observed between two variables within a sample may merely be due to chance variations within that particular sample and may not generalize to the entire population. Considerations of Type 1 error refer to the likelihood that an observed relationship between variables is unreliable. Within many social science disciplines, the "95% level of confidence" or threshold – i.e. probability less than 0.05 or less than 1 chance in 20 of being wrong – must be achieved or surpassed before one can conclude that the relationship is statistically "significant." This concept should be quite familiar to survey researchers. However, Type 2 error does not have its direct equivalent in traditional survey research literature. It refers to the statistical power of a research design to help researchers avoid drawing an erroneous conclusion (i.e. false negative) based on the finding that no reliable covariation exists between variables within a sample, when in fact such a relationship does exist in the population.

Related to the possibility of Type 2 error is the tradition within experimental research of not "accepting" a null hypothesis (cf. Cook et al. 1979). Instead, a prudent researcher concludes that there exists no empirical basis for accepting the alternative hypothesis if the results do not achieve statistical significance. This is more than mere semantics, as it represents the *explicit acknowledgement on the part of the researcher that any research study based on a sample, regardless of how that sample was chosen, is fallible* and that any certainty that one has about one's results within the sample is probabilistic and thus less than 100% (i.e. less than absolute certainty).[5]

Coverage error and nonresponse error are clearly related to external validity in that the three concepts address the issue of whether research findings gathered in a sample of respondents are representative of some larger, known population. Since the use of a sample, rather than a census, is almost always a survey researcher's practical preference, it should be paramount that the sample be drawn from a sampling frame that accurately "covers" the entire population of

---

3 Of note, often it is not recognized – nor understood – that the concepts of Type 1 error, Type 2 error, and sampling error are irrelevant (i.e. they do not exist) within a true census of a population. That is, those concepts apply only to research that is based upon a sample.

4 Of course, sampling error also applies to univariate sample statistics whenever the researcher's purpose is to make a point estimate on the level of a single variable within the target population. For example, 45% of a random sample of 500 likely voters may report their intention to vote for candidate A, yet when sampling error is taken into consideration this finding falls within the 95% confidence interval of 41–49%.

5 This also holds for Type 1 error. And, in addition to the effects of sampling, statistical conclusion validity refers to making correct assumptions, e.g. not assuming independence in cluster samples.

interest. However, avoiding coverage error by utilizing a representative sampling frame is not a sufficient condition for external validity. Instead, if one wants strong external validity, one also must avoid nonresponse error so that the data gathered from the respondents do not differ to a nonignorable extent from data that would have been gathered from nonrespondents (i.e. those who were sampled but from whom no data were collected due to refusals, noncontacts, etc.). In sum, the avoidance of both nonignorable coverage error and nonresponse error is a necessary condition for external validity, and together they likely comprise the sufficient condition.

Measurement error and construct validity both concern whether the tools of measurement yield reliable and valid data. Construct validity has traditionally focused on the manner in which a construct is operationalized as a specific variable. This often has meant merely how a question is worded in a survey. Measurement error within the TSE typology is a much broader concept as it explicitly includes the full data collection environment encompassing the instrument (or tool) by which the data are captured – including item wording and ordering if a questionnaire is used – and extends to the role of the interviewer when one is present. It also includes the mode of data collection and the role of the respondent herself/himself in providing reliable and valid data.

The concept of internal validity, however, especially as manifested via the use of true experimentation, is not resident within traditional survey research nomenclature. As we have noted, it is not part of the traditional steps in conceptualizing a survey research design. And, this is something we hope to change with this volume.

1.3 Internal Validity

Cook and Campbell (1979) define internal validity as "the approximate validity with which [one infers] that a relationship between two variables is causal or that the absence of a relationship implies the absence of cause" (p. 37).[6] There are three conditions for establishing that a relationship between two variables (X and Y) is a causal one, as in "X causes Y." The researcher must demonstrate (i) covariation (that there is a reliable statistical relationship between X and Y), (ii) temporal order (X must occur before Y), and (iii) an attempt to eliminate other plausible explanations than changes in X for observed changes in the dependent variable (Y). The use of a true experimental design with random assignment to conditions is of special value for the last of these three conditions, i.e. eliminating plausible alternative explanations. Figure 1.1 shows the simple formula often used to depict covariation between two variables, X and Y, and is read, "Y is a function of X."

$$Y = f(X)$$

Figure 1.1 Simple mathematical representation of the interrelationship of two variables.

The concept of internal validity essentially addresses the *nature of the equal sign (=)* in the equation, i.e. is the relationship between X and Y a causal one? For internal validity to exist, there must be covariation demonstrated between X and Y, and, therefore, X must predict Y to some statistically reliable extent. But the equal sign (and the covariation it implies) in itself does not provide evidence that cause and effect between X and Y has been demonstrated.

If the relationship shown in Figure 1.1 is causal, then it presupposes that X causes Y, and thus that X precedes Y in a temporal sense. This is what distinguishes this specification from one

6 Cook and Campbell (1979) note that "internal validity has nothing to do with the abstract labeling of a presumed cause or effect; rather it deals with the *relationship* between the research operations *irrespective of what they theoretically represent.*" (p. 38)

that says Y is the cause of X, or that each cause the other, or that each are caused by some other unspecified variable (Z) – any of which could be the interpretation of the correlation between two variables. Only by the use of a controlled experiment or series of experiments can the nature and direction of these interrelationships be parceled out through the application of the hypothesized independent variable (under the control of the researcher) followed by the measurement of the dependent variable. This overcomes the typical problem of a cross-sectional survey where the independent and dependent variables are measured simultaneously.

The essential design feature of an experiment is the use of random *assignment* of subjects/respondents to different experimental conditions. The logic here is that with random assignment of respondents/subjects to different conditions (i.e. the different levels of the independent variable, X), all other factors will be equivalent except for the differences that the researcher directly controls in implementing the independent variable of interest. If statistically significant differences in the "average value"[7] of the dependent variable (Y) then are found between the randomly assigned groups, these differences then can be attributed to the independent variable that the researcher has controlled. Then the researcher usually will have a solid basis to conclude that it was the controlled differences in the independent variable (X) that caused the observed differences between the groups in Y.

A simple survey-based example of this occurs when a group of respondents are randomly assigned to two conditions: often called a "Control" condition and a "Treatment" condition. For this example, assume that a questionnaire employs a so-called "split-half" experiment whereby one randomly assigned half of respondents are exposed to the standard wording of a question (e.g. the wording used in the 2000 Census to measure whether someone is Hispanic; "Are you Spanish, Hispanic or Latino?"). The group receiving this standard wording is the Control Group. The other random half of respondents would be asked the question with some altered version of the wording (e.g. "Are you of Spanish, Hispanic or Latino origin?"). In this example, the group of respondents seeing or hearing the word "origin" in their questionnaire is the Treatment Group (sometimes called the "Experimental" Group), as they are receiving a different wording of the question to investigate what effect adding the word "origin" will have on the proportion in the Treatment Group that answers "Yes." The researcher has controlled the administration of the question wording (i.e. X, the independent variable) in order to learn whether the change in wording causes a change in the proportion of people who say "Yes" (i.e. Y, the dependent variable, whether they are Hispanic). And, this control over administration of the independent variable (X) is exercised via random assignment so that, in theory, nothing else is dissimilar between the two groups except for the slight change of wording between the control question and the treatment question. Thus, random assignment is the design equivalent to holding "all other things equal." Due to the strong internal validity of the experimental design in this example – which includes the expectation that there is no differential nonresponse between the control and treatment groups – the researcher can conclude with great confidence that any statistically significant difference between the two groups in the proportion that answered "Yes" to being Hispanic is associated with (i.e. caused by) the presence or absence of the word "origin" (the independent variable).[8]

7 In this use, "average value" generally means the mean, median, or mode for a group.
8 Of note, in fact it has been found that using the word "origin" when asking about one's Hispanic ethnicity does cause significantly more respondents to say "Yes" that they are Hispanic than when the question wording does not include "origin" (cf. Lavrakas et al. 2002, 2005).

1.4 Threats to Internal Validity

To better appreciate the power of an experimental design and random assignment, it is worth a brief review of various reasons why cause-and-effect inferences that are drawn from nonexperimental research designs are subject to many threats to their validity. For a much more detailed discussion of these and other reasons see Campbell and Stanley (1966) and Cook and Campbell (1979).

1.4.1 Selection

Too often the "selection" of the respondents (or subjects) that constitute different comparison groups turns out to be the main threat to a study's internal validity. For example, if a survey researcher sampled two different municipalities and measured the health of residents in each community, the researcher would have no statistical or methodological grounds on which to base any attributions about whether living in one community or the other "caused" any observed differences between the average health within the respective communities. In this example, no controlled effort was built into the study to make the two groups equivalent through random assignment, and in this example that would not be possible. As such, any observed differences between the health in each community could be due to countless other reasons than the place of residence, including a host of demographic and behavioral differences between the residential populations of each community.

Thus, any time two (or more) groups have been selected for comparison via a process other than random assignment, the researchers most often will have no solid empirical grounds on which to draw causal inferences about what may have caused any observed difference between the two groups.[9] Unfortunately, this does not stop many researchers from making unfounded causal attributions. This is especially the case in the field of epidemiological research when the health of large samples of volunteer patients is tracked over time. Such a panel survey may find many significant correlations between behavior and health (e.g. eating a lot of carrots is associated with better eyesight), but this is mere covariation and the study design, with its lack of random assignment to comparison groups, provides no internal validity to support causal inferences between the measured behaviors and health.

Furthermore, any study that allows respondents to "self-select" themselves into different comparison groups will suffer from selection as a threat to its internal validity. However, this point is different from the common way that self-selection into a sample is thought of. So-called convenience samples suffer from a self-selection sampling bias, as the researcher has no means of knowing whether a larger population is represented by a self-selected sample (this is an issue that affects external validity as discussed later in this chapter). However, the researcher could legitimately build a valid experiment into a survey that uses a convenience sample, simply by randomly assigning the self-selected respondents to different comparison groups. Thus, as long as the respondents do not self-select themselves into the treatment and control groups, it does not threaten the internal validity of the experiment, even if they have self-selected themselves into the larger sample. We also note that in experiments embedded in a survey, selection also includes differences in nonresponse rates (including break-off rates) between the experimental and control groups.

9 There is the possibility that a researcher deployed a quasi-experimental design, one without random assignment, but one that may have design features that avoid some of the potential threats to internal validity (cf. Cook and Campbell 1979).

1.4.2 History

This potential threat to internal validity refers to the possibility that something other than the independent variable may have taken place between the time respondents were first exposed to the independent variable and time of the measurement of the dependent variable. If so, then a differential "history" effect may have caused any observed differences among respondent groups in the dependent variable.

To illustrate this, consider a survey of attitudes toward local police being administered in two different, yet socioeconomically similar, communities to establish a baseline (i.e. a pretest measure). Imagine that the survey found that these two communities held essentially similar "pretest" attitudes. Then imagine that in one of the two communities, local police implemented a foot patrol program putting many more police officers on neighborhood streets. After this program is implemented for several months, both communities are resurveyed (i.e. a posttest measure), and the community with the new foot patrol program is now found to hold significantly more positive attitudes than the other community.

Could the researchers conclude with confidence (i.e. with strong internal validity) that the foot patrol program *caused* the improvement in attitudes? The answer is "No" for many reasons, including that there was no way for the researcher to control for whatever else may have occurred locally between the time that the pretest and posttest surveys were conducted that may have led to the attitudes in one community to change compared to the other. For example, a major crime may have been solved in one community in the intervening period, but not in the other. Was this the cause of more positive attitudes toward the police? Or was it the foot patrols?[10] This is how the differential history of two groups can confound any interpretation of cause when a true experiment is not used.

Furthermore, even if a research study starts out as a true experiment, subsequent uncontrolled history between randomly assigned groups can undermine the experiment and, thereby, its internal validity. For example, imagine a study in which the interviewers at a survey organization were randomly assigned into two groups to be trained, separately, to administer one of two different introductory spiels to randomly selected households in order to determine the differential effects on response rates of the two introductions. If something eventful happened at one of the training sessions other than the difference in the content related to the respective introductory spiels – e.g. an interviewer and the trainer got into a heated argument about the wording of the introductory spiel thereby lowering the confidence of the rest of the interviewers in that group regarding the effectiveness of that introductory wording – then this differential "history" could pose a serious threat to the internal validity of this research study, despite it being originally designed as a true experiment. If this were to happen, then the researchers would have a weakened basis on which to conclude that it was the content of the different introductions *and only that content* that caused any observable differences in response rates between the two groups of respondents.

All this notwithstanding, in many survey-based experiments, history is not a likely threat to internal validity because the dependent variable often is measured immediately after the administration of the independent variable (e.g. most wording experiments built into a questionnaire require that the respondent answers the question immediately after being exposed to the wording), but in other instances, the researcher must be very conscious of the possibility that history may have invalidated the integrity of the experimental design.

---

10 The posttest survey in each community could be used to gather information to determine if there were any differential history effects in the communities. Although this may appear to be a prudent approach to take, it is not possible to do this in such a comprehensive manner as to be able to rule out all potential history effects.

1.4.3 Instrumentation

Any time a measurement instrument, e.g. a survey question, is changed between a pre- and postperiod, any observed changes in the dependent variable of interest may be due solely to the change in instrumentation as opposed to real changes between the two groups due to a treatment or stimulus. For example, take a panel survey with two waves of data collection in which all respondents were asked, "Do you support or oppose the President's new plan to reduce taxes?" in Wave 1 data collection. Suppose that after Wave 1, a random half of the respondents were exposed to a direct mail campaign touting the popularity of the new tax plan. Suppose also that after Wave 1, the President began actively campaigning on behalf of the new tax plan and received consistently positive press coverage. After some passage of months, another wave of data is gathered from the same respondents, but using the following question, "Do you support or oppose the President's popular plan to reduce taxes?" Imagine that at Wave 2, a sizably larger proportion of respondents who were exposed to the direct mail campaign said that they supported the plan than had supported it at Wave 1 and that this increase was larger than the increase in support among the nontreatment group. Would this mean that the direct mail campaign exposure caused the apparent growth within that portion of the sample exposed to it?

The answer is "No, not necessarily," because although the small change in the wording of the measure at Wave 2 may appear innocuous – and given the positive press coverage might appear to be an appropriate wording change – the use of the word "popular" in the Wave 2 version of the questions could by itself have prompted (i.e. caused) more people to "conform" with apparent public opinion and say "Yes" to the question than otherwise would have happened had the exact Wave 1 wording been used. This could especially be true for the respondents exposed to the direct mail campaign. In particular, the treatment (the direct mail campaign) may have interacted with the wording change in the posttest question to cause the disproportionate shift in expressed support of the new tax plan among the group exposed to the mail campaign. Thus, it is possible that the change in support among the treatment group would have been no different in size than the change among the group that did not receive the direct mail campaign had the question wording not been altered.

1.4.4 Mortality

Imagine that an experimental test of a new remedial science curriculum is implemented so that a large random *sample* of inner-city high students is randomly *assigned* to a treatment group or a control group. The control group does not receive the remedial curriculum. The treatment group receives the remedial instruction during special 30-minute class sessions held only for them at the end of their regular school days. After six months of being exposed to the remedial curriculum, researchers find that the treatment group actually scores lower in science knowledge than does the control group. Does this mean that the curriculum actually caused the treatment group to do more poorly on their science knowledge test?

Although that is possible, imagine instead that receiving the remedial education curriculum caused more students in the treatment group to remain in school after six months because they were receiving the special attention. However, in the control group, more students dropped out of school during the six months, with students having the lowest knowledge of science being the most likely to drop out. In this case, differential "mortality" (i.e. differential attrition) would render the two groups no longer equivalent when the comparison was made between each group's average science knowledge score after six months. As such, researchers must guard against respondent/subject mortality threatening the internal validity of their experiment. And, even if the researchers cannot foresee and/or control against differential mortality, the possibility that this might occur must be measured and its possible effects taken into account before one can interpret experimental results with confidence. In particular, any survey-based experiment in which the experimental treatment causes differential response rates, but the dependent variable is something other than the response rate, is subject to the effects of differential mortality.

There are other threats to the internal validity that may undermine a research design's ability to support cause-and-effect reasoning (cf. Campbell and Stanley 1966). However, by using a true experiment with random assignment, the researcher is on a much firmer ground in making valid causal attributions than without an experimental design. In addition, if threats to internal validity are suspected, using a pretest–posttest design provides additional information that can be used to test for such threats and to some extent control for them (cf. Cook et al. 1979).

1.5 External Validity

Now that we have explained how *random assignment* is the cornerstone of experimentation and establishment of the interval validity of a research design, it is worth clarifying a common point of confusion, namely, the difference between random assignment and *random sampling*. Random sampling is very much a cornerstone of external validity, especially when it is done within the context of a probability sampling design. In fact, the beauty and strength of high-quality survey research is that a researcher can meld *both* random assignment and random sampling, thereby having strong internal validity *and* strong external validity whenever the survey mode of data collection is appropriate for the needs and resources of the researcher.

Many researchers who use the survey mode of data collection are much more familiar with the science of sampling than they are with the science of experimentation. Although they may not have prior familiarity with the terminology of external validity, they likely are quite familiar with the principals underlying the concerns of external validity: If one wants to represent some known target population of interest accurately, then one best utilize a sampling design that (i) well represents that population via a properly chosen sampling frame, (ii) uses a random probability sampling scheme to select respondents from the frame, thereby allowing one to generalize research findings from the sample to the population with confidence and within a known degree of sampling error, and (iii) uses successful recruitment methods to achieve a final sample that is well representative (unbiased) of the initially drawn sample. As stated earlier in this chapter, "avoidance of coverage error and nonresponse error each are necessary conditions for external validity, and together they likely comprise the sufficient condition." Thus, survey researchers need to use sampling frames that fully "cover" the target population they purport to represent and need to achieve adequate cooperation from the sampled respondents that avoids nonignorable differential nonresponse and nonresponse bias.

The *linkage* between internal validity and external validity concerns whether any cause-and-effect relationship that has been observed in a research experiment can be generalized beyond the confines of the particular sample (subjects/respondents) on which the experiment was conducted. For example, the field of psychology has a long and honored history of using experimentation with strong internal validity. However, it also has the well-known (and not so honorable) history of questionable external validity for many findings related to too often using unrepresentative samples of college undergraduates (cf. Visser et al. 2013).

To achieve strong external validity in a survey, one needs to use a probability sample and avoid both coverage error and nonresponse error. As a reminder, a probability sample is one in which each element in the sampling frame has a *nonzero* and *known* probability of selection (the reader should note that having an *equal* probability of selection is not a necessary attribute of a probability sample). Coverage error is avoided if the researcher's sampling frame "covers" (i.e. represents) the target population to which the researcher wants to generalize the study's findings. Nonresponse error is avoided as long as the group of sampled elements (i.e. respondents) chosen from the sampling frame, but from whom no data are gathered, does not differ in meaningful ways from the group of elements that are sampled and from whom data are gathered. All in all, using a probability sample that adequately covers the target population and achieves a high response rate typically has strong external validity.

1.6 Pairing Experimental Designs with Probability Sampling

We cannot understate the importance of the research "power" that is afforded by an experimental design in allowing a researcher to test the causal nature of the relationship between variables with confidence. We strongly encourage survey researchers to challenge themselves in thinking about how to utilize experimentation more often within their surveys so as to achieve this power, and the resulting internal validity, in advancing theory throughout the social sciences and in explaining human behaviors and cognitions more fully. As we noted earlier, this often can be done at little or no additional cost in the data collection process, and sometimes can even save costs as it may reduce the amount of data that must be gathered from any one respondent. We believe that the value of experimentation will be realized more often if survey researchers build into their survey planning process a decision stage where they explicitly consider how, if at all, to use experimentation in their survey. Similarly, we challenge social scientists who are familiar with the strengths of experimentation to improve their research studies by deploying probability sample surveys whenever those are appropriate for the topic of study. (Of note, Chapter 21 of this book provides a discussion of the usage of nonprobability sampling in survey research, including when experiments are used in the survey).

Thus, it is our sincere hope that survey researchers will heed the call of Don Campbell's (1988) vision of an "Experimenting Society" and help build it within the field of survey research. The chapters that follow provide many examples of how both strong internal validity and strong external validity can be achieved by using experimentation within surveys.

1.7 Some Thoughts on Conducting Experiments with Online Convenience Samples

Historically, there have been a great many social science research studies that have been conducted using nonprobability samples and that has included the vast majority of experimental studies. Reasons for this include the lower cost and less time that it requires to use a nonprobability sample for one's experiment. But it was also due to a greater concern among many experimenters about internal validity than external validity. Nowadays, many research studies are conducted using respondents/subjects from the myriad of readily available online convenience samples, which have been created using nonprobability methods. Primarily due to concerns with costs and timing, many of these experimenters seem to either assume or at least act as though, the quality of their research will not suffer by ignoring external validity concerns. Online convenience platforms include panel surveys, intercept surveys, and crowdsourced labor markets. Relative to surveys using probability-based designs, online convenience surveys tend to be substantially less expensive and faster to conduct, and thus their considerable attraction to many researchers.

Proponents of online convenience sources point to their greater representativeness relative to in-person convenience samples (e.g. Hauser and Schwarz 2016; Levay, Freese, and

Druckman 2016; Mullinix et al. 2015), which have been the modal subject pool for decades in the fields of experimental psychology and experimental social psychology, and more recently in the field of experimental political science (Berinsky, Huber, and Lenz 2012). Additionally, several studies report that estimated treatment effects in online convenience samples are similar to those observed in probability-based studies (Berinsky, Huber, and Lenz 2012; Coppock forthcoming; Mullinix et al. 2015; Weinberg, Freese, and McElhattan 2014). However, even advocates of online convenience sources note that they do not replace the need for probability-based population samples, as the latter provide a critical baseline allowing researchers to assess the conditions under which convenience samples provide useful or misleading inferences about causal attribution (Levay et al. 2016; Mullinix et al. 2015).

For researchers with limited training in survey methods, this can be difficult terrain to navigate. It may not be clear to them how convenience samples differ from probability-based ones or what steps one might take to reduce the risk of making erroneous inferences with either kind of sample. As such, and in the tradition of Donald T. Campbell, we offer what we consider to be some suggested best practices for researchers conducting experiments with any kind of survey data, but especially data coming from a nonprobability online convenience source.

1. *Consider how the nature of the sample might interact with the experiment*: Participants in a study might differ from their counterparts in the broader population in some fundamental way that is related to their participation in the survey platform. For example, online marketplace workers presumably all have access to the Internet as well as a desire to earn supplemental income from their participation in an online convenience panel. If these characteristics are related to key variables in the experiment, but are not shared with the study's target population, then such a subject pool may yield biased results. For example, if the research goal is to study the behavior of senior citizens and there is a reason to believe that the seniors in a given sample will react to the experimental stimulus differently than seniors in the population, further investigation is warranted, as is enhanced caution on behalf of the researchers in drawing conclusions about their findings. The researcher should consider embedding checks for that possibility into the survey, as well as whether other (e.g. offline) sample sources should be used to guard against this issue.
2. *Apply weighting designed to address sample skews that could bias estimates*: Many convenience survey platforms have demographic and ideological skews different from the broader population that their creators and users hope they represent. For example, online opt-in samples tend to have disproportionately high shares of adults who live alone, collect unemployment benefits, do not have children, and are low income (Pew Research Center 2016). Laborers on Mechanical Turk skew younger and more liberal than the U.S. public (Berinsky, Huber, and Lenz 2012). Weighting is an important tool for attempting to address such skews when they threaten survey estimates, as they often do. For online convenience, surveys in particular, weighting just on core demographics (e.g. gender, age, race) tends to be insufficient. Adjusting on demographics plus additional variables associated with both propensity to be interviewed and survey outcomes can help to reduce bias (e.g. Pew Research Center 2018; Schonlau, van Soest, and Kapteyn 2007; Wang et al. 2015). That said, even careful weighting does not always reduce bias on substantive variables. A meta-analysis by Tourangeau, Conrad, and Couper (2013) found that weighting removed at most up to three-fifths of the bias in online samples and that large biases on some variables persisted. Therefore, we recommend applying weights when analyzing the effects of experiments embedded in convenience surveys and contrasting results based on the weights with those results ignoring the weights entirely. Elliott and Valliant (2017) provide more technical details about the use of weights when analyzing data from nonprobability samples.

3. *Attempt to estimate variances and discuss/disclose the assumptions involved*: Professional guidance on this point has evolved in recent years, at least as it concerns using online convenience samples to make claims about public opinion (e.g. the percentage of Americans that favor privatizing social security). The early advice was for researchers to simply refrain from reporting variance estimates (e.g. a margin of sampling error) with opt-in samples. However, as Baker et al. (2013) noted that guidance ignored decades of variance estimation work by statisticians in clinical trials, evaluation research, and other domains. Arguably, it also had the inadvertent effect of encouraging survey consumers to simply ignore sampling error. Updated guidance encourages researchers working with opt-in samples to report measures of precision for their estimates (AAPOR 2015; Baker et al. 2013). In doing so, it is important to note that probability-based, simple random sample (SRS) assumptions are inappropriate. The precision of estimates from opt-in samples is a *model-based* measure, not the average deviation from the population value over all possible samples. Researchers using opt-in data should, therefore, provide/disclose a detailed description of how their underlying model was specified, its assumptions validated, and the measure(s) calculated. The AAPOR provides guidance for doing this (2016).
4. *Exercise particular caution when making inferences about non-Whites*: One acute concern is the possibility that subgroups at elevated risk of negative outcomes (e.g. on health, employment, crime victimization, etc.) also happen to be subgroups that are not well-represented in online convenience platforms. Studies documenting that online convenience samples tend to underrepresent lower income, less educated, Black, and/or Hispanic adults provide some evidence to this effect (e.g. Baker et al. 2010; Berinsky, Huber, and Lenz 2012). Critically, this problem appears to persist even after weighting. A recent study found that *weighted* online survey estimates for Blacks and Hispanics were off by over 10 percentage points on average, relative to benchmark values for these subpopulations computed from federal studies (Pew Research Center 2016). In addition, the online samples rarely yielded accurate estimates of the marginal effects of being Hispanic or Black on substantive outcomes, when controlling for other demographics. These results suggest that researchers using online nonprobability samples for their experiments are at risk of drawing erroneous conclusions about the effects associated with race and ethnicity. At the very least, when considering models for experimental effects on measures of interest in convenience samples, researchers should consider the possibility of treatment effect heterogeneity (addressed in Chapter 22 of this volume) and test interactions between indicators for non-Whites and the treatment variable of interest.
5. *Attempt to replicate findings with other sources*: If possible, experimental researchers using nonprobability platforms should attempt to replicate their experiments using random samples of the U.S. population or whatever else may be their target population similar to the program Time Sharing Experiments in the Social Sciences (Mutz 2011). If that is cost prohibitive, a potentially useful though less informative strategy would be to use a different convenience sample source. Knowing, for example, that an estimated treatment effect observed on Mechanical Turk replicated on an opt-in survey panel would provide some support for the robustness of the finding, though this assumes that the selection mechanisms and associated biases on the two platforms are different. Given that selection mechanisms for online convenience samples are neither controlled nor well documented, this strategy is far from infallible.

Although we consider these steps as a good practice, we would emphasize that they do not eliminate the risk of significant problems when using nonprobability samples, including online convenience ones, for experimentation. We note that many of these recommendations also apply to low response rate probability-based surveys, not just convenience samples. Techniques for addressing noncoverage and/or nonresponse biases tend to be equally applicable, if not equally effective, for both probability and nonprobability experimental designs.

1.8 The Contents of this Book

There are 23 substantive chapters that follow in this book, and these are organized into nine topical areas, many of which are components of the TSE framework:

- Coverage
- Sampling
- Nonresponse
- Questionnaire
- Interviewers
- Special surveys
- Trend data
- Vignettes
- Analysis.

For each area, there is a brief introduction written by the coeditors responsible for the chapters in that area. Most of the chapters were chosen by the editors through a "competitive" process by which a widely publicized Call for Chapters was distributed throughout the world. A few of the chapters were invited from the authors by the editors.

In each chapter, authors were asked to state the focus of the experimental research topic they would be addressing, provide a critical literature review of past experimentation in this topic area, present at least one original case study of an experiment that they conducted within a probability-based survey, address the strengths and limitations of their case study/studies, make proscriptive recommendations for how to best carry out experimentation in this topic area, and identify areas where additional research is needed. Most of the substantive sections of the book have multiple chapters. No effort was made to reconcile redundancies across chapters or disagreements. Instead, the editors thought that it would benefit readers to learn where there is an agreement on a topic and where it does not yet exist.

Finally, we six editors would like to very much thank the 61 authors throughout the world who contributed the 23 substantive chapters in the book. We thank them most for their intellectual contributions, including showcasing in their respective chapters at least one previously unpublished survey-based experiment that they planned, conducted, and analyzed within a probability-based survey sampling design. We also would like to thank them for their considerable commitment and patience during the four years they worked closely and very cooperatively with us.

References

American Association for Public Opinion Research (AAPOR) (2015). Code of ethics. https://www.aapor.org/Standards-Ethics/AAPOR-Code-of-Ethics.aspx (accessed 1 March 2019).

American Association for Public Opinion Research (AAPOR) (2016). AAPOR guidance on reporting precision for nonprobability samples. https://www.aapor.org/getattachment/Education-Resources/For-Researchers/AAPOR_Guidance_Nonprob_Precision_042216.pdf.aspx (accessed 1 March 2019).

Aronson, E. and Carlsmith, J.M. (1968). Experimentation in social psychology. In: *The Handbook of Social Psychology*, 2e, vol. 2 (ed. G. Lindzey and E. Aronson), 1–79. Reading, MA: Addison-Wesley.

Baker, R., Blumberg, S.J., Brick, J.M. et al. (2010). Research synthesis: AAPOR report on online panels. *Public Opinion Quarterly* 74: 711–781.

Baker, R., Brick, M.J., Bates, N.A. et al. (2013). Summary report of the AAPOR task force on non-probability sampling. *Journal of Survey Statistics and Methodology* 1: 90–143.

Berinsky, A.J., Huber, G.A., and Lenz, G.S. (2012). Evaluating online labor markets for experimental research: Amazon.com's mechanical turk. *Political Analysis* 20: 351–368.

Biemer, P.B., Groves, R.M., Lyberg, L.E. et al. (2004). *Measurement Error in Surveys*. New York: Wiley.

Biemer, P., de Leeuw, E., Eckman, S. et al. (2017). *Total Survey Error in Practice*. Hoboken, NJ: Wiley.

Campbell, D.T. (1988). The experimenting society. In: *Methodology and Epistemology for the Social Sciences: Selected Papers of Donald T. Campbell* (ed. S. Overman). Chicago: University of Chicago Press.

Campbell, D.T. and Stanley, J.C. (1966). *Experimental and Quasi-Experimental Designs for Research*. Chicago: Rand McNally.

Cook, T.D. and Campbell, D.T. (1979). *Quasi-Experimentation: Design and Analysis Issues for Fields Settings*. Cambridge, MA: Houghton Mifflin.

Cook, T.D., Gruder, C.L., Hennigan, K.M., and Flay, B.R. (1979). The history of the sleeper effect: some logical pitfalls in accepting the null hypotheses. *Psychological Bulletin* 86: 662–679.

Coppock, A. (2018). Generalizing from survey experiments conducted on mechanical turk: a replication approach. *Political Science Research and Methods* 1–16.

Crano, W.D. and Brewer, M.B. (1973). *Principles of Research in Social Psychology*. New York: McGraw-Hill, Inc.

Elliott, M.R. and Valliant, R. (2017). Inference for nonprobability samples. *Statistical Science* 32 (2): 249–264.

Fienberg, S.E. and Tanur, J.M. (1987). Experimental and sampling structures: parallels diverging and meeting. *International Statistical Review* 55 (1): 75–96.

Fienberg, S.E. and Tanur, J.M. (1989). Experimentally combining cognitive and statistical approaches to survey design. *Science* 243: 1017–1022.

Fienberg, S.E. and Tanur, J.M. (1996). Reconsidering the fundamental contributions of Fisher and Neyman on experimentation and sampling. *International Statistical Review* 64: 237–253.

Gaziano, C. (2005). Comparative analysis of within-household respondent selection techniques. *Public Opinion Quarterly* 69: 124–157.

Groves, R.M. (1989). *Survey Errors and Survey Costs*. New York: Wiley.

Groves, R.M. and Couper, M.P. (1998). *Nonresponse in Household Interview Surveys*. New York: Wiley.

Groves, R. and Lyberg, L. (2010). Total survey error past, present, and future. *Public Opinion Quarterly* 74 (5): 849–879.

Kalaian, S.A. and Kasim, R.M. (2008). External validity. In: *Encyclopedia of Survey Research Methods* (ed. P.J. Lavrakas), 254–256. Thousand Oaks, CA: Sage.

Hauser, D.J. and Schwarz, N. (2016). Attentive Turkers: MTurk participants perform better on online attention checks than do subject pool participants. *Behavioral Research* 48: 400–407.

Lavrakas, P.J., Courser, M., and Diaz-Castillo, L. (2002) Differences between Hispanic 'origin' and Hispanic 'identity' and their implications. 2002 American Association for Public Opinion Research Conference, St. Petersburg, FL.

Lavrakas, P.J., Courser, M.W., and Diaz-Castillo, L. (2005). What a difference a word can make: new research on the differences between hispanic 'origin' and hispanic 'identity' and their implications. 2005 American Association for Public Opinion Research; Miami, FL.

Levay, K.E., Freese, J., and Druckman, J.N. (2016). The demographic and political composition of mechanical turk samples. *SAGE Open* 6: 1–17.

Mullinix, K.J., Leeper, T.J., Druckman, J.N., and Freese, J. (2015). The generalizability of survey experiments. *Journal of Experimental Political Science* 2: 109–138.

Mutz, D. (2011). *Population-Based Experiments.* Princeton, NJ: Princeton University Press.

Orwin, R.G. and Boruch, R.F. (1982). RRT meets RDD: statistical strategies for assuring response privacy in telephone surveys. *Public Opinion Quarterly* 46 (4): 560–571.

Pew Research Center (2016). Evaluating Online Nonprobability Surveys. http://www.pewresearch.org/methods/2016/05/02/evaluating-online-nonprobability-surveys (accessed 1 March 2019).

Pew Research Center (2018). For Weighting Online Opt-in Samples, What Matters Most? http://www.pewresearch.org/methods/2018/01/26/for-weighting-online-opt-in-samples-what-matters-most (accessed 1 March 2019).

Schonlau, M., van Soest, A., and Kapteyn, A. (2007). Are "webographic" or attitudinal questions useful for adjusting estimates from web surveys using propensity scoring? *Survey Research Methods* 1: 155–163.

Tourangeau, R., Conrad, F.G., and Couper, M.P. (2013). *The Science of Web Surveys.* Oxford: Oxford University Press.

Traugott, M.W. and Lavrakas, P.J. (2016). *The Voter's Guide to Election Polls,* 5e. Lulu Online Publication.

Visser, P., Krosnick, J.A., and Lavrakas, P.J. (2013). Survey research. In: *Handbook of Research Methods in Personality and Social Psychology,* 2e (ed. H.T. Reis and C.M. Judd). Cambridge: Cambridge University Press.

Wang, W., Rothschild, D., Goel, S., and Gelman, A. (2015). Forecasting elections with non-representative polls. *International Journal of Forecasting* 31: 980–991.

Weinberg, J.A., Freese, J., and McElhattan, D. (2014). Comparing data characteristics and results of an online factorial survey between a population-based and a crowdsource-recruited sample. *Sociological Science* 1: 292–310.

Wikipedia (2018a). Replication crisis. https://en.wikipedia.org/wiki/Replication_crisis (accessed 1 March 2019).

Wikipedia (2018b). Ecological validity. https://en.wikipedia.org/wiki/Ecological_validity (accessed 1 March 2019).

Part I

Introduction to Section on Within-Unit Coverage

Paul J. Lavrakas[1] *and Edith D. de Leeuw*[2]

[1] *NORC, University of Chicago, 55 East Monroe Street, Chicago, IL 60603, USA*
[2] *Department of Methodology & Statistics, Utrecht University, Utrecht, the Netherlands*

There are many surveys for which there is not an accurate and comprehensive frame that provides names and other information about the people who live at sampled residences or who work at sampled businesses and other institutions. Therefore, for many probability samples of households, businesses, or other organizations, it is very important that a researcher consider how the designated respondent(s) at each unit will be chosen so as to minimize errors of representation of the target population. There is a fairly large literature on the topic of selecting one designated respondent from within a household in general population surveys when recruitment is done via an interviewer (cf. Gaziano 2005). However, there is not much literature on techniques for the selection of a designated respondent from businesses or other organizations. In the case of the latter, it is most often that the person holding a certain position within the company or organization (e.g. CEO, CFO, Owner, Principal, Chief of Staff, Head of HR, etc.) is the one person at the sampled unit from whom the researchers seek to gather data. However, in some instances, there are multiple people within a residence or an organization who qualify as being eligible to complete the survey questionnaire and thus a systematic selection technique must be deployed. For probability surveys that do not use interviewers to recruit respondents, the issue of within-unit selection is just as important as it is in interviewer-administered surveys. However, these surveys must rely on other means than interviewers to explain to someone in the sampled unit how to determine who within the unit the designated eligible respondent becomes (cf. Olson et al. 2014). Regardless of the technique, it must be known to work reliably in leading to an accurate within-unit selection of the designated respondent, otherwise why would a researcher want to use it?

In less common surveys, more than one person is recruited to provide data from within the same sampling unit. Examples of these studies include surveys that strive to gather data from dyads, such as a patient and her/his health-care provider, a parent and one of her/his children, a supervisor and one of her/his employees, a school teacher and one of her/his administrators, or an accused perpetrator and the arresting officer. In these instances, the accuracy of the techniques used to select the eligible *respondents* at the sampled unit is also crucial to the success of the survey.

Experimental Methods in Survey Research: Techniques that Combine Random Sampling with Random Assignment, First Edition.
Edited by Paul J. Lavrakas, Michael W. Traugott, Courtney Kennedy, Allyson L. Holbrook, Edith D. de Leeuw, and Brady T. West.

Companion Website: www.wiley.com/go/Lavrakas/survey-research

There are several systematic techniques that have been devised to select one eligible respondent within a sampled household. The most rigorous of these was the first reported by Kish (1949). That technique, unlike those that followed, closely simulated a true random selection of a designated respondent within the household. Since then, several other quasi-random systematic methods have been devised in part because it has long been feared that the rigor of the Kish technique increases nonresponse when the person who is completing the selection technique for her/his household or business turns out not to be the designated respondent (cf. Gaziano 2005).

When experiments are conducted within surveys to assess the effects of the within-unit eligibility/selection technique used to identify who within the sampled unit should provide data, there are three types of dependent variables that researchers should measure. One type is variables that reflect the extent to which coverage of the target population is affected by a given selection technique. The second type of dependent variables is the metrics used to indicate how the selection techniques being tested in the experiment affect the response rates to the survey. The third type is variables that can show the level of accuracy that is achieved via the deployment of a given selection technique. Thus, experiments in this domain aim to test selection/eligibility techniques that minimize within-unit coverage errors without inflating unit nonresponse and unit nonresponse bias.

This section contains two chapters. The first chapter by Smyth, Olson, and Stange addresses experiments comparing different techniques to select one eligible respondent within each household sampled for a general population survey. As noted, there has been a fair amount of research reported on this topic related to surveys that use interviewers to select and recruit a designated respondent from the household. Using a total survey error perspective, Smyth et al. provide a broad review of the types of selections techniques available to researchers, addressing their use both within interviewer-administered telephone surveys and self-administered mail surveys. Their chapter goes on to report their original experiment testing of two theory-guided techniques (incentives and targeted wording) to try to improve the accuracy of the selection process within a probability-based mail survey of the general population. Their dependent variables were the response rate, representation of the resulting sample, and accuracy of the selection under the different experimental treatment conditions.

The second chapter is from O'Muircheartaigh, Smith, and Wong and addresses the domain of surveys that need to gather data from more than one person in the sampled unit. The focus is on how experiments can be used to try to (i) improve response rates which otherwise may be lower than when data need be gathered from only one household member, (ii) improve the quality of all the data from a household by avoiding direct contamination through discussion among within-unit respondents, including that stemming from anxiety about possible breaches of confidentiality, and (iii) address the statistical concern that a positive intra-household correlation renders second, and subsequent, interviews in a household less informative than the first. In addition, their chapter addresses many operational concerns that researchers face when they must implement and control the integrity of a complex experiment within such a study.

With the increased use of addressed-based samples, the topic of how well a given within-unit selection technique accomplishes the purposes it meant to serve becomes even more important. Apart from these chapters, issues that merit more future experimentation within this domain include whether within-unit coverage error or nonresponse error is the greater threat to the accurately of a survey. Furthermore, experiments in this domain in which interviewers play a major role in implementing the experimental treatments (those selection techniques that are being compared in the experiment) must use analytic techniques that model the variance that occurs across interviewers in the manner that they implement all the selection techniques to which they are assigned (cf. Chapter 12 in this book by Lavrakas, Kelly, and McClain).

References

Gaziano, C. (2005). Comparative analysis of within-household selection techniques. *Public Opinion Quarterly* 69 (1): 124–157.

Kish, L. (1949). A procedure for objective respondent selection within the household. *Journal of the American Statistical Association* 44 (247): 380–387.

Olsen, K., Stange, M., and Smyth, J. (2014). Assessing within-household selection methods in household mail surveys. *Public Opinion Quarterly* 78 (3): 656–678.

2

Within-Household Selection Methods: A Critical Review and Experimental Examination

Jolene D. Smyth[1], *Kristen Olson*[1], *and Mathew Stange*[2]

[1] *Department of Sociology, University of Nebraska-Lincoln, Lincoln, NE, USA*
[2] *Mathematica Policy Research, Ann Arbor, MI, USA*

2.1 Introduction

Probability samples are necessary for making statistical inferences to the general population (Baker et al. 2013). Some countries (e.g. Sweden) have population registers from which to randomly select samples of adults. The U.S. and many other countries, however, do not have population registers. Instead, researchers (i) select a probability sample of households from lists of areas, addresses, or telephone numbers and (ii) select an adult within these sampled households. The process by which individuals are selected from sampled households to obtain a probability-based sample of individuals is called within-household (or within-unit) selection (Gaziano 2005). Within-household selection aims to provide each member of a sampled household with a known, nonzero chance of being selected for the survey (Gaziano 2005; Lavrakas 2008). Thus, it helps to ensure that the sample represents the target population rather than only those most willing and available to participate and, as such, reduces total survey error (TSE).

In interviewer-administered surveys, trained interviewers can implement a prespecified within-household selection procedure, making the selection process relatively straightforward. In self-administered surveys, within-household selection is more challenging because households must carry out the selection task themselves. This can lead to errors in the selection process or nonresponse, resulting in too many or too few of certain types of people in the data (e.g. typically too many female, highly educated, older, and white respondents), and may also lead to biased estimates for other items. We expect the smallest biases in estimates for items that do not differ across household members (e.g. political views, household income) and the largest biases for items that do differ across household members (e.g. household division of labor).

In this chapter, we review recent literature on within-household selection across survey modes, identify the methodological requirements of studying within-household selection methods experimentally, provide an example of an experiment designed to improve the quality of selecting an adult within a household in mail surveys, and summarize current implications for survey practice regarding within-household selection. We focus on selection of one adult out of all possible adults in a household; screening households for members who have particular characteristics has additional complications (e.g. Tourangeau et al. 2012; Brick et al. 2016; Brick et al. 2011), although designing experimental studies for screening follows the same principles.

Experimental Methods in Survey Research: Techniques that Combine Random Sampling with Random Assignment, First Edition. Edited by Paul J. Lavrakas, Michael W. Traugott, Courtney Kennedy, Allyson L. Holbrook, Edith D. de Leeuw, and Brady T. West.

Companion Website: www.wiley.com/go/Lavrakas/survey-research

2.2 Within-Household Selection and Total Survey Error

Inaccurate within-household selection can contribute to TSE in multiple ways. First, every eligible member of the household has to be considered by the household informant during the within-household selection process. The household informant needs to identify a "list" (written down or not) of eligible household members. If eligible members are excluded from the list, undercoverage occurs. If certain people tend to be systematically excluded from household lists (e.g. young men) and their characteristics are related to constructs measured in the survey (e.g. health-care expenditures), their exclusion will result in increased coverage bias of survey estimates. Second, assuming that the list of eligible household members is complete, the interviewer, if there is one, has to accurately administer, and the informant has to correctly follow, the selection instructions. Mistakenly or intentionally selecting the wrong household member from the (conceptual) list of household members can affect sampling error, especially sampling bias (e.g. if there is similarity across households in the characteristics of those erroneously selected and these characteristics are related to measured survey constructs). Finally, nonresponse error can result if the within-household selection procedure dissuades certain types of households or certain types of selected household members from completing the survey. The joint effects of the within-household selection procedure on any one of these three error sources (coverage, sampling, and nonresponse) may bias survey estimates. As a result, a number of different selection procedures have been developed, some of which prioritize obtaining true probability samples and some of which relax this criteria to potentially reduce coverage, sampling, and nonresponse errors.

2.3 Types of Within-Household Selection Techniques

Researchers can sample individuals within households using various probability, quasi-probability, and nonprobability, and convenience methods (Gaziano 2005). The Kish (1949), age-order (Denk and Hall 2000; Forsman 1993), and full enumeration (and variations of these) techniques obtain a *probability* sample of individuals from within households by ensuring that each eligible member of a sampled household has a known, nonzero chance of becoming the selected survey respondent. Probability methods of within-household selection require the most information about household members, including the number of people living in the household, and often more intrusive information such as household members' sex and age. The interviewer asks the household informant for the requisite information about the household and then follows systematic procedures to select (or the interviewer's computer selects) a respondent from the household. To our knowledge, full probability sample procedures are rarely used in self-administered surveys because they are so complex; even in interviewer-administered modes, they pose some challenges from a TSE framework. While full enumeration procedures are intended to reduce coverage error, by requesting sensitive information upfront, they may increase nonresponse error.

The last birthday and next birthday within-household selection techniques are quasi-probability methods because household members' birthdates identify who should be the respondent rather than a truly random selection mechanism. In the birthday techniques, the researcher uses an interviewer (in interviewer-administered modes) or the cover letter (in self-administered modes) to ask the household member who has the birthday that will occur next (next birthday) or who most recently had their birthday (last birthday) relative to a reference date to respond to the survey. Birthday techniques assume that birthdates are functionally random for the purposes of identifying a member of the household to respond. For many topics, this assumption seems warranted; however, for topics where the variables of interest are related to birthdays (i.e. voting at age 18), this method may not be appropriate. These techniques are popular in both interviewer- and self-administered questionnaires because of their ease of implementation, although the selection process is often inaccurately completed in any mode.

Variations that aim to reduce the intrusiveness of the probability methods by combining probability and quasi-probability methods and accounting for household size also exist (e.g. Rizzo method – Rizzo et al. 2004; Le et al. 2013). These methods first obtain information about the number of people in a household and then use different methods for households with two adults (unobtrusive random selection of the informant or the other adult) and households with three or more adults (more obtrusive requests for enumeration, using a birthday method, asking by age position and possibly sex of the adults in the household). These methods reduce the proportion of households subjected to more intrusive methods; for example, most U.S. households have only one or two adults (Rizzo et al. 2004).

Quota or targeted techniques identify a respondent based on demographic criteria, such as the youngest male or oldest female from the selected household, or simply select any adult from the household. These methods are nonprobability methods, meaning the researcher loses the statistical theory linking the sample to the target population, thus undermining the representation side of the TSE framework. However, they are less costly, less intrusive, and easier to implement accurately. Nonprobability methods can be used in any data collection mode. In a telephone survey, the interviewer may ask for a knowledgeable respondent or take the phone answerer as the respondent; in a mail survey, the instructions will appear in a cover letter, if at all.

2.4 Within-Household Selection in Telephone Surveys

In a telephone survey, the interviewer (typically assisted by a computer) selects and encourages the sampled household member to participate in the survey using one of the methods described above. In telephone surveys conducted up to the early 2000s, less invasive techniques (i.e. birthday and nonprobability techniques) demonstrated the tradeoff across error sources. They tended to have higher response rates and lower cost but less representative demographic compositions than more invasive probability techniques such as the Kish method (Gaziano 2005; Yan 2009).

More recently, probability-based within-household selection methods continue to result in lower *response rates* than quasi-probability birthday techniques and nonprobability techniques (Marlar et al. 2014; Longstreth and Shields 2005; Beebe et al. 2007). For example, in a comparison of probability, quasi-probability, and nonprobability methods, Marlar et al. (2014) found that the probability-based Rizzo et al. (2004) method garnered response rates that were roughly 2.5 percentage points lower than selecting a respondent based on age/sex criteria and being at home, with quasi-probability and nonprobability methods selecting among all people in the household (not just those at home) in the middle. Longstreth and Shields (2005) had a similar magnitude difference in response rates comparing the last birthday method to the Rizzo method. Beebe et al. (2007) compared the Rizzo method with the next birthday method, finding response rates for the next birthday method about 4 percentage points higher than the Rizzo method.

In these studies, the *composition* of completed samples did not differ unless demographic characteristics were part of the selection method. For example, Beebe et al. (2007) and Longstreth and Shields (2005) both found no differences in demographic characteristics such as sex, age, race, education, income, and number of people in the household in the completed samples produced by the Rizzo selection procedure and either birthday selection technique (last or next birthday). On the other hand, Marlar et al. (2014) found that the youngest male/oldest female technique resulted in more males in the sample, while selecting the youngest person in the household produces a sample that contained more females. In an international context, Le et al. (2013) found no difference in composition of the respondent pool across sex or age characteristics comparing the Kish method to a new household-size dependent procedure. Moreover, none of the studies found differences in substantive estimates by within-household selection procedure.

Other outcomes used to assess within-household selection methods include the *accuracy of selection* and *cost information*. Among the existing telephone research, only Marlar et al. (2014) examined the accuracy of selection, finding that roughly 20–30% of respondents were inaccurately selected in the quasi-probability birthday methods, which is similar to earlier research (O'Rourke and Blair 1983; Troldahl and Carter 1964; Lavrakas et al. 2000; Lind et al. 2000). By comparison, the nonprobability methods they tested had considerably lower inaccuracy rates (youngest person – 20.1%, youngest male/youngest female – 1.8%, multiquestion youngest male/youngest female – 0.5%) (Marlar et al. 2014). For cost, little information is available. Longstreth and Shields (2005) found that the completion time for interviewers to implement the Rizzo and last birthday methods did not significantly differ, and Beebe et al. (2007) found that the mean number of call attempts to interview was the same across the Rizzo and next birthday methods.

2.5 Within-Household Selection in Self-Administered Surveys

Unlike telephone surveys, self-administered surveys cannot rely on trained interviewers to administer the selection procedures. In mail surveys, the household informant opens the mail, reads the selection technique typically described in a survey's cover letter or on the questionnaire, and determines which member of the household should complete the survey. The household informant must then complete the survey if they are selected or must convince the selected person to complete the survey. Problems can arise at any of these steps.

One fundamental difference between mail and telephone surveys is that true probability methods such as the Kish selection method are considered too complex for households to implement in mail surveys (Battaglia et al. 2008; Reich et al. 1986). As such, researchers have most often employed the quasi-probability birthday methods or nonprobability techniques to try to reduce coverage, sampling, and nonresponse errors at the expense of true probability methods.

Unlike telephone surveys, there are few significant differences in *responses rates* by type of within-household selection method in mail surveys (Battaglia et al. 2008; Olson et al. 2014). For example, despite the any-adult technique being minimally burdensome, it yields response rates similar to the next birthday method (Battaglia et al. 2008). Across two studies of Nebraskans, the next and last birthday selection procedures had statistically identical response rates as the oldest adult procedure, but the youngest adult method had a significantly lower response rate, likely driven by lower response rates among younger adults in general (Olson et al. 2014).

Similarly, the demographic *composition* of respondent pools do not significantly differ across the within-household selection techniques in mail surveys and all the methods result in samples that significantly differ from demographic benchmarks in similar ways (Battaglia et al. 2008; Hicks and Cantor 2012; Olson et al. 2014). The any adult, all adult, and next and last birthday techniques all tend to underrepresent younger people and overrepresent non-Hispanic whites, adults with higher education, and married people. Studies also find no significant differences in substantive survey estimates across within-household selection techniques (Battaglia et al. 2008; Hicks and Cantor 2012; Olson et al. 2014).

Selection accuracy is the primary focus of within-household selection evaluations in self-administered surveys. Across studies, up to 30% of within-household selections are inaccurate, with (substantially) higher rates when excluding one-adult households that have accurate selections by default (Stange et al. 2016; Olson and Smyth 2014; Olson et al. 2014; Battaglia et al. 2008; Schnell et al. 2007; Gallagher et al. 1999). Moreover, inaccuracy rates do not significantly differ by selection technique (Olson et al. 2014).

Few studies have examined within-household selection methods in web surveys. In part, this is because web surveys are often used to survey named people, using individualized emails to deliver the survey invitation. When researchers want to send a web survey to a household and administer a within-household selection procedure, they typically do so using a mixed-mode design in which the invitation letter is delivered by postal mail (e.g. Smyth et al. 2010). In this case, researchers have largely adopted mail within-household selection procedures, most often including a quasi-probability selection instruction in the invitation letter. In the only study of which we are familiar that assessed selection accuracy in web surveys (using postal mail invitations), the inaccuracy rate in web was about 20% and did not significantly differ from the inaccuracy rate in mail-only, or in conditions mixing mail and web data collection modes (Olson and Smyth 2014).

2.6 Methodological Requirements of Experimentally Studying Within-Household Selection Methods

The goal of within-household selection of a single adult is to produce a (quasi-)probability-based sample that mirrors the target population (i.e. minimizes coverage and nonresponse error from a TSE perspective) on characteristics being measured in the survey. As such, there are three general methods for assessing the quality of the within-household selection:

1. comparing the characteristics of the completed sample to benchmark measures for the target population,
2. comparing survey estimates across the experimental treatments, and
3. evaluating how well the completed samples followed the within-household selection instructions by measuring the accuracy of selection.

For example, for state, regional, or national surveys, one can compare the demographic makeup of the completed sample to official statistics for the same geographic region, such as from the American Community Survey (ACS). Of course, this requires that benchmark outcomes be available for the target population. Comparison of estimates in the survey across experimental treatments should be guided by the mechanisms for what might differ across experimental treatments (e.g. age in the youngest adult method). Accuracy requires obtaining external information about household composition from a rich sampling frame or incorporating methods to assess the accuracy of selection (at least among the respondents) in the survey itself. With these three methods in mind, it is possible to identify the appropriate experimental design for studying within-household selection methods.

Comparing characteristics of the completed sample to benchmark data requires minimizing other sources of survey error that might affect the composition of the final sample. Thus, an experimental study of within-household selection should start with **a sample frame with good coverage** so that coverage error in the sample frame is excluded as an explanation for differences between the completed sample characteristics and benchmark measures. It also means that within-household selection experiments need to **start with a probability sample of housing units** from the sample frame. A probability sample of housing units will produce a sample that mirrors the target population so that any differences between the final completed sample of individuals and the benchmark measures can be attributed to measurable sampling error and the within-household selection techniques after accounting for probabilities of selection. The sample does not have to be a simple random sample as long as information about strata, clusters, and unequal probabilities of selection are maintained and incorporated in the analyses. A sample frame with poor coverage or a nonprobability sample of housing units will make it impossible to tell how much of the difference between the respondent pool and benchmark outcomes is due to coverage and sampling from the household frame versus coverage and sampling of individuals within households. Statistically, it is also necessary to ensure that the **sample size is sufficiently large** to allow for enough power to detect significant differences across treatments. This decision will be driven by a power analysis that accounts for the number of experimental treatments, the outcome of interest, the type of analysis used for evaluating the experiment, and the expected effect size that will result from the experiment for that outcome. Thus, the first requirement of such a study under a TSE framework is a sufficiently large probability sample of a known population from a good sample frame.

Of course, any experiment needs experimental treatments or factors. The **selection of the experimental factors and items included in the questionnaire should be informed by theory** to anticipate possible effects of the experimental factors. For example, Olson and Smyth (2014) theorized that there were three reasons for inaccurate within-household selections: confusion, concealment, and commitment. They were able to test these theories using a limited set of questions included in their questionnaire as proxies for each reason: size of household, education, and presence of children in the household were proxies for confusion; gender, age, race, income, concern with identity theft, and fear of crime for concealment; and previously reported mode preference (a variable on the sample frame) for commitment. Thus, using theory to guide the selection of experimental factors and including measures that allow researchers to test theoretical reasons for the success or failure of the selection methods can help advance knowledge of why certain methods work or fail and ways to improve them.

The next requirement of an experimental study of within-household selection methods is that **the sampled housing units be randomly assigned to the alternative experimental treatments**. Randomly assigning housing units to the within-household selection method treatments ensures that each treatment is assigned a representative subset of the sample of housing units. Thus, the composition of the respondent pool in each treatment can be attributed to the within-household selection method used in the treatment, not to differences in the composition of housing units assigned to each treatment. Using both a probability sample of households from a known population and then randomly assigning sampled households to treatments ensures that the sampling design and experimental assignments are not confounded with the experimental treatments.

In addition, **design differences other than the factors being tested between the experimental treatment versions should be eliminated** (i.e. eliminate confounding factors). For example, if incentives are to be used, they should be used in exactly the same way (type, amount, timing, etc.) in all treatments. Likewise, the response device type (e.g. cell phone versus landline phone; computer versus mobile web, etc.) should not differ across treatments nor should the number, type, and timing of contacts or the information communicated in those contacts, other than changes needed for the factor being tested. If testing methods to try to improve the quality of a single within-household selection method, then all other features of the within-household selection method should be held constant. For example, to compare the effects of the selection instruction wording for the next birthday method on selection accuracy, all treatments should use the same selection method (next birthday) and the same wording for all other aspects of the cover letter other than the relevant part being manipulated. Essentially, the only thing that should differ across the treatments is the within-household selection method or elements that are modifying a single within-household selection method.

Another possible confounding factor for within-household selection method experiments in interviewer-administered modes is the interviewer themselves. Interviewers pose two types of threats to the integrity of these experiments. First, they may be differentially skilled at administering within-household selection methods and/or obtaining cooperation from selected household members. If more skilled interviewers are disproportionately assigned to a particular experimental treatment, that treatment may end up performing better because of the interviewers, not because it is the better method. To solve this problem, **interviewers should be randomly assigned to experimental treatments** so that interviewer characteristics (both observable and unobservable) are equally distributed across the treatments. If interviewers cannot be randomly assigned to treatments, then observable interviewer characteristics such as demographic characteristics, interviewer experience or tenure, and even measures of interviewer skill, such as cooperation rates on previous studies, should be collected so that they can be used to statistically control for potential differences in interviewers across the experimental treatments. Analyses of experiments in interviewer-administered surveys should use multilevel models that can account for the nesting of selection procedures and sample cases within interviewers (e.g. Raudenbush and Bryk 2002; Hox et al. 1991). Likewise, interviewer assessments of the characteristics of each method (e.g. ease, sensitivity) should be evaluated. In addition to randomly assigning interviewers to treatments, **the way that cases are assigned to interviewers should be the same across all treatments** to ensure that the assignment of cases to interviewers does not confound results across the treatments.

The second type of challenge that interviewers pose is that their knowledge of the experiment itself may lead them to change their behaviors, either intentionally or unintentionally, in ways that undermine the integrity of the experiment. For example, an interviewer may prefer the ease of the next or last birthday method to a full probability method such as the Kish method. Interviewer expectations can affect response rates (Durrant et al. 2010), and thus these preferences or expectations confound the experiment itself with interviewer preferences. This suggests that **each interviewer should only be assigned to one treatment**; the same interviewer(s) should not work on multiple treatments. This topic is covered in more depth in Chapter 12 of this volume by Lavrakas, Kelly, and McClain.

In addition to considering how the sample of households is drawn and assigned to treatments and interviewers, considerable thought should be given to whether there are **variables on the frame or that can be measured in the survey that can help assess the quality of each selection treatment**. For example, to assess the accuracy of the birthday and oldest/youngest adult methods, Olson et al. (2014) included a household roster in their questionnaire that collected relationship to the respondent, age, date of birth, and sex of each person living or staying in the household. They could then check whether or not the respondent actually had the next or last birthday or were the youngest or oldest adult in the household, depending on the assigned selection method. They were also able to examine whether selection accuracy differed by factors such as the size of the household or whether a household member had a birthday during the field period, both of which positively predicted selection inaccuracies.

In sum, the methodological requirements of experiments for studying within-household selection methods under a TSE framework are:

- Identifying analytic outcomes that will be used to evaluate the methods (e.g. benchmarks and/or ways to assess accuracy);
- A sample frame with good coverage from which a probability sample of housing units will be selected;
- Theoretically driven experimental treatments;
- Random assignment of the selected housing units to experimental treatments;
- Elimination of design differences across the treatments that are not the focus of the comparison;
- In interviewer-administered surveys, random assignment of interviewers to experimental treatments, separate interviewer corps for each treatment, and consistent assignment of cases to interviewers across treatments;
- Inclusion of covariates from the frame or measured in the survey to better understand why differences occur between the treatments.

2.7 Empirical Example

The process of implementing within-household instructions in mail surveys can break down if the informant does not read the instructions, understand them, enumerate a full list of eligible household members, believe in the importance of the selection process, feel motivated to follow the instructions, and/or have the ability to recruit the sampled household member. In our early studies, we found that proxies for confusion such as complexity of the instruction, number of adults in the household, children in the household (Olson and Smyth 2014), and a member of the household having a birthday during the field period (Olson et al. 2014) were associated with higher inaccuracy rates. However, our research designed to reduce confusion (i.e. providing a calendar to help informants place household birthdays in time and providing explanatory instructions to help informants understand why the selection instructions should be followed) failed to improve the quality of sample pools (Stange et al. 2016). The motivation of the informant to implement the instructions and of the selected household member to complete the survey, a factor also discussed by Battaglia et al. (2008), had not been tested. As a result, we designed a new experiment to target motivation. That is, this experiment focuses on the commitment part of the confusion, concealment, and commitment framework theorized by Olson and Smyth (2014). We discuss the theoretical motivation for the experimental treatments, the design of the experiment, and its results here.

One technique previously shown to be effective at encouraging survey participation among unmotivated sample members is providing prepaid (i.e. noncontingent) cash incentives. Numerous studies show that incentives significantly increase response rates (e.g. Church 1993; James and Bolstein 1992; Singer 2002; Singer and Ye 2013; Trussell and Lavrakas 2004). Importantly, Baumgartner and Rathbun (1996) and Groves et al. (2006) found that incentives also encourage participation among sample members who are less interested in the survey topic. These findings suggest that incentives might improve within-household selection in several ways. First, incentives may increase the likelihood that the letter opener will read the cover letter in the first place and the importance they attribute to the survey (Dillman et al. 2014), thus increasing the likelihood that they see and subsequently follow the within-household selection instruction rather than simply doing the survey themselves (i.e. reducing the potential for sampling error). Second, incentives may increase the likelihood that otherwise reluctant household members are included in the household list (i.e. reducing undercoverage), either because they themselves are the selected respondent and want to receive the incentive or because another informant believes that the reluctant household member would want it. Third, the incentive may increase the otherwise reluctant household members' willingness to respond if selected. Thus, we examine whether providing a prepaid, noncontingent incentive improves the performance of the next birthday within-household selection method.

For the incentive to have the largest impact within the context of within-household selection, the selected respondent should receive the incentive. Their receiving the incentive should reduce resistance to being included in the selection process and increase the likelihood that they respond if selected, thereby improving coverage and response rather than simply increasing response rates from the household more generally. As such, in addition to examining the effects of providing an incentive versus no incentive, we also experimentally varied whether or not wording about the incentive in the cover letter was targeted to the selected respondent.

The experiment had three treatments:

1. No incentive
2. $1 incentive with standard letter wording
3. $1 incentive with targeted letter wording.

The standard letter wording was, "We have enclosed a small token of appreciation to thank you for your help," and the targeted letter wording was, "We have enclosed a small token of appreciation to thank the adult with the next birthday for their help." The no incentive condition necessarily omitted all mention of incentives. To eliminate confounds in the experiment and ensure that we could attribute all differences across treatments to either the provision of the incentive or the standard versus targeted wording, the remaining content of the letter in all three treatments was identical.

We expected that the incentive would encourage households to notice and follow the selection instruction, include all household members in the list of eligible household members, and encourage participation when the selected household member was uninterested in the survey, thus targeting coverage, sampling, and nonresponse errors. Thus, we hypothesized that the incentive would lead to:

1. a higher response rate,
2. a completed sample that more closely matched ACS benchmarks for the area under study, and
3. a higher rate of accurate selections, determined through the use of information from a household roster.

We hypothesized that the effect of the incentive on response rates would be attenuated somewhat in the targeted letter wording as this wording reinforces the idea that the incentive and therefore the survey is for the specifically selected person in the household. Thus, we thought it was more likely in this condition that if the selected person refused, the survey would be discarded rather than returned by another adult. Because of this, however, we expected the incentive and targeted wording treatment to most closely match ACS benchmarks and to have the highest rate of accurate selections (i.e. we expected a tradeoff between response rates and selection accuracy in this treatment).

2.8 Data and Methods

We embedded the incentive and cover letter wording experiment in the 2014 Nebraska Annual Social Indicators Survey (NASIS), which is an annual, omnibus mail survey of Nebraska adults aged 19 and older (Bureau of Sociological Research 2014). NASIS 2014 included 93 questions (some with multiple prompts) across 11 pages about natural resources, underage drinking, vaccinations, the Affordable Care Act, invasive plant species, household characteristics, finances, and demographics. The surveys were administered in English only. After obtaining institutional review board approval for the study within our university, the questionnaires were mailed on 20 August 2014. A postcard reminder was sent one week later, and a replacement survey packet was sent to nonrespondents on 18 September 2014. The survey cover letter instructed (with bolded text) that the household member with the next birthday after August 1, 2014, should complete the survey.

The sample consisted of a simple random sample of $n = 3500$ addresses from across Nebraska drawn by Survey Sampling International (SSI) from the USPS computerized delivery sequence (CDS) file. NASIS 2014 was an ideal survey for this experiment because of its use of an address-based sample frame with excellent coverage of US households (Iannacchione 2011) and a probability sampling method. Thus, with frame-based confounds minimized, differences between characteristics of our completed sample and ACS estimates for the state of Nebraska (i.e. a key outcome) can be attributed to coverage, sampling, and response *within* households rather than coverage and sampling *of* households.

The NASIS 2014 sample size was also sufficiently powered to allow us to test our hypotheses. For example, previous years' NASIS surveys, which did not use incentives, yielded response rates around 25% (Bureau of Sociological Research 2013). At the planning stage, we assumed a similar response rate for NASIS 2014 would yield 875 completes or roughly 291 completes per treatment. Based on these assumptions, Table 2.1 shows the effect sizes we anticipated being able to detect with a given level of power across treatments with an alpha of 0.05. If our assumptions held, we would be able to detect effect sizes of 10.7 percentage points with power 0.8 (a typical minimum power level). Table 2.1 also shows the effect sizes we would be able to detect by power level if we compared one treatment to two others combined, to determine the overall effects of the incentive (i.e. the no incentive treatment compared to the two incentive treatments). Ultimately, a total of $n = 1018$ sampled households completed NASIS 2014 for a 29.1% response rate (AAPOR RR1) (for effect sizes by post hoc power level with actual response rates, see the right half of Table 2.1).

To ensure that we could attribute differences across the three experimental treatments to the features of the treatments themselves and not other factors (i.e. different types of households assigned to different treatments), each sampled household was randomly assigned to one of

Table 2.1 Detectable effect sizes (proportions) by power level for anticipated and actual response rates ($\alpha = 0.05$).

	Anticipated response rate = 25%; n = 875		Actual response rate = 29.1%; n = 1018	
Power	Effect size comparing any two treatments	Effect size comparing one treatment to the other two combined	Effect size comparing any two treatments	Effect size comparing no incentive treatment to both incentive treatments combined
0.4	0.064	0.056	0.048	0.056
0.5	0.074	0.064	0.055	0.064
0.6	0.083	0.072	0.062	0.073
0.7	0.094	0.082	0.070	0.082
0.8	0.107	0.092	0.080	0.093
0.9	0.124	0.107	0.093	0.107

the three experimental treatments. This resulted in 1166 households assigned to the no incentive treatment and 1167 households assigned to each of the $1 incentive with standard letter wording and $1 incentive with targeted letter wording treatments. Comparisons of household characteristics provided with the sample (e.g. FIPS [Federal Information Processing Standard] code – a geographic code identifying counties and county equivalents, Census tract, delivery type, race of population in Census tract, age, children, homeowner versus renter, length of residence, and gender) revealed that the randomization worked; there were virtually no significant differences in the types of households assigned to each treatment. The exception is that the no incentive treatment was assigned to slightly more black households and slightly fewer white households than the other treatments. Both of these differences were small in magnitude – less than 2 percentage points and likely attributable to Type I error (Type I error refers to a statistical test being significant by chance alone – that is, a false positive; results available from the authors).

In addition to using a sample frame, probability sampling method, and random assignment to treatments to allow for comparison to the ACS benchmark, NASIS 2014 included a household roster (see Figure 2.1) that we used to determine whether the person answering was the adult in the household with the next birthday (i.e. accuracy/inaccuracy of selection). Following Olson and Smyth (2014), the questionnaire also included a set of covariates designed to reflect theoretically guided correlates of confusion, concealment, or commitment in within-household selection. However, because we had more control over questionnaire content in this experiment, a more extensive set of proxies were included. The confusion proxies included respondent education, children in the household, respondent's marital status, number of adults in the household, and whether the respondent lived in the same household as they did two years ago. These variables capture aspects of cognitive ability (Krosnick 1991; Narayan and Krosnick 1996) or complexity of the household makeup (Martin 1999, 2007; Martin and Dillman 2008;

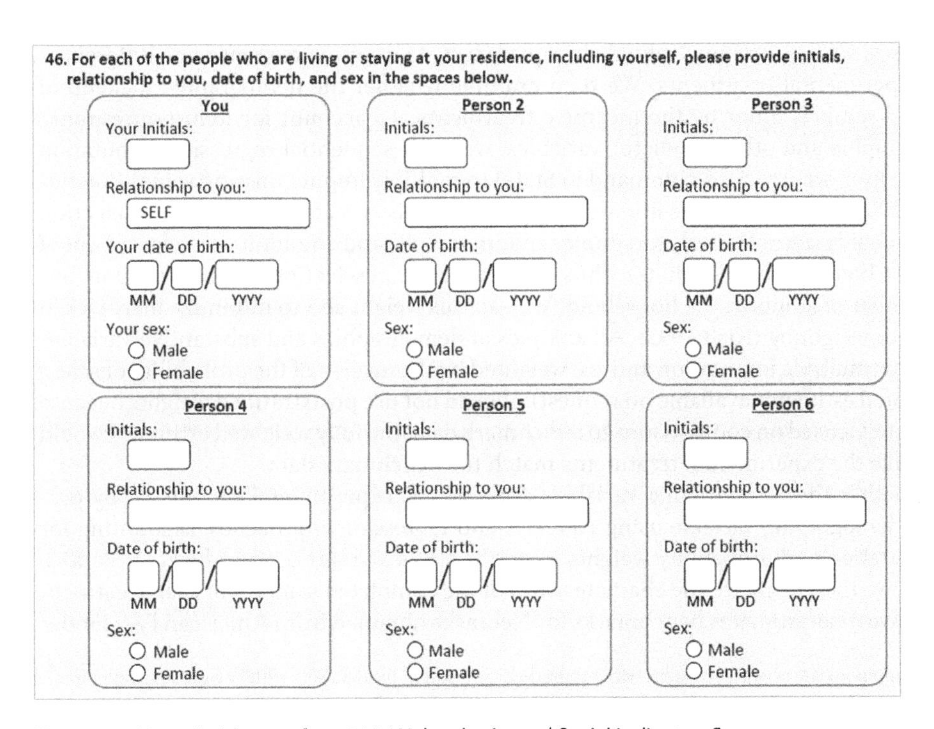

46. For each of the people who are living or staying at your residence, including yourself, please provide initials, relationship to you, date of birth, and sex in the spaces below.

You
Your Initials:
Relationship to you: SELF
Your date of birth: MM / DD / YYYY
Your sex: Male / Female

Person 2
Initials:
Relationship to you:
Date of birth: MM / DD / YYYY
Sex: Male / Female

Person 3
Initials:
Relationship to you:
Date of birth: MM / DD / YYYY
Sex: Male / Female

Person 4
Initials:
Relationship to you:
Date of birth: MM / DD / YYYY
Sex: Male / Female

Person 5
Initials:
Relationship to you:
Date of birth: MM / DD / YYYY
Sex: Male / Female

Person 6
Initials:
Relationship to you:
Date of birth: MM / DD / YYYY
Sex: Male / Female

Figure 2.1 Household roster from 2014 Nebraska Annual Social Indicators Survey.

Olson and Smyth 2014), both of which are expected to increase confusion and thus increase inaccurate selections. The concealment proxies included sex, age, income, and race because previous research has shown that young black men are underrepresented in surveys, with a hypothesis that the household is concealing household members (Tourangeau et al. 1997; Valentine and Valentine 1971). They also include a measure of how often respondents are concerned with identity theft (never to always) and measures of whether the respondent believes most people cannot be trusted, is suspicious of others, is concerned about personal privacy and the number of days the respondent felt sad or hopeful in the past seven days. These measures all reflect respondents' openness to the outside world; those who worry about intrusions from others, feel sad, or lack hopefulness are expected to be more hesitant to engage with the outside world and thus more likely to conceal themselves or family members (Caplan 2003; Kim et al. 2011; Malhotra et al. 2004; McKenna et al. 2002; Olson and Smyth 2014; Phelps et al. 2000; Segrin 2000), leading to inaccurate within-household selections. Finally, the commitment proxies include a set of items measuring who controls entrance into the household (the household gatekeeper) and thus would be the one to initially handle an incoming mail survey. We hypothesized that this person would be more likely to erroneously complete the survey because they are the household member who introduces it to the household, but that this effect would be diminished in the incentive treatments, especially with the targeted letter wording. The gatekeeper covariates included measures of who in the household opens the mail, answers the landline telephone (if available), opens the door for friends and relatives, and opens the door for strangers. These were recoded into dichotomous variables indicating whether the respondent was the person most likely to do each task (0 = no, 1 = yes). Under commitment, we also included an item measuring how likely the respondent is to answer surveys "like this one."

2.9 Analysis Plan

For the analyses, we first use unweighted chi-square tests to examine response rate differences across the experimental treatments. We then examine whether the demographic makeup of the completed samples differ by the incentive treatments. To account for item nonresponse in the demographic and other predictor variables, we use a sequential regression imputation approach (the user-written `ice` command in Stata) to multiply impute missing values (Raghunathan et al. 2001). We created five imputed datasets.[1] We also created probability of selection weights; households were selected as a simple random sample, and one adult was selected out of all adults in the households (1/# adults). Thus, the probability of selection weight is proportionate to the number of adults in the household. We cap this weight at 3 to minimize increases in variance due to weighting (Kish 1992). All analyses of demographics and substantive variables account for this multiple imputation and are weighted by the inverse of the probability of selection (unweighted estimates available on request). We did not use poststratified weights because our analyses are focused on comparisons to benchmark data; the fully weighted estimates would artificially make the experimental treatments match the benchmark data.

We test whether the demographic variables differ across experimental treatments by predicting each demographic variable using ANOVA and regression approaches, accounting for multiple imputation and probability weights using the `mi estimate` procedures in Stata13. Using *t*-tests, we then compare the characteristics of the completed samples in each treatment to ACS 2014 five-year estimates benchmarks for Nebraska obtained from American Fact Finder

1 Creating five imputed data sets is consistent with established convention for data sets with low missing data rates and small fractions of missing information (Rubin and Schenker 1987; Raghunathan et al. 2001). More data sets are needed when the fraction of missing information is high, but our overall low item nonresponse rate (maximum <10%) and low fraction of missing information (maximum <0.18) suggest that five is adequate.

(factfinder.census.gov). For these analyses, we look at respondent's sex, education, whether there are children in the household, age, family income, and race.

We then use birthdate information from the household roster to examine if the household member who completed the survey was the household member with the next birthday following August 1 (i.e. accurate versus inaccurate selection). We examine this for all households and those households with two or more adults because one-adult households automatically have accurate within-household selections. We then test for differences in accuracy by the incentive treatments and examine associations between our proxy measures for confusion, concealment, and commitment and accuracy of selections using logistic regression. In these analyses, we also include control variables for whether the household was located on a farm, open country (not a farm), or a town or city; whether the home was owned, and whether it was a single family dwelling to account for any potential household composition differences across these characteristics. We look at these predictors overall, and whether there are any differences across the experimental treatments using interaction terms between the treatment indicators and proxies.

For all analyses, consistent with our power analysis, we adopt a $p < 0.050$ cutoff for determining statistical significance. However, consistent with the American Statistical Associations statement on p-values (Wasserstein and Lazar 2016), we recognize more than a p-value has to be considered in assessing the importance of statistical results. Therefore, we also discuss results with p-values ranging from 0.050 to 0.100 where effect sizes are also large enough to be meaningful.

2.10 Results

2.10.1 Response Rates

As hypothesized, the incentive increased response rates. The response rate (AAPOR RR1) for the no incentive condition was 22.3% compared to 32.5% for the two incentive conditions combined ($\chi^2(1) = 39.05, p < 0.001$). Also consistent with expectations, among the two incentive conditions, the response rates were 34.3% with the standard letter wording and 30.7% with the targeted wording. Both incentive conditions significantly differed from the no incentive condition (standard $\chi^2(1) = 41.25, p < 0.001$; targeted $\chi^2(1) = 21.03, p < 0.001$), and the 3.6 percentage point difference (a 10.4% reduction) between the standard and targeted incentive conditions approached significance ($\chi^2(1) = 3.45, p = 0.060$).

2.10.2 Sample Composition

As Table 2.2 shows, the sample composition only differed significantly across the three treatments on sex ($F = 8.38, p < 0.001$). The incentive with the standard letter wording treatment yielded a sample that was 62.9% female, which was about 11 percentage points higher than the no incentive treatment ($t = 2.68, p = 0.007$) and 15 percentage points higher than the incentive with targeted wording treatment ($t = 3.97, p < 0.001$). This is also about 12 percentage points higher than the ACS estimate ($t = 4.67, 0 < 0.001$). Thus, the incentive on its own resulted in an overrepresentation of women but using the targeted wording with the incentive appears to have corrected for this overrepresentation.

The distribution of age was moderately significantly different across the three treatments (design-adjusted $F = 1.89$, $p = 0.079$). The addition of the incentive reduced the percent of respondents in the oldest age group (65+) by 6.4 percentage points in the targeted wording treatment ($t = -1.57, p < 0.117$) and 11.5 percentage points in the standard wording treatment ($t = -2.95, p < 0.010$). In the standard wording treatment, this reduction was accomplished primarily through an 8.3 percentage point increase in the percent of respondents in the next

Table 2.2 Demographic composition and comparison to 2014 five-year ACS estimates by treatment.

	Composition (%)					Significance tests across treatments				Significance tests NASIS vs. ACS estimates			
	All	No $ n = 260	$ + stand. n = 400	$ + target n = 358	ACS	Overall F	t stand. vs. no $	t target vs. no $	t stand. vs. target	All vs. ACS	No $ vs. ACS	Stand. vs. ACS	Target vs. ACS
Sex													
Female	54.6	51.7	62.9	47.6	50.9	8.38***	−2.68**	0.93	3.97***	2.21*	0.23	4.67***	−1.17
Male	45.4	48.3	37.1	52.4	49.1					−2.21*	−0.23	−4.67***	1.17
Education													
HS or less	24.6	24.1	22.9	26.7	37.3	1.34	−0.33	0.71	1.15	−8.93***	−4.68***	−6.55***	−4.21***
Some college	35.5	37.0	32.7	37.5	36.2		−1.05	0.12	1.30	−0.44	0.24	−1.39	0.47
BA+	40.0	39.0	44.4	35.8	26.4		1.30	−0.76	−2.25*	8.17***	3.91***	6.72***	3.44***
Children in HH													
No kids	66.4	68.4	66.6	64.6	68.1	0.43	0.44	0.92	0.54	−1.06	0.11	−0.55	−1.27
Has kids	33.6	31.6	33.4	35.4	31.9					1.06	−0.11	0.55	1.27
Age													
19–34	13.3	12.0	13.1	14.3	20.7	1.89+	0.39	0.77	0.43	−6.60***	−4.02***	−4.17***	−3.30***
35–54	30.2	28.4	30.6	31.0	25.4		0.54	0.63	0.11	3.01**	1.01	2.02*	2.09*
55–64	25.1	21.3	29.6	22.9	12.2		2.14*	0.43	−1.89+	8.71***	3.30***	6.95***	4.39***
65+	31.5	38.2	26.7	31.8	13.8		−2.95**	−1.57	1.46	11.48***	7.73***	5.49***	6.96***
Family income													
<$50k	38.0	39.8	36.3	38.5	35.1	0.42	−0.82	−0.31	0.57	1.78	1.45	0.47	1.22
$50–99k	40.3	41.4	40.6	39.1	38.4		−0.17	−0.51	−0.38	1.11	0.90	0.81	0.25
$100k+	21.7	18.8	23.0	22.4	26.4		1.15	0.97	−0.18	−3.22**	−2.77**	−1.41	−1.66
Race													
Nonwhite	9.5	9.2	10.7	8.3	18.8	0.59	−0.63	0.33	1.06	−9.69***	−5.09***	−4.99***	−6.82***
Non-Hispanic white	90.5	90.8	89.3	91.7	81.2					9.69***	5.09***	4.99***	6.82***

Notes: +$p<0.100$, *$p<0.050$, **$p<0.010$, ***$p<0.0001$. Data are weighted for probability of selection and adjusted for multiple imputations.

highest age category (55–64), but in the targeted wording treatment, the increase was spread among all the younger age categories.

With the exception of sex and age, none of the other demographic variables differed significantly across the treatments. Moreover, the overall pattern is that all three treatments significantly differed from the ACS estimates on a number of the demographic characteristics, especially education (overrepresented high education), age (underrepresented the young and overrepresented the old), and race (overrepresented non-Hispanic whites). The no incentive treatment and the incentive with targeted wording treatment did not differ from the ACS on sex or children in the household and the two incentive conditions did not differ from the ACS on family income.

Because the treatments differed in how their estimates compared to the ACS, it is difficult to say that one treatment is better than another from these analyses. One way to assess the overall performance of the treatments is to examine the average absolute differences between the estimates produced by each treatment for each demographic and the corresponding ACS estimate. Looking across all characteristics, the treatment with the incentive and targeted wording had the lowest average absolute difference from the ACS estimates at 6.5 percentage points versus 6.8 percentage points for the no incentive treatment and 8.1 percentage points for the treatment with incentive and standard wording. Taken altogether, the sample composition results suggest that while the differences are not large in magnitude, the treatment with the incentive and targeted wording produced demographic estimates that most closely matched the ACS estimates.

2.10.3 Accuracy

Sufficient information about household members had to be provided in the household roster to determine whether or not the within-household selection was done accurately for each responding household. Accuracy could be determined for 92.6% of households; the accuracy analyses thus are limited to the 943 cases where accuracy could be determined. Households with complete versus incomplete roster information did not differ on any characteristic other than the likelihood to answer surveys – those for whom accuracy could not be determined rated their likelihood of answering surveys like this one significantly lower (2.92 on a 4 point scale) than those for whom accuracy could be determined (3.34; $t = -3.15, p = 0.002$).

Table 2.3 shows accuracy rates overall and by treatment for both the full sample ($n = 943$) and the sample limited to households with two or more adults ($n = 660$). In the full sample, 63.2% of respondents were selected accurately with accuracy rates ranging from 59.9% in the no incentive condition to 66.2% in the incentive condition with targeted letter wording (a 6.3 percentage point difference); the overall difference in accuracy by treatment was not significant ($F = 1.07$, $p = 0.343$). The accuracy rate in the sample limited to households with two or more adults was lower because households with only one adult can only get the selection correct – 55.2% overall and ranging from 50.4% in the no incentive condition to 59.5% in the incentive condition with targeted letter wording. Among this sample in which errors of selection could occur, the overall difference across treatments was not significant ($F = 1.58$, $p = 0.207$), but there was a 9 percentage point difference between the no incentive and incentive with targeted wording treatments ($t = 1.75, p = 0.081$). With more statistical power, this sizable difference would likely reach statistical significance.

While there were a few demographic differences across the treatments as discussed above (Table 2.2), none of the estimates of the theoretically driven concealment or commitment proxies (e.g. concern over identity theft, trust, mail opener, likelihood to answer surveys, etc.) significantly differed across treatments (results available from the authors).

Table 2.4 shows the results of logistic regression models predicting accuracy by experimental treatment; the proxy measures for confusion, concealment, and commitment; and control

Table 2.3 Selection accuracy rates overall and by treatment for the full sample and for households with at least two adults.

	All households	Two+ adult households
All sample($n = 943/n = 660$)	63.2%	55.2%
No incentive treatment ($n = 243/n = 165$)	59.9%	50.4%
Incentive + standard wording treatment ($n = 371/n = 261$)	62.5%	54.3%
Incentive + targeted wording treatment ($n = 329/n = 234$)	66.2%	59.5%
Overall F	1.07	1.58
t Incentive + standard wording vs. no incentive	0.61	0.76
t Incentive + targeted wording vs. no incentive	1.44	1.75+
t Incentive + standard wording vs. incentive + targeted wording	0.93	1.12

Note: +p<0.100.

Table 2.4 Logistic regression predicting accurate selection by experimental treatment, confusion, concealment, commitment, and control variables.

	Complete sample ($n = 943$)						Two+ adult households ($n = 660$)					
	Model 1 $F(2, 940) = 1.07$			Model 2 $F(31, 932.3) = 4.26^{***}$			Model 1 $F(2, 657) = 1.58$			Model 2 $F(31, 653.7) = 1.28$		
	Coeff.	SE	Odds ratio	Coeff.	SE	Odds ratio	Coeff.	SE	Odds ratio	Coeff.	SE	Odds ratio
Constant	0.40**	0.141	1.49	2.51***	0.877	12.25	0.02	0.159	1.02	1.71+	0.886	5.51
Experimental treatments												
No incentive (omitted)												
Incentive + standard	0.11	0.184	1.12	0.25	0.203	1.28	0.16	0.204	1.17	0.22	0.213	1.25
Incentive + targeted	0.27	0.190	1.31	0.49*	0.215	1.63	0.37+	0.209	1.44	0.42+	0.218	1.52
Confusion proxies												
Education												
High school or less (omitted)												
Some college				−0.12	0.239	0.89				−0.08	0.239	0.92
BA+				0.02	0.247	1.02				0.01	0.248	1.01
Kids in household				0.18	0.215	1.20				0.13	0.212	1.14
Marital status												
Married (omitted)												
Never married				0.46	0.322	1.58				−0.21	0.382	0.81
Div./sep./widow				1.13***	0.351	3.09				0.01	0.431	1.01
Number of adults in HH				−1.02***	0.197	0.36				−0.55**	0.189	0.58
Same residence 2 yr				0.05	0.253	1.05				0.04	0.259	1.04

(*Continued*)

Table 2.4 (Continued)

	Complete sample (n = 943)						Two+ adult households (n = 660)					
	Model 1 $F(2, 940) = 1.07$			Model 2 $F(31, 932.3) = 4.26^{***}$			Model 1 $F(2, 657) = 1.58$			Model 2 $F(31, 653.7) = 1.28$		
	Coeff.	SE	Odds ratio	Coeff.	SE	Odds ratio	Coeff.	SE	Odds ratio	Coeff.	SE	Odds ratio
Concealment proxies												
Sex (male)				0.27	0.204	1.31				0.22	0.206	1.25
Age												
19–34 (omitted)												
35–54				0.64*	0.282	1.89				0.56+	0.291	1.76
55–64				0.47	0.323	1.60				0.30	0.334	1.35
65+				0.43	0.331	1.54				0.30	0.344	1.36
Family income												
<50k (omitted)												
50–99k				−0.02	0.245	0.98				−0.02	0.242	0.98
100k+				−0.02	0.300	0.98				−0.04	0.290	0.96
Race (white)				0.48	0.299	1.61				0.41	0.298	1.51
Worry about identity theft				−0.04	0.104	0.97				−0.06	0.107	0.95
Trust (be careful)				0.01	0.197	1.01				0.00	0.197	1.00
Suspicious of others				0.20	0.203	1.23				0.24	0.200	1.27
Privacy concern				−0.16	0.131	0.86				−0.14	0.130	0.87
Days sad				0.06	0.060	1.06				0.06	0.060	1.06
Days hopeful				0.00	0.036	1.00				−0.01	0.036	0.99
Commitment proxies												
Opens mail				−0.24	0.229	0.79				−0.25	0.216	0.78
Answers phone				0.29	0.192	1.34				0.21	0.190	1.23
Answers door for strangers				−0.01	0.231	0.99				−0.07	0.229	0.93
Answers door for friends/family				−0.44+	0.236	0.64				−0.46*	0.224	0.63
Likelihood to answer surveys				0.10	0.100	1.10				0.10	0.105	1.10
Control variables												
Rural/urban												
Farm (omitted)												
Open country, not farm				0.10	0.376	1.10				0.03	0.363	1.03
Town or city				−0.18	0.291	0.83				−0.20	0.286	0.82
Own home				−0.15	0.328	0.86				−0.18	0.330	0.84
Single family dwelling				−0.59+	0.301	0.56				−0.39	0.321	0.68

Notes: $+p<0.100$, $^{*}p<0.050$, $^{**}p<0.010$, $^{***}p<0.001$.

variables. Contrary to hypotheses, the full models indicate that the incentive with standard wording treatment was no more effective at producing accurate selections than the no-incentive treatment (full sample: $t = 1.22, p = 0.224$; 2+ adults sample: $t = 1.05, p = 0.295$), but, consistent with hypotheses, the incentive with targeted wording treatment was 63% more likely to produce accurate selections than the no incentive treatment in the full sample ($t = 2.27, p = 0.024$) and 52% more likely in the two or more adult households ($t = 1.92, p = 0.055$).

The results for the other predictors of accuracy were fairly consistent across the full and 2+ adult samples, indicating that survey estimates of these predictors differ for accurately and inaccurately selected households. Larger households were 42–64% less likely to make accurate selections (full sample: $t = -5.18$, $p < 0.0001$; 2+ adults sample: $t = -2.90$, $p = 0.004$). Households where the respondent was the household member who is most likely to answer the door for friends or family were about 35% less likely to make accurate selections (full sample: $t = -1.86$, $p = 0.063$; 2+ adults sample: $t = -2.06$, $p = 0.040$). Respondents age 35–54 were 76–89% more likely than their younger counterparts to be accurately selected (full sample: $t = 2.26$, $p = 0.024$; 2+ adults sample: $t = 1.94$, $p = 0.053$). In the full sample, those who were divorced, widowed, or separated were over three times as likely as their married counterparts to be selected accurately ($t = 3.22, p = 0.001$), and those living in a single family dwelling were 44% less likely than those in other types of housing to be selected accurately ($t = -1.94$, $p = 0.052$). These were not significant in the 2+ adult sample. No interactions between the proxies for confusion, concealment, or commitment, and the experimental treatments were statistically significant.

2.11 Discussion and Conclusions

Within-household selection is an important step for maintaining a probability sample of individuals. Unlike sampling housing units or households, within-household selection requires household members to identify who are members of the household and follow rules to garner a (quasi-)random selection of adults. Thus, within-household selection has implications for coverage, sampling, and nonresponse survey errors. Although this process is fairly straightforward when interviewers are present, it is much more difficult in self-administered surveys when no interviewer is present to assist the household. Previous research and the experimental results presented here suggest that households get this selection wrong at high rates. In fact, in households with more than one adult, the chance that the correct adult is selected is roughly equivalent to a coin flip. Thus, understanding how well different within-household selection methods work, why they may fail, and how to improve them is important. This kind of understanding is facilitated by the use of experimental methods.

Experimental tests of within-household selection methods are the strongest when they have good external validity through a sample frame with good coverage of households and a probability sample of households from that frame and strong internal validity through unconfounded experimental treatments and outcomes identified prior to data collection with requisite information collected in the questionnaire. This requires paying close attention to the design and its implementation. For instance, although implementing a within-household selection technique is easier overall in an interviewer-administered survey, implementing an experiment to test alternative within-household selection methods is more difficult with interviewers because they can introduce (unobserved) confounding factors through their attitudes or expectations about a given method.

One challenge in implementing within-household selection field experiments on a probability sample of the general population is that multiple errors of nonobservation can be impacted

Table 2.5 Outcome rates for sampled households ($n = 3500$) by experimental treatment.

	Nonresponding household	Responding household: correct selection	Responding household: incorrect selection	Responding household: unknown correctness
Percent within each treatment				
No incentive	77.7	14.0	6.9	1.5
Incentive and standard wording	65.7	21.9	9.9	2.5
Incentive and targeted wording	69.3	20.1	8.1	2.5
Differences between treatments				
Incentive and standard wording minus no incentive	−12.0	8.0	3.0	1.0
Incentive and targeted wording minus no incentive	−8.4	6.2	1.2	1.0
Incentive and targeted wording minus incentive and standard wording	3.6	−1.8	−1.8	0.0

by the experimental treatments. We do not know exactly what was happening inside sampled households as they processed the survey materials. Thus, we cannot be fully certain whether it was coverage, sampling, or nonresponse, or a combination of these errors that produced the differences we observed across the selection methods. Yet, Table 2.5 is suggestive about possible mechanisms. It shows that adding the incentive increased response rates by increasing the percentage of households responding with the correctly selected household member between 6.2 and 8 percentage points but also increasing the percentage responding with an incorrectly selected household member by between 1.2 and 3 percentage points. This finding suggests that the incentive may not only have improved the coverage and response propensity of reluctant household members but also may have slightly increased errors in the sample selection (perhaps due to informants selecting themselves to get the incentive). Likewise, among the two incentive conditions, the targeted wording decreased response rates by about 3.6 percentage points with half of the decrease (1.8 percentage points) coming from responding households with correct selections and the other half from those with incorrect selections. Thus, the incentive with targeted wording resulted in a lower response rate and a lower percentage of households that responded with correct selection when looking at the entire sample (Table 2.5) but a higher percentage of households with correct selection when looking at only the respondent pool (Table 2.3). Multiple error sources were clearly at play in this treatment. Again, while not definitive, we believe that we have a tradeoff between nonresponse and coverage/sampling errors occurring in this treatment. Overall, we believe that the increased accuracy rate outweighs the decrease in response rate because, even though the response rate was lower, this treatment had better alignment with the target population on important demographic characteristics. For experimental design, this example shows the importance of identifying multiple outcomes of interest prior to conducting the field experiment so that these different effects, and their relative importance, can be jointly weighed. For survey practice, if incentives are to be used in self-administered surveys with within-household selection of an adult, targeted wording about who should receive the incentive should be used in the cover letters as such wording improved the composition of the final sample compared to standard letter wording, especially on the characteristic of sex.

In this chapter, we provided an example of an experimental study of the effects of incentives on sample composition, variables theoretically measuring the mechanisms of confusion, concealment, and commitment, and accuracy of selection. Our results suggest that the incentive with the targeted wording yielded slightly better representation relative to official benchmarks and more accurate selection than the other two approaches. Even with these improvements, roughly 40% of respondents in households with two or more adults were not the correct respondent. Thus, there is ample room for improvement. The research here should be replicated with different types of samples and survey topics and additional strategies for improving the accuracy for within-household selection should be tested.

Designing an experiment to evaluate and, potentially, improve within-household selection methods requires careful planning and thoughtful consideration of theory, design, and implementation challenges. With a theoretically guided set of experimental factors, implemented to minimize any other confounding features, and a thorough set of outcomes examining the multiple possible error sources, within-household selection experiments can yield useful and important insights. These experiments are even more necessary as self-administered surveys continue to grow in use and importance.

References

Baker, R., Brick, J.M., Bates, N.A. et al. (2013). *Report of the AAPOR Task Force on Non-probability Sampling*. American Association for Public Opinion Research. Retrieved 29 January 2016 from http://www.aapor.org/AAPOR_Main/media/MainSiteFiles/NPS_TF_Report_Final_7_revised_FNL_6_22_13.pdf.

Battaglia, M.P., Link, M.W., Frankel, M.R. et al. (2008). An evaluation of respondent selection methods for household mail surveys. *Public Opinion Quarterly* 72: 459–469.

Baumgartner, R.M. and Rathbun, P.R. (1996). Prepaid monetary incentives and mail survey response rates. Paper presented at the Annual Conference of the American Association of Public Opinion Research. Norfolk, VA.

Beebe, T.J., Davern, M.E., McAlpine, D.D., and Ziegenfuss, J.K. (2007). Comparison of two within-household selection methods in a telephone survey of substance abuse and dependence. *Annals of Epidemiology* 17 (6): 458–463.

Brick, J.M., Andrews, W.R., and Mathiowetz, N.A. (2016). Single-phase mail survey design for rare population subgroups. *Field Methods* https://doi.org/10.1177/1525822X15616926.

Brick, J.M., Williams, D., and Montaquila, J.M. (2011). Address-based sampling for subpopulation surveys. *Public Opinion Quarterly* 75 (3): 409–428.

Bureau of Sociological Research (2013). *NASIS 2012–2013 Methodology Report*. Lincoln, NE: Department of Sociology, University of Nebraska-Lincoln.

Bureau of Sociological Research (2014). *NASIS 2013-2014 Methodology Report*. Lincoln, NE: Department of Sociology, University of Nebraska-Lincoln.

Caplan, S.E. (2003). Preference for online social interaction: a theory of problematic Internet use and psychosocial well-being. *Communication Research* 30 (6): 625–648.

Church, A.H. (1993). Estimating the effect of incentives on mail survey response rates: a meta-analysis. *Public Opinion Quarterly* 57: 62–79.

Denk, C.E. and Hall, J.W. (2000). Respondent selection in RDD surveys: a randomized trial of selection performance. Paper presented at the annual meeting of the American Association for Public Opinion Research, Portland, OR.

Dillman, D.A., Smyth, J.D., and Christian, L.M. (2014). *Internet, Phone, Mail, and Mixed-Mode Surveys: The Tailored Design Method*. Hoboken, NJ: Wiley.

Durrant, G.B., Groves, R.M., Staetsky, L., and Steele, F. (2010). Effects of interviewer attitudes and behaviors on refusal in household surveys. *Public Opinion Quarterly* 74 (1): 1–36.

Forsman, G. (1993). Sampling individuals within households in telephone surveys. Paper presented at the Annual Meeting of the American Association for Public Opinion Research, St. Charles, IL, USA.

Gallagher, P.M., Fowler, F.J. Jr., and Stringfellow, V.L. (1999). Respondent selection by mail obtaining probability samples of health plan enrollees. *Medical Care* 37: MS50–MS58.

Gaziano, C. (2005). Comparative analysis of within-household respondent selection techniques. *Public Opinion Quarterly* 69: 124–157.

Groves, R.M., Couper, M.P., Presser, S. et al. (2006). Experiments in producing nonresponse bias. *Public Opinion Quarterly* 70 (5): 720–736.

Hicks, W. and Cantor, D. (2012). Evaluating methods to select a respondent for a general population mail survey. Paper presented at the Annual Meeting of the American Association for Public Opinion Research, Orlando, FL, USA.

Hox, J.J., de Leeuw, E.D., and Kreft, I.G.G. (1991). The effect of interviewer and respondent characteristics on the quality of survey data: a multilevel model. In: *Measurement Errors in Surveys* (ed. P.P. Biemer, R.M. Groves, L.E. Lyberg, et al.). New York: Wiley.

Iannacchione, V.G. (2011). The changing role of address-based sampling in survey research. *Public Opinion Quarterly* 75 (3): 556–575.

James, J.M. and Bolstein, R. (1992). Large monetary incentives and their effect on mail survey response rates. *Public Opinion Quarterly* 56: 442–453.

Kim, J., Gershenson, C., Glaser, P., and Smith, T.W. (2011). The polls-trends: trends in surveys on surveys. *Public Opinion Quarterly* 75: 165–191.

Kish, L. (1949). A procedure for objective respondent selection within the household. *Journal of American Statistical Association* 44: 380–387.

Kish, L. (1992). Weighting for unequal Pi. *Journal of Official Statistics* 8 (2): 183–200.

Krosnick, J.A. (1991). Response strategies for coping with the cognitive demands of attitude measures in surveys. *Applied Cognitive Psychology* 5: 213–236.

Lavrakas, P.J. (2008). Within-household respondent selection: how best to reduce total survey error? Unpublished report prepared for the Media Rating Council, Inc. Retrieved 29 January 2016 from http://www.mediaratingcouncil.org/MRC%20Point%20of%20View%20-%20Within%20HH%20Respondent%20Selection%20Methods.pdf

Lavrakas, P.J., Stasny, E.A., and Harpuder, B. (2000). A further investigation of the last-birthday respondent selection method and within-unit coverage error. In: *JSM Proceedings, Survey Research Methods Section*, 890–895. Alexandria, VA: American Statistical Association. Retrieved from http://www.asasrms.org/Proceedings/papers/2000_152.pdf.

Le, K.T., Brick, J.M., Diop, A., and Al-Emadi, D. (2013). Within-household sampling conditioning on household size. *International Journal of Public Opinion Research* 25: 108–118.

Lind, K., Link, M., and Oldendick, R. (2000). A comparison of the accuracy of the last birthday versus the next birthday methods for random selection of household respondents. In: *JSM Proceedings, Survey Research Methods Section*, 887–889. Alexandria, VA: American Statistical Association. Retrieved from http://www.asasrms.org/Proceedings/papers/2000_151.pdf.

Longstreth, M. and Shields, T. (2005). A comparison of within household random selection methods for random digit dial surveys. Paper presented at the annual meeting of the American Association For Public Opinion Association, 12–15 May 2005, Fontainebleau Resort, Miami Beach, FL.

Malhotra, N.,.K., Kim, S.S., and Agarwal, J. (2004). Internet users' information privacy concerns (IUIPC): the construct, the scale, and a causal model. *Information Systems Research* 15: 336–355.

Marlar, J., Jones, J., Manas, C., et al. (2014). Within-household selection for telephone surveys: an experiment of eleven selection methods. Paper presented at the Midwest Association for Public Opinion Research Annual Conference. 21–22 November 2014, Chicago, IL.

Martin, E. (1999). Who knows who lives here: within-household disagreements as a source of survey coverage error. *Public Opinion Quarterly* 63: 220–236.

Martin, E. (2007). Strength of attachment: survey coverage of people with tenuous ties to residences. *Demography* 44: 427–440.

Martin, E. and Dillman, D.A. (2008). Does a final coverage check identify and reduce census coverage errors? *Journal of Official Statistics* 24: 571–589.

McKenna, K.Y.A., Green, A.S., and Gleason, M.E.J. (2002). Relationship formation on the Internet: what's the big attraction? *Journal of Social Issues* 58 (1): 9–31.

Narayan, S. and Krosnick, J.A. (1996). Education moderates some response effects in attitude measurement. *Public Opinion Quarterly* 60: 58–88.

O'Rourke, D. and Blair, J. (1983). Improving random respondent selection in telephone surveys. *Journal of Marketing Research* 20: 428–432.

Olson, K. and Smyth, J.D. (2014). Accuracy of within-household selection in web and mail surveys of the general population. *Field Methods* 26 (1): 56–69.

Olson, K., Stange, M., and Smyth, J.D. (2014). Assessing within-household selection methods in household mail surveys. *Public Opinion Quarterly* 78 (3): 656–678.

Phelps, J., Nowak, G., and Ferrell, E. (2000). Privacy concerns and consumer willingness to provide personal information. *Journal of Public Policy & Marketing* 19: 27–41.

Raghunathan, T.E., Lepkowski, J.M., van Hoewyk, J., and Solenberger, P. (2001). A Multivariate technique for multiply imputing missing values using a sequence of regression models. *Survey Methodology* 27: 85–95.

Raudenbush, S.W. and Bryk, A.S. (2002). *Hierarchical Linear Models: Applications and Data Analysis Methods*, 2e. Newbury Park, CA: Sage.

Reich, J., Yates, W., and Woolson, R. (1986). Kish method for mail survey respondent selection. *American Journal of Public Health* 76: 206.

Rizzo, L., Brick, J.M., and Park, I. (2004). A minimally intrusive method for sampling persons in random digit dial surveys. *Public Opinion Quarterly* 68 (2): 267–274.

Rubin, D.B. and Schenker, N. (1987). Interval estimation from multiply-imputed data: a case study using census agriculture industry codes. *Journal of Official Statistics* 3 (4): 375–387.

Schnell, R., Ziniel, S., and Coutts, E. (2007). Inaccuracy of birthday respondent selection methods in mail and telephone surveys. Presentation at the European Survey Research Association Conference, 29 June, Prague.

Segrin, C. (2000). Social skills deficits associated with depression. *Clinical Psychology Review* 20 (3): 379–403.

Singer, E. (2002). The use of incentives to reduce nonresponse in household surveys. In: *Survey Nonresponse* (ed. R.M. Groves, D.A. Dillman, J.L. Eltinge and R.J.A. Little), 163–178. New York: Wiley-Interscience.

Singer, E. and Ye, C. (2013). The use and effects of incentives in surveys. *Annals of the American Academy of Political and Social Science* 645 (1): 112–141.

Smyth, J.D., Dillman, D.A., Christian, L.M., and O'Neill, A.C. (2010). Using the Internet to survey small towns and communities: limitations and possibilities in the early 21st century. *American Behavioral Scientist* 53: 1423–1448.

Stange, M., Smyth, J.D., and Olson, K. (2016). Using a calendar and explanatory instructions to aid within-household selection in mail surveys. *Field Methods* 28 (1): 64–78.

Tourangeau, R., Kreuter, F., and Eckman, S. (2012). Motivated underreporting in screening interviews. *Public Opinion Quarterly* 76 (3): 453–469.

Tourangeau, R., Shapiro, G., Kearney, A., and Ernst, L. (1997). Who lives here? Survey undercoverage and household roster questions. *Journal of Official Statistics* 13: 1–18.

Troldahl, V.C. and Carter, R.E. Jr. (1964). Random selection of respondents within households in phone surveys. *Journal of Marketing Research* 1: 71–76.

Trussel, N. and Lavrakas, P.J. (2004). The influence of incremental increases in token cash incentives on mail survey response: is there an optimal amount? *Public Opinion Quarterly* 68 (3): 349–367.

Wasserstein, R.L. and Lazar, N.A. (2016). The ASA's statement on p-values: context, process, and purpose. *The American Statistician* 70 (2): 129–133.

Valentine, C.A. and Valentine, B.L. (1971). *Missing Men: A Comparative Methodological Study of Underenumeration and Related Problems*. Washington, DC: U.S. Census Bureau. Retrieved 5 February 2016 from https://www.census.gov/srd/papers/pdf/ex2007-01.pdf.

Yan, T. (2009). A meta-analysis of within-household respondent selection methods. Paper presented at the Annual Meeting of the American Association for Public Opinion Research, Hollywood, FL, USA.

3

Measuring Within-Household Contamination: The Challenge of Interviewing More Than One Member of a Household

Colm O'Muircheartaigh[1], *Stephen Smith*[2], *and Jaclyn S. Wong*[3]

[1] *Harris School of Public Policy, University of Chicago and NORC at the University of Chicago, Chicago, IL, United States*
[2] *NORC at the University of Chicago, Chicago, IL, United States*
[3] *Department of Sociology, University of South Carolina, Columbia, SC, United States*

Survey researchers are aware of three possible negative effects of attempting to interview more than one member of a household: (i) that the response rate may be lowered for the household; (ii) that the response quality of all the interviews in the household may be reduced because of direct contamination through discussion among respondents or anxiety about possible breaches of confidentiality; and (iii) the statistical concern that a positive intra-household correlation renders second and subsequent interviews in a household less informative than the first (and less informative than independent interviews in other households). For these reasons, standard protocol in survey research for surveys of individuals tends to dictate that a single respondent be chosen at random within each selected household.[1]

The chapter describes the design and implementation of an experiment to evaluate the support for this received wisdom. We describe an experiment in which we measure the impact of seeking to interview (and interviewing where possible) in Wave 2 the *co-resident spouse or romantic partner*[2] of each prime respondent from Wave 1 of the National Social Life, Health and Aging Project (NSHAP; see Lindau et al. 2007), an NIH-funded panel study designed to increase our understanding of the role that social support and personal relationships play in healthy aging. We show that the inclusion of a second respondent from within the household does not negatively affect data quality and has several advantages for data collection and analysis efforts. It is noteworthy that this result arises in a context where the potential for contamination and distortion might be expected to be particularly high.

We discuss the practical issues and considerations that arise in embedding a randomized experiment in a longitudinal study.

3.1 Literature Review

The basic building block of sample design for demographic surveys is typically the household, as almost all the major sampling frames are frames of households, dwelling units, or addresses.

1 This does not mean that information should be collected on only one member of a household. There are important surveys where a single respondent is seen as qualified to report on all members of the household; the Current Population Survey (CPS) is an example.
2 Following discussion with our colleagues in the field division, we had decided not to attempt interviews with the partners of Wave I nonrespondents. During the fieldwork, having had a positive response from some of those nonrespondents, we did approach a subset of their partners.

Experimental Methods in Survey Research: Techniques that Combine Random Sampling with Random Assignment, First Edition. Edited by Paul J. Lavrakas, Michael W. Traugott, Courtney Kennedy, Allyson L. Holbrook, Edith D. de Leeuw, and Brady T. West.

Companion Website: www.wiley.com/go/Lavrakas/survey-research

Thus, the samples that are allocated to field interviewers are samples of households. A number of factors play a role in the decision on how many interviews to attempt within each selected household: the *target unit*, the *informant*, and the *nature of the data* being collected. A survey may have as its target: (i) the whole household as a unit, without regard to disaggregation of the data, (ii) an individual without reference to the relationship of that individual to others in the household; or (iii) both the individual and the household. In choosing informant(s), we may decide that information on each individual should be collected only from that individual (*self-reporting*), or we may accept reports from others (*proxy reporting*). And in collating the data, we may simply aggregate the data, or we may be interested in using reports on different individuals as a cross-check on each other. There are important examples of each of these situations in the survey literature.

The Current Population Survey (CPS) collects information on each member of the household and accepts reports on all members from a single informant. The focus of the data collection is the individual's labor force status in a particular week; the principal output is the unemployment rate, and the individual is the unit of analysis. The person who responds is called the "reference person" and usually is the person who either owns or rents the housing unit. If the reference person is not knowledgeable about the employment status of the others in the household, attempts are made to contact those individuals directly. The data are analyzed at the individual level; no results are produced at the household level.

The US Consumer Expenditure Survey (CE), essentially a household budget survey, collects data on expenditures by all members of the household; any eligible household member who is at least 16 years old can serve as the respondent covering the whole household. The analysis is at the level of the household.

On the other hand, the National Crime Victimization Survey (NCVS) and the National Survey of Drug Use and Health (NSDUH) interview all persons 12 and older in the household individually. For these surveys, the analysis is at the level of the individual.

In the National Health Interview Survey (NHIS), a household adult reports for a randomly selected child less than 18 years of age (in the Child Core questionnaire) and a randomly selected adult reports for him/herself (in the Adult Core questionnaire). The data are analyzed at the level of the individual.

The NSDUH provides national- and state-level estimates on the use of tobacco products, alcohol, illicit drugs (including nonmedical use of prescription drugs), and mental health in the United States. Following a face-to-face screening interview, up to two residents of the household may be selected for interview. The analysis is carried out at the level of the individual.

In the General Social Survey (GSS), as in most attitude surveys, a single respondent is selected randomly from within each selected household. In the first wave of the NSHAP, a survey of social networks, health, and sexuality among older adults, one adult aged between 57 and 84 years was selected in each household.

Kish (1949) developed a practical field procedure for objective selection of one respondent from within each household in a sample for the Survey of Consumer Finances being carried out by the Survey Research Center at the University of Michigan's Institute for Social Research. The procedure was proposed to overcome two potential problems: (i) it was desired to take no more than one interview in any household, in order to obtain each interview before the respondent had a previous opportunity to discuss the questions and (ii) furthermore, multiple interviews were considered to be potentially statistically inefficient because of the expected correlation of attitudes within the household.

The more objective survey questions are thought to be, the more likely proxy reports are to be accepted. Thus, for unemployment and expenditures (CPS and CE) reports from a single household respondent on all household members are considered appropriate. The more subjective the topics, the less likely proxy reports are to be sought or accepted (NSDUH and NCVS). And the more sensitive the topics, the less likely it is that more than one respondent will be recruited in a single household (NHIS). Where multiple respondents are interviewed in the same household (NSDUH and NCVS, for instance), the main justification is the saving in costs.

There is considerable evidence in the literature that the introduction of sensitive (or potentially threatening) material in a questionnaire can affect the quality of responses to subsequent questions; the most severe effect is of course termination of the interview (Anderson et al. 1966; Bradburn et al. 2004; Kornhauser and Sheatsley 1959; Landon 1971; Podsakoff et al. 2012; Tourangeau et al. 2000). This has led to the recommendation that difficult or challenging questions be postponed until late in a survey interview. Requesting an interview with a second member of a household may also be seen as imposing unreasonably on a respondent; this was the rationale in National Comorbidity Study – Replication (NCS–R) for waiting until the main interview was completed before recruiting a second respondent to the survey.

However, there is a concern among practitioners that even this postponement may not remove the effects of such questions; it has been hypothesized that the fact that the questions are to be asked affects the interviewer (perhaps by making the interviewer anxious, and communicating the anxiety). The NSHAP experiment is designed to measure the impact on data quality of adding a second respondent within a household, assessing not only measurement errors but also nonresponse.

An unavoidable result of selecting and interviewing a second individual from a household is to introduce an additional element of clustering into the sample; cases selected from the same household share characteristics that may be related to the target variables in the survey, and may thus increase the homogeneity of the sample and the standard error of the estimates. Consequently, survey researchers generally interview only one individual from each household (Groves 1989; Groves et al. 2009; Kish 1949, 1965).

The method developed by Kish in 1949 to select objectively a single respondent from a household spawned numerous alternatives (Gaziano 2005; Salmon and Nichols 1983). Originally designed for face-to-face interviewing, these methods have been adapted for use in mail/Internet and telephone surveys (Dillman 1978; Dillman et al. 2014).

Primarily for cost reasons some surveys collect information from multiple household members. The CPS collects data about all household members more than 16 years old, often from a single respondent. In the NCVS, every person aged 12 and older in the household provides self-reports; and in the NSDUH every person aged 12 and older reports for self; respondents may allow a more knowledgeable person to complete the health insurance and income sections for them. None of these surveys carries out any experimental assessment of the impact of the field strategy.

One major study did design and carry out a formal assessment of one aspect of the impact of collecting data from a second household member. The NCS–R selected a second respondent for interview *after the primary respondent had been interviewed* (Kessler et al. 2004). This decision was made to avoid compromising the response rate for the primary respondent. The response rate among primary respondents was 70.9%; the conditional response rate among the secondary respondents was 80.4%.

NCS–R compared the design effects for the sample of primary respondents with the design effect for the sample including both primary and secondary respondents; as the design effect is independent of the overall sample size, this provides a measure of the statistical efficiency of each design (Groves and Heeringa 2006). The evidence is inconclusive, though the authors suggest that selecting the second adult may have led to an increase in the variance of sample estimates.

In NSHAP, most of the analyses are carried out separately for men and for women, and most partners are of the opposite gender to each other, and therefore the relevance of this within-household design effect is less than it would be for a general household survey where the analyses would include both men and women. Nevertheless, by selecting both members of the household, we can obtain estimates of the impact that our design would have in such surveys. Therefore, the data provide an opportunity to estimate the intra-household correlation coefficient for key survey variables, providing a measure of the impact on the standard errors of the additional selection even in the absence of measurement or nonresponse effects.

3.2 Data and Methods

NSHAP is a longitudinal study of a nationally representative probability sample of adults born between 1920 and 1947. NSHAP collects a wide range of information, including quality of relationships, social support, functional health, sexual behavior and function, mental health, cognitive function, and financial status. The study's multimode data collection combines in-home CAPI with the collection of a wide-ranging set of biomeasures, and a postinterview Leave-Behind Questionnaire (LBQ). The response rate (AAPOR RR3) to the initial Wave (Wave 1) was 75%. In Wave 1 (2005–2006), 3005 community-dwelling individuals age 57–85 were interviewed. The second Wave of data collection took place five years later (2010–2011). Wave 2 returned to all surviving Wave 1 respondents and approached again all nonhostile nonrespondents from Wave 1 (Jaszczak et al. 2014; O'Muircheartaigh, Eckman and Smith 2009; O'Muircheartaigh et al. 2014; Smith et al. 2009).

Besides returning to Wave 1 nonrespondents, the other major modification at Wave 2 was the plan to add interviews with cohabitating spouses or romantic partners of the original sample members. For the Wave 2 partner-study design, those who are reinterviewed following their participation in Wave 1 together with Wave 1 nonrespondents who responded in Wave 2 constitute the *prime* respondents; these are members of the original NSHAP sample. Because the hypotheses in NSHAP focus on the importance of the intimate partner relationship, the Wave 2 design planned to seek to interview also the current partner of each prime respondent. For Wave 1, we had decided to include only one respondent per household for the reasons described above; we were particularly concerned in NSHAP about the sensitivity of some of the subject matter and the fact that respondents are asked to rate aspects of their lives that directly relate to their partner's intimate behavior and attitudes.

The addition of partners to the study design allows us to examine if and how romantic or sexual relationships promote better health trajectories. Analyses that require such data include (i) paired analyses, involving within-couple comparisons to examine gender differences or control for unobserved household-level characteristics; (ii) the estimation of the within-couple correlation in various health outcomes, controlling for individual-level factors; and (iii) analyses in which the couple is the unit of analysis, using parallel data from both partners and/or derived variables capturing features of the couple (measures of similarity or difference between partners are examples). Returning to prime respondents and their spouses or romantic partners in

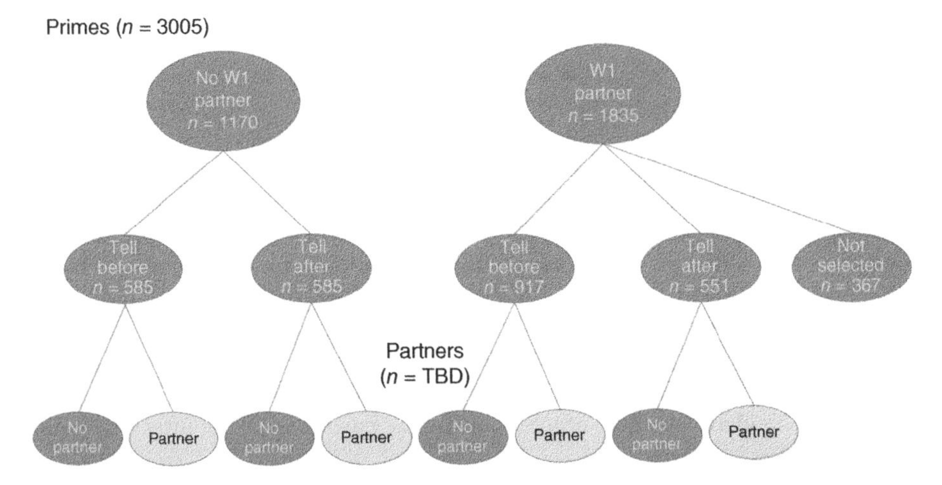

Figure 3.1 Partner study experimental design.

the third Wave of NSHAP (2015–2016) would also allow longitudinal analysis of partnership characteristics and relationship quality alongside health trajectories.

A second (and independent) advantage to the inclusion of partners is the effect on sample size. Projections based on data from Wave 1 suggested that reinterviewed respondents would yield an estimated 1827 coresident romantic partners. *Age-eligible partners* (ages 57–85 at the time of Wave 1) constitute additional respondents for cross-sectional analyses and increase the power of the research design. Potentially, the sample size could be augmented by more than 1500 age-eligible individuals. With a response rate of 75%, this would provide an additional 1100 cases for cross-sectional analysis in Wave 2. The impact on the precision of estimates, particularly for gender-specific estimates, would be considerable. For gender-specific estimates, the sample of women (and the sample of men) would be some 40% larger; as most partners are of the opposite gender, these additional respondents would represent additional households, and the intra-household correlation coefficient would be effectively zero. Consequently, the effective sample size would be increased proportionately, reducing sampling variances by 30% (and standard errors by 15%). For comparisons between men and women (where both partners would be included in the analyses), the within-household correlation would in fact reduce the variance of the estimates below that for a comparison of samples of the same size of men and women from different households (Kish 1965; Verma et al. 1980).

NSHAP Wave 2 provided an opportunity to examine the general issue of interviewing multiple members of a household. The circumstances were favorable in one way; the prime respondent had already agreed to respond in Wave 1 of the study, so access to the household was promising. On the negative side, the prime, whose permission we sought, was already aware of the nature of the interview and of the questions, and might in some circumstances be expected to be less positively inclined to accede to our gatekeeper request.

We were however concerned that the inclusion of partners might influence the results in two ways: (i) the response rate for the prime respondents might be decreased, thereby negatively affecting the panel analyses and (ii) the responses of the primes themselves to (sensitive) questions might be affected. These concerns had been influential in the original decision not to include partners in the design of Wave 1. At Wave 2, the introduction of unmeasured biases into the NSHAP data (by including partners) could threaten the validity of the NSHAP panel. Consequently, an evaluation experiment was incorporated into the fieldwork.

3.2.1 The Partner Study Experimental Design and Evaluation

A randomized experiment was designed to test a number of basic questions. We wished to test whether seeking to interview the prime respondent's partner would reduce the response rate for the primes. Furthermore, we wished to establish whether, for primes successfully interviewed in Wave 2, the responses of the primes would be affected in this circumstance. Finally, we wished to establish whether the effect would be different depending on when we informed the prime respondents of our intention. There is a general belief, illustrated by the design decision in NCS–R, that making such a request at the end of the interview does not affect the interview itself; however, we considered that knowing that the request would be made could influence the interviewer and hence the respondent by affecting the rapport/tone of the interview. To test this hypothesis, two separate protocols were developed; in the first, we informed the prime at the beginning of the interview that we would be requesting an interview from the partner; in the second, we postponed any mention of interviewing the partner until the interview with the prime and been successfully completed.

Consequently, we randomized the prime respondents across two experimental conditions and a control.

A (control): No partner interviews were attempted; no mention of partner interviews (the CONTROL group);
B: We informed the Wave 1 respondent (prime) before, or at the beginning of, the Wave 2 interview that we wished to interview his/her partner (the TELL BEFORE group);
C: We informed the Wave 1 respondent (prime) only after his/her interview was completed that we would seek to interview the partner (the TELL AFTER group).

In making these assignments, we took into consideration whether or not the Wave 1 respondent (the prime) had a partner in Wave 1. Because we thought the likelihood of people in this age group acquiring a partner was small, in the event that a new partner was identified, we decided always to interview the partner. Therefore, all Wave 1 respondents without a partner in Wave 1 were assigned to one of the two experimental conditions, using a 50–50 split. For Wave 1 respondents with a partner in Wave 1, we assigned cases to Tell Before: Tell After: Not Selected in the ratios 50 : 30 : 20. A diagrammatic representation of the allocation of the 3005 cases is given in Figure 3.1. These decisions and their implementation illustrate the complexity of integrating randomized experiments into surveys.

Prior to beginning Wave 2, the project field operations' staff worked closely with the methodologist and sampling staff to design all elements of the experiment, including the addition of questions and scripts to the survey questionnaire that were needed to manage their subsequent routing within the questionnaire, programming of the case management system, sample allocation and monitoring, and field training and operations.

Using the information from Wave 1 that indicated whether they had a partner at that time, all primes for Wave 2 were randomly preassigned in the case management system to meet the approximate distribution needed in each of the three experiment groups (i.e. Tell Before, Tell After, No Partner Interview needed). When a prime was allocated to the Tell Before group, the interviewer knew in advance of commencing the interview that the prime should be told that if they had a partner we would be seeking to interview the partner too. Field interviewers were not able to see if the Tell After or the No Partner Interview experiment group had been assigned until the questionnaire made that information available to them at the appropriate point in the interview. Scripts were included at the appropriate point in the questionnaire, in the case of the Tell Before and Tell After cases, for the interviewers to inform the prime about the desire to interview their partners.

Prior to beginning field work, the field interviewers were briefed during their in-person project training that when completing the network roster sometimes the questionnaire would indicate that the partner, if one was identified, would be eligible to be interviewed and they would be instructed to tell the prime (i.e. Tell Before). They also understood that in some cases the questionnaire would indicate after the prime's interview had been completed that the partner was eligible to be interviewed (i.e. Tell After) or no partner interview was required (No Partner Interview – the control).

In general, partners eligible for the experiment were identified by the prime during the completion of a social network roster at the start of the interview. A series of questions were asked that elicited information about the prime's marital status and whether they considered anyone to be their romantic partner. When a potentially eligible partner was identified, additional questions were asked to confirm the partners' age (they needed to be 18 or over) and their current cohabiting status (they needed to be living primarily in the same household as the prime). Once a partner in Wave 2 who was assigned to the Tell Before or Tell After group was identified in the network roster, the case management system generated a new questionnaire that the interviewer was to administer to the partner after the primes interview had been completed.

Investigators

There is invariably a wish on the part of the investigators/substantive researchers to maximize the number of cases available for subject matter analysis; methodologists on the other hand require sufficient power to identify effects should they occur. In our case, restricting the control group to 20% of the cases was deemed an acceptable price to pay to assess whether the introduction of partners would undermine the validity of the data.

Field/Project Directors

The primary responsibilities of field and project directors are to balance the sometimes competing requirements of successfully meeting the targeted number of completed interviews and maintaining the appropriate level of quality and completion, while staying on schedule and within budget. Consequently, field directors and project managers can be prone to avoiding methodological experiments that might interfere with the smooth operation of the fieldwork. This can lead to a generalized resistance to the introduction of any perturbations into the field operation process. Among the arguments put forward to resist incursions into the standard process are the following:

1. Interviewers will be confused and demotivated by unexplained variation in instructions
2. Questionnaire programming and systems support staff will be unable to program and implement the necessary changes in the data collection protocol reliably and in time
3. Methodological experiments in the field can be challenging to monitor
4. Experiments compete for and use valuable resources – such as project, technical, and field staff – that likely distract from the primary purpose and goals of the study
5. No one else does this.

These are not unreasonable concerns, and it behooves methodologists to address these in the design.

1. One way that will often be suggested to overcome the difficulty of interviewer training is to randomize treatments across interviewers rather than across cases, i.e. use a group randomized design. However, this would lead to a very substantial reduction in the power of the experiment to identify an effect, as the number of degrees of freedom to estimate the variance between treatments would be reduced to the number of interviewers rather than the number of cases; the variance of the estimate of the effect would be inversely proportional to the number of interviewers. In order to implement our approach successfully, a significant training effort is necessary to prepare the interviewers.
2. Programming implementation for real-time execution in the field can be a problem, in terms of both cost and reliability. In this case, the second wave of a longitudinal survey where we know in advance all the potential cases to be interviewed, the randomization can be done in advance, and cases will automatically generate the appropriate interviewer instructions during the introductory stage of the interview. This means that the interviewer does not have to implement the allocation procedure and can be alerted in advance to the type of case with which s/he is dealing.
3. Monitoring of field instructions and how they are actually implemented in the field is always a challenge and can be addressed by communicating the experiment protocol and developing tools and reports to monitor its implementation. It is impossible to overstate the importance of briefing the field supervisors on the nature of the intervention and its relevance to the success of the survey. The briefing should ideally be carried out jointly by the project director and a senior member of the methodology team/investigator to ensure that all parties are aware of and on board with its significance. Once implemented in the field, it is imperative to have the appropriate tools and reports available to monitor the intervention. Often only standard reports for tracking usual field activities are available, so custom reports are needed by the project staff and methodologist to manage and control the experiment.
4. Concern over cost is always a source of tension between practitioners and methodologists. In most cases, timely planning and preparation can minimize the actual cost of the experimental intervention. Maintaining a close working relationship is essential; it is a mistake to think that just because the researcher sees the importance of the experiment that this is clear to others. Responsiveness to the concerns of the field staff is essential.
5. By emphasizing the contribution of the experiment to the skills of the workforce and the reputation of the survey organization, the fact that "no one else does this" can be turned into an advantage, but only if the benefits can be shown to accrue not just to the methodologist.

The experiment was designed to test the following hypotheses:

3.2.1.1 Effects on Response Rates and Coverage

For Primes

Hypothesis 1: The introduction of partner interviews will reduce the response rates of the primes, both to the in-person interview and the LBQ

1a. A subsidiary question is whether the timing of the request to interview the partner affects the response rates of the primes. The direction of the effect is not clear *a priori.*

For Partners

Hypothesis 2: The early introduction of partner interview requests (TELL-BEFORE) will reduce the number of partners found relative to postinterview requests (TELL-AFTER).

Effect on Responses

Hypothesis 3a: The introduction of partner interview requests will increase the amount of missingness to sensitive questions for primes, both for the interview and the LBQ.

Hypothesis 3b: The introduction of partner interview requests will increase the social desirability effect in responses by primes, both for the interview and the LBQ.

We do not address the contamination of partner reports; indeed only the differential effect by timing can be addressed in this study.

3.3 The Sequence of Analyses

In order to understand the ways in which the inclusion of partner in the sample could influence the results, we considered all the stages of the process.

1. The effect on both response rate and response errors of having the interviewer be aware that s/he will have to inform the prime we wish to interview the prime's partner with the same interview and biomeasures.
2. The effect on both response rates and response errors of informing the prime respondent that we wish to interview his/her partner with the same interview and biomeasures.
3. The effect on the prime of knowing that we will be obtaining some of the same information from the partner on both response rates and response error. This could manifest itself either in differences across conditions in the current reports or in the change over time across conditions.

We carried out a series of analyses to investigate these issues. We assessed primes' response rates and the results of the experimental group assignment to establish whether there were any systematic and/or consequential demographic differences in primes across the experimental categories. We examined whether experimental group assignment affected the number of partners found, and whether the experimental condition affected partners' response rates. We compared response rates and substantive answers to in-person and LBQ items across experimental categories for primes and partners. Finally, we conducted a longitudinal comparison of primes' answers across experimental conditions.

3.4 Results

3.4.1 Results of Experimental Group Assignment

Analyses assessing the results of experimental group assignment showed few demographic differences across experimental categories. However, we did find some differences in experimental condition by education. College attendance is significantly associated with experimental group, with a higher proportion of those who attended college being assigned to the Tell After group (56.8% vs. 49.2%, $p = 0.008$). Significantly more females who attended college were assigned to the Tell After group; the pattern is the same for males, but does not reach statistical significance (Table 3.1). There is also a significant difference among the youngest primes (born between 1940 and 1947); a significantly higher proportion of youngest primes who attended college was assigned to the Tell After group (Table 3.2). There were no other differences in marginal distributions on demographic variables. These discrepancies illustrate the difficulty of equalizing the distribution of experimental condition across all potentially relevant characteristics.

Table 3.1 Results of experimental group assignment by education, gender, and age group.

	Not Selected	Tell Before	Tell After	Total
Males				
Attended college	105	270	197	572
	49.5%	51.2%	56.8%	52.7%
Did not attend college	107	257	150	514
	50.5%	48.8%	43.2%	47.3%
Total	212	527	347	1086
	100%	100%	100%	100%
$\chi^2 = 3.62(2), p = 0.164$				
Females				
Attended college	72	180	116	368
	46.5%	46.4%	56.9%	49.3%
Did not attend college	83	208	88	379
	53.5%	53.6%	43.1%	50.7%
Total	155	388	204	747
	100%	100%	100%	100%
$\chi^2 = 6.48(2), p = 0.039$				
All respondents				
Attended college	177	450	313	940
	48.2%	49.2%	56.8%	51.3%
Did not attend college	190	465	238	893
	51.8%	50.8%	43.2%	48.7%
Total	367	915	551	1833
	100%	100%	100%	100%
$\chi^2 = 9.72(2), p = 0.008$				

Table 3.2 Results of experimental group assignment by education and age group.

	Not Selected	Tell Before	Tell After	Total
Youngest (1940–1947)				
Attended college	94	214	156	464
	55.0%	53.2%	68.1%	57.9%
Did not attend college	77	188	73	338
	45.0%	46.8%	31.9%	42.1%
Total	171	402	229	802
	100%	100%	100%	100%
$\chi^2 = 14.01(2), p = 0.001$				

3.5 Effect on Standard Errors of the Estimates

The impact of departures from simple random sampling is generally summarized using the design effect (deff), which is the ratio of the variance of the estimate based on a particular design to the variance of the estimate for a simple random sample of the same size. For a clustered sample, the design effect can be expressed as $1 + \rho(b - 1)$, where ρ is the intracluster correlation coefficient and b is the number of elements selected in each cluster.

By analogy here, we can consider the impact of a sample of size n, based on selecting one member of the household at random, and a sample of the same size, n, where we select two people per household; thus, we would have $n/2$ households and n people: $b = 2$. To the extent that the members of a household have positively correlated characteristics – health or attitudes say – the additional household member will contribute less to the precision of the estimate than would be the case with an individual selected randomly from a different household.

For NSHAP analyses this is not an issue, as all our analyses are carried out separately for men and women, and our procedure did not generate cases where we had more than one respondent of either sex from the same household in our sample. We present below the implications for other studies where this might not be the plan of analysis.

We calculated the standard errors and design effects for all the variables in the study. These are complex design effects, reflecting the geographical clustering of sample and the effects of a variety of other aspects of the design, including weighting. We have chosen 15 key variables that were measured on both partners in the household. For the purposes of this chapter, we confine ourselves to the effect of clustering within households due to selection of partners. The measure we present is the within-household (intraclass) correlation coefficient, ρ_{hh}; this measures the similarity, for each variable, between the two partners in the same household relative to the similarity between two people taken at random from the same population. If the within-household correlation coefficient is ρ_{hh}, the variance is multiplied by $1 + \rho_{hh}\ (2 - 1) = 1 + \rho_{hh}$; the standard error is increased by the square root of $\sqrt{(1 + \rho_{hh})}$ (Table 3.3).

Our analysis provides structural information on homogeneity in the population for the variables we consider here. The table illustrates the fact that the implications of choosing more than one respondent from within a household will be quite different depending on the target variables in the survey. The values of ρ_{hh} range widely across the 15 variables; the values of ρ_{hh} range from 0.033 (for level of *C-Reactive Protein*) to 0.872 (for *sex in last three months*). For the most homogeneous of these variables, there is almost no additional value in obtaining a response from the partner in addition to the prime. For the least homogeneous, the partner data are almost as informative as an additional independent observation.

If we consider the median of the 15 values of ρ_{hh} ($\rho_{hh} = 0.193$), the impact of interviewing the partner is to increase the variance by a factor of about 1.2; this is equivalent to reducing the effective sample size by the same factor. For the 1548 respondents included in Table 3.2, this implies that, for analyses that combined all the cases, adding the 774 partners produced equivalent precision to adding 516 additional independent respondents.

Ideally, we would compare the costs of two competing strategies: adding 774 partners from households where we already had cooperation from the primes, compared to recruiting 516 additional respondents. However, the result implies that, for a value of $\rho_{hh} = 0.2$, if the additional cost of adding and interviewing a partner is less than two-thirds (516/774) the cost of recruiting and interviewing an additional independent case, recruiting the partner is the more efficient approach. A great deal of the cost of obtaining a respondent is in getting access to the household, screening for eligibility, and obtaining agreement to be interviewed. Though we do not have detailed cost estimates for this cost trade-off, for NSHAP, the cost of additional recruitment of independent case would have been many times the marginal cost of recruiting a partner.

Table 3.3 Within-household intraclass correlation coefficients[a] among those households with two age-eligible NSHAP respondents ($n = 774$ households; 1548 respondents).

Variable	Household (Within PSU)
Self-rated health	
Good or better	0.231
Very good or better	0.193
BMI	0.196
Diabetes	0.076
Cognitive function	0.357
Blood pressure	
Systolic	0.117
Diastolic	0.058
Log (CRP)	0.033
Frailty	0.323
CESD (Depressive sympt.)	0.143
PSS (Perceived stress)	0.217
HADS (Anxiety)	0.187
UCLA loneliness	0.256
Smoking (Cigarettes)[b]	0.594
Sex in last 3 mo[b]	0.872

a) Unweighted.
b) Model including strata would not converge.

3.6 Effect on Response Rates

3.6.1 Response Rates Among Primes

Wave 1 respondents were first categorized according to whether they had a partner at the time of the Wave 1 interview. Those who had a partner at Wave 1 were assigned to one of three experimental groups – (i) Control group – partners not selected; (ii) Tell Before group – partner selected and prime informed at the beginning of the Wave 2 interview (In the "Tell Before" condition, the prime was informed that we would wish to interview her/his partner as soon as the interviewer had confirmed that the prime had (or still had) a partner. This usually occurred during or at the end of the *Social Network Roster* section of the interview, quite early in the visit.); and (iii) Tell After group – partner selected but prime not informed until the end of the Wave 2 interview. The design is illustrated in Figure 3.1; the allocation of sample sizes is given in Tables 3.4 and 3.5.

Tables 3.4 and 3.5 compare the response rates of the primes across the different experimental groups. As the assignment rules differed according to whether the prime had reported a partner in Wave 1, the comparison is made separately for different Wave 1 partner status. A difference between the response rates for the "Not Selected" cases and the "Partner" cases would indicate that knowing the partner would be approached could color the interviewer's expectations (and therefore behavior) in a way that influenced their effectiveness in inducing cooperation from the prime. A difference between the "Tell Before" and "Tell After" cases would indicate that the interviewer could have a different level of concern (approaching the case) about the stage

Table 3.4 Case counts and response rates for primes with Wave 1 partner.

Wave 1 partner	Not Selected	Tell Before	Tell After	Total
Wave 2 respondents with no Wave 2 partner	44	98	53	195
Wave 2 respondents with Wave 2 partner	256	624	392	1272
Wave 2 nonrespondents	33	79	47	159
Wave 2 out-of-scope	34	116	59	209
Total	367	917	551	1835
Response rate	90.1%	90.1%	90.4%	90.2%

Table 3.5 Case counts and response rates for primes with no Wave 1 partner.

No Wave 1 partner	Tell Before	Tell After	Total
Wave 2 respondents with no Wave 2 partner	380	371	751
Wave 2 respondents with Wave 2 partner	32	11	43
Wave 2 nonrespondents	55	77	132
Wave 2 out-of-scope	118	126	244
Total	585	585	1170
Response rate	88.2%	83.2%	85.7%

Note: These cases (all those without partners in Wave 1) generated very few partners, only 42 in all. The small number of partners involved makes the difference substantively insignificant in terms of partner analysis.

in the prime interview that the topic of interviewing the partner would be broached with the respondent. In presenting these comparisons, we use unweighted response rates for simplicity; the weighted response rates are almost identical.

Each of the 1835 respondents who had reported a partner in Wave 1 was allocated to one of three conditions – "Tell Before," "Tell After," and a "Control group" (i.e., Not Selected). Table 3.4 shows the response rates for the groups; these are the response rates for the primes themselves. There is no effect either of being invited to include one's partner or of the timing of the invitation. The response rates are all essentially identical. This allays a major concern, encapsulated in hypothesis 1, that the decision to recruit partners in Wave 2 would depress the response rates of the primes. We also reject hypothesis 1a for this set of respondents, that the timing of the request has an effect.

Each of the 1170 respondents who had reported no partner in Wave 1 was allocated to one of only two conditions – "Tell Before" and "Tell After" as we wished to be able to interview any new partners who were identified during the interview. Table 3.5 shows the results. This provides a separate test of hypothesis 1a, for this distinct set of primes. There was a statistically significant difference in the response rates; interestingly, the response rate was higher for the "Tell Before" group. There are two competing forces that might be at play here: on the one hand, earlier disclosure of our intention might reduce the willingness of the prime to be interviewed, but on the other hand, the effect on the interviewer of knowing that our intention would be revealed later might diminish the interviewer's job performance.

When we separate Primes who had a partner in Wave 1 from Primes who did not have a partner in Wave 1, we found that Primes without a partner in Wave 1 had higher response rates in the Tell Before condition than in the Tell After condition. These Primes also generated more partners than those in the Tell After group. Perhaps Primes with new partners were more willing or excited to talk about their partners.

3.6.2 Response Rates Among Partners

The overall (conditional) response rate for partners who were identified (or confirmed) during the Wave 1 interview was 86.5%, substantially higher than the 70% we had assumed in our design. The response rates were essentially identical for men and women. There was some variation by age, though not as pronounced as the variation for primes; again the oldest respondents had the lowest response rate, though only 3 percentage points below the mean. There were some partners who were outside the eligible age range, mostly younger women and older men; their response rates were slightly lower than those for the age-eligible partners. The biggest difference was for partners of Wave 1 nonrespondents. Only two-thirds of these responded. This may be explained by the fact that the field effort for those partners was considerably shorter than that for the others, due to (i) the late decision to pursue them and (ii) the fact that Wave 1 nonrespondents (the primes corresponding to those partners) were in general interviewed later in the field period. Excluding these exceptional cases from the calculations the response rates are about a point and a half higher.

Table 3.6 gives the detailed results for the categories of the experiment; the table includes all partners who were selected for the interview (a control group of partners was not targeted for interview).

Table 3.6 Response rates for partners of primes identified during the fieldwork by experimental treatment.

	Prime Tell Before	Prime Tell After	Total
All partners			
Complete	568	339	907
NIR	81	61	142
OOS	7	3	10
Total	656	403	1059
Response rate	87.5%	84.7%	86.5%
"Old partners" (Prime had Wave 1 partner)			
Complete	540	329	869
NIR	78	61	139
OOS	6	2	8
Total	624	392	1016
Response Rate	87.4%	84.4%	86.2%
"New partners" (Prime did not have Wave 1 partner)			
Complete	28	10	38
NIR	3	0	3
OOS	1	1	2
Total	32	11	43
Response rate	90.3%	100%	92.7%

NIR, Non-interview Response; OOS, Out of Scope Cases.

Interestingly, there was a difference of 3% (in favor of the "Tell Before" group) between the response rates for the partners in the "Tell Before" and "Tell After" conditions. This, though not significant ($p = 0.18$), parallels the evidence of a difference in favor of the "Tell Before" condition among primes who did not have a partner at Wave 1. A difference of this magnitude in response rate, were it to be confirmed by future studies, would be of considerable substantive significance. The marginal cost per case of the final 3% in response rate is extremely high relative to the average cost per case, and indeed relative to the cost per case of the preceding 3%.

3.7 Effect on Responses

3.7.1 Cases Considered in the Analyses

Our analyses include 1272 Primes with partners at Wave 1 and Wave 2 and 869 Partners of a prime with a Wave 1 partner. All other categories were excluded (we excluded the 43 Primes with partners at Wave 2 but no partner at Wave 1). The number of Primes falls to 1145 when we analyze data from the LBQ because 127 did not return the LBQ; the sample falls to 1132 when we analyze LBQ questions related to partnership because 13 Primes received the wrong version of the LBQ (one version of the LBQ includes questions about respondents' relationships and the other does not). Similarly, the number of Partners falls to 781 when we analyze data from the LBQ because 88 Partners did not return the LBQ; the number of Partners in the sample falls to 747 when we examine LBQ questions about partnership because 34 Partners received the wrong version of the LBQ.

3.7.2 Sequence of Analyses

Following the logic of the preceding sections, we first examined item nonresponse; reluctant respondents anxious about the inclusion of their partners might choose not to answer particular questions even though they had consented to the interview as a whole.

3.7.3 Dependent Variables of Interest

We chose 80 interview items to examine; these were chosen to represent the most significant analyses carried out on the NSHAP data. Many are sexuality measures presented in Lindau et al.'s (2007) article "A study of sexuality and health among older adults in the United States" (22 variables); the remaining variables include items from the CES-D scale (12 variables), questions about activities of daily living (30 variables), and measures of incontinence (6 variables) and physical health (4 variables). Seven of the sexuality variables come from a check-all-that-apply list, and these variables were collapsed into a single variable for some of our analyses, resulting in 74 variables in total. In other analyses, the seven sexuality variables were kept separate. Additionally, three sexuality variables were gender-specific; analyses with these variables were completed only where appropriate.

We chose 10 LBQ items for analysis based on whether they, at face value, might be affected by the partner experiment. Seven of the 10 items are related to relationship quality and sexual behavior. The other three items are questions assessing felt loneliness.

3.7.4 Effect of Experimental Treatment on Responses of Primes

3.7.4.1 Missingness

There was no evidence of any systematic difference across the experimental conditions, though there was a handful of significant results. We compared results across experimental condition in two ways: (i) we tested item nonresponse (0 = answered, 1 = did not answer) and unit nonresponse for the LBQ (0 = returned LBQ, 1 = did not return LBQ), by experimental condition (Tell Before, Tell After) using χ^2 and (ii) we carried out multinomial logistic regressions predicting outcome (0 = answered, 1 = don't know, 2 = refused) and binomial logistic regressions predicting unit nonresponse for the LBQ, using experimental condition.

3.7.4.2 Substantive Answers

We carried out a number of analyses of the substantive responses to investigate whether experimental treatment had an effect. Again, we found no evidence of systematic variation across experimental conditions. Though there were some statistically significant results, the pattern of differences was haphazard (inconsistent in terms of the direction and impact of the experimental treatments) and did not differ materially from that expected by chance.

3.7.5 Changes in Response Patterns Over Time Among Primes

We were able to compare changes in response patterns over time for 49 of the 80 CAPI variables (25 variables are new to Wave 2 and an additional 6 variables in Wave 1 did not match the Wave 2 variables in form, so could not be appropriately analyzed) and 5 of the 10 LBQ variables (5 variables are new to Wave 2).

3.7.5.1 Missingness

To examine the effect of the experimental treatment on missingness over time, we condition on the level of missingness in Wave 1 and look at the change in missingness between the two Waves. Comparing Wave 1 and Wave 2 missing answers shows that, on average, missingness increased from Wave 1 to Wave 2; the average number of missing answers was greater in both the CAPI and the LBQ in Wave 2 (1.63 and 0.26, respectively) than in Wave 1 (0.74 and 0.17, respectively) ($t = -8.9571, p = 0.0000$ for CAPI items; $t = -3.5420, p = 0.0004$ for LBQ items). However, the increase in missingness did not differ across experimental groups (ANOVA, $F = 0.96, p = 0.3845$ for CAPI items; $F = 0.88, p = 0.4155$ for LBQ items).

3.7.5.2 Substantive Answers

For the analysis of substantive answers, the number of comparable variables from the CAPI dropped from 49 to 48 between Wave 1 and Wave 2.

Comparing the average number of affirmative answers to CAPI and LBQ questions in Wave 1 to that in Wave 2 among all Primes shows no significant difference. For example, in Wave 1, the average for CAPI items is 14.50 and in Wave 2 the average is 14.67 ($t = -1.28, p = 0.2018$). ANOVA and post hoc analyses show no differences in the change in affirmative answers across experimental groups ($F = 0.41, p = 0.6623$).

3.7.5.3 Social Desirability

We were interested in whether knowing one's partner would be interviewed might cause a respondent to give answers that might be seen as more socially desirable. We analyzed 47 key variables in the CAPI only and checked for differences across groups in the number of affirmative (socially desirable) answers. There were 36 variables rated as negative in social desirability, and 11 questions were rated as positive.

3.7.5.4 Negative Questions

There was no significant difference across experimental groups in the number of affirmative answers given to the 36 negative questions in Wave 2. Furthermore, the change in the number of affirmative answers to negative questions across Waves did not differ across experimental groups ($F = 0.07, p = 0.9306$).

3.7.5.5 Positive Questions

There was weak evidence of an effect on answers to positive questions. The change in the number of affirmative answers to positive questions is marginally significantly different across experimental category ($F = 2.42$, $p = 0.0893$); a post hoc test shows that a marginally significant difference lies between the Tell Before (change of −0.51) and Tell After (change of −0.24) groups ($p = 0.086$).

3.7.5.6 Effect on Responses of Partners

We performed the same analyses for partners. In summary, the analyses provided no evidence of a significant difference across experimental treatments.

3.7.6 Conclusion

The experience of integrating a field experiment into the second wave of data collection for NSHAP highlighted the importance of collaboration among all those involved in the study. There are three major actors in the research design and implementation process, each with a different perspective. The investigators are anxious to maximize the amount of usable data produced for analysis; the project managers wish to complete the fieldwork on time and on budget; and the methodologists would like to use the project to learn about the design aspects that determine, and can be manipulated to improve, data quality.

The issue for the investigators was whether we could recruit and interview the partners of our panel members; doing so would augment the sample size (of individuals) while providing data on a range of new research questions about couples. For the project managers, the challenge was to develop a practical protocol and training process that could be used by the interviewers. For the methodologists, the question was whether recruiting partners would undermine the quality of the data being collected, not just from the partners but also from the original panel members (the primes).

The different responsibilities and consequent priorities of the investigators, the project management, and the field interviewers could easily lead to dysfunctional competition. It is critical that the rationale for the experiment be communicated to all concerned and that the metrics used to evaluate the outcomes should give appropriate incentives and credits to all parties, especially the operational staff. We describe the protocol for the experiment in some detail and emphasize that the specifics of the protocol and its implementation are critical to the interpretation of the outcomes.

Our results are of three kinds: implications for design of field experimentation, technical considerations for sampling and field strategy, and substantive results in relation to NSHAP variables.

3.7.6.1 Design

In order to provide a basis for evaluation, it was necessary to exclude a portion of the sample (the control group) from the experimental treatments. This was one of the more contentious issues. If one assumes success for the process (and in general within an ongoing panel there is generally a high a priori confidence that the treatment will be successful before it is sanctioned), then any untreated cases are a wasted opportunity. On the other hand, to be able to establish results with an acceptable level of power requires an adequate number of cases in the control group. After intense debate, 20% of the cases were allocated to the control.

The NSHAP approach was to involve all three interests in the design of the experiment from the outset of the discussions. The details of the protocol are given above. By instituting detailed consultation with the project manager and the field staff from the very beginning of the process, we established that it is possible to introduce, implement, and monitor a fairly complex field experiment without disrupting a major ongoing panel survey field operation.

3.7.6.2 Statistical Considerations

The values of the within-household correlation coefficient that we found for most of the survey variables were sufficiently low that the benefit of adding an additional respondent (the partner) within a household easily exceeds the cost for face-to-face surveys where the cost of identifying and establishing contact with a household is a large fraction of the total cost.

For our study, NSHAP, there were three additional factors that favored adding partners to the sample (i) only about 40% of households contain age-eligible respondents, and (ii) the cost of finding additional independent cases would consequently be even higher than in most surveys and (iii) we had a particular interest in analyzing couples (both partners as a unit), and this would have been impossible without including both partners in a household.

3.8 Substantive Results

Though these are not the focus of this chapter, it is reassuring to note that we found no evidence of deleterious effects of recruiting partners in the household. First, there was no evidence that including partners led to any reduction in response rates, whether for the existing panel members (prime respondents) or their partners. Second, there was no evidence that adding respondents in the same household contaminated the responses. This was a critical factor for our future analyses; contamination would have ruled out the possibility of following a partner-recruitment strategy in the future. On the basis of our results, NSHAP decided that all eligible members of a household would be included in Wave 1 of the NSHAP second cohort, recruited in 2015–2016.

References

Anderson, L.K., Taylor, J.R., and Holloway, R.J. (1966). The consumer and his alternatives: An experimental approach. *Journal of Marketing Research* 62–67.

Bradburn, N.M., Sudman, S., and Wansink, B. (2004). *Asking Questions: The Definitive Guide to Questionnaire Design–For Market Research, Political Polls, and Social and Health Questionnaires.* Wiley.

Dillman, D.A. (1978). *Mail and Telephone Surveys: The Total Design Method.* Wiley https://books.google.com/books?id=hDhHAAAAMAAJ.

Dillman, D.A., Smyth, J.D., and Christian, L.M. (2014). *Internet, Phone, Mail, and Mixed-Mode Surveys: The Tailored Design Method.* Wiley https://books.google.com/books?id=fhQNBAAAQBAJ.

Gaziano, C. (2005). Comparative analysis of within-household respondent selection techniques. *Public Opinion Quarterly* 69 (1): 124–157.

Groves, R.M. (1989). *Survey Errors and Survey Costs.* New York, NY: Wiley.

Groves, R.M. and Heeringa, S.G. (2006). Responsive design for household surveys: tools for actively controlling survey errors and costs. *Journal of the Royal Statistical Society: Series A (Statistics in Society)* 169 (3): 439–457.

Groves, R.M., Fowler, F.J. Jr., Couper, M.P. et al. (2009). *Survey Methodology.* Hoboken, NJ: Wiley.

Jaszczak, A., O'Doherty, K., Colicchia, M. et al. (2014). Continuity and innovation in the data collection protocols of the second Wave of the National Social Life, Health, and Aging Project. *The Journals of Gerontology Series B: Psychological Sciences and Social Sciences* 69: S4–S14.

Kessler, R.C., Berglund, P., Chiu, W.T. et al. (2004). The US National Comorbidity Survey Replication (NCS-R): design and field procedures. *International Journal of Methods in Psychiatric Research* 13 (2): 69–92.

Kish, L. (1949). A procedure for objective respondent selection within the household. *Journal of the American Statistical Association* 44 (247): 380–387.

Kish, L. (1965). *Survey Sampling*. New York, NY: Wiley.

Kornhauser, A. and Sheatsley, P. (1959). Questionnaire Construction and Interview Procedure Research Methods in Social Relations, C. New York, NY: Holt, Rinehart and Winston.

Landon, E.L. (1971). Order bias, the ideal rating, and the semantic differential. *Journal of Marketing Research* 8 (3): 375–378.

Lindau, S.T., Schumm, L.P., Laumann, E.O. et al. (2007). A study of sexuality and health among older adults in the United States. *New England Journal of Medicine* 357 (8): 762–774.

O'Muircheartaigh, C., Eckman, S., and Smith, S. (2009). Statistical design and estimation for the National Social Life, Health, and Aging Project. *The Journals of Gerontology: Series B* 64B (Suppl 1): i12–i19. https://doi.org/10.1093/geronb/gbp045.

O'Muircheartaigh, C., English, N., Pedlow, S., and Kwok, P.K. (2014). Sample design, sample augmentation, and estimation for Wave 2 of the NSHAP. *The Journals of Gerontology Series B: Psychological Sciences and Social Sciences* 69 (Suppl 2): S15–S26.

Podsakoff, P.M., MacKenzie, S.B., and Podsakoff, N.P. (2012). Sources of method bias in social science research and recommendations on how to control it. *Annual Review of Psychology* 63: 539–569.

Salmon, C.T. and Nichols, J.S. (1983). The next-birthday method of respondent selection. *Public Opinion Quarterly* 47 (2): 270–276.

Smith, S., Jaszczak, A., Graber, J. et al. (2009). Instrument development, study design implementation, and survey conduct for the National Social Life, Health, and Aging Project. *The Journals of Gerontology Series B: Psychological Sciences and Social Sciences* 64 (Suppl 1): i20–i29.

Tourangeau, R., Rips, L.J., and Rasinski, K. (2000). *The Psychology of Survey Response*. Cambridge University Press.

Verma, V., Scott, C., and O'Muircheartaigh, C. (1980). Sample designs and sampling errors for the World Fertility Survey. *Journal of the Royal Statistical Society. Series A (General)* 431–473.

Part II

Survey Experiments with Techniques to Reduce Nonresponse

Edith D. de Leeuw[1] *and Paul J. Lavrakas*[2]

[1] *Department of Methodology & Statistics, Utrecht University, Utrecht, the Netherlands*
[2] *NORC, University of Chicago, 55 East Monroe Street, Chicago, IL 60603, USA*

Nonresponse in surveys can occur at the unit level and/or the item level. Nonresponse and the bias it can cause are two of the greatest challenges that survey researchers have faced for more than a century (de Leeuw 1999). Thus, it is no surprise that experiments to reduce survey nonresponse – especially nonresponse at the unit level – are among the most commonly conducted experimental studies to appear in the survey research literature. The independent variables that researchers manipulate in their studies to reduce the effects of unit nonresponse include a host of respondent recruitment techniques, including various contingent and noncontingent incentives (Singer and Ye 2013), advance contacts (de Leeuw et al. 2007; Edwards et al. 2009), multiple contact attempts, multiple modes of contact, reducing the burden of the survey task, and appealing to altruism and other prosocial motivations. The dependent variables in these studies include the resulting response rates, the final sample's unweighted representation of the target population, changes in nonresponse bias, resulting quality of data, and changes in total survey costs.

In survey-based studies to reduce item nonresponse, researchers have experimented with incentives and communication about the importance of data quality to try to reduce missing data (de Leeuw et al. 2003). The dependent variables in such experiments include item nonresponse rates and the nature of item-nonresponse bias. The independent variables serve as the treatments meant to prevent and thereby reduce item nonresponse, which researchers have controlled via experimentation include changing mode of data collection, altering the nature of the questionnaire including its layout/format, motivational respondent instructions about data quality in the cover letter and/or the questionnaire, and noncontingent incentives to motivate higher data quality.

This section of the book contains two chapters on unit nonresponse. The first is by Bianchi and Biffignandi and addresses the issue of incentives and their effects in ongoing online probability-based panels. Their literature review of survey-based incentives provides an up-to-date summary of the state of knowledge that has accrued via hundreds of controlled experiments on incentives that have been embedded in probability sample surveys during the past nine decades. In their original case study, the authors use two experiments within the UK's Society Innovation panel to simultaneously study the effects of incentives and data collection

Experimental Methods in Survey Research: Techniques that Combine Random Sampling with Random Assignment, First Edition.
Edited by Paul J. Lavrakas, Michael W. Traugott, Courtney Kennedy, Allyson L. Holbrook, Edith D. de Leeuw, and Brady T. West.

Companion Website: www.wiley.com/go/Lavrakas/survey-research

mode and illustrate how such experiments can be set up and analyzed. Results are presented regarding the effects on participation, bias, data quality, and costs. Their first experiment tested how different levels of noncontingent incentives worked in a single-mode face-to-face design versus a sequential mixed-mode design. In their second experiment, the authors tested the effects on contingent incentives in enhancing web response in the mixed-mode group. In their discussion, they address the issue of "lurking variables" that may lead to spurious results when conducting these survey-based experiments.

The second chapter comes from Vogl, Parsons, Owen, and Lavrakas and addresses experimentation within the domain of advance contacts. The authors provide a comprehensive literature review of this domain of survey-based experimentation, including detailed discussion of the dependent variables that should be considered when conducting such an experiment. The chapter presents information about two original case studies the authors carried out. Within a survey on violence against men in intimate relationships, the first of the experiments tested the effects of personalized advance letters on outcome rates and reasons for refusals, recruitment effort, and the reporting on sensitive topics. In a second case study within a survey of neighborhood crime and justice issues, an original experiment was carried out that incorporated experiments with both advance letters and envelope branding. Furthermore, it explored the effect of envelope contents by including a letter only vs. letter plus study brochure treatment. Of note, this case study is unique in its use of focus on face-to-face data collection instead of telephone data collection.

Despite the considerable survey-based experimentation that has been carried out for more than 80 years, there remains a great deal of experimentation that is needed in order to improve the cost efficiency of conducting surveys. This is especially important in an era of diminishing budgets to fund scientific research. There is a need for more and better surveys to understand incentives, especially the interaction of contingent and noncontingent incentives in the same study on the success of initially recruiting respondents and maintaining their cooperation in the case of panel studies. More knowledge is needed on how to effectively personalize advance contacts, especially when email is a recruitment mode. But most important is future experimentation with adaptive/responsive design, including those that use response propensity modeling to tailor recruitment strategies either prior to the start of the field period or during the field period (cf. Tourangeau et al. 2017).

References

Edwards, P.J., Roberts, I., Clarke, M.J. et al. (2009). Methods to increase response to postal and electronic questionnaires. *Cochrane Database of Systematic Reviews* 3, Article number MR000008.

de Leeuw, E. (1999). Preface." Special issue editor. *Journal of Official Statistics* 15 (2): 127–128.

de Leeuw, E., Hox, J., and Huisman, M. (2003). Prevention and treatment of item nonresponse. *Journal of Official Statistics* 19 (2): 153–176.

de Leeuw, E., Callegaro, M., Hox, J. et al. (2007). The influence of advance letters on response in telephone surveys: a meta-analysis. *Public Opinion Quarterly* 71: 413–443.

Singer, E. and Ye, C. (2013). The use and effects of incentives in surveys. *The Annuals of the American Academy of Political and Social Science* 645 (1): 112–141.

Tourangeau, R., Brick, J.M., Lohr, S., and Li, J. (2017). Adaptive and responsive survey designs: a review and assessment. *Journal of the Royal Statistical Society: Series A (Statistics in Society)* 180: 203–223.

4

Survey Experiments on Interactions and Nonresponse: A Case Study of Incentives and Modes

A. Bianchi and S. Biffignandi

Department of Management, Economics and Quantitative Methods, University of Bergamo, via dei Caniana 2, 24127 Bergamo, Italy

4.1 Introduction

Three major trends are requiring studies for updating survey best practices and understanding the impact of new survey protocols. These trends are (i) increasing Internet diffusion and, as a consequence, increasing the use of web as a survey tool, and of mixed-mode designs that include web (de Leeuw 2005). Mixing modes are needed in probability-based sampling when a web-only design would result in undercoverage of the target population and/or an unacceptably low response rate (RR); and (ii) decreasing response rates. Respondent incentives can reduce the occurrence of nonresponse, and their role should be reevaluated in the context of lower response rates and mixed modes; and (iii) growing attention to survey quality evaluation that goes beyond the response rate toward a total survey error perspective and broader issues such as ethical evaluation.

Understanding how these trends can be better managed to improve surveys, surveys results, and data quality is one of the greatest challenges of survey methodological research nowadays. The scenario is rather complex as many factors act both separately and interactively. For this reason, experiments are needed to extend empirical-based evidence, and researchers need to understand how to analyze such experiments. In this chapter, we present a case study to illustrate how these issues can be tackled through a survey-based experiment. The case study raises generic issues that would also be relevant to the experimental study of other issues.

Knowledge of the effects of incentives in mixed-mode surveys, and whether these effects differ from those in single-mode surveys, is lacking. The literature on incentives is extensive and dates back many years. However, it mainly refers to traditional data collection modes. Studies on the use of incentives in web surveys and mixed-mode surveys are scarce. Little is known about the cost–benefit impact of incentives in a mixed-mode context or whether targeting the use of incentives to specific groups could lead to cost savings.

Our case study follows from experiments first reported by Jäckle et al. (2015) and focuses on the comparison of a mixed-mode design with a web component and a computer assisted personal interviewing (CAPI)-only design and on the use of incentives. Outcome measures include response rate and other quality indicators. We investigate these factors in a single wave of a longitudinal panel.

Experimental Methods in Survey Research: Techniques that Combine Random Sampling with Random Assignment, First Edition.
Edited by Paul J. Lavrakas, Michael W. Traugott, Courtney Kennedy, Allyson L. Holbrook, Edith D. de Leeuw, and Brady T. West.

Companion Website: www.wiley.com/go/Lavrakas/survey-research

Section 2 presents a critical review of the literature on randomized experiments to study the effects of incentives with special reference to different modes. We try to disentangle the main questions faced by the literature on incentives in the context of the key issues stated above. We also report findings of experimental studies on incentives carried out by other existing panels.

In Section 3, we present our case study based on experiments carried out in Wave 5 of the Understanding Society Innovation Panel (USIP) in 2012. Outcomes from the experiments are presented using statistical tests and models. Our case study provides an example of how to design such experiments and how to handle the analysis of such experiments from a statistical perspective, while also providing evidence of the impact of incentive strategies and modes. Impacts are assessed not only on response rates but also on data quality and indicators of costs, going toward a total survey error perspective.

Section 4 provides suggestions for future research, especially in the context of mixed-mode and longitudinal surveys. Our findings also speak to the debate on the role of incentives in socioeconomic surveys. Some general guidance about the design and analysis of survey-based experiments to estimate interaction effects is also presented.

4.2 Literature Overview

Results in the literature regarding the effects of respondent incentives on response rates are generally consistent and point to a positive effect of incentives in increasing response rates (Singer and Kulka 2000), especially when the incentive is monetary in form and offered unconditionally in advance (Singer and Ye 2013; Singer et al. 1999a). Incentives have been found to be effective for both mail surveys (Singer and Ye 2013) and interviewer-administered surveys (Cantor et al. 2008; Singer et al. 1999a). The size of the effect generally appears to be larger for mail surveys (Singer and Ye 2013). The effect of different levels of incentives have been tested as well (Trussell and Lavrakas 2004).

Even though highly relevant for the future, the literature on the effects of incentives on web surveys is smaller. Göritz (2006, 2010, 2015) and Brown et al. (2016) suggest a generally positive effect of incentives in web surveys. Millar and Dillman (2011) report on two experiments conducted to evaluate several strategies for improving response to web and web/mail mixed-mode surveys.

Most evidence of differences between modes in the effect of incentives refers to comparisons of separate studies. This approach does not allow consideration of interactions between mode and incentive levels. Other than Jäckle et al. (2015), only a few studies are known to the authors where comparisons of the effects of incentives in different modes are made. Ryu et al. (2006) compare the effects of different types of incentives (monetary and gifts) in a mixed-mode postal and face-to-face survey.

As for the aspects considered in the evaluation of the effects of incentives, past literature has focused mostly on response rate as the key dependent variable. Little is known about possible effects on bias, data quality, or costs.

Nonresponse bias is caused by differential nonresponse across sample groups. Some studies found that incentives increase participation of typically underrepresented respondents: those with less education, single people and those not in paid employment (Ryu et al. 2006), black or Indian minority ethnic groups; those living in larger households or households with dependent children, aged 0–20, or single (Stratford et al. 2003); and those with less education (Singer et al. 2000), black (Mack et al. 1998) or poor (Mack et al. 1998; James 1997). As a consequence, if these respondent characteristics are related to substantive variables that are measured in the questionnaire, the result of using incentives will be less nonresponse bias. Regarding the effects of incentives on data quality, the literature presents mixed evidence. A common measure of data quality is item nonresponse. This is a critical aspect, as many analysts only use complete cases to perform analyses. The main concern is that the effect of incentives may be to include more respondents who are not diligent about answering survey questions and thus to increase item nonresponse, thereby lowering data quality. However, existing studies either found that incentives lead to less item nonresponse (Singer et al. 2000; Mack et al. 1998) or found no relationship between item nonresponse and the use of incentives (Ryu et al. 2006; Davern et al. 2003; Singer et al. 1999a; Teisl et al. 2005; Tzamourani and Lynn 1999; Willimack et al. 1995). Studies found that item nonresponse rates tend to be higher in (paper) self-completion surveys than interviewer-assisted face-to-face (Bowling 2005; de Leeuw 2005; Nicolaas et al. 2000) and computer-assisted telephone interviewing – CATI (de Leeuw 2005; Fricker et al. 2005). Other measures of data quality have also been used in studies of the effects of incentives, such as indicators of the extent of satisficing behavior (Grauenhorst et al. 2016).

With reference to costs, Gajic et al. (2012) conducted a randomized experiment to test the cost-effectiveness of incentives in a general community population, in a sequential mixed-mode approach, with invitation using a traditional mailed letter request for web completion. Individuals were randomized to four incentive groups: no incentive, prepaid cash incentive ($2), a low value lottery (10 prizes of $25), and a higher value lottery (2 prizes of $250). Looking at the incremental cost-effectiveness ratio (ICER) per completed survey, it turned out that the high value lottery was the most cost-effective incentive for obtaining completed surveys. This is consistent with other experiments on web surveys (Bosnjak and Tuten 2003; Duetskens et al. 2004).

Most studies on the effect of incentives are based on cross-sectional surveys. Few studies have been conducted in the longitudinal context. The longitudinal context introduces distinct considerations. If panel members have completed the questionnaire in a previous wave, they are already used to it and may have better comprehension of the questions. Further, respondents who might fear repercussions as a result of giving a particular answer might be aware of the absence of such repercussions. Incentives seem to be at least as effective in longitudinal surveys as they are in cross-sectional surveys (Laurie and Lynn 2009; Schoeni et al. 2013). Laurie and Lynn (2009) review the use of incentives in longitudinal surveys, describing common practices and the rationale for these practices.

The use of unconditional versus conditional incentives in longitudinal studies warrants some discussion. Social Exchange Theory (Dillman et al. 2014) advises that unconditional incentives build trust in the researcher, whereas conditional incentives used by themselves do not. However, once trust is built it likely remains, unless the researcher does something later to violate that trust. So it might be expected that there is no need to use unconditional incentives for continuing panel members. However, the use of incentives has been found to raise the expectations of respondents in future surveys (Singer et al. 1998). Thus, once incentives are given, withdrawing them in a subsequent wave could have adverse effects on response rates. Overall, there is limited evidence of the role of these expectations in terms of respondent behavior in longitudinal surveys. Little is known about the effects of ceasing to provide incentives when respondents had previously received them. Some studies suggest that they may not be significant (Singer et al. 1999b). Laurie and Lynn (2009) report that an incentive sent unconditionally in advance of the interview appears to be most effective in increasing response rates. Jäckle and Lynn (2008) report results from an experiment comparing the use of unconditional and conditional incentives. Their experiment was carried out on cohort 10 of the Youth Cohort Study (YCS) of England and Wales, which used a combination of postal and telephone modes. No incentives were provided at Wave 1. In Wave 2, within the sample for each mode, a random subset was sent a £5 voucher while the remainder received no incentive. In the postal sample, the incentive treatment group was further divided into two subgroups. To one subgroup, incentives were given unconditionally, to the other subgroup, incentives were provided conditionally on response. At Waves 3 and 4, all incentives were paid unconditionally. Unconditional incentives were found to have a greater effect in reducing attrition than conditional incentives (but similar effects on nonresponse and attrition bias).

Turning attention to the mixed-mode (with web component) longitudinal context, very few studies of incentives have been carried out. Jäckle and Lynn (2008) found that incentives had a stronger effect on attrition and item nonresponse in postal than telephone mode, and no effect on attrition bias in either mode. We can speculate that web surveys, another form of self-administered survey, could also benefit from the use of incentives.

In the same specific context that we consider of a longitudinal survey in which previous waves were interviewer-administered, Jäckle et al. (2015) study the effects of introducing mixed-mode in Wave 5 of USIP and provide some considerations about the use of incentives. With reference to more recent panel members, they find a suggestion that higher (unconditional) incentive levels might counterbalance the negative effect of the mixed-mode design on individual and household response rates. However, these differences were not statistically significant. Further, they find that complete household response by web depends on the level of unconditional incentives. Higher unconditional incentive levels increased the probability of households participating fully by web for those who had been in the panel for fewer waves. A marginally significant increase in the proportion of households fully participating by web was found for those who had been in the panel longer and had responded at the previous wave. The conditional incentive (conditional on household fully responding by web) increased the proportion of households fully responding by web for panelists with longer tenure in the panel. The effect was found to be stronger among households in which sample members received higher levels of unconditional incentives. No effect of conditional incentives was found for more recent panel members.

With reference to the same experiment, but considering outcomes over three waves, Bianchi et al. (2017) investigate some aspects related to survey costs, taking several components into account. The analysis highlights that the mixed-mode design appears to have the potential to deliver substantial cost savings.

In summary, most of the knowledge about the effects of incentives on response comes from the cross-sectional context and points to a positive effect of incentives in increasing response rates. The available evidence comes mostly from mail and interviewer-administered surveys, while evidence from web surveys is much smaller. Further, available evidence of differences between modes in the effect of incentives refers mostly to comparisons of separate studies, thus not allowing investigation of possible interactions between mode and incentive levels. The aspect that has been investigated the most in the literature is the response rate, little is known about possible effects on nonresponse bias, data quality, and costs. Some studies found that incentives have a positive effect in increasing participation of typically underrepresented groups, while mixed evidence was found on the effect of incentives on item nonresponse. In general, item nonresponse rates tend to be higher in (paper) self-completion surveys than interviewer-assisted face-to-face and CATI. As for costs, high value lotteries were found to be the most cost-effective incentive for obtaining completed surveys. Literature on the effects of incentives in longitudinal studies is very scarce, particularly if the focus is on the mixed-mode longitudinal context. However, in general, incentives seem to be at least as effective in longitudinal studies as they are in cross-sectional surveys. In this respect, some evidence was found that incentives had a stronger effect on attrition and item nonresponse in postal than telephone mode. In the same context that we consider of a longitudinal survey in which previous waves were interviewer-administered, positive effects were found but restricted to specific subgroups of respondents, depending on the number of waves they have been in the panel and on previous wave response outcome.

4.3 Case Study: Examining the Interaction Between Incentives and Mode

In this section, we provide an example of how an experiment can be set up and analyzed to simultaneously study the effects of incentives and data collection mode. Results are presented regarding effects on participation, bias, data quality, and costs.

First, we introduce the probability survey within which the experiment was carried out (Section 3.1); next, we describe the experiment (Section 3.2). In Section 3.3, we describe the data obtained from the experiment and the statistical methods we adopted for analyzing them. We present our results in Section 3.4.

4.3.1 The Survey

USIP is a longitudinal panel survey of individuals in the United Kingdom. Annual interviews have been carried out since 2008 (Buck and McFall 2012). For methodological experiments and testing that could lead to improvements in the main panel see Uhrig (2011) and Lynn and Jäckle (2019).

The target population for the USIP is all individuals aged 16 or over and living in England, Scotland, or Wales. The USIP is based on a stratified clustered probability-based sample of addresses. All residents at selected addresses are included in the panel. Primary sampling units are postal sectors, secondary sampling units are residential addresses selected from the Postcode Address File (Lynn and Lievesley 1991), and sample elements are persons. An initial sample of 2760 addresses was included from USIP1, which forms the original sample. In 2011, 960 addresses were added at USIP4 by means of a refreshment sample. Further details on the USIP sample design are found in Lynn (2009).

The USIP involves interviews at 12-month intervals with the initial sample and all members of the current household of each sample person. Only sample members who were in participating households at the first wave for that sample were reapproached for interview at each subsequent wave. Interviews cover a wide range of topics, such as household dynamics, economic activity, income, health, housing, and political attitudes. Household participation can be complete or partial as follows: household response is considered complete if all household members answer the questionnaire and partial if only some of the household members participate. From USIP2 onward, nonresponse at one wave did not preclude an interview attempt at the next wave. Households in which no person responded at two successive waves are no longer issued to the field. Thus, in the sample issued to the field at USIP5, the original sample included individuals who had responded at USIP4 and a number of individuals and whole households that had not been contacted or had refused at USIP4. The refreshment sample, added in 2011, only included individuals in households that had responded at IP4.

4.3.2 The Experiment

At USIP5, a data collection mode experiment and two incentives experiments were carried out. The mode experiment considered two modes: mixed-mode and single-mode face-to-face interview (Jäckle et al. 2015; Bianchi et al. 2017). Mixed-mode was sequential: web first (first step) and face-to-face in the follow up (second step) among nonrespondents to the web step. Single mode face-to-face was fielded parallel to the mixed-mode second step. Note that for the mixed-mode group, the web option for the questionnaire remained available throughout the fieldwork period, so that it was possible to respond online at any time. Sample members were randomly allocated to one of the two mode treatments. The allocation was at the household level, so that all individuals in the same household received the same treatment. One-third was allocated to the single-mode face-to-face design and two-thirds to the sequential mixed-mode design.

The first incentive experiment considered unconditional prepaid incentives in both modes. The second concerned conditional incentives to enhance web response in the mixed-mode group. The experiment on unconditional incentives was fully crossed with the mode experiment and is focused on the different amounts of incentives. Allocation is at the household level, so all individuals in the same household received the same incentive. The incentive was in the form of a High Street gift voucher, a voucher which is accepted by over 40 leading Retail Groups and a range of Leisure Groups in the United Kingdom. Incentives were provided unconditionally in an advance/prenotification letter. The value of the voucher was manipulated experimentally. Treatment allocation was conditional on the type of sample (original/refreshment) and previous wave incentive allocation. Original sample members were allocated to receive either £5 or £10, based on their randomized allocation to incentive levels at previous waves. All those who had received either £5 or £10 at USIP4 again received the same amount, while those who at USIP4 had received an initial £5 with a promise of an additional £5 if all household members participated were randomly allocated to receive either £5 or £10 at USIP5. Members of the refreshment sample had been randomly allocated to receive £10, £20, or £30 at USIP4. Each received the same value incentive again at USIP5. Notice that all eight possible combinations of the levels of the two factors (mode and level of unconditional incentive) are present. Thus, this is a 2×4 full factorial experiment. Higher levels of incentives in the refreshment sample are justified by expectations of lower response rates for this group, as explained in Section 3.3.

The second incentives experiment, on conditional incentives, was fully crossed with the unconditional incentive experiment. The purpose was to test ways of increasing web response rates. Thus, this experiment was restricted to the mixed-mode group and, if considered in combination with the unconditional incentives experiment, is again a 2×4 full factorial design. Half of the households in the mixed-mode group were offered an additional £10 per person conditional on all eligible household members completing the web survey within two weeks. This was mentioned in the advance letters to all household members in this treatment group. Table 4.1 summarizes the experimental design.

Some other details of USIP5 survey procedures are relevant. The face-to-face treatment involved standard USIP procedures. Each adult sample member (aged 16 or over) was sent an advance letter with an unconditional incentive, after which interviewers visited to attempt face-to-face interviews. In each household, one person was asked to complete the household enumeration grid and the household questionnaire. All household members aged 16 or over were asked for an individual interview, which included a CASI (computer-assisted self-interviewing) self-completion component.

In the mixed-mode treatment group, sample members were sent a letter with an unconditional incentive, inviting them to take part by web. The letter included the URL and a unique user ID, which was to be entered on the welcome screen. A version of the letter was additionally sent by email to all sample members for whom an email address was available (around one-third of the sample). For people who had indicated at previous waves that they do not use the Internet regularly for personal use, the letter mentioned that they would also have the opportunity to do the survey with an interviewer. Up to two email reminders were sent at three-day intervals.[1] Sample members who had not completed the web interview after two weeks were sent a reminder by post and interviewers then started visiting them to carry out face-to-face interviews. The web survey remained open throughout the fieldwork period.

1 The impact of email reminders on panel members who had indicated at previous waves that they do not use the internet regularly for personal use was very limited.

Table 4.1 Design of the mode and incentive experiments.

	Factors		
	Unconditional incentives		Conditional incentives
Mode	Original sample	Refreshment sample	
Mixed-mode	£5	—	£10
			Nothing
	£10	£10	£10
			Nothing
	—	£20	£10
			Nothing
	—	£30	£10
			Nothing
Face-to-face	£5	—	
	£10	£10	
	—	£20	
	—	£30	

The first household member to log on to do the web survey was asked to complete the household grid, which collects information on who is currently living in the household. The web grid included an additional question to identify who is responsible for paying bills. The household questionnaire could be completed by either this person or their spouse/partner. For these sample members, the household questionnaire was displayed first, then leading on to the individual questionnaire. Once one partner had completed the household questionnaire, it would not appear for the other partner. The web questionnaire was based on the face-to-face one, with some adaptations, such as incorporating interviewer instructions into question wording or removing references to show cards.

4.3.3 Data and Methods

We use data from the USIP, and we restrict attention to Wave 5 of the panel, where the incentive experiment under consideration was carried out. We study outcomes at USIP5 only, though we use some variables (especially socio-demographics) from earlier waves as covariates. Analysis of participation and cost indicators refer to eligible households ($n = 1566$) and individuals aged 16 or over ($n = 3040$). Analysis of data quality relates to USIP respondent individuals ($n = 1995$). We consider results for the overall sample and for the original and refreshment samples separately, as these two groups may show different behavior. Indeed, it is commonly found in longitudinal studies that wave-on-wave attrition is highest at the second wave and then declines over time. A reason for attrition at the initial waves of a panel is related to "absence of commitment" (Laurie et al. 1999). According to this theory, some respondents really never wanted to participate in the survey but were persuaded in the first wave. If participation itself does not change their commitment, then these respondents are very likely to drop out in Waves 2 or 3. Thus, as individuals are member of the panel for longer time they are expected to be more faithful respondents. In our case, the wave under study was the fifth wave for individuals in the original sample. Individuals in the refreshment sample entered the panel at USIP4, so the wave under study was the second wave for them.

As for correlates of nonresponse, there is also evidence that they may change over waves of a survey. Farrant and O'Muircheartaigh (1991), in a study based on the British Election Panel Survey, find that the strongest predictors of nonresponse were sociodemographics at Wave 1 (age and level of education), interest in politics at Wave 2 and, at subsequent waves, characteristics associated with the likelihood to move. Bianchi and Biffignandi (2017a), in a study based on four waves of the UK Household Longitudinal Study, report that under- and overrepresentation of subgroups in the panel tends to decrease and be more and more similar at each subsequent wave.

Furthermore, those who have been in the panel longer have more experience with the questionnaire in another mode and prior knowledge of its content than those who have entered the panel more recently. Thus, it is expected that those who have joined the panel more recently will show higher levels of attrition/nonresponse. Using the same data as us, Jäckle et al. (2015) found that, for those who have been longer in the panel, the proportion of interviews of any form was lower with mixed-mode, while there was no difference by the mode of data collection for those who joined the panel more recently. Thus, the two samples show different mode effects in response. This result may be explained by the fact that original sample members are more familiar with face-to-face mode and thus more reluctant to change the mode of the interview. Bianchi et al. (2017) found no difference between mixed-mode and face-to-face designs with respect to the cumulative response rate over Waves 5, 6, and 7, regardless of the number of waves that panelists have been in the panel.

Table 4.2 summarizes the distribution of the issued sample of households across samples, mode, and unconditional incentive experimental allocation. Table 4.3 shows the allocation of households across samples and conditional incentive experimental groups in the mixed-mode group.

Table 4.2 The number of households allocated to the unconditional incentive and mode experimental groups by sample.

	Original sample		Refreshment sample		Total
Treatment level	**F2F**	**MM**	**F2F**	**MM**	
£5	199	399	—	—	598
£10	162	323	42	90	617
£20	—	—	63	111	174
£30	—	—	63	114	177
Total	361	722	168	315	1566

F2F, face-to-face; MM, mixed-mode.

Table 4.3 The number of households allocated to the conditional incentive experimental groups by sample (mixed-mode group).

Treatment level	Original sample	Refreshment sample	Total
Yes	363	157	520
No	359	158	517
Total	722	315	1037

As the data are from a so-called population-based survey experiment (Mutz 2011), to perform proper analyses, one needs to take into account both the survey design and the experimental design. To study the effects of treatments on outcome variables (response, item nonresponse, and number of interviewer visits per household), we apply two-way ANOVA, adjusting standard errors and p-values for sample design. ANOVA is a typical technique in the analysis of experiments (e.g. Maxwell and Delaney 1990). ANOVA is more statistically efficient than multiple two-group t-tests. In ANOVA, each factor can be tested while controlling for all other factors. Further, with ANOVA, interaction effects between variables can be detected and, therefore, it is possible to test more complex hypotheses. The experiment that we analyze is a randomized design. Randomization controls for the effects of extraneous variables: on average, extraneous factors will affect treatment conditions equally; so any significant differences between conditions can fairly be attributed to the independent variable. This ensures internal validity. As stated above, the probability sample design is nationally representative, and we take into account in our analysis the survey nature of the data by adjusting standard errors and p-values for the sample design in terms of clustering and stratification. These features add to external validity.

Where a treatment effect is found to be significant, we perform multiple comparisons among treatment levels by means of t-tests, adjusted for both sample design and multiplicity of the comparisons. We use Tukey's honest significant difference (HSD) post hoc test to take into account the multiplicity of the comparisons. Analyses are carried out in SAS version 9.4, using PROC SURVEYREG. We used the statement LSMEANS to perform multiple comparisons with the option ADJUST to adjust p-values for multiplicity of comparisons. Finally, to test the effect of incentives on attrition bias, we use logit regression models including incentive treatment, a number of covariates, and their interaction with treatment. We adjust standard errors and p-values for sample design.

4.3.4 Results of the Experiment: Incentives and Mode Effects

4.3.4.1 Effects on Participation

All our analyses are conditional on being issued to the field at USIP5. This means that all USIP1 nonresponding households and some explicit refusers or persistent nonrespondents at USIP2 to USIP4 have been dropped from the sample. For this reason, response rates do not meet standard definitions. Our individual-level response rate is defined as the percentage of individuals giving a full interview among those issued to the field and not found to be ineligible. Our household-level response rate is the percentage of complete or partial households among those issued to the field and (assumed to be) eligible.

Table 4.4 presents results from a two-way ANOVA to study the effects of the unconditional incentive, mode and their interaction on both individual and household response, and multiple comparisons for effects found to be significant in the ANOVA analysis. At the individual level, only the incentive has an effect on response rates ($p < 0.001$). Looking at multiple comparisons, only £30 is significantly different from the other levels of incentives. The response rate difference between £30 and £5 is 19.7 pp ($p < 0.05$) and between £30 and £10 is 16.4 pp ($p < 0.05$). The interaction between mode and level of incentive is not significant, meaning that there is no evidence of a difference in effect of incentive treatment between the mode treatments.

Looking separately at the two samples, in the original sample, no significant effect is found, while the incentive effect is significant for the refreshment sample ($p < 0.001$). Household response shows a similar pattern.

Mode has no significant effect overall or in either the original or refreshment samples.

Table 4.4 ANOVA results and multiple comparisons for the effects of unconditional incentive, mode and their interaction on response.

Overall sample								
	Individuals				Households			
Effects	**Model**	**Incentive**	**Mode**	**Incentive × mode**	**Model**	**Incentive**	**Mode**	**Incentive × mode**
p-Value	**0.00**	**0.00**	0.88	0.24	**0.00**	**0.00**	0.36	0.78
Incentives	**Mean RR**	**Multiple comparisons**			**Mean RR**	**Multiple comparisons**		
		£5	**£10**	**£20**		**£5**	**£10**	**£20**
£5	60.9				75.3	—		
£10	64.2	3.3			76.7	1.4		
£20	72.5	11.6	8.3		79.9	4.6	3.2	
£30	80.6	**19.7****	**16.4****	8.1	91.0	**15.7****	**14.3****	**11.1****
Original sample								
Effects	**Model**	**Incentive**	**Mode**	**Incentive × mode**	**Model**	**Incentive**	**Mode**	**Incentive × mode**
p-Value	0.27	0.29	0.14	0.38	0.47	0.75	0.25	0.47
Refreshment sample								
Effects	**Model**	**Incentive**	**Mode**	**Incentive × mode**	**Model**	**Incentive**	**Mode**	**Incentive × mode**
p-Value	**0.01**	**0.00**	0.54	0.60	**0.01**	**0.01**	0.40	0.44
Incentives	**Mean RR**	**Multiple comparisons**			**Mean RR**	**Multiple comparisons**		
		£10	**£20**			**£10**	**£20**	
£10	64.0	—			76.5	—		
£20	72.5	8.5			79.9	3.4		
£30	80.6	**16.6****	8.1		90.1	**14.5****	**11.1****	

Note: p-Values are adjusted for sample design and multiplicity of comparisons. RR, response rate. $^{**}p<0.05$ In the ANOVA part of the table, p-values for F-tests are shown, while in the multiple comparison part of the table, response rates differences are displayed.
Bold values correspond to significant values.

4.3.4.2 Effects on Subgroups

To investigate whether incentives had different effects on attrition for different subgroup characteristics, we fitted a logit model predicting individual full response (versus proxy or nonresponse) using individual characteristics and interactions of those characteristics with treatment as predictors.

Individual characteristics were measured in USIP4 (or at the last available interview before USIP5). Results for the original sample and for the refreshment sample are summarized in

Table 4.5 Average predicted probabilities of giving full interview in the original and refreshment samples – based on a logit model including the allocated incentive level, the allocated mode, characteristics of the sample members, and interactions between the incentive and the other variables as predictors.

	Original sample				Refreshment sample				
	£5	£10	*p*-Value	*p*-Value (joint test)	£10	£20	£30	*p*-Value	*p*-Value (joint test)
F2F	65.1	64.8	0.93		66.4	69.3	73.97	0.49	
Mixed-mode	59.1	64.3	0.11	0.32	63.2	75.3	83.84	0.00	0.33
Female	66.0	67.2	0.65		65.7	73.9	85.55	0.00	
Male	55.8	61.4	0.10	0.20	62.7	71.7	75.10	0.07	0.13
Rural	60.1	60.4	0.96		66.6	74.7	79.38	0.27	
Urban	61.4	65.5	0.16	0.55	63.8	72.4	80.91	0.00	0.33
Age 16–20	60.0	49.7	0.25		60.4	65.3	78.29	0.36	
Age 21–30	40.3	54.6	0.04		37.8	56.0	53.92	0.22	
Age 31–40	56.3	64.1	0.24		58.4	71.0	79.63	0.17	
Age 41–50	65.3	68.5	0.50		73.8	59.8	87.21	0.01	
Age 51–60	66.2	73.2	0.15		68.2	79.7	89.44	0.01	
Age 61–70	69.5	76.7	0.21		68.6	83.4	91.12	0.08	
Age 71+	60.7	56.3	0.58	0.31	77.1	89.4	74.48	0.20	0.90
Single	75.7	74.9	0.90		65.9	68.6	81.56	0.33	
Single, children	45.1	83.4	0.00		86.5	89.8	79.39	0.80	
Couple	66.4	70.7	0.37		65.3	67.6	77.57	0.16	
Couple, children	58.2	57.7	0.94		66.5	84.1	85.73	0.14	
2+ unrelated adults	56.8	59.0	0.72		57.2	66.8	77.05	0.10	
2+ unrelated, children	50.1	52.7	0.78	0.02	58.1	66.4	77.77	0.22	0.27

Table 4.5, which shows average predicted probabilities[2] from the model, together with *p*-values of Wald tests (adjusted for sample design) of the effect (contrast) of incentive within levels of the covariates and *p*-values of joint tests of whether the contrast estimates for incentive are (jointly) equal for different levels of the variable. The *margins* command in Stata 14 has been used for computing predicted probabilities. In the original sample, the joint test is significant only for household type ($p = 0.02$), with single respondents with children having the lowest predicted response rate with a £5 incentive and the highest predicted response rate with £10. A £10 incentive was more effective than £5 for single respondents with children ($p = 0.00$) and for the 21–30 age group ($p = 0.04$).

For the refreshment sample, the joint test is not significant for any variable. Differences in the treatment effect are found for the mixed-mode group, males, females, people leaving in urban areas, and age groups 41–50, 51–60, and 61–70.

Mode is jointly not significant for either sample. For the refreshment sample only, in the mixed-mode group, higher incentives correspond to significantly higher probability of response.

2 Average predicted probabilities are averages of predicted probabilities of giving full response from the model, given the observed levels of the covariates.

4.3.4.3 Effects on Data Quality

To investigate possible effects on data quality, we fitted a two-way ANOVA to two measures of item nonresponse. The first one is an overall item nonresponse rate, based on over 1000 items in the individual questionnaire and including 6 items about unearned income sources. The rate is the proportion of items for which the respondent was eligible, to which they answered "don't know" or "refused." The second measure refers to the proportion of employed respondents who did not provide an answer for their last gross pay. This item is important for the derivation of income measures, and it is prone to high levels of item nonresponse (Al Baghal and Lynn 2015). Results are shown in Table 4.6, together with multiple comparisons for significant effects.

Table 4.6 ANOVA results and multiple comparisons for item nonresponse indicators (overall measure and Gross pay indicator), by sample and unconditional Incentive.

				Overall sample				
Indicator	**Overall measure**				**Gross pay**			
Effects	**Model**	**Incentive**	**Mode**	**Incentive × mode**	**Model**	**Incentive**	**Mode**	**Incentive × mode**
p-Value	**0.00**	**0.01**	**0.00**	0.94	**0.03**	0.61	**0.00**	0.47
Incentives	**Mean INR**	**Multiple comparisons**			**Mean RR**	**Multiple comparisons**		
		£5	**£10**	**£20**		**£5**	**£10**	**£20**
£5	1.57	—			16.5	—		
£10	1.66	0.09			13.3	3.3		
£20	0.99	0.58	0.67		15.5	1.0	2.3	
£30	1.20	0.37	0.46	0.21	13.7	2.8	0.4	1.8
Mode	**Mean INR**	**Comparison**			**Mode**	**Mean INR**	**Comparison**	
		F2F					**F2F**	
F2F	1.04	—			F2F	9.4		
MM	1.73	**0.69****			MM	17.7	**8.3****	
				Original sample				
Effects	**Model**	**Incentive**	**Mode**	**Incentive × mode**	**Model**	**Incentive**	**Mode**	**Incentive × mode**
p-Value	**0.01**	0.52	**0.00**	0.66	**0.08**	0.18	**0.02**	0.75
Mode	**Mean INR**	**Comparison**			**Mode**	**Mean INR**	**Comparison**	
		F2F					**F2F**	
F2F	1.2	—			F2F	10.0		
MM	1.9	**0.75****			MM	17.6	**7.5****	

(Continued)

Table 4.6 (Continued)

Refreshment sample								
Effects	Model	Incentive	Mode	Incentive × mode	Model	Incentive	Mode	Incentive × mode
p-Value	**0.00**	0.10	**0.00**	0.92	**0.00**	0.97	**0.01**	**0.04**
Mode	Mean INR	Comparison F2F			Mode	Mean INR	Comparison F2F	
F2F	0.82	—			F2F	8.0		
MM	1.38	**0.56****			MM	18.0	**10.0****	

Note: p-Values are adjusted for sample design and multiplicity of comparisons. INR, item nonresponse.
**$p < 0.05$. In the ANOVA part of the table, p-values for F-tests are shown, while in the multiple comparison part of the table, response rates differences are displayed.
Bold values correspond to significant values.

In the complete sample and considering the overall measure of item nonresponse, the incentive effect is significant. Looking at multiple comparisons, at the 5% level of significance, no difference in item nonresponse can be detected among different incentive levels. No effect of incentive is found for the gross pay measure. There is a significant effect of mode ($p < 0.001$) for both item nonresponse measures. Mixed-mode showed a higher level of item nonresponse rate overall (1.73% vs. 1.04%) and for the gross pay item (17.7% vs. 9.4%). The interaction of mode with incentives is not significant for either item nonresponse measure.

Looking at the original sample and refreshment sample separately, for both samples, the only significant factor is mode both for the overall item nonresponse measure and the gross pay measure, with higher levels of item nonresponse for the mixed-mode group.

4.3.4.4 Effects on Costs

Web data collection brings potential cost savings relative to mail surveys and interviewer-administered surveys. When adopting an incentive strategy and choosing a particular incentive among several possible ones, the possible impact on response and retention should be weighed against the cost of the incentive itself and its distribution to respondents. A thorough analysis of costs would require careful consideration of all factors affecting costs: sample size, sample design, interviewers' fees, interview length, etc. We do not perform such an analysis here. Rather, we examine two cost indicators: (i) the number of interviewer visits to households and (ii) the percentage of households fully responding by web in the mixed-mode group. The latter is justified as data collection costs are very much influenced by the proportion of households requiring a face-to-face visit at the second step of the sequential field design. Since all household members are required to respond to the survey, an interviewer visit to a household can only be avoided if all household members respond by web.

Table 4.7 reports results from a two-way ANOVA to study the effect of unconditional incentive, mode, and their interaction on the number of interviewer visits per household, for the overall sample, the original sample, and the refreshment sample separately. In all cases,

Table 4.7 ANOVA results and multiple comparisons for the number of interviewer visits per household, by sample.

Overall sample							
ANOVA					Multiple comparisons		
Effects	**Model**	**Incentive**	**Mode**	**Incentive × mode**	**Mode**	**Mean no. of visits**	**F2F**
p-Value	**0.00**	0.11	**0.00**	0.28	F2F	3.7	—
					MM	2.8	**0.9****
Original sample							
ANOVA					Multiple comparisons		
Effects	**Model**	**Incentive**	**Mode**	**Incentive × mode**	**Mode**	**Mean no. of visits**	**F2F**
p-Value	**0.00**	0.93	**0.00**	0.11	F2F	3.8	—
					MM	3.9	**0.9****
Refreshment sample							
ANOVA					Multiple comparisons		
Effects	**Model**	**Incentive**	**Mode**	**Incentive × mode**	**Mode**	**Mean no. of visits**	**F2F**
p-Value	**0.00**	0.29	**0.00**	0.51	F2F	3.7	—
					MM	2.5	**1.2****

Note: *p*-Values are adjusted for sample design and multiplicity of comparisons. **$p < 0.05$. In the ANOVA part of the table, *p*-values for *F*-tests are shown, while in the multiple comparison part of the table, response rates differences are displayed.
Bold values correspond to significant values.

unconditional incentives do not have an effect on the number of interviewer visits. Mode is the only significant effect, with mixed-mode requiring a lower number of interviewer visits per household. The interaction is not significant.

Looking at the percentage of households fully responding by web in the mixed-mode treatment group, we test the effect of both unconditional and conditional incentives and their interaction. Table 4.8 shows that all effects are significant in the complete sample. Higher unconditional incentive levels (£20 or £30 rather than £5 or £10) improve the probability of participating fully by web. The conditional incentive increases the proportion of households fully responding by web (26.4 vs. 21.5, $p < 0.01$). The interaction is also significant, meaning that conditional incentives have a different effect at different levels of the unconditional incentive. Figure 4.1 shows that the most effective combination of unconditional and conditional incentive is a £30 prepaid unconditional incentive with a conditional incentive. Effects differ depending on the number of waves members have been in the panel. In the original sample, only the conditional incentive seems to be effective ($p = 0.01$). On the other hand, in the refreshment sample, the unconditional incentive ($p < 0.001$) and the interaction of conditional and unconditional incentive ($p = 0.01$) are effective. These findings are in line with those of Jäckle et al. (2015).

Table 4.8 ANOVA results and multiple comparisons, for the effects of unconditional incentive, conditional incentives, and their interaction on complete response by web.

Overall sample

ANOVA					Unc.	RR	Multiple comparisons			Con.	RR	Comparison
Effects	**Model**	**Unc.**	**Con.**	**Unc. × con.**			**£5**	**£10**	**£20**			**No**
p-Value	**0.00**	**0.01**	**0.00**	**0.03**	£5	17.3	—			No	21.5	—
					£10	21.3	4.0			Yes	26.4	4.9**
					£20	37.8	20.5**	16.5**				
					£30	43.0	25.7**	21.7**	5.1			

Original sample

ANOVA					Unc.	RR	Multiple comparisons			Con.	RR	Comparison
Effects	**Model**	**Unc.**	**Con.**	**Unc. × con.**			**£5**	**£10**	**£20**			**No**
p-Value	**0.02**	0.17	**0.01**	0.45	£5	17.3	—			No	15.6	—
					£10	21.1	3.8			Yes	22.3	**6.7****

Refreshment sample

ANOVA					Unc.	RR	Multiple comparisons			Con.	RR	Comparison
Effects	**Model**	**Unc.**	**Con.**	**Unc. × con.**			**£5**	**£10**	**£20**			**No**
p-Value	**0.00**	**0.00**	0.98	**0.01**	£10	22.2				No	34.8	—
					£20	37.8	15.6			Yes	35.7	0.9
					£30	43.0	**20.8****	5.1				

Note: p-Values are adjusted for sample design and multiplicity of comparisons. Unc, unconditional incentive; Con, conditional incentive. $^{**}p < 0.05$. In the multiple comparison part of the table, response rates differences are displayed. Bold values correspond to significant values.

4.4 Concluding Remarks

The design and analysis of experiments to study best practices for survey design features that may interact with each other must be done well if appropriate inferences are to be made. The main suggestion is to try to implement experimental designs that allow for clear identification of the factor effects, trying to avoid spurious results depending on lurking variables[3] and risks of confounding (Lynn and Jäckle 2019). Lurking variables have an effect on both the explanatory and the response variables, creating the illusion of a causal link between them. Lurking variables are neither measured nor incorporated in the design of the study. Confounding variables are measured, so their association with the explanatory and response variables can be determined. The best approach to experimental design is to clearly define the task of the experiment, to thoroughly review existing literature on similar tasks, and possibly activate a peer review process

3 A lurking variable is a variable that is not considered in a research study that could influence the relations between the variables in the study.

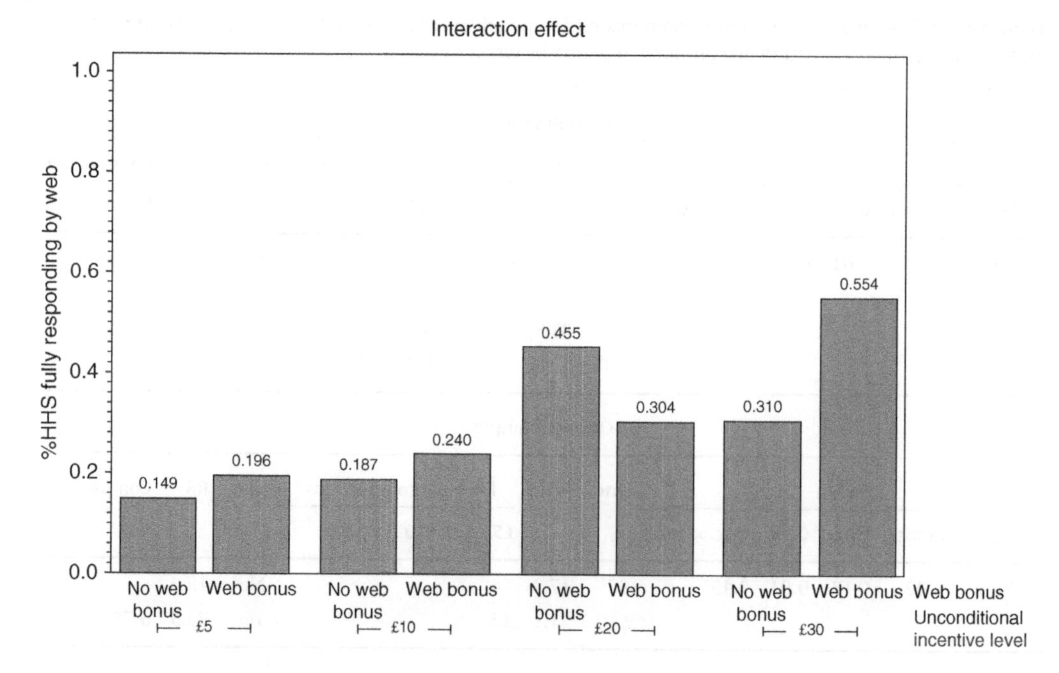

Figure 4.1 Interaction effect of conditional and unconditional incentives for the overall sample.

to get assistance in removing/avoiding confounding and selecting most appropriate treatment levels. The best available defense against the possibility of spurious results due to confounding is often to conduct a randomized study of a sufficiently large sample, such that all potential confounding variables (known and unknown) will be distributed by chance across all study groups and hence will be uncorrelated with the binary variable for inclusion/exclusion in any group. Furthermore, the external validity of findings can be greatly strengthened by basing the experiment on a probability-based representative sample.

Our case study illustrates a possible approach to the statistical analysis of this type of experiment (multiple interacting treatments, mounted on a survey with a complex design). We propose the use of ANOVA followed by multiple comparisons, taking into account multiplicity of comparisons by Tukey's HSD post hoc adjustments. Furthermore, we suggest that the sample design should be taken into account in the form of standard error adjustment for the effects of sample clustering and stratification.

The substantive focus of our case study was the interaction between the choice of a respondent incentive strategy and the choice of survey mode, specifically including mixed-mode web surveys. These design choices have important implications both for survey quality in the broad perspective of total survey error and for costs. Our findings demonstrate that a well-designed and carefully analyzed survey experiment can extend scientific knowledge with respect to complex issues such as these. One has to bear in mind, however, that our empirical results apply to panel surveys and not necessarily to other research contexts. Some of the key implications of our findings are:

- Higher unconditional incentives (£30) are effective in stimulating participation in both face-to-face and mixed-mode in ongoing longitudinal panels: for individual response, £30 unconditional incentive is effective only when compared to £5 and £10, while for household response, a significant difference in response is registered with respect to all incentive levels (£5, £10, and £20).

- We have identified some characteristics that could contribute to the implementation of targeted designs (Groves and Heeringa 2006; Luiten and Schouten 2013; Wagner 2008; Lynn 2015) in longitudinal studies, where the targeted feature is incentive and/or mode. For example, among those who had joined the panel more recently, some groups show higher predicted probabilities to provide full response for higher levels of incentives: mixed-mode (with web component), females, people living in urban centers, and age classes ranging from 41 to 70 years.
- Our findings contribute to the debate on the role of incentives in the context of survey errors as we have evaluated outcomes beyond the response rate. We did not find evidence of an effect of different incentive levels nor of an interaction between mode and incentive level. In the context of total survey error, other aspects related to the quality of the data could also be evaluated, such as effects on "representativity" (Bianchi and Biffignandi 2017a) or on estimates (Jäckle and Lynn 2008).
- Costs evaluation is an additional element for deciding incentive strategies. We did not find evidence that the level of unconditional incentive affected the number of interviewer visits to households in either of the modes studied. We did, however, find that in the mixed-mode design a conditional incentive (additional incentive for households fully responding via web) improved the proportion of households fully responding by web. Taking into account that lesser number of face-to-face visits reduces cost and that more web surveys is cheaper, this finding could imply that the use of conditional incentives leads to reduced costs. However, in order to have a complete picture, a more thorough analysis should be carried out, including the costs of the incentives. This would provide information on the net costs per complete questionnaire. This analysis is beyond the scope of this chapter. However, Bianchi et al. (2017) provide an indication of the scale of the data collection cost differential between the two mode treatments, taking into account incentives costs, in the same experiment we are considering and with reference to Waves 5–7 of the USIP. The mixed-mode design was found to bring cost savings of around 15% at USIP5.

In conclusion, some suggestions emerge from our case study both with regard to the methodological approach to the experimental study of complex survey design features and with respect to empirical evidence to support targeted strategies and further research needs.

With regard to targeted incentive/mode strategies, the usefulness of conditional incentives, higher levels of unconditional incentives, and mixed-mode designs with respect to participation and quality emerge in this study. Some evidence on the possibility of differential impact of incentives on subgroups is found, too. More refined strategies with respect to targeted subgroups might be evaluated taking into account a variety of specific survey features and studies (Bianchi and Biffignandi 2014, 2017b).

For future research, it could be fruitful to carry out an experiment to investigate the effect of using incentive and modes targeted to the characteristics of sample members. The implementation of such a design in the context of a longitudinal study, where a lot of information is available on panel members, could be very beneficial. Further, another strand of research in the longitudinal context would be to study the longitudinal effects (over many waves) of incentives and the effect of changing incentive strategies over waves.

Acknowledgments

The authors would like to acknowledge the support by the COST Action IS1004 and by the ex 60% University of Bergamo, Biffignandi grant. We are grateful for the comments from Peter Lynn and the Editors.

USIP is an initiative by the UK Economic and Social Research Council, with scientific leadership by the Institute for Social and Economic Research, University of Essex, and survey delivery by NatCen Social Research and Kantar Public. The Innovation Panel data are freely available from the UK Data Service (data collection S.N. 6849: "Understanding Society: Innovation Panel, Waves 1-8, 2008-2015").

References

Al Baghal, T. and Lynn, P. (2015). Using motivational statements in web-instrument design to reduce item-missing rates in a mixed-mode context. *Public Opinion Quarterly* 79 (2): 569–579.

Bianchi, A. and Biffignandi, S. (2014). Responsive design for economic data in mixed-mode panels. In: *Contribution to Sampling Statistics* (ed. F. Mecatti, P.L. Conti and M.G. Ranalli), 85–102. Springer.

Bianchi, A. and Biffignandi, S. (2017a). Representativeness in panel surveys. *Mathematical Population Studies* 24 (2): 126–143.

Bianchi, A. and Biffignandi, S. (2017b). Targeted letters: effects on sample composition and item non-response. *Statistical Journal of the International Association for Official Statistics* 33: 459–467.

Bianchi, A., Biffignandi, S., and Lynn, P. (2017). Web-face-to-face mixed-mode design in a longitudinal survey: effects on participation rates, sample composition and costs. *Journal of Official Statistics* 33 (2): 385–408.

Bosnjak, M. and Tuten, T. (2003). Prepaid and promised incentives in web surveys. *Social Science Computer Review* 21 (2): 208–217.

Bowling, A. (2005). Mode of questionnaire administration can have serious effects on data quality. *Journal of Public Health* 27: 281–291.

Brown, J.A., Serrato, C.A., Hugh, M. et al. (2016). Effect of a post-paid incentive on response rates to a web-based survey. *Survey Practice* 9 (1): https://www.surveypractice.org/article/2821-effect-of-a-post-paid-incentive-on-response-rates-to-a-web-based-survey.

Buck, N. and McFall, S. (2012). Understanding society: design overview. *Longitudinal and Life Course Studies* 31: 5–17.

Cantor, D., O'Hare, B., and O'Connor, K. (2008). The use of monetary incentives to reduce non-response in random digit dial telephone surveys. In: *Advances in Telephone Survey Methodology* (ed. J. Lepkowski, C. Tucker, J. Brick, et al.), 471–498. Wiley.

Davern, M., Rockwood, T.H., Sherrod, R., and Campbell, S. (2003). Prepaid monetary incentives and data quality in face-to-face interviews: data from the 1996 survey of income and program participation incentive experiment. *Public Opinion Quarterly* 67: 139–147.

De Leeuw, E. (2005). To mix or not to mix data collection modes in surveys. *Journal of Official Statistics* 21: 233–255.

Dillman, D.A., Smyth, J.D., and Christian, L.M. (2014). *Internet, Phone, Mail and Mixed-Mode Surveys: The Tailored Design Method*, 4e. Hoboken, NJ: Wiley.

Duetskens, E., Ruyter, K.D., Wetzels, M., and Oosterveld, P. (2004). Response rate and response quality of internet-based surveys: an experimental study. *Marketing Letters* 15 (1): 21–36.

Farrant, G. and O'Muircheartaigh, C. (1991). Components of nonresponse bias in the British election surveys. In: *Understanding Political Change* (ed. A. Heath, J. Curtice, R. Jowell, et al.), 235–249. London: Pergamon Press.

Fricker, S., Galesic, M., Tourangeau, R., and Yan, T. (2005). An experimental comparison of web and telephone surveys. *Public Opinion Quarterly* 69: 370–392.

Gajic, A., Cameron, D., and Hurley, J. (2012). The cost-effectiveness of cash versus lottery incentives for a web-based, stated-preference community survey. *The European Journal of Health Economics* 13 (6): 789–799.

Göritz, A. (2006). Incentives in web studies: methodological issues and a review. *International Journal of Internet Science* 1: 58–70.

Göritz, A. (2010). Using lotteries, loyalty points, and other incentives to increase participant response and completion. In: *Advanced Methods for Conducting Online Behavioural Research* (ed. S. Gosling and J. Johnson), 219–233. Washington, DC: American Psychological Association https://doi.org/10.1037/12076-014.

Göritz, A. (2015). Incentive effects. In: *Improving Survey Methods: Lessons from Recent Research* (ed. U. Engel, B. Jann, P. Lynn, et al.), 339–350. London: Routledge.

Grauenhorst, T., Blohm, M., and Koch, A. (2016). Respondent incentives in a national face-to-face survey: do they affect response quality? *Field Methods* 28 (3): 266–283. https://doi.org/10.1177/1525822X15612710.

Groves, R.M. and Heeringa, S.G. (2006). Responsive design for household surveys: tools for actively controlling survey errors and costs. *Journal of the Royal Statistical Society: Series A* 169: 439–457.

Jäckle, A. and Lynn, P. (2008). Respondent incentives in a multi-mode panel survey: cumulative effects on nonresponse and bias. *Survey Methodology* 34: 105–117.

Jäckle, A., Lynn, P., and Burton, J. (2015). Going online with a face-to-face household panel: effects of a mixed mode design on item and unit non-response. *Survey Research Methods* 9 (1): 57–70.

James, T.L. (1997). Results of wave 1 incentive experiment in the 1996 survey of income and program participation. In: *JSM Proceedings of the Survey Research Methods Section*, 834–839. Alexandria, VA: American Statistical Association.

Laurie, H. and Lynn, P. (2009). The use of respondent incentives on longitudinal surveys. In: *Methodology of Longitudinal Surveys* (ed. P. Lynn). Chichester: Wiley https://doi.org/10.1002/9780470743874.ch12.

Laurie, H., Smith, R., and Scott, L. (1999). Strategies for reducing nonresponse in a longitudinal panel survey. *Journal of Official Statistics* 15: 269–282.

Luiten, A. and Schouten, B. (2013). Tailored fieldwork design to increase representative household survey response: an experiment in the survey of consumer satisfaction. *Journal of the Royal Statistical Society: Series A* 176: 169–189.

Lynn, P. (2009). Sample design for understanding society. Understanding Society working paper 2009-01, ISER, University of Essex, Colchester. www.understandingsociety.ac.uk/research/publications/working-paper/understanding-society/2009-01.

Lynn, P. (2015). Targeted response inducement strategies on longitudinal surveys. In: *Improving Survey Methods: Lessons from Recent Research* (ed. U. Engel, B. Jann, P. Lynn, et al.), 322–338. New York: Routledge.

Lynn, P. and Jäckle, A. (2019). Mounting multiple experiments on longitudinal social surveys: design and implementation considerations. In: *Experimental Methods in Survey Research: Techniques that Combine Random Sampling with Random Assignment* (ed. P.J. Lavrakas and E. De Leeuw). Wiley.

Lynn, P. and Lievesley, D. (1991). *Drawing General Population Samples in Great Britain*. London: SCPR.

Mack, S., Huggins, V., Keathley, D., and Sundukchi, M. (1998). Do monetary incentives improve response rates in the survey of income and programme participation? In: *Proceedings of the Survey Research Methods Section*, 529–534. Alexandria, VA: American Statistical Association.

Maxwell, S.E. and Delaney, H.D. (1990). *Designing Experiments and Analyzing Data: A Model Comparison Perspective*. Pacific Grove, CA: Brooks/Cole Publishing.

Millar, M.M. and Dillman, D.A. (2011). Improving response to web and mixed-mode surveys. *Public Opinion Quarterly* 75: 249–269.

Mutz, D.C. (2011). *Population-Based Survey Experiments*. Princeton, NJ: Princeton University Press.

Nicolaas, G., Thomson, K., and Lynn, P. (2000). *The Feasibility of Conducting Electoral Surveys in the UK by Telephone*. London: Centre for Research into Elections and Social Trends.

Ryu, E., Couper, M.P., and Marans, R.W. (2006). Survey incentives: cash vs. in-kind; face-to-face vs. mail; response rate vs. nonresponse error. *International Journal of Public Opinion Research* 18: 89–106.

Schoeni, R., Stafford, F., McGonagle, K., and Andreski, P. (2013). Response rates in national panel surveys. *Annals of the American Academy of Political and Social Science* 645: 60–87.

Singer, E. and Kulka, R.A. (2000). Paying respondents for survey participation. Survey Methodology working paper, N. 092, Survey Research Center, Institute for Social Research, University of Michigan, Ann Arbour.

Singer, E. and Ye, C. (2013). The use and effects of incentives in surveys. *Annals of the American Academy of Political and Social Science* 645: 112–141.

Singer, E., Van Hoewyk, J., and Maher, M.P. (1998). Does the payment of incentives create expectation effects? *Public Opinion Quarterly* 64: 152–164.

Singer, E., Gebler, N., Raghunathan, T. et al. (1999a). The effect of incentives in interviewer-mediated surveys. *Journal of Official Statistics* 15: 217–230.

Singer, E., Groves, R.M., and Corning, A.D. (1999b). Differential incentives: beliefs about practices, perceptions of equity, and effects on survey participation. *Public Opinion Quarterly* 63: 251–260.

Singer, E., Van Hoewyk, J., and Maher, M.P. (2000). Experiments with incentives in telephone surveys. *Public Opinion Quarterly* 64: 171–188.

Stratford, N., Simmonds, N., and Nicolaas, G. (2003). *National Travel Survey 2002: Report on Incentives Experiment*. London: National Centre for Social Research.

Teisl, M.F., Roe, B., and Vayda, M. (2005). Incentive effects on response rates, data quality, and survey administration costs. *International Journal of Public Opinion Research* 18: 364–373.

Trussell, N. and Lavrakas, P.J. (2004). The influence of incremental increases in token cash incentives on mail survey response is there an optimal amount? *Public Opinion Quarterly* 68: 349–367.

Tzamourani, P. and Lynn, P. (1999). The effect of monetary incentives on data quality – results from the British social attitudes survey 1998 experiment. CREST working paper No. 73. University of Oxford, Oxford.

Uhrig, S.C.N. (2011). Using experiments to guide decision making in Understanding Society: Introducing the Innovation Panel. In: *Understanding Society: Early Findings from the First Wave of the UK's Household Longitudinal Study* (ed. S.L. McFall and C. Garrington). Colchester: University of Essex. http://research.understandingsociety.org.uk/findings/early-findings.

Wagner, J. (2008). Adaptive survey design to reduce non-response bias. PhD thesis. University of Michigan.

Willimack, D.K., Schuman, H., Pennell, B.-E., and Lepkowski, J.M. (1995). Effects of a prepaid nonmonetary incentive on response rates and response quality in a face-to-face survey. *Public Opinion Quarterly* 59: 78–92.

5

Experiments on the Effects of Advance Letters in Surveys

Susanne Vogl[1], Jennifer A. Parsons[2], Linda K. Owens[3], and Paul J. Lavrakas[4]

[1] Department of Education, Department of Sociology, University of Vienna, Vienna, Austria
[2] Survey Research Laboratory, University of Illinois at Chicago, Chicago, IL, USA
[3] Stephens Family Clinical Research Institute, Carle Foundation Hospital, 611 W. Park Street, Urbana, IL 61801, USA
[4] NORC, University of Chicago, 55 East Monroe Street, Chicago, IL 60603, USA

5.1 Introduction

5.1.1 Why Advance Letters?

Sending advance letters prior to the first contact by an interviewer has long been a strategy to increase response rates in both telephone and face-to-face surveys. However, there is little published research examining the effect of advance letters on the outcomes in face-to-face studies; the literature is predominantly focused on telephone surveys, as evident by the de Leeuw et al.'s (2007) meta-analysis. Furthermore, the rapidly changing landscape of survey recruitment and data collection, with less emphasis on interviewer-mediated data collection (especially via telephone) makes the examination of the effect of advance letters and other forms of advanced communications in other survey recruitment modes more urgent (cf. Lavrakas et al. 2017a). At the same time, in their last updated volume, Dillman et al. (2014) posit that as people are increasingly saturated with survey requests, the efficiency of prenotification letters may be in decline. Thus, continuous assessment of the effect and efficiency of advance letters has to be evaluated.

Advance letters are thought to stimulate interest (boosting both intrinsic and extrinsic motivation to participate), while conveying the legitimacy and credibility of the survey and the survey organization and providing assurances about confidentiality (Sangster 2003; Collins et al. 2001; Traugott et al. 1987; Dillman et al. 1976). For interviewer-administered surveys, it has been suggested that advance letters may communicate the value of the survey and evoke principles of reciprocity, as well give the interviewer support in their recruitment efforts (Groves and Snowden 1987, p. 633). Though, how interviewers actually use these letters to their advantage in their field work has largely been untested (cf. Lavrakas et al. 2017b). There are indications that interviewers' confidence is increased by an advance letter (Groves and Snowden 1987). Interviewers found advance letters helpful in allaying respondents' initial suspicions and value them as a more professional way of seeking an interview (Collins et al. 2001). Letters are helpful in the initial contact with households (Collins et al. 2001) and eliminate the surprise of an unexpected (aka "cold") contact by the interviewer (Dillman et al. 1976).

Experimental Methods in Survey Research: Techniques that Combine Random Sampling with Random Assignment, First Edition.
Edited by Paul J. Lavrakas, Michael W. Traugott, Courtney Kennedy, Allyson L. Holbrook, Edith D. de Leeuw, and Brady T. West.

Companion Website: www.wiley.com/go/Lavrakas/survey-research

5.1.2 What We Know About Effects of Advance Letters?

Research on the effect of advance letters usually focuses on:

- *outcome rates* (cooperation, response, and refusal rates in particular);
- the *cost effectiveness* of advance mailings based on these outcome rates; and
- possible reduction of *nonresponse bias*.

In addition, these studies often assess which specific properties of advance letters cause the observed differences in the dependent variable(s). Covariates typically include *sociodemographic characteristics* of the target groups. *Sampling strategies* (random digit dialing [RDD] vs. list-based samples) and potentially survey mode also can be independent variables if cases are randomly assigned to different recruitment modes. This relates to an important issue: regardless of data collection mode, advance letters require a sampling frame that includes a mailing address either postal or email (the latter is notoriously more difficult to obtain and thus hardly employed). This challenges the applicability of advance letters because sampling frames often contain only information on one contact mode, for example, telephone number or postal address and not both together, or no prior contact information is available (e.g. random-route strategies). Consequently, interview mode, sampling frame, and recruitment strategy determine the applicability and mode of sending advance letters. If sampling and advance letter require different information, then datasets might have to be matched, for example, telephone numbers have to be matched with postal addresses. Thus, potential biases can multiply when frames systematically differ and groups of respondents are excluded.

The study by Luiten (2011) illustrates the problem: "The sampling frame, derived from the municipal registries, contains the names of the inhabitants. This information is used by Statistics Netherlands for finding telephone numbers [...]. The telephone numbers are the ones that can be found by automated search in the records of the Royal Dutch telephone company (KPN), owner of the landlines" (p. 13). Around 35% of addresses could be linked to telephone numbers by this method. By intensive manual search, numbers could be retrieved from other providers, for an additional 25% of addresses. The other 40% has either a shielded/unlisted landline number, or an unregistered cell phone. In this study, the number of households with an unlisted landline (62%) exceeds the number of listed households (38%) (Luiten 2011, p. 13).

First, we briefly summarize empirical evidence on effects of advance letters on outcome rates, nonresponse bias, and effectiveness of properties of advance letters. Then, we focus on experimental designs for researching the effects of advance letters. This helps in understanding the research questions commonly pursued with experimental research on advance letters.

5.1.2.1 Outcome Rates

In general, the literature on advance letters tends to show positive effects on *outcome statistics* (e.g. Link and Mokdad 2005; Robertson et al. 2000; Camburn et al. 1995; Smith et al. 1995; Traugott et al. 1987), though null findings have also been reported (Woodruff et al. 2006; Singer et al. 2000; Sykes and Hoinville 1985). In a comprehensive meta-analysis of advance letters *in telephone surveys,* de Leeuw et al. (2007) found that advance letters have a positive effect on the response rate, both with RDD and with list-based address samples; the effect was greater when the sample was based on a list of known addresses. Overall, their analysis showed that cooperation rates improved by 11 percentage points and the response rate by 8 percentage points, when advance letters were mailed. One of the primary challenges in using advance letters, in any mode, is that unless the respondent is a specifically named person (as in a list frame), there is no way to tailor the advance letter to the selected respondent.[1] In fact, the respondent typically

1 However, Lavrakas et al. (2016) reported the beneficial response rate effects of tailoring the addressee line of an envelope to a hard-to-reach demographic cohort by using "Resident Actual" in a mailings to high density Hispanic

is not even selected at the time the advance mailings are sent. For *face-to-face interviews*, we are not aware of a similar meta-analysis.

Interpreting differences in outcome rates should be supplemented with other information, such as the topic of the study, survey method, the age of the target group, combined with incentives, properties of the advance letter like wording, layout, etc. because these could alter the effectiveness of advance letters (von der Lippe et al. 2011). In fact, heterogeneous results in research on advance letters are not surprising, considering the vast variety of survey topics, different target groups and sampling methods (von der Lippe et al. 2011). These mixed results call for further investigation with factorial designs in which multiple variables and interactions can be formally examined.

5.1.2.2 Nonresponse Bias

Although advance letters can potentially reduce nonresponse, they may also have *differential effects on subgroups* by raising participation rates disproportionally among certain demographic segments, thereby contributing to a worsening of the representation of the unweighted final sample, which in turn could contribute to nonresponse bias (Link and Mokdad 2005; Goldstein and Jennings 2002). Thus, there can be a systematic difference between respondents and nonrespondents in some characteristics under study. Nonresponse error may systematically distort descriptive and inferential statistics and thus make unbiased estimates of population characteristics almost impossible (Berinsky 2008; Groves et al. 2009). At the same time, advance letters could counterbalance some selection effects by being more effective with subgroups that would have higher refusal rates without the advance mailings.

Goldstein and Jennings (2002) used a listed sample that contained key demographic and political characteristics for all sample units. Thus, they could analyze the effect of advance letters in telephone surveys by comparing basic demographic characteristics of a letter and nonletter respondent with all interviewees, all contacted sample units, and the entire sample frame (p. 612). Their results showed that the notification letter improved both response rates as well as survey accuracy. The advance letter improved the representation of men, older respondents, and Democrats. Groves and Snowden (1987) also found that advance letters were most successful among older participants who generally exhibit low response rates. Besides this positive effect on the demographic accuracy of a sample completing telephone interviews, de Leeuw et al. (2007) could not find any indication that advance letters have differential effects on population subgroups.

5.1.2.3 Cost Effectiveness

The idea behind the use of advance letters is to increase response rates while decreasing interviewer effort, thus reducing costs. Thus, from the perspective of a survey research organization, the question of *cost-effectiveness* of advance letters arises. Do advance letters reduce the recruitment effort and the time needed for data collection? Cost-effectiveness calculations are not always straightforward because complete accounting information related to various cost components is required. Principal cost elements are interviewer and supervisor labor, telephone charges, printing and postage costs, and labor associated with mailings (Hembroff et al. 2005). One of the biggest costs associated with data collection is the number of call or contact attempts necessary to finalize a case, whether it is a completed interview or not. The assumption would be to have a lower number of contact attempts when advance letters are sent. Thus, the average

---

neighborhoods. Also, Lavrakas et al. (2017b) report that tailoring the addressee line to a class of eligible respondents – such as addressing it to "parent/guardian" in a survey of children – can have a beneficial effect on response rates for such surveys.

number of call attempts should reduce the costs more than what printing and postage for advance letters adds.

However, results on the cost effectiveness of advance letters are rarely tested or reported, and when they are, they do not show a clear picture. Groves and Snowden (1987) compared call back protocols of an experimental group that had received an advance letter and control groups without prenotifications and did not find any effect of the advance letter on the amount of calling effort required to complete a sample case. This result calls the cost efficiency of advance letters into question. However, Link and Mokdad (2005) found that advance letters paid for themselves "in that the cost for obtaining a fixed number of completed surveys from an address-matched sample using advance letters was lower than without the letters" (p. 584). Also, Hembroff et al. (2005) compared a postcard, an advance letter and a no-mailing group. They found differences in the average number of calls required for one case to receive a final disposition and also in the percentage of calls that led to a completed interview. They judged the overall efficiency "by comparing the costs of the alternative efforts required to accomplish the same results" (p. 240). With an advance letter, the number of call attempts to reach a final disposition code was significantly lower than with postcards or without any advance mailing. Similarly, the number of call attempts to complete an interview was lowest in the advance letter group, similar to the postcard group but the control group required most call attempts. They conclude: "the overall efficiency of the calling was greater with the use of letters than postcards, and the calling was more efficient using letters or postcards than using no mailing at all" (p. 241). In face-to-face interviews, the cost for interviewers and contacting a potential respondent is much higher and therefore it seems reasonable to assume that advance mailing would be even more cost-effective.

Advance letters can only have an effect if the letter is *received, opened, and read* (ideally, by the household member who is sampled; see Groves and Snowden 1987). Letters are of little utility if they are discarded without being opened, or read and discarded by someone other than either the informant who completes the household screener or the selected respondent. To determine whether a respondent has actually received the advance letter, corresponding questions should be implemented in the survey interview (cf. Lavrakas et al. 2017b) – being aware that some responses to these questions might be unreliable. Ideally, data on the exposure to the advance letters among refusers would be collected, which is certainly a very difficult if not impossible task.

Twenty years ago, results from the Survey of Census Participation showed that in 65% of households mail was thrown away without opening it (Couper et al. 1995). Other data suggests that only 50–60% of respondents in telephone surveys recall seeing the advance letter (Parsons et al. 2002), although Lavrakas et al. (2017b) found that less than 25% of their respondents reported seeing the advance letter. Results indicate that women and lower income groups read the letter more frequently than higher income groups and men (Groves and Snowden 1987; Groves 2004). Recall of the letter content was lowest among younger adults, lower income groups, and larger households (Link and Mokdad 2005), whereas a recent study found recall to be lower among households with three or more adults (Lavrakas et al. 2017b). Cannell and Fowler (1965) report that 44% of the respondents in an National Health Interview Survey (NHIS)-like interview claimed that they had not received the letter and brochure describing the survey – although an estimated 73% should have received the letter. Still, it does not appear that most rigorous examinations on the effect of advance letters have looked at the subject's recall of seeing the letter, only on the response and/or cooperation rates. What happens to advance letters before and after they are delivered is a major knowledge gap.

5.1.2.4 Properties of Advance Letters

To increase the likelihood that recipients open advance letter mailings, researchers have also experimented with the effect of different *layout* features, for example, including a short message or graphic on the envelope to increase the distinctiveness of the mailing. As this is a strategy employed more commonly by direct marketers, there is the possibility that *branding* might work against researchers as respondents could mistake the envelope for a commercial mailing rather than a scientific research-related one (Dillman et al. 2014). Finn et al. (2004) concluded that a brand reduced their response rates by 6 percentage points in a mail survey of New Zealand's national identity (although the difference was not significant) and in experiments with household surveys in Wisconsin, use of envelope messaging had no effect (Dykema et al. 2015). McLean et al. (2014) in a study on eating disorders found that the brand on their mail questionnaire interacted with gender: males who received the branded envelope had lower response rates than males who received an unbranded envelope, while females had a slightly higher response rate with the brand, but took longer to return the questionnaire. Again, these experiments usually use outcome rates as dependent variable. Overall, the research on branding appears to have been tested more often with questionnaire packets in mail surveys than with advance letters for face-to-face or telephone surveys.

5.2 State of the Art on Experimentation on the Effect of Advance Letters

To determine the effects of advance letters, experimental studies have great promise. An experimental design allows for a causal interpretation of differences. The advance letter as such is considered the independent treatment variable that the researchers control. In the simplest design form, researchers compare a randomly assigned advance letter group with a random control group that is not sent an advance letter. The effect on outcome statistics can be determined with a single treatment and control group. However, the reach of the results is limited. For a more complex analysis, for example, regarding the effect of advance mailing formats (e.g. postcard, email, letter, and brochure) or content, a factorial design is necessary. In more complex factorial designs, researchers also test the effects of multiple advance letter properties (e.g. wording, branding, postcard vs. letter, layout, delivery service, etc.) as one of other recruitment features (e.g. incentive amounts, reminder phone calls, use of a refusal conversation protocol, etc.) being tested.

In the following section, we illustrate how such experiments have been designed and results analyzed.

Key questions regarding the effect of advance letters mainly evolve around the effect on outcome rates and cost effectiveness, potential nonresponse bias, and how an advance letter should be designed to be most effective. These research questions partly require different types of information and designs. Beyond the experimental set-up, additional information might be required, particularly when nonresponse bias or bias due to the requirement of matching data sources in order to send an advance letter and later conduct the interview is to be assessed. For evaluating nonresponse bias, researchers face the challenge that information on nonrespondents is usually not available. Bias is not only a function of the mere number of people who are not contacted or refuse. Rather, nonresponse bias is a function of sampling units interviewed (response rate) and the difference between respondents and nonrespondents. Thus, even low response rates can generate robust results (e.g. Keeter et al. 2000). But assessing the difference between respondents and nonrespondents is a difficult task because we rarely have precise information on the nonrespondents in a specific study. However, different sampling frames offer different possibilities, some sampling strategies allow for more information than others do. Also the national context can influence the amount and accessibility of information. Address-based sampling frames can allow for the analysis of nonresponse bias by matching rich data onto each address, e.g. municipal registries contain demographic information for respondents and nonrespondents and the local areas in which they live (see, for example Luiten 2011; Dykema et al. 2011; Montaquila and Olson 2012). Other sampling frames usually do not contain information on (all) members of the sample – respondents and nonrespondents (e.g. RDD samples or telephone lists). Another form of nonresponse bias analysis that can be used in advance letter experiments is to determine whether the level of effort (e.g. how many follow-up contacts does it take to achieve a completed questionnaire) among respondents is correlated with a survey's key statistics, with the hypothesis being that those who take the most effort may be similar to nonrespondents (Montaquila and Olson 2012). Beyond these means we are left with the alternative of comparing respondents to a survey to the population they represent using auxiliary data, such as decennial Census data or American Community Survey data. These comparisons are typically limited to sociodemographics characteristics, such as, age, race, gender, and education. However, this procedure gives a limited idea of nonresponse bias, because it only addresses nonresponse that is related to demographics and can be addressed for weighting by those characteristics. It is also difficult if not impossible with mere demographic comparisons to determine whether any observed differences between ones final sample and its population parameters are due to nonresponse, noncoverage, or both.

In sum, the *nonresponse or recruitment bias* in relation to advance letters can be difficult to assess and depends to a great extent on sampling frames and available information on nonrespondents. In an experimental design with a random assignment to treatment and control group a comparison of respondents can help (assuming a sufficient sample size) to estimate differential effects of the advance letter on certain segments of the population by comparing variables of interest, for example, with ANOVAs.

To assess the effect of advance letters on *outcome rates*, a split-ballot design with a random assignment of participants to the advance letter and nonadvance letter condition can suffice but could of course be extended to factorial designs when different types of advance mailings are tested. According to de Leeuw et al. (2007) most studies included in their meta-analysis (23 out of the 29) have only one experimental condition. "Two studies investigated the effect of two different forms of advance letters, three studies used three different versions of advance letters in the experiment, and one study investigated four different types of advance letters" (de Leeuw et al. 2007, p. 417). However, Lavrakas et al. (2017a) conducted a 3 × 3 factorial design with nine different versions of their advance mailing.

The comparison of outcome rates is usually only a first step in the analysis of advance letter effects. Typically eligibility screening, response, contact, refusal, and cooperation rates in the treatment and control groups are compared (with χ^2-statistics; and/or logistic regression and logit models, as ANOVAs are not preferred and at the individual level the dependent variable is a binary one). Goldstein and Jennings (2002), for example, implemented a fully randomized telephone experiment to test the effectiveness of advance letters. They compare outcome rates in the advance letter and nonadvance letter group and regressed cooperation against gender, party identification, and receipt of a prenotification letter (p. 612). In a next step, they analyze differential effects of advance letters on social groups and estimate a logit model with interaction effects.

To summarize, experimental research to determine the effect of advance letter focuses on a comparison of treatment and control groups to infer on the effect of the treatment. The number of treatment groups varies according the elaboration of the treatment, (e.g. branded letter vs. branded postcard vs. nonbranded letter vs. nonbranded postcard). Despite the number of existing experiments, there are still some knowledge gaps. There is little research on the assumed interaction of advance letter (properties), sampling strategy, topic of the study, and survey method, which is due to the practice of often conducting methodological research on the back of substantive surveys. Although most research focuses on outcome rates, this can only be a first step.

Analytical strategies presented in the literature on advance letter experimentation are limited to only a few approaches. A common approach is to compare survey outcomes (e.g. eligibility rates when screeners are used, response rates, percent of respondents who refused, etc.) by advance letter condition, using cross-tabulation and χ^2 tests to determine if the differences in rates and percentages are statistically significant (e.g. Luiten 2011). A second approach is to calculate rates and ratios for different advance letter conditions and compare the confidence intervals around them. Dykema et al. (2011), for example, tested for nonresponse bias by comparing proportion of respondents with various characteristics with the distribution of these characteristics in administrative data with one sample z-tests for proportions (p. 439). A third approach has been to examine percentages in outcome categories, by advance letter condition, using ANOVA. But, instead of using ANOVA, *logistic regression* would be better to use in light of the binary nature of the dependent variables that are tested in most advance letter experiments (e.g. Luiten 2011). Logistic regressions are used for binary dependent variables and focus on the prediction of experiencing an event or having a characteristic – in this case completing an interview or not. Dykema et al. (2011) used logistic regression models for their comparisons between experimental groups, as did Lavrakas et al. (2017b) for their 3×3 experimental design.

Understanding how and why advance letters can make a difference (but sometimes do not) demands a more complex design, taking a total survey error perspective, including sampling error, nonresponse, and measurement error. This requires a rigorous comparison of advance letter conditions not only on basis of cost or outcome rates but also in terms of total survey error. Thus, a combination of experimental designs with appropriate analytical strategies is necessary. To detect sampling error, we need some frame information and be aware of distortions if a matching process (e.g. telephone numbers as sampling frame has to be matched with addresses) is required because no sampling frame offers a perfect match between population and sampling frame(s), commonly referred to as problems of over- and undercoverage. To assess nonresponse error, we require information on nonrespondents or the population. Measurement error is even more difficult to assess because it assumes a true value which is generally unknown in most cases. Under certain circumstances a measurement error experiment could be built into an advance letter experiment to investigate how, if at all, the measurement error differs under different advance letter conditions. However, a nonexperimental approximation can be doing a comparison between treatment and control groups regarding their reporting – but without knowing with certainty if more or less reporting is closer to the true value.

5.3 Case Studies: Experimental Research on the Effect of Advance Letters

In the following section, we describe and reflect on the research designs and analysis of our own studies, which we utilize to exemplify the use of experiments in the research of advance letters. The original studies included represent telephone and area probability surveys, and address the following topics:

(1) design and analytic features of advance letters;
(2) cost effectiveness (outcome rates, recruitment effort, respondents recall, and awareness of advance letter content);
(3) reporting: does the advance letter affect reporting (e.g. on sensitive topics).

5.4 Case Study I: Violence Against Men in Intimate Relationships

In the CATI-study "Violence against Men" conducted in Germany by the Catholic University of Eichstaett-Ingolstadt in 2007, we implemented a split-ballot experiment on the effect of advance letters. More specifically, we researched (i) the effect of personalized advance letters on outcome rates and reasons for refusals, (ii) recruitment effort, and (iii) the reporting on sensitive topics (Vogl, 2018). Sensitive topics are notoriously difficult to research. Technically speaking, we compared the advance letter and no-advance letter group regarding the outcome statistics, the propensity to complete an interview or refuse, and the number of call attempts required to come to a final disposition code. Furthermore, we compared respondents of the advance letter and nonadvance letter group to determine the effects on the reporting of violence. The assumption was that contents of the advance letter (AL) could emphasize the importance of the study and increase trust and credibility. As a result, the response rates and the reported victimization could be higher.

In preparation of the data collection, we sent personalized advance letters to half of our sample members. The other half did not receive any advance mailing. The data collection mode was CATI and we interviewed just over 900 Bavarian men.

5.4.1 Sample Design

The target population were men aged 21–70 with German citizenship. For CATI studies, in principle, two strategies for random sampling exist: telephone directories and RDD. Similar to the United States, German public telephone directories are an inadequate source of telephone samples because they are neither complete nor up-to-date, thus resulting in a great deal of differential noncoverage. Now, a growing share of telephone numbers is not published. Although in RDD designs potentially every household with a telephone can be captured, a problem is that addresses for a large and nonrandom proportion of respondents are usually unknown. This mostly prohibits the mailing of advance letters.

Thus, a different sampling strategy was employed: Our sample was based on addresses from official registration databases. For random samples, registration office data seem to be the most complete and up-to-date source. German residents are legally obligated to register their home address at a local registration office. Then individuals rather than households are selected. Thus, addressing an advance letter personally to a specific person is therefore straightforward and the probability that the advance letter will be read improves.

Because a complete list of inhabitants does not exist in Germany, we had to employ a complex sample design. In a first step, all Bavarian administrative units were categorized into either "big city" (100 000 and more inhabitants), "rural area" (on average less than 3000 inhabitants per municipality), and "town" (between 40 000 and 100 000 inhabitants). We randomly selected one big city, three towns, and four (rural) administrative districts in which we selected five municipalities. Altogether, 20 rural municipalities within Bavaria were randomly selected.

Next, for each of the three categories of size of municipality (big city, town, and rural area), 1500 addresses of men of German citizenship aged 21–70 per strata were randomly selected from official registration data bases. Information obtainable from registration offices for scientific research is name, title, address, year of birth, nationality, and sex. The most important information for telephone surveys – telephone numbers – is *not* available. In a third step, we matched telephone number entries in public directories with the selected names (and if necessary addresses).

For 2539 out of the 4473 sample members, we could identify telephone numbers. From the list of telephone numbers, we randomly selected every second number and assign the

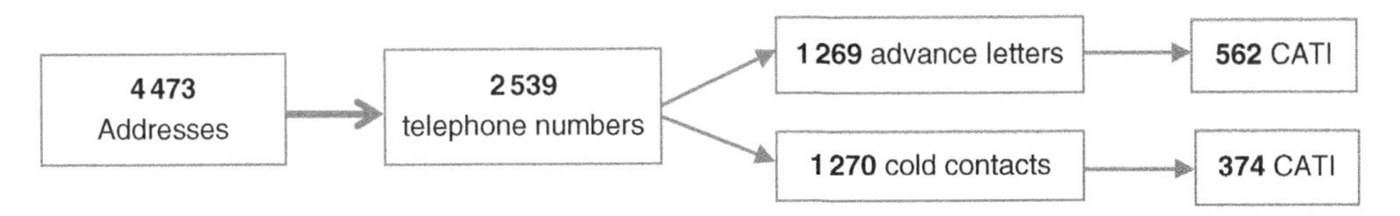

Figure 5.1 Sampling procedure.

potential respondent to the advance letter group. Thus, half the sample members received a personalized advance letter with information about the study, the conducting institution, the data collection, and an assurance of confidentiality. The non-AL group did not receive any prenotification.

5.4.2 Fieldwork

In the last stage, 936 men aged 21–70 completed the survey on their experience of intimate partner violence. In total, 374 of our respondents had not received an advance letter, 562 had (see Figure 5.1). Up to 20 call attempts were made per number, with an average of 3.7. Callback times alternated in terms of weekdays/weekend and afternoon/evening and flexible appointments were made. The field phase lasted for three weeks. The data were collected through CATI in spring 2007 at the telephone survey facilities at the Catholic University of Eichstaett-Ingolstadt. The calls were made without telephone number suppression and respondents could associate the numbers with the University – if they had received the advance letter. The interviewers knew about the advance letter experiment and referred to the advance letter during recruitment, if the case was marked to have received a prenotification.

5.4.3 Sampling Bias

Due to our sampling strategy, a coverage bias was introduced in that people without published telephone numbers could not participate in the study. People who make their telephone number publicly available differ in certain characteristics from those that do not – the so-called "nonpubs" (e.g. Deutschmann and Häder 2002; Lavrakas, 1993; Otte 2002; Parsons et al. 2002). For our sample, nonpubs tended to be younger, compared to people with telephone numbers published in directories and compared to the target population. Furthermore, the number of nonpubs increases with the size of place of residency: in cities 54.3% were nonpubs, 45.3% in towns, and 29.6% in rural areas.

5.4.4 Analytical Strategies

To evaluate the effect of advance letters, we used two approaches. First, we compared outcome statistics as a dependent variable of the advance letter and nonadvance letter group and the proportion of respondents with various sociodemographic characteristics, testing for statistically significant differences. Second, we estimated logistic regression equations that predicted completion of an interview for eligible sample units with theoretically sensible predictors, testing for interactions of advance letters with demographic variables to gauge whether the predictors were more strongly related to the dependent variable under the advance letter-condition than the nonadvance letter condition. We took the same analytical steps for reporting on sensitive topics.

5.4.5 Results

5.4.5.1 Effect of Advance Letters on Outcome Rates and Reasons for Refusals

Our analysis shows that while advance letters did not have any impact on Refusal Rates (non-AL: 39.4%; AL: 38.8%) they positively and significantly influenced Contact (non-AL: 71.5%; AL: 88.1%), Cooperation (non-AL: 43.0%; AL: 53.2%), and overall Response Rates (non-AL: 30.8%; AL: 46.9%). All in all, the Refusal Rate was similar in both groups and the reported reasons for refusal remained the same. In contrast, the Contact Rate[2] was higher under the advance letter condition. Furthermore, the Cooperation Rate was also higher with an advance letter. In other words, more potential respondents who had received an advance letter could be reached[3] and they were then also more likely to participate in the survey. As a result, also the Response Rate is higher for the advance letter group. The difference between the outcome rates in the advance letter versus nonadvance letter group are statistically significant ($\chi^2 = 164.698, p < 0.000$).

Logistic regression models were developed to assess the likelihood of completing an interview for eligible persons sending an advance letter versus not using an advance letter.[4] Although the advance letter significantly improved the odds to complete an interview for eligible sample members the effect was rather small (Nagelkerke Pseudo-$R^2 = 0.011$). Without an advance letter, the odds for completing an interview decreases by 0.7 ($\beta = -0.360$; SE = 0.091; Wald = 15.730; d.f. = 1; $p < 0.000$; Exp(B) = 0.698; 95% CI = 0.584–0.834; reference category: "eligible, noninterview"). Furthermore, we introduced demographic variables into the regression model. The outcome (interview vs. eligible, noninterview) was the dependent variable while advance letter was the main independent variable of interest. In another step, we took an interaction effect between advance letter and respondents age into account. However, there was no change in the explained variance.

5.4.5.2 Recruitment Effort

Recruitment effort is measured by the number of call attempts. The number of contact attempts required to achieve a final disposition was significantly higher for the advance letter group compared to the nonadvance letter group ($F = 48.4$; $p < 0.000$; $\eta^2 = 0.019$). On average, sample units who had received an advance letter required 3.8 call attempts, whereas those without required only 2.9 attempts to receive a final disposition code (e.g. complete interview, final refusal, and noneligible). This seems counterintuitive. However, the call attempts in the advance letter group resulted in a significantly higher number of completed interviews. It is likely that a higher willingness to cooperate after receiving an advance letter resulted in a higher number of callbacks, which increased the total number of attempts, because of fewer refusals at early contact attempts. In the logistic regression model, advance letter and number of call attempts, and the interaction between these two, did not improve the prediction of whether eligible people completed an interview or not.

5.4.5.3 Effect of Advance Letters on Reporting on Sensitive Topics

To investigate a potential effect of an advance letter on reporting on sensitive topics, we compared substantive responses to items on victimization in intimate relationships in the advance letter and nonadvance letter groups. The hypotheses were that an advance letter increases the trust in and credibility of the survey institution and should thus increase the self-report of

2 The cases in which the lines were always busy (non-AL: 2.7 %; AL: 0.7 %), nobody ever answered (non-AL: 13.0 %; AL: 4.6 %), or only an answering machine responded (non-AL: 9.8 %; AL: 2.3 %) were less frequent for men who had received an AL. This result does not make immediate sense, unless respondents could identify the displayed telephone number as being related to a scientific study rather than an unknown caller. However, this explanation remains a speculation.

3 Because they could not be reached their eligibility could not be checked and they are subsumed under "unknown eligibility (UH), noninterview (Category 3)"-code according to the AAPOR definition.

4 We only included eligible people in the model. Noneligible persons and those with unknown eligibility were excluded. We predict the dichotomous dependent variable "interview" (Yes/No).

Table 5.1 Reporting on questions on experience of domestic violence.[a)]

My partner ...	AL	Non-AL
... insulted or verbally abused me	3.25	3.32
... shouted at me	3.05	3.12
... ignored me	3.15	3.11
... accused me of being a bad lover	3.78	3.76
... threatened me with leaving me	3.67	3.73
... deliberately damaged my belongings	3.93	3.91
... pushed or shoved me	3.83	3.87
... slapped me	3.98	3.94*
... threatened me with a knife or a weapon	4.00	4.00
... hit me with an object that could have injured me	3.99	3.99

*$p < 0.05$.
a) Mean values, results from ANOVA, 1 = often, 2 = sometimes, 3 = rarely, 4 = never.

experiences with intimate partner violence. But, our results imply that advance letters do not affect the overall self-reports on sensitive topics in any significant way (see Table 5.1).

5.4.6 Discussion

This experiment showed that potential bias could be introduced through the use of advance letters in different steps of the survey lifecycle. In our specific context, the requirement of obtaining both telephone numbers and mail addresses introduced a coverage bias excluding nonpubs. However, having registration office data allowed for a more detailed analysis of nonresponse because we had some sociodemographic information on nonrespondents. This facilitates a nonresponse error analysis as well as creating survey weights. Although there is general acceptance that advance letters are a positive means, in telephone surveys the use of advance letters is becoming increasingly difficult as the coverage of phone books decreases. Consequently, reducing the sample on sample units with a published telephone number or attempting to match RDD with addresses could introduce bias.

Furthermore, it seems worthwhile to investigate more aspects than just cooperation, response, and refusal rates but take a Total Survey Error perspective and thus look at the survey lifecycle from a procedural perspective. Recruitment effort and its implications for costs do play an important role in survey research and deserve more attention when it comes to experiments on the effect of advance letters. So does measurement error. Effects of advance letters on reporting (or recruiting respondents for a specific topic) have mostly been neglected. However, future research should investigate differences in reporting further, taking into account the possibility that it is not a measurement error that could lead to differences but potentially also sampling bias – for example, when more respondents with a specific behavior or experience are recruited after they have received an advance letter.

What we could include in a new study: We would advise to include questions on whether the advance letter has been received and read – although the reliability might be hampered. This information can serve as an important control variable for the effect of advance letters. We also recommend gathering information on interviewer experience in if and how they made reference to an advance letter while recruiting respondents. However, interviewers could potentially confound some findings in the experiment, if interviewers have differential (recruiting) behavior the treatment is not only the advance letter but also the interviewer behavior (see Chapter 12). Furthermore, interviewer variation could be included in statistical models.

5.5 Case Study II: The Neighborhood Crime and Justice Study

An area probability survey fielded by the Survey Research Laboratory (SRL) at the University of Illinois at Chicago incorporated experiments with both advance letters and envelope branding; furthermore, it explored the effect of envelope contents by including a letter only vs. letter plus study brochure. This case study is unique in its focus on face-to-face data collection instead of telephone surveys. Advance letters are commonly used in area probability surveys, but there has been little published research examining their effectiveness in increasing rates of contact and cooperation and our goal, given the cost associated with these mailings, was to assess whether they would have any effect on either the level of effort it took to finalize the sample (in terms of the number of contact attempts) or response rates. This was of particular interest in this study because of the considerable amount of sample we had in restricted access buildings, which in an urban area such as Chicago are increasingly difficult to access.

5.5.1 Study Design

The Neighborhood Crime and Justice Study was an area probability survey of Chicago residents aged 16 and over about their perceptions of and encounters with the Chicago Police Department. It was a two-wave panel study with in-person interviews conducted approximately 13 months apart. The sample was drawn using Address Based Sampling (ABS) (Iannacchione et al. 2003). ABS relies on the U.S. Post Office's (USPS) Delivery Sequence File (DSF) as the sample frame. The DSF is a file of all deliverable addresses in the United States, used by the USPS for delivering mail. After a sample of 289 blocks was drawn with probabilities proportionate to size, SRL interviewers used the DSF as the basis for block listing, adding addresses that did not appear on the DSF and removing those that were no longer there. After the block listing was completed, a simple random sample of 28 households from each sampled block was drawn. Prior to sampling from the block, blocks that did not have the minimum number of households were clustered with geographically proximate blocks until the cluster size reached the minimum of 28. Thus, the sampled clusters may consist of a single census block or multiple census blocks.

5.5.2 Experimental Design

All sampled households were randomly assigned to one of the 11 conditions. Only one condition involved no mailing whatsoever (no envelope, brochure not included); seven condition mailings were in an unbranded envelope and another three in a branded envelope, with varying combinations of letters and brochures in each. Condition 1 (letter and brochure) is SRL's standard protocol, and the one we hypothesized would have the highest response rate. Therefore, we assigned half of the sample to this condition. The remaining sample was equally split among the other experimental conditions.

The "regular" letter used a study-specific return address label with the study logo. The branded envelope included the same return address label, along with a stamp in the lower left corner that read, RESEARCH STUDY ABOUT THE CHICAGO POLICE. The study brochure included information on the SRL, the Principal Investigator, the purpose of the study and the study sponsor, sample survey questions, and explained that their household was randomly selected for inclusion. These letters, and the study brochure, can be found at www.wiley.com/go/Lavrakas/survey-research.

Table 5.2 Condition number of conditions assigned to each condition.

Condition	Sample size	Envelope	Brochure
1	540	Plain	Yes
2	572	Plain	Yes
3	527	Plain	Yes
4	611	Plain	Yes
5	636	Plain	Yes
6	586	None	No
7	467	Plain	No
8	489	Branded	No
9	478	Branded	Yes
10	467	Plain	No
11	440	Branded	No

A one-way ANOVA was conducted to assess the differences in completion rates by condition. Tukey[5] tests were run to determine which conditions differed significantly from each other. Table 5.2 shows the factorial design and the conditions assigned to each condition.

5.5.3 Fieldwork

Sample was assigned to interviewers so that the same interviewer was responsible for making all contact attempts. Their schedules and daily production were closely monitored by supervisory staff to ensure that interviewers attempted contacts at different times of the day and on different days of the week to maximize the chance of finding some at home. In weekly meetings with supervisory staff, contact histories were reviewed and sample was sometimes allocated to different interviewers if the assigned interviewer could not make a contact on a needed day or was not keeping up with contact attempt demands. The majority of interviewing was conducted during evening and weekend hours, as many households are vacant during the day. If an English-only interviewer reached a Spanish-speaking household, field coordination staff transferred the case to a bilingual interviewer for the duration of the data collection period.

To gain access to hard-to-reach households, interviewers used a variety of techniques, including contacting neighbors or other key informants. Up to 10 personal contact attempts were made at different times of the day and on different days of the week to contact each household. Interviewers documented each attempt in the electronic Case Management System (CMS) installed on their laptops. In all, 1450 completed interviews were attained across all the conditions. The mean number of contact attempts for completed interviews was 4.34; notably, 111 completes came after 10 or more contact attempts. The CMS also was used to record notes regarding any contacts the interviewer made with the household or neighbors. Field coordinators reviewed the CMS with interviewers during their regular in-person meetings when progress and sample production are reviewed.

Field interviewers were told about the advance letter experiments, but were not told which condition their sampled housing units were in. Since, we asked all respondents whether or not

5 http://www.statisticshowto.com/tukey-test-honest-significant-difference

they recalled receiving the letter (regardless of what condition they were in) it was not necessary to divulge this information to interviewers. While interviewers had a copy of the study brochure and advance letter in their set of materials, they were not trained to systematically use them as leverage in gaining cooperation with households. However, interviewers were advised to wear their UIC identification card on a lanyard and to use the advance letter if they needed to demonstrate their legitimacy while canvassing.

Baseline data collection began in December 2014 and concluded in February 2016.[6] A total of 5813 housing units were sampled, and 1450 interviews were completed, corresponding to a response rate of 28%[7] and a cooperation rate of 52% according to the AAPOR (2016) standard definitions.

5.5.4 Analytical Strategies

The purpose of the experiment was to test the effects of various advance-mailing features on survey outcomes. Thus, the independent variables are the binary/dummy variables indicating the presence of an advance letter, branded versus plain envelope, and brochure versus no brochure. The dependent variables were:

- Completion rate (completed interviews divided by starting sample size)
- Percent no answer (housing units in which no potential respondent ever answered the door)
- Number of contact attempts
- Percent contact to screener (interviewer made contact with the household to screen for eligible respondent)
- Percent cooperation to screener (household provided information on eligible household members)
- Percent contact to final (interviewer made contact with selected eligible respondent)
- Percent cooperation to final (interviewer completed interview with eligible respondent)
- Response rate (AAPOR Response Rate 3).

We included multiple outcome rates because they reflect different components of overall response. Thus, we can tease out the effects of advance letters on contact and cooperation with the household as well as contact and cooperation with the selected respondent.

The specific hypotheses being tested included:

- H1: Advance letters in unbranded plain envelopes and without a brochure will boost outcome rates compared to no letters (outcome rates of conditions 7 and 10 compared to those of condition 6).
- H2: Advance letters in branded envelopes and without a brochure will boost outcome rates compared to unbranded plain envelopes without a brochure (the outcome rates of conditions 8 and 11 compared to those of conditions 7 and 10).
- H3: The inclusion of a brochure in an advance letter will boost outcome rates compared to an advance letter without a brochure (outcome rates of conditions 1 through 5 compared to those of conditions 7 and 10; and outcome rates of condition 9 compared to those of conditions 8 and 11).

6 The second wave of data collection began in January 2016; the experiment described here was only implemented at baseline.

7 The response rate is the proportion of eligible respondents who completed the interview. SRL uses the American Association of Public Opinion Research's (AAPOR) Standard Definitions for response rate calculation; the rate reported here is RR3. See The American Association for Public Opinion Research (2016).

5.5.5 Results

Outcome rates by condition are shown in Table 5.3. Because the outcomes are rates for each of five different groups, we tested for statistically significant differences by first calculating 95% confidence intervals around the outcome estimates. Rates are significantly different if there is no overlap in their confidence intervals. Figure 5.2 shows the 95% confidence intervals for the response rates, by condition. For all significant differences indicated by the 95% confidence intervals, we ran statistical tests using the Summary TTEST procedure in SPSS, which tests for difference in rates.[8] The response rates are higher in the conditions that received a brochure in a plain envelope ($M = 0.32$, SD = 0.47) compared to those who received a branded envelope with only a letter ($M = 0.23$, SD = 0.42); $t = 2.59, p = 0.01$. Table 5.3 shows the various sample rates by advance mailing condition. For all of the outcome rates except cooperation to final, outcomes are significantly better for those who received a brochure in a plain envelope compared to other outcomes.

- The completion rates are higher in conditions 1–5 ($M = 0.28$, SD = 0.45) than in conditions 8 and 11 ($M = 0.20$, SD = 0.40); $t = 2.41, p = 0.016$.
- The no answer rate is lower in conditions 1–5 ($M = 0.16$, SD = 0.37) than in conditions 8 and 11 ($M = 0.26$, SD = 0.44); $t = -2.954$, $p = 0.003$ and conditions 7 and 10 ($M = 0.23$, SD = 0.42); $t = -2.058, p = 0.04$.
- The contact to screener rate is higher in conditions 1–5 ($M = 0.71$, SD = 0.45) than in condition 6 ($M = 0.60$, SD = 0.49); $t = 3.809, p = 0.000$; conditions 7 and 10 ($M = 0.65$, SD = 0.48); $t = 2.67$, $p = 0.008$; conditions 8 and 10 ($M = 0.64$, SD = 0.48); $t = 3.113$, $p = 0.002$; and condition 9 ($M = 0.61$, SD = 0.49); $t = 3.194, p = 0.002$.
- The cooperation to screener rate is higher in conditions 1–5 ($M = 0.62$, SD = 0.49) than in condition 6 ($M = 0.53$, SD = 0.43); $t = 2.527$, $p = 0.012$; conditions 7 and 10 ($M = 0.54$, SD = 0.43); $t = 2.845$, $p = 0.005$; and conditions 8 and 10 ($M = 0.46$, SD = 0.43); $t = 5.015$, $p = 0.000$.
- The cooperation to final rate is lower in conditions 1–5 ($M = 0.83$, SD = 0.38) than in conditions 8 and 11 ($M = 0.89$, SD = 0.31); $t = -2.29, p = 0.023$.

Given the overall response rate is highest in the condition with the brochure in the plain envelope, the advantage of this condition in household screening rates outweighs its disadvantage in respondent contact and cooperation rates.

We also examined the effort required across the experimental conditions (in terms of mean number of contact attempts needed to finalize sample). Table 5.4 shows the mean number of contact attempts for the entire sample; the highest mean contact attempts was among the households in the condition that did not receive an advance letter ($p < 0.001$). When effort is examined just for the completed interviews (Table 5.5), we again see that the mean number of contact attempts was highest in the no advance letter condition; these differences were marginally significant. Together, these results suggest that the use of an advance letter reduces the level of interviewer effort required. While these differences are significant, they are marginal, and it is difficult to assess whether the added costs of advance letters are enough to offset the costs of additional contact attempts, because there are too many operational complexities to factor in.

5.5.6 Discussion

Advance letter mailings in this study appear to have the biggest impact on contact and cooperation at the screening stage of the interview. The conditions in which the advance letter included

8 https://developer.ibm.com/predictiveanalytics/2010/02/01/a-new-extension-command-spssinc-summary-ttest

Table 5.3 Sample rates by condition.

	Condition 6 no advance mailing ($n = 586$)	Conditions 7 and 10 plain envelope, letter only ($n = 934$)	Conditions 8 and 11 branded envelope, letter only ($n = 929$)	Conditions 1–5 plain envelope, brochure ($n = 2886$)	Condition 9 branded envelope, brochure ($m = 478$)	Total ($n = 5813$)
Completion rate	20% ($n = 120$)	24% ($n = 221$)	20% ($n = 190$)	28% ($n = 800$)	24% ($n = 115$)	25% ($n = 1446$)
Percent no answer	21% ($n = 112$)	23% ($n = 208$)	26% ($n = 231$)	16% ($n = 444$)	24% ($n = 107$)	20% ($n = 1102$)
Percent Contact screener	60% ($n = 329$)	65% ($n = 575$)	64% ($n = 574$)	71% ($n = 1945$)	61% ($n = 274$)	67% ($n = 3697$)
Percent coop screener	53% ($n = 173$)	54% (313)	46% ($n = 266$)	62% ($n = 1201$)	55% ($n = 151$)	57% ($n = 2104$)
Percent contact final	88% ($n = 147$)	88% ($n = 256$)	85% ($n = 213$)	85% ($n = 967$)	90% ($n = 131$)	86% ($n = 1714$)
Percent coop final	82% ($n = 120$)	86% ($n = 221$)	89% ($n = 190$)	83% ($n = 800$)	88% ($n = 115$)	84% ($n = 1446$)
Response rate	24% ($n = 120$)	28% ($n = 221$)	23% ($n = 190$)	32% ($n = 800$)	28% ($n = 115$)	29% ($n = 1446$)

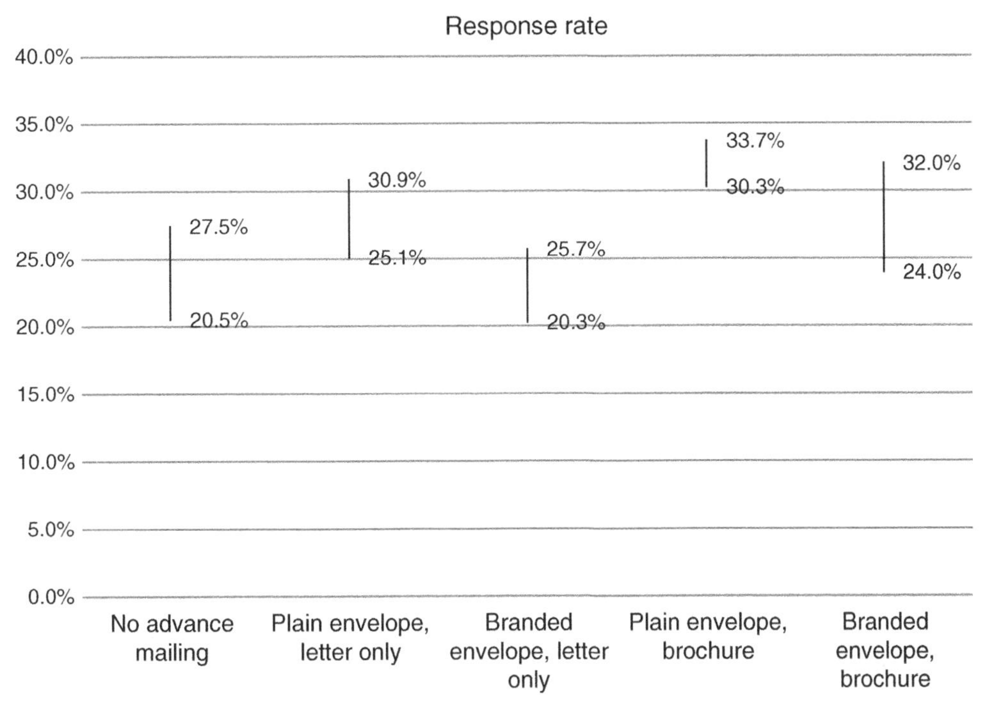

Figure 5.2 Response rate 95% confidence interval by condition.

Table 5.4 Mean number of contact attempts by condition, all cases.

Advance letter/brochure combo	Mean	*N*	Std. deviation
No advance mailing	8.0	586	3.7
Plain envelope, letter only	7.5	934	3.6
Branded envelope, letter only	7.8	929	3.7
Plain envelope, brochure	7.1	2886	3.9
Branded envelope, brochure	7.2	478	3.6
Total	7.4	5813	3.8

$F(4, 5808) = 11.477, p < 0.001$.

a brochure in a plain envelope had the highest rates of contact to screener and cooperation to screener. There were fewer differences in final contact and cooperation rates with eligible respondents. The biggest difference in these rates is that respondents who received a letter in a branded envelope had higher rates of contact and cooperation to the final interview than those who received a brochure in a plain envelope. Overall, these results suggest that advance letters have a significant impact in response rates, mostly via their effects on household screening rates (cf. Lavrakas et al. 2017a). Further, a plain envelop appears to be more effective than a branded one.

A limitation of the Neighborhood Crime and Justice Study is that the assignment of conditions was confounded with the timing of when those conditions were released during the field period. For example, the results show consistently better outcomes among those who received

Table 5.5 Mean number of contact attempts by condition, completes only.

Advance letter/brochure combo	Mean	*N*	Std. deviation
No advance mailing	5.3	120	3.3
Plain envelope, letter only	4.7	221	3.2
Branded envelope, letter only	4.3	190	2.9
Plain envelope, brochure	4.4	800	3.1
Branded envelope, brochure	4.6	115	3.3
Total	4.6	1446	3.1

$F(4, 1441) = 2.331, p > 0.06.$

a brochure in a plain envelope. However, that condition was assigned to conditions 1 through 5, which were also the first five conditions to be released. It is possible that the outcomes had more to do with interviewer enthusiasm in the early part of the study rather than advance mailing conditions. Interviewer burnout and turnover may have contributed to lower rates in later conditions. Thus, we cannot determine precisely that the nature of the advance mailing treatment that a household received was entirely responsible for the observed decrease in nonresponse. Future experiments should assign advance mailing conditions randomly throughout the data collection period, and perhaps test fewer conditions than the number included in this study.

5.6 Discussion

At least in telephone surveys, advance letters seem to be a general tool for increasing response rates (e.g. de Leeuw et al. 2007). Generally, there is a clear tendency toward a *positive impact* of advance letters on response rates. Taking a total survey error perspective, the mere (non)response rate is of secondary interest: even low levels of response can yield good data and allow valid inferences. It is the nonresponse bias that is of greater interest. Whether *nonresponse bias* can be reduced with advance letters is still an open question. The same holds for effects on measurement. Do advance letters only have an effect on the recruitment or also on the reporting? Nevertheless, it also should be noted that response rates are important as they are related to survey costs; lower response rates mean that the average cost per completion is greater.

There are also indications for interdependence of *interviewer strategies/training* and effects of advance letters. Beyond this interviewer and respondent level, we know even less about the way advance letters work on a respondent level. What we do know is that a considerable amount of advance mailing is not read or noticed. In an attempt to raise the awareness of advance letters and their content, there are some studies on design and layout features that could influence the probability that an advance letter is read.

Altogether, advance letters can have an indirect or direct impact on data quality. Advance letters could (positively or negatively) affect data quality in the following key steps in the survey lifecycle: sample procedure and selection bias, nonresponse bias, and measurement error. The rationale of using advance letters is to reduce nonresponse or increase response rates and thereby reducing nonresponse-related total survey error. However, it is possible that advance letters increase nonresponse bias and survey costs (Camburn et al. 1995). So, the practice of using advance letters anyway, just in the hope that they will increase response rates, might be counterproductive from a total survey error perspective.

Therefore, there is need for research using experiments to study the effects of advance letters: The current knowledge gap includes the interdependence of sampling requirements for conducting the survey and requirements for sending advance letters which could cause coverage bias, differential effects of advance letters on subgroups, and the effect of branding and the content of the advance letter (both in terms of layout and text). Furthermore, we know little about what happens with advance letters before and after they are delivered, that is, the receipt of letters and by whom. With changing survey modes, modes of sending prenotifications also change. In the future, experiments on the effects of advance letters will also have to focus on the applicability and specific effects in combinations with new recruitment and data collection modes, e.g. online surveys.

5.7 Research Agenda for the Future

We identify the following knowledge gaps regarding how and under what conditions advance letters might affect:

- Nonresponse bias (is there less such bias when advance letters are used?)
- Cost effectiveness (is the number of contact attempts affected?)
- "Mode" effects (are there any differences between face-to-face and telephone surveys and the effect of advance letters?)
- Interviewer strategies (how does an interviewer utilizes advance letters in their recruitment effort?)
- Measurement error (are advance letters only relevant in the recruitment phase or do they also have an impact on reporting?)
- Respondents' perception (how do respondents perceive different features of advance letters? What do they notice and what influences their decision to participate in a survey interview? Which component of the content is most effective [reciprocity?])
- Reliability of respondent recall about receiving advance letters and question wording.

And how these effects might vary depending on:

- Topic of the study (does, for example, sensitivity of the survey topic increase the effect of advance letters?)
- Sampling frame/target population (different data collection methods require different sample information. Do coverage biases multiply when more contact information might be needed?)
- Envelope layout and design features (what motivates the recipient of an advance letter to open and read the letter? What type of advance mailing has an effect on what?)
- Interaction of advance letters with other strategies such as incentives (is there an additive effect of incentives and advance letters?)

To study these effects, researchers should conduct experiments in either a split-ballot or a factorial design and analyze results with analysis of variance or regression models. Limitations of experiments to test effects of advance letter definitely relate to sampling issues and topic-specific surveys. Regarding the latter: Methodological research is usually not conducted for its own sake but on the back of substantive survey. Thus, the study topic cannot be manipulated experimentally – however, the content and layout of an advance letter can be, at least to some degree.

The sampling frame has crucial effects on the applicability of advance letters. Depending on the survey mode, different options of sending advance mailings occur with different advantages and disadvantages. For good quality data and confidence in results, random assignment to experimental and treatment groups have to be warranted. The same prerequisites (e.g. a published telephone number) have to be met for all participants.

Furthermore, it has been shown to be difficult to assess how many and who (in a household) opens and maybe reads or discards advance letters. All experiments on advance letters we are aware of suffer from this unknown quantity. However, the effect of advance letters can only be reliably judged if this information can be estimated. So far, this has been included in a limited number of studies and to actually obtain the information reliable can be challenging.

Also, advance letter experiments on mere outcome rates can only offer superficial insights but hardly and deeper understanding of processes and biases in sending, receiving, opening advance letters, and responding to survey requests. As for most research on nonresponse biases, it would be insightful to have a nonrespondent survey combined with factorial designs or at least split-ballot experiments. Additionally, to ascribe potential effects to advance letters, other factors have to be controlled, most importantly interviewer behavior. If the interviewers are not kept blind regarding the experimental condition, they may use different behaviors that could lead to observed findings rather than sending an advance letter. Thus, interviewers should ideally be assigned to only one experimental condition and kept blind to the experiment as such.

Overall, we advocate a stronger conceptualization of advance letters within the total survey error framework, an interview mode sensitive discussion and a more qualitative inquire in the motives of respondents to participate or to refuse after an advance letter was sent to them. Furthermore, as the case studies in our chapter demonstrated, there are a host of dependent variables that can be used to evaluate the effects of advance letters, and we recommend that future research not be limited to only the standard response rate dependent variables that have been used in most past advance letter experimentation.

References

Berinsky, A.J. (2008). Survey nonresponse. In: *Public Opinion Research* (ed. W. Donsbach and M.W. Traugott), 309–321. Los Angeles: Sage.

Camburn, D., Lavrakas, P.J., Battaglia, M.P. et al. (1995). Using advance respondent letters in random-digit-dialing telephone surveys. In: *Proceedings of the 50th Annual Conference of the American Association for Public Opinion Research* (ed. American Statistical Association), 969–974.

Cannell, C.F. and Fowler, F.J. Jr., (1965). Comparison of hospitalization reporting in three survey procedures. In: *Vital and Health Statistics. National Center for Health Statistics.* DHEW Publication No. 1000, Series 2, No. 8. Washington, D.C.: U.S. Government Printing Office.

Collins, M., Sykes, W., Wilson, P., and Blackshaw, N. (2001). Nonresponse: the UK experience. In: *Wiley Series in Survey Methodology. Telephone Survey Methodology* (ed. R.M. Groves, P. Biemer, L. Lyberg, et al.), 213–231. New York: Wiley.

Couper, M.P., Mathiowetz, N., and Singer, E. (1995). Related households, mail handling, and returns to the 1990 U.S. Census. *International Journal of Public Opinion Research* 7: 172–177.

de Leeuw, E., Callegaro, M., Hox, J. et al. (2007). The influence of advance letters on response in telephone surveys: a meta-analysis. *Public Opinion Quarterly* 71 (3): 413–443.

von der Lippe, E., Schmich, P., and Lange, C. (2011). Advance letters as a way of reducing nonresponse in a National Health Telephone Survey: differences between listed and unlisted numbers. *Survey Research Methods* 5 (3): 103–116.

Dillman, D.A., Gallegos, J.G., and Frey, J.H. (1976). Reducing refusal rates for telephone interviews. *Public Opinion Quarterly* 40: 66–78.

Dillman, D.A., Smyth, J.D., and Christian, L.M. (2014). *Internet, Phone, Mail, and Mixed-Mode Surveys: The Tailored Design Method*. Wiley.

Dykema, J., Stevenson, J., Day, B. et al. (2011). Effects of incentives and prenotification on response rates and costs in a national web survey of physicians. *Evaluation & the Health Professions* 34 (4): 434–447.

Dykema, J., Jaques, K., Cyffka, K. et al. (2015). Effects of sequential prepaid incentives and envelope messaging in mail surveys. *Public Opinion Quarterly* 79 (4): 906–931.

Finn, A., Gendall, P., and Hoek, J. (2004). Two attempts to increase response to a mail survey. *Marketing Bulletin* 15: 1–5.

Goldstein, K.M. and Jennings, M.K. (2002). The effect of advance letters on cooperation in a list sample telephone survey. *Public Opinion Quarterly* 66: 608–617.

Groves, R.M. (2004). *Survey Errors and Survey Costs*. Hoboken: Wiley Interscience.

Groves, R.M. and Snowden, C. (1987). The effects of advance letters on response rates in linked telephone surveys. In: *Proceedings of the Section on Survey Research Methods of the American Statistical Association*, 633–638. Alexandria, VA: American Statistical Association.

Groves, R.M., Fowler, F.J. Jr., Couper, M. et al. (2009). *Survey Methodology*. Hoboken: Wiley Interscience.

Deutschmann, M. and Häder, S. (2002). Nicht-Eingetragene in CATI-Surveys (unlisted numbers in CATI surveys). In: *Telefonstichproben: Methodische Innovationen und Anwendungen in Deutschland* (Hg. S. Gabler und S. Häder), 68–84. Münster: Waxmann.

Hembroff, L.A., Rusz, D., Rafferty, A. et al. (2005). The cost-effectiveness of alternative advance mailings in a telephone survey. *Public Opinion Quarterly* 69 (2): 232–245.

Iannacchione, V.G., Stabb, J.M., and Redden, D.T. (2003). Evaluating residential mailing lists. *Public Opinion Quarterly* 67: 202–210.

Keeter, S., Miller, C., Kohut, A. et al. (2000). Consequences of reducing nonresponse in a national telephone survey. *Public Opinion Quarterly* 64 (2): 125–148.

Lavrakas, P.J. (1993). *Telephone Survey Methods: Sampling, Selection, and Supervision*, 2e. Thousand Oaks, CA: Sage Publications.

Lavrakas, P.J., Dirksz, G., Lusskin, L., and Ponce, B. (2016). Experimenting with the addressee line in a mail survey of hispanic households. *Survey Practice* 9 (4). http://www.surveypractice.org/index.php/SurveyPractice/article/view/380/pdf_80.

Lavrakas, P.J., Benson, G., Blumberg, S. et al. (2017a). *The Future of General Population Telephone Survey Research*. Deerfield, IL: AAPOR. https://www.aapor.org/Education-Resources/Reports/The-Future-Of-U-S-General-Population-Telephone-Sur.aspx.

Lavrakas, P.J., Skalland, B., Ward, C. et al. (2017b). Testing the Effects of Envelope Features on Survey Response in a Telephone Survey Advance Letter Mailing Experiment. *Journal of Survey Statistics and Methodology* 6 (2): 262–283.

Link, M.W. and Mokdad, A. (2005). Advance letters as a means of improving respondent cooperation in random digit dial studies. *Public Opinion Quarterly* 69 (4): 572–587.

Luiten, A. (2011). Personalisation in advance letters does not always increase response rates. Demographic correlates in a large scale experiment. *Survey Research Methods* 5 (1): 11–20.

McLean, S.A., Paxton, S., Massey, R. et al. (2014). Prenotification but not envelope teaser increased response rates in a bulimia nervosa mental health literacy survey: a randomized controlled trial. *Journal of Clinical Epidemiology* 67 (8): 870–876.

Montaquila, J.M. and Olson, K.M. (2012). Practical tools for nonresponse bias studies, SRMS/AAPOR Webinar, 24 April 2012.

Otte, G. (2002). Erfahrungen mit zufallsgenerierten Telefonstichproben in drei lokalen Umfragen. In: *Telefonstichproben. Methodische Innovationen und Anwendungen in Deutschland* (Hg. S. Gabler und S. Häder), 85–110. Münster: Waxmann.
Parsons, J., Owens, L., and Skogan, W. (2002). Using advance letters in RDD surveys: results of two experiments. *Survey Research* 33 (1): 1–2.
Robertson, B., Sinclair, M., Forbes, A. et al. (2000). The effect of an introductory letter on participation rates using telephone recruitment. *Australian and New Zealand Journal of Public Health* 24 (5): 552.
Sangster, R.L. (2003). *Do Current Methods Used to Improve Response to Telephone Surveys Reduce Nonresponse Bias?* United States Department of Labour.
Singer, E., van Hoewyk, J., and Maher, M.P. (2000). Experiments with incentives in telephone surveys. *Public Opinion Quarterly* 64: 171–188.
Smith, W., Chey, T., Jalaludin, B. et al. (1995). Increasing response rates in telephone surveys: a randomised trial. *Journal of Public Health Medicine* 17 (1): 33–38.
Sykes, W. and Hoinville, G. (1985). *Telephone Interviewing on a Survey of Social Attitudes: A Comparison with Face-to-Face Procedures.* Social and Community Planning Research.
The American Association for Public Opinion Research (2016). *Standard Definitions: Final Dispositions of Case Codes and Outcome Rates for Surveys*, 9e. AAPOR.
Traugott, M.W., Groves, R.M., and Lepkowski, J.M. (1987). Using dual frame designs to reduce nonresponse in telephone surveys. *Public Opinion Quarterly* 51: 522–539.
Vogl, S. (2018). Advance Letters in a Telephone Survey on Domestic Violence: Effect on Unit-Nonresponse and Reporting. *International Journal of Public Opinion Research.* https://doi.org/10.1093/ijpor/edy006
Woodruff, S.I., Mayer, J.A., and Clapp, E. (2006). Effects of an introductory letter on response rates to a teen/parent telephone health survey. *Evaluation Review* 30 (6): 817–823.

Part III

Overview of the Section on the Questionnaire

Allyson Holbrook[1] *and Michael W. Traugott*[2]

[1] *Departments of Public Administration and Psychology and the Survey Research Laboratory, University of Illinois at Chicago, Chicago, IL, USA*
[2] *Center for Political Studies, Institute for Social Research, University of Michigan, Ann Arbor, MI, USA*

Perhaps the largest proportion of survey experiments to date has been conducted on questionnaire-related issues, including question wording, response formats and scales, and question order and context effects. One somewhat underexplored aspect of experiments involving the questionnaire is the way in which experimental conditions may disproportionately affect respondents' willingness to participate fully in the experiment, or expressed another way, the extent to which specific treatments can produce high levels of unusable responses or complete withdrawal from the study. Such unequal (i.e. differential) attrition may threaten the validity of studies if the characteristics of respondents who do not participate in the experiment are different across conditions. All of these topics are covered in the five chapters in this section.

In the first chapter, Beatty, Cosenza, and Fowler address the tradeoffs and problems with grammatically complex questions designed to obtain detailed and specific information. Using examples, they demonstrate the utility of extensive pretesting and cognitive interviewing involving split-ballot questionnaires, coupled with a recording of each interviewer interaction with a respondent, improved the eventual design of the questions. In addition to exploring the differences that alternative question wordings and forms can make, they argue for the use of carefully designed experiments to estimate accurately how much effect question wording can have.

Keusch and Yan report on experiments to assess the difference that alternative response scales can make. Focusing on frequency scales, they explore the impact of three distinct attributes of such scales in an online survey experiment: direction, alignment on the screen, and the vagueness of the scale labels. Another important research question they address is how these scale attributes might affect satisficing behavior in relation to certain respondent characteristics. As a form of validity test, they also compare responses to the experimental scale treatments with those to open-ended questions about the frequency of the same behaviors. They find that vague quantifiers produce data of lower validity, but none of their three treatments are related to satisficing behavior.

The chapter by de Leeuw, Hox, and Scherpenzeel looks at the interaction between question effects and mode in a multi-mode study conducted online and by telephone in the Netherlands.

Experimental Methods in Survey Research: Techniques that Combine Random Sampling with Random Assignment, First Edition.
Edited by Paul J. Lavrakas, Michael W. Traugott, Courtney Kennedy, Allyson L. Holbrook, Edith D. de Leeuw, and Brady T. West.

Companion Website: www.wiley.com/go/Lavrakas/survey-research

The experimental treatments involved two different scale attributes: fully labeled *vs.* labeling only of the end points and an unfolding two-question sequence vs. all response categories deployed in a single question. They were used on eight questions in which the expected responses were equally divided between acceptable and unacceptable practices. They find small but consistent mode effects and question format effects as would be expected, but only small interaction effects between data collection mode and question format, which are encouraging results for those who conduct mixed-mode data collection.

Cuik and Yost addresses issues of noncompliance and the problems they can create. They explore the role of unexpected policy positions on high and low salience issues of candidates from different political parties in relation to the party identification of randomly assigned respondents. The rate of "Don't Knows" in the conflicting cue condition was large (27%) and twice as high as in the neutral or consistent cue conditions. The main body of the chapter focuses on a set of alternative approaches to data analytic approaches for dealing with unequal sample sizes that suggest biased responses to the treatments.

The chapter by Creighton, Brenner, Schmidt, and Zavala-Rojas reports on an application of the item count technique (ICT) or list experiment in the context of Dutch attitudes toward immigration. In addition to a detailed description of the method, the authors provide a discussion of the strengths and weaknesses of the method especially as it relates to social desirability patterns of response and the inability of it to provide information at the individual level because of the way in which it guarantees anonymity.

Experiments that manipulate elements of the questionnaire represent a large portion of experiments conducted in surveys with probability-based sampling methods, in part because it is relatively easy to manipulate questionnaire characteristics, particularly when surveys are conducted using computer-assisted technology.

The chapters in this section represent the diversity of questionnaire characteristics that have been examined, but there are several areas in which further experimentation would be useful. One is the effect of question order broadly. Such order experiments have been examined somewhat narrowly within a small set of items, but there has been less experimental research into question order overall, in part because the number of possible orders is often high and some orders are illogical or obviously problematic. A second area where future experimentation seems useful is in the effects of technological advancements in survey design and delivery. Technology increasingly makes it possible to deliver surveys in new ways (e.g. SMS) or to ask questions in new ways (e.g. with slider bars). Further experimental studies are needed to examine the implications of new technologies as they develop for questionnaire design and survey measurement more broadly.

6

Experiments on the Design and Evaluation of Complex Survey Questions

Paul Beatty[1]*, Carol Cosenza*[2]*, and Floyd J. Fowler Jr.*[2]

[1] *Center for Behavioral Science Methods, U.S. Census Bureau, Washington, DC, USA*
[2] *Center for Survey Research, University of Massachusetts, Boston, MA, USA*

A fundamental tenet of questionnaire design, dating back to some of the earliest volumes on the subject, is to write simple and straightforward questions (Payne 1951). Nevertheless, a casual examination of a variety of questionnaires, including those administered by government agencies, academic institutions, and private firms, reveals that survey questions are often long and detailed, using many words to specify particular attitudes or behaviors of interest. Some questions use elaborate definitions or examples to illustrate concepts, sometimes followed by particular prompts to help stimulate recall. Qualifying clauses may be used to instruct respondents to consider or exclude particular situations and timeframes when responding. Some questions also provide specific response options. As a result, questions used to produce a single response are often long, detailed, and grammatically sophisticated.

Complex questions may appeal to survey researchers because they maximize the specificity of data while minimizing the number of distinct data items collected. Short and simple questions may be easier for respondents, but analysts may need to combine multiple responses to reach specifically desired data points. The advantages of acquiring a desired data point through a single question are clear – doing so can reduce steps needed to conduct analysis, and can avoid expanding the risk of item nonresponse through multiple questions. The key issue is whether such questions can be feasibly administered and accurately answered.

From a Total Survey Error perspective, complex questions may introduce measurement error in several ways. The level of detail provided may overwhelm respondents' working memory, making it difficult to initially comprehend questions or to remember pertinent details when completing the response task (Just and Carpenter 1992; Tourangeau et al. 2000, p. 45). Complex questions may also be difficult for interviewers to administer as written, potentially reducing the standardization of measurement. The measurement error risks of complex questions may very well outweigh their benefits – but it is hard to know for sure. In reality, survey researchers often resort to complex questions because the benefits are easily understood, while the risks of error are difficult to quantify.

Experimentation has the potential to shed more light on those risks, and in fact has long played an important role in understanding the consequences of questionnaire design decisions on survey results. The approach is fairly straightforward, typically entailing random assignment of alternative question versions (or alternative conditions in which questions are presented) to comparable samples; the effect of the variation is determined by comparing response distributions, or associations of responses with some external variable of interest (Krosnick 2011).

Experimental Methods in Survey Research: Techniques that Combine Random Sampling with Random Assignment, First Edition.
Edited by Paul J. Lavrakas, Michael W. Traugott, Courtney Kennedy, Allyson L. Holbrook, Edith D. de Leeuw, and Brady T. West.

Companion Website: www.wiley.com/go/Lavrakas/survey-research

For example, experiments by Fowler (2004) provided several examples of how parsing complex questions into multiple, shorter questions significantly affected response distributions. Numerous other experiments over several decades have shown how specification of response categories, time frames, question contexts, and framing of issues affect behavioral reports (Sudman et al. 1996), and attitudinal expression (Schuman and Presser 1981). Collectively, experimental findings serve as a basis for many widely adopted questionnaire design principles (see, for example, Fowler 1995; Bradburn et al. 2004).

As useful as this collective guidance is, survey practitioners may often find that there is still insufficient guidance to decide among various question alternatives. For example, consider the following question:

> During the past 12 months, how many times have you seen a doctor or other health care professional about your own health at a doctor's office, a clinic, or some other place? Do not include times you were hospitalized overnight, visits to a hospital emergency room, home visits, or telephone calls.

There are numerous ways that one could construct a question, or questions, to obtain this information. Does this represent the optimal approach? Note that it includes various components, including a time frame to be considered, a specification of what to report (a number of times), a key behavior of interest (seeing a doctor) and details about what is relevant (visits regarding one's own health, in particular settings) and what should be excluded (hospitals and home visits, and telephone calls). As alternatives, a questionnaire designer might consider whether the level of detail is too complicated for a single item, warranting multiple questions. Alternatively, if a single question is used, the designer might consider whether its components are optimally structured to maximize comprehension and accurate recall.

Although questionnaire design principles may help to navigate such decisions, more commonly they are made through a combination of experience and common sense. Pretesting, therefore, plays a critical role in evaluating the performance of questions prior to fielding. Cognitive interviewing in particular is useful for identifying problems respondents are experiencing, in addition to which attributes of questions are responsible for difficulties (Beatty and Willis 2007). However, cognitive interviewing does not provide evidence regarding how statistically significant problems will actually be among representative samples (Fowler 2004); furthermore, cognitive interviewing may reveal tradeoffs to alternative versions of questions, without being able to quantify the magnitude of the respective pros and cons (Beatty 2013; Willis 2015). Still, such results can lead to hypotheses that a particular question form or format should be superior to an alternative. These hypotheses often lend themselves very well to further experimentation, which in turn has the potential to lead to more specific questionnaire design principles.

In this chapter, we provide examples of experiments designed to inform particular questionnaire design decisions in which the available guidance was limited or ambiguous. Each of the experiments we present began with a dilemma involving draft questions proposed for national surveys (e.g. at the National Center for Health Statistics), in which there was more than one plausible way to craft the question, and limited guidance regarding which was preferable. Initially, the dilemmas involved the optimal structure of questions that were long and grammatically complex, potentially including detailed descriptions and definitions, examples, explanations, multiple clauses, qualifiers, or including specific response categories. As our research progressed, it moved from matters of structural complexity to conceptual complexity – for example, considering the optimal way to manage questions containing disparate concepts, or ambiguity of question intent arising from context. Our results, therefore, explore question complexity in various forms.

We constructed alternative question versions to explore each questionnaire dilemma and embedded them in split-ballot questionnaire instruments. These questionnaires were administered via random digit dialing (RDD) telephone surveys, based on probability samples of U.S. households with landlines, conducted by the University of Massachusetts – Boston, in which respondents were randomized to one of two or one of three forms of questions designed to achieve the same objective. The basic design was to have comparable samples of respondents answering different versions of the questions, in order to identify the effect of the variation on results. Some experiments were embedded in an initial survey, and the remainder was embedded in a second survey conducted about a year later. A total of 454 interviews were completed in the first survey, and 425 interviews were completed in the second survey. Several shortcuts were taken in the data collection phases of both: interviewers were instructed to arrange interviews with any available adult, and both callbacks and attempts to convert reluctant respondents were limited. We made these decisions to maximize response yield given limited resources, and a primary concern of maintaining comparability of the samples rather than making inferences to the population. Basic tabulations indicated that the randomized approach we used produced comparable samples with respect to gender, age, ethnicity, and education, so that differences in responses to the question alternatives could be confidently attributed to the effects of the question forms.

In each experiment, we were interested in whether the survey responses differed across question versions. In addition, we were interested in whether interactions between respondents and interviewers differed across question versions. In some cases, differences in how often interviewers read questions exactly as written, or how often they needed to probe to obtain a codeable response, provide evidence regarding the extent of problems with the questions. We therefore audio-recorded the interviews, and behavior coded them using procedures summarized in Fowler and Cannell (1996). In this study, accuracy of coders' work was evaluated as part of their training before production coding began, but we did not collect reliability statistics. In the first sample, usable recordings were obtained from 378 (83%) of the respondents. In the second sample, usable recordings were obtained from 314 (73%) of interview respondents.

A number of different codes were used, including the number of times that the initial response to the question was inadequate; the number of times the respondent interrupted the question before it was fully read; the number of times some sort of probing was required to get a response; and the number of times the respondent asked for clarification, repeat of question, or similar assistance. The next few sections of the chapter illustrate our approach and the evidence some of our randomized experiments produced regarding differences in question performance.

6.1 Question Construction: Dangling Qualifiers

When questionnaire designers determine that it is necessary to convey complex concepts within a single question, they generally have several options about how to structure this information. One common question construction is to provide explanatory or qualifying information following the question mark, as in the following example:

> In the past 12 months, how many times have you talked to any health professional about your own health? Include in-person visits, telephone calls, or times you were a patient in a hospital.

The *dangling qualifier* in this case asks respondents to include information that they might not have automatically considered. Presumably, this construction keeps the core question as simple as possible, and encourages respondents to recall additional details after absorbing the overall meaning of the question. However, from the respondent's perspective, the question may end with the question mark – at which point they begin responding and pay relatively little attention to the material that follows. This is one basis for Fowler's (1995, pp. 86–87) recommendation to provide qualifiers and definitions prior to the core question. In this case, such an alternative might be:

> People talk to health professionals in person, over the phone, or as patients in a hospital. Including any of those, in the past 12 months, how many times have you talked to any health professional about your own health?

The potential downside of this construction is that respondents must make sense of qualifying information before they fully understand what it pertains to. However, upon reaching the question mark, the respondent will have heard all of the information that the researcher intended to present. We know of no prior research that provides guidance on which approach is preferable; furthermore, since survey sponsors often insist upon presenting all of the details to respondents, it is a common dilemma for survey designers.

We administered both versions of the question in a split-ballot experiment, with two key hypotheses. The first was that reports of contact with a health professional would increase in the alternative version of the question, because respondents would uniformly hear the qualifier that was designed to prompt full reporting. Table 6.1 shows that the data do not support this hypothesis. In fact, responses to the original version of the question, with the dangling qualifier, produced somewhat higher reports, although the difference is not significant. The second hypothesis was that any evidence of problems responding (e.g. through inadequate initial responses, or requests for clarification) would be roughly equivalent across versions. That turns out to be the case – in fact, the alternative version seems to require clarification slightly less often – but it is worth noting that problems are undesirably common in *both* versions. A common rule of thumb is that problem rates of 15% or higher are worthy of notice (Oksenberg et al. 1991), and both versions exceed that mark on one or more measures.

We conducted numerous other experiments based on similar structural variations (data not shown), with similar results. Although original versions (with dangling qualifiers) are more commonly interrupted than the alternatives, the two versions rarely produce significantly different responses – perhaps indicating that respondents interrupt only after realizing that qualifying information could not reasonably affect their response. Similarly, there was rarely evidence that one construction created significantly more response problems than the other. In fact, it often appeared that *both* versions posed notable degrees of difficulty, and manipulating the structure did not make a fundamental difference in whatever attributes of the question were causing problems. (For further discussion on such overarching problems, see Fowler 2004; Beatty et al. 1999.)

Table 6.1 Effect of location of qualifiers in the question.

Qualifier	V1 (after Q)	V2 (beginning of Q)	Significance
Mean times talked to health professional	6.59 ($n = 214$)	5.93 ($n = 205$)	n.s.
Inadeq initial response	52 (32.5%)	39 (25.5%)	n.s.
R requests help	32 (20.0%) ($n = 160$)	20 (13.1%) ($n = 153$)	$p < 0.1$

Table 6.2 Effect of location of definition within question.

Definition	V1 (after Q)	V2 (beginning of Q)	Significance
Relative has diabetes	89 (42.6%) ($n = 209$)	74 (34.4%) ($n = 215$)	$p < 0.1$
Inadeq initial response	11 (7.2%)	4 (2.5%)	$p < 0.1$
Interrupted	25 (16.5%)	1 (0.6%)	$p < 0.01$
Interviewer intervention needed for response	14 (9.2%) ($n = 152$)	5 (3.1%) ($n = 159$)	$p < 0.05$

One experiment produced somewhat stronger evidence against a definition dangling after the question mark. The original question was:

> Have any of your immediate blood relatives ever been told by a doctor that they have diabetes? By "immediate blood relatives", we mean your parents, your children, and your brothers and sisters, whether or not they are still living.

Note that, because the term "immediate blood relatives" could reasonably be interpreted in different ways, it *must* be defined to fully convey the intent of the question. If interrupted, respondents could reasonably interpret the question more broadly than intended. An alternative that avoids that possibility would be:

> The next question is about immediate blood relatives – by that, we mean your parents, your children, and your brothers and sisters, whether or not they are still living. Have any of your immediate blood relatives ever been told by a doctor that they have diabetes?

Evidence from an experimental comparison systematically favors the alternative version: responses to V1 and V2 differ as hypothesized. V1 also performs less favorably on a number of behavior codes – respondents provide immediately adequate responses less often, it is interrupted before being completely read much more often, and interviewers need to take action (other than simply reading the question) to get a response more often (Table 6.2).

On the whole, results from these experiments suggest that the tradeoffs involved in restructuring questions are fairly modest. However, at least under some circumstances, dangling qualifiers and definitions have the potential to be ignored. It seems reasonable to present important definitions up front, as long as doing so does not present additional comprehension problems for respondents. Experiments such as these provide an effective means for evaluating such tradeoffs.

6.2 Overall Meanings of Question Can Be Obscured by Detailed Words

As noted earlier, one concern about long and detailed questions is that respondents may lose track of important details; or, in listening to the details, may incorrectly interpret the overall question. Such questions are often included on surveys anyway, because the benefits of obtaining data through a concise question are clear, but evidence of statistical error is elusive. Consider the level of detail in the following question:

> In the past 12 months, how many times have you seen or talked on the telephone about your physical or mental health with a family doctor or general practitioner?

The purpose of the question is to capture a wide variety of contacts with general practitioners – this could include *seeing* or *talking on the phone*; it could be about *physical* or *mental* health; and the terms *family doctor* and *general practitioner* are both used in case respondents are more familiar with one or the other. The question is long and rather wordy, but if successful would produce a desired data point quite efficiently.

When tested through cognitive interviews, several interviewers independently observed an interesting phenomenon. Participants sometimes answered "zero" even though answers to other questions on the same survey suggested that they *had* recently seen a general practitioner. When probed about their responses, several participants confirmed that they had seen their doctors, but answered "zero" to this question because they *had not talked to them on the telephone*. In other words, several respondents had misunderstood the question, thinking it was *only* about talking on the telephone.

We believed that these errors could be attributed to the overall level of detail in the question, and in particular, an unbalanced emphasis on words. In listening to the central section of the question ("seen or talked on the telephone about physical or mental health"), it is easy to imagine how the first part ("seen or") could be missed given the number of words devoted to other details, especially when immediately followed by still more details ("with a family doctor or general practitioner"). If true, there could be clear advantages in asking a distinct question about *seeing* a general practitioner, and if necessary asking a follow-up question about talking on the telephone.

Of course, the cognitive interview findings could not predict the extent of this error in actual statistics. In contrast, the costs of asking a second question were more easily quantifiable, and results from two questions would not necessarily be comparable to single-question results from other surveys. The following experiment was designed to test the hypothesis that even a few extra words could be responsible to significant misunderstanding, and if so to demonstrate the magnitude of the effect. We constructed simplified versions of the question, as follows:

> V1: The next question is specifically about primary care doctors. In the past 12 months, how many times have you seen or talked on the telephone with a primary care doctor about your health?
>
> V2: The next question is specifically about primary care doctors. In the past 12 months, how many times have you seen or talked with a primary care doctor about your health?

Both versions strip some of the details from the original question, and differ in only one minor respect: whereas V1 includes the words "seen or talked on the telephone," V2 simply says "seen or talked with," to reduce the imbalance in words. The experimental comparison allows us to explore whether three additional words are sufficient to throw respondents off from the intended meaning of the question. Results from tabulations and some behavior codes appear below (Table 6.3).

Table 6.3 Effect of detailed wording (seen or talked on the telephone).

	V1 telephone	V2 no telephone	Significance
Mean doctor contacts in 12 months	3.41	3.64	n.s.
Resp reporting no dr. contact in 12 months	48 (24.7%) ($n = 194$)	18 (9.2%) ($n = 195$)	$p < 0.01$
Inadequate initial response	21 (14.9%)	30 (21.1%)	n.s.
Resp requests help	8 (5.7%) ($n = 141$)	16 (11.3%) ($n = 142$)	$p < 0.1$

At first glance, the difference between response distributions appears to be modest: although reports of contact with doctors did rise slightly in the "no telephone" version, the difference was not significantly different. However, there was one notable difference in responses to the two versions: respondents were much more likely to report "zero" when the words "on the telephone" are included in the question. Whereas 24.7% of respondents reported "zero" to the original question, only 9.2% reported the same when the words "on the telephone" were dropped. The difference is both large and consistent with the hypotheses emerging from cognitive interviews.

On the other hand, behavior coding results suggest that the alternative "no telephone" version of the question is somewhat more ambiguous than the original, resulting in a higher number of respondent requests for clarification from the interviewer. Since the only difference between the two questions is the inclusion of the words "on the telephone," respondents most likely queried regarding whether telephone contact was consistent with the intent of the question.

This experiment is valuable because it highlights the nature and extent of the tradeoffs involved. Although the inclusion of the additional words may reduce some ambiguity, it arguably creates a much larger problem of what appears to be significant misreporting. Furthermore, although occasional requests for help can be resolved, invisible misunderstandings of the question cannot, and lead to error. Ironically, the additional words intended to produce more complete reporting actually have the opposite effect.

6.3 Are Two Questions Better than One?

One of the earliest established principles of survey question design is to ask one question at a time. The rationale for this wisdom is that respondents have difficulty thinking about two things at once, and may also have different responses to different components. When presented with two questions at once, they may concentrate on answering one of the questions and essentially ignore the other one. Yet researchers often seem to ignore this principle, hoping that it does not apply in all situations, perhaps to minimize the number of distinct questions to be administered.

For example, being poisoned and being injured severely enough to require medical care are two relatively infrequent events. Given their rarity, researchers thought they might save some interview time and space by combining them as follows:

> Version A: During the last 6 months, that is since (six months ago date) were you or anyone else who lives in your household injured or poisoned seriously enough that you or they got medical advice or treatment?

We wondered if there was a price to pay in accuracy for this double question, and split-ballot experimentation provided a means to explore this. We, therefore, created an alternative that asked one question at a time:

> Version B: During the past 6 months, that is since (since months ago date) did you or anyone else who lives in your household have an injury where any part of the body was hurt – for example, with a cut or wound, sprain, burn, or broken bone?
>
> During the last 6 months, that is since (date 6 months ago), were you or anyone else who lives in your household poisoned in any way?

Respondents who answered "yes" were asked if the injured or poisoned person had seen a medical professional about any of the events (Table 6.4).

Table 6.4 Poisonings or injuries requiring medical care per 100 households.

Version	Rates of injuries or poisonings requiring medical care
One question	13.1% ($n = 206$)
Two questions $p < 0.1$	20.6% ($n = 214$)

It can be seen that asking about each problem separately yielded over 50% more reports of events. The split-ballot test results are consistent with our hypothesis and the conventional wisdom that asking more than one question at a time leads respondents to not fully consider both parts of the question.

We then turned our attention to a subtler form of a double barreled question. Consider the following series:

- **To lower your risk of heart problems or stroke**, has a doctor or other health professional advised you to:
  - Cut down on salt or sodium in your diet
  - Eat fewer high fat foods
  - Get more exercise
  - Control your weight or lose weight.

At first, this may not seem like two questions blended into one. However, a "yes" answer to these questions means that a respondent is saying both that a doctor advised, for example, cutting down on salt *and* that the reason for this was to lower the risk of heart problems or stroke. Would respondents attend to both parts of this question when they answered? Again, a split-ballot experiment seemed to be an effective way to address that question. An example of the two-question version looked like this:

- A. Has a doctor or other health professional ever advised you to cut down on salt?
- B. (IF YES) Did the doctor recommend this for your general health or specifically to lower your risk of heart problems or stroke? (Table 6.5)

It can be seen that the estimates from the two-question approach are much lower than from the one-question version. Table 6.6 shows why: it compares the answers to the test question that included the idea of lowering the risk of heart problems or stroke as the reason for the advice with the results from the first question of the two-question series in which the rationale was not mentioned.

The percentage of respondents reporting that their doctors advised various health behaviors were practically identical regardless of whether or not the rationale of reducing risks of heart

Table 6.5 Percent reporting doctor advising health behavior to reduce risk of heart problems or stroke.

Healthy behavior	Single question	Separated questions	Significance
Reduce salt	32 (22.3%) $n = 143$	22 (16.3)% $n = 135$	n.s.
Fewer high-fat foods	64 (44.8%) $n = 143$	23 (17.0%) $n = 135$	$p < 0.01$
Get more exercise	71 (49.7%) $n = 143$	12 (8.9%) $n = 135$	$p < 0.01$
Control or lose weight	56 (39.1%) $n = 143$	6 (4.4%) $n = 135$	$p < 0.01$

Table 6.6 Percent reporting doctor advising health behavior by whether or not "to lower your risk of heart problems or stroke" was included in the question.

Healthy behavior	Single question (Reduce risk of heart problems/ stroke included in question)	First of two questions (Reduce risk of heart problems/ stroke NOT included in question)	Significance
Reduce salt	32 (22.3%) $n = 143$	38 (28.1%) $n = 135$	n.s.
Fewer high-fat foods	64 (44.8%) $n = 143$	59 (43.7%) $n = 135$	n.s.
Get more exercise	71 (49.7%) $n = 143$	58 (43.0%) $n = 135$	n.s.
Control or lose weight	56 (39.1%) $n = 143$	40 (30.0%) $n = 135$	$p < 0.1$

problems or stroke was included in the question. Table 6.6 clearly demonstrates that respondents to the original question ignored the issue of why the advice was given and simply focused on the single issue of whether or not they had gotten a particular kind of advice from a doctor. Table 6.6 is a clear-cut exhibit for the case that it is important to ask one question at a time, and the data were produced through a small split-ballot experiment.

6.4 The Use of Multiple Questions to Simplify Response Judgments

In some instances, the case for asking multiple questions is not that the single question mixes truly distinctive concepts, but rather poses a response task too broad to be reasonably completed. Consider the following question that was proposed as part of a module on consumption of various foods:

> During the past 30 days, how many times did you eat cheese, including cheese as snacks, and cheese in sandwiches, burgers, lasagna, pizza, or casseroles? Do not count cream cheese.

Cognitive interviewing suggested that this was a difficult question to answer accurately. Many respondents indicated that their answers were essentially guesses, guided by very crude estimation strategies. Furthermore, when probed to think about more specific eating situations, respondents often remembered additional information and revised their initial answers upwards. Over time, it became clear that respondents ate cheese in several distinct ways: as a part of dishes, such as a casserole; as an addition to food, such as on a sandwich; and as a snack. Respondents expressed more confidence when answering more specific questions along these lines than they had to a single question. While suggestive, these findings did not provide information about how different responses would be to simpler "decomposed" questions.

A subsequent split-ballot experiment compared the original question with the following multiple questions, which were developed based on the cognitive interview results:

- During the last 30 days, how many times have you eaten cheese on a sandwich, including burgers?
- During the last 30 days, how many times have you eaten cheese in lasagna, pizza, casseroles, or mixed in with other dishes?
- During the last 30 days, how many times have you eaten cheese as a snack or appetizer?

Table 6.7 Cheese consumption in the past 30 days, single and multiple questions.

	V1 (single)	V2 (multiple)			Significance
Mean times eating	13.9 ($n = 218$)	17.5 ($n = 228$)			$p < 0.01$
	V1	V2a	V2b	V2c	V1 and V2a/b/c
Inadequate initial response	33.3%	23.4	12.0	4.7	$p < 0.05$ except*
Probes used to obtain resp.	13.7%	7.8	6.3	2.1	
Resp. requested help/repeat	19.1%*	15.1*	3.1	2.1	

*$p = 0.3$.

Based on cognitive interviewing results, we believed that the original question produced an undercount, and that decomposed questions would produce higher reports. We also believed that the decomposed questions would be easier to administer based on some key behavior coding indicators. Results of the experiment appear to be consistent with those expectations: not only do the decomposed questions produce significantly higher reports, but they also produce fewer requests for clarification or repeats of the question, and fewer responses that are initially inadequate or require probing.

Closer analysis muddies that picture somewhat. One significant disadvantage of the decomposed questions is time (not presented in Table 6.7): whereas the original version was administered in a mean 28 seconds, the multiple questions required a mean of 51 seconds. Furthermore, some of the apparent advantages seen in behavior coding proved to be illusory. Whether asking one question or three, the questions still produce one data point – the objective is to produce a measure of overall cheese consumption. In that sense, it might be more appropriate to consider the *cumulative* rates of undesirable behavior codes to arrive at that data point. When comparing the cumulative behavior codes across the decomposed questions with the original, the rates are not significantly different. For example, it takes about the same amount of probing to yield a response for both the single original and the three decomposed questions.

Of course, these data do not reveal which responses are actually more accurate. Several studies have suggested that decomposing questions can actually lead to overreporting (e.g. Belli et al. 2000, and see also Tourangeau et al. 2000, pp. 95–96). Subsequent research (Beatty and Maitland 2008; Beatty 2010), using a food diary as a validation measure, suggested that actual consumption often fell somewhere between single and decomposed responses; results also suggested that decomposed questions can be less accurate than single questions if the questions are not parsed in a manner that corresponds to memories of the behavior in question.

Results of this experiment do not indicate that these decomposed questions are uniformly better or worse than global alternatives. Rather, they provide quantitative data that illustrate the nature of the tradeoffs involved in the alternative versions. Researchers can use the data on responses, response behaviors, and time to judge which alternative is likely to meet their objectives – or, to craft additional alternatives with the potential to further improve efficiency and accuracy.

6.5 The Effect of Context or Framing on Answers

The role of social desirability in the answers people give to surveys has long been known. There is an extensive literature that shows that behaviors or characteristics that may be seen as devalued are underreported, while overreporting occurs for answers that may be seen as socially desirable. The fact that these tendencies have been found to be more pronounced when an interviewer asks the questions than when respondents are self-administering surveys or providing answers to a computer supports the notion that the potential evaluation from another person is a key part of the dynamic.

These generalizations largely come from observing answers to questions that are thought to be potentially stigmatizing, such as using illegal drugs or frequenting prostitutes. However, many things we ask about are not so clearly labeled as good or bad. In fact, some behaviors can be good or bad depending on the context and how they are framed. Two such examples are drinking alcohol and owning guns. While drinking alcohol to excess is stigmatized, alcohol consumption in moderation has been shown to provide some health benefits. Gun ownership can lead to accidents, but it also can be an approach to home protection. How respondents answer questions like this might be a function of how they think their answers will be interpreted. If we explicitly manipulated the apparent context of the questions, we might be able to see more clearly whether or not the answers would be affected by the implied interpretation. Again, a split-ballot experimentation seemed to be a good way to examine that hypothesis.

We created two versions of a series of questions about drinking alcohol and gun ownership.

6.5.1 Alcohol

There are a number of things researchers have found that can affect your risk of serious illness or death. The next few questions are about things you can do that might after your risks.

Version 1: In the last 30 days, on how many days did you ....

a. Smoke any cigarettes
b. Ride in a car or truck without wearing a seatbelt
c. Have at least one drink of any alcoholic beverage
   i. On days when you drank, how many drinks of alcohol did you most often have.

Version 2: In the last 30 days, on how many days did you ....

a. Take a vitamin pill or some kind of vitamin supplement
b. Have at least one glass of milk
   i. On days when you drank milk, how many glasses of milk did you most often drink.
c. Have at least one drink of any alcoholic beverage
   i. On days when you drank, how many drinks of alcohol did you most often have.

6.5.2 Guns

The next few questions are about things you can do to affect the safety of your home.

Version 1

a. Do you keep any pesticides inside your home
b. Do you keep any cans filled with gasoline inside your home
c. Do you keep any guns inside your home or in your basement
   i. How many do you have
   ii. Are any of them/Is it loaded and ready for use.

Version 2

a. Do you have a security or burglar alarm system in your home?
b. Do you have deadbolt locks on all the doors that open to the outside of your home.
c. Do you keep any guns inside your home or in your basement
   i. How many do you have
   ii. Are any of them/Is it loaded and ready for use.

Table 6.8 Framing differences, alcohol and gun ownership questions.

	Version 1 (Negative)	**Version 2 (Positive)**	**Significance**
Alcohol use			
Mean number of days had at least one drink of alcohol	4.3 ($n = 229$)	6.0 ($n = 223$)	$p < 0.05$
Mean number of drinks on days had anything to drink	2.2 ($n = 115$)	2.2 ($n = 127$)	n.s.
Gun ownership and readiness			
Per cent have any guns in home or basement	78 (36.8%) $n = 212$	77 (38.3%) $n = 201$	n.s.
Per cent of those with guns who have then loaded and ready for use	23 (29.5%) $n = 78$	30 (40.0%) $n = 75$	n.s. ($p = 0.12$)

In each case, Version 1 was designed to communicate that drinking alcohol or having guns was perhaps not such a good thing, while Version 2 implied each might be reasonable. The question that interested us was whether or not the context and implied evaluation would affect reporting (Table 6.8).

In each experiment, we found one measure that seemed to be affected in the expected direction and one that was not. Moreover, the estimate that was affected in both cases was the one that was most closely tied to the framing in the question. The alcohol framing might suggest that having an alcoholic drink was good for health, but probably did not imply that having a lot of alcoholic drinks was a good idea. Hence, it is consistent that the number of days with at least one alcoholic drink was higher in the positive version, but not the number of drinks per day. In a similar way, gun ownership per se may or may not seem to make you safer unless it is loaded and ready to serve as protection. Although the effect is not quite significant at the $p < 0.1$ level, more gun owners reported that their guns were loaded in the context of protection than in context of safety hazards in the home. The split-ballot experiments provided an excellent opportunity to explore whether the framing of questions can have a significant effect on reporting and to provide some nuanced insights into which estimates were and were not affected by the framing.

6.6 Do Questionnaire Effects Vary Across Sub-groups of Respondents?

The preceding analyses show a variety of response effects attributable to the form of the questions, but only consider these effects on the overall samples. A reasonable follow-up hypothesis is that such effects do not operate in a uniform manner, but rather, are stronger among some groups of respondents than others. In particular, we wondered if education was a mitigating factor. Several prior studies have suggested this possibility – for example, both Narayan and Krosnick (1996) and Holbrook et al. (2007) found that education moderated the magnitude of numerous response effects in attitudinal surveys. We know of no studies exploring such effects among behavioral self-reports, but similarly expected stronger response effects among less-educated respondents.

Sample sizes for our experiments were small, and the experiments were not designed with such comparisons in mind. However, to explore the possibility of differential effects, we subdivided our samples into two groups: one with relatively low education (no more than a high school diploma) and one higher (any education beyond high school). We reran all of our analyses on the subdivided samples and evaluated whether the question effects operated differently among the high and low education subsamples.

Most of these comparisons did not reveal any differences. However, several were suggestive of different effects among higher- and lower-educated respondents. For example, one comparison revealed that the unintentional influence of a few words was limited to lower-educated respondents. Recall the question in which the words "on the telephone" seemed to restrict some respondents' interpretations, such that they only included telephone calls to doctors. Initially, it appeared that the effect only changed the proportion of respondents who reported having zero visits to the doctor (original data in Table 6.3). However, when broken down by education, it appears that less-educated respondents also reported significantly fewer visits to the doctor overall when the words "on the telephone" were included in the question. This provides further evidence of the hypothesized response effect, while suggesting that the influence of these additional words might be markedly stronger among the less-educated subset, who are perhaps more easily distracted from the overall intention of the question (Table 6.9).

In addition, further analysis of the experiment on poisoning and injury questions (original data in Table 6.4) suggest that the benefit of asking separate questions may center on lower-educated respondents. Asking two questions significantly increases their reports, whereas the effect is not significant among higher-educated respondents (Table 6.10).

Finally, some context effects also appeared to center on lower-educated respondents. For example, the effect of reframing the question on alcohol consumption in positive terms (original data in Table 6.8) seems to affect the lower-educated, but not the higher-educated respondents (Table 6.11).

We are hesitant to overgeneralize from these findings, which were based on fairly crude comparisons, and did not materialize in many of our experiments (data not shown). However, these findings are at least suggestive of the possibility that the response effects we observed operate somewhat differently on respondents at different education levels (e.g. with lower-educated

Table 6.9 Effect of detailed wording (seen or talked on the telephone), divided by education.

Mean doctor contacts in 12 months	V1 (telephone)	V2 (no telephone)	Significance
Overall	3.41 ($n = 194$)	3.64 ($n = 195$)	n.s.
Low education	1.78 ($n = 40$)	3.70 ($n = 54$)	$p < 0.05$
High education	3.86 ($n = 152$)	3.64 ($n = 140$)	n.s.

Table 6.10 Poisonings or injuries requiring medical care per 100 households, divided by education.

	V1 (one question)	V2 (two questions)	Significance
Low education	4 (8.5%) $n = 47$	15 (23.4%) $n = 64$	$p < 0.05$
High education	23 (14.5%) $n = 159$	29 (19.3%) $n = 150$	n.s.

Table 6.11 Framing differences, alcohol (divided by education).

Mean number of days had at least one drink of alcohol	V1 (Negative)	V2 (Positive)	Significance
Low education	2.7 $n = 74$	6.2 $n = 67$	$p < 0.05$
High education	5.2 $n = 152$	6.0 $n = 156$	n.s.

respondents being more susceptible to context effects, and higher-educated respondents drawing more heavily upon detailed definitions). The specific mechanisms that may be at play are also unclear. In any case, it seems reasonable that subsequent questionnaire design research should look more closely at how effects vary across subgroups, in addition to results across all respondents.

6.7 Discussion

The experiments described in this chapter were designed to bridge troublesome gaps between theory and practice in questionnaire design. Design principles are helpful for creating first-draft survey questions, but are often not specific enough to navigate decisions between various alternatives. For example, few would argue against the "enduring counsel for simplicity" of survey questions (Converse and Presser 1986, p. 9), but it is not always obvious whether a particular question has crossed an acceptable threshold. Researchers constantly grapple with decisions regarding whether one question is too complex; or, if it is not, how to optimally structure the words and clauses within; or, whether the benefits of various details outweigh the costs in terms of cognitive burden.

Cognitive testing and related qualitative methods can provide more specific insights into the sort of problems likely to result from a particular rendering of a question. It is highly unlikely that design principles will ever be specific enough to fully inform questionnaire design on their own. Questions pose such unique combinations of concepts and response tasks that we cannot imagine anything replacing the process of trying out questions on research participants, and observing how well they actually function. However, such testing has clear limitations: typical sample sizes are too small to cover very many permutations of responses and respondent characteristics, and results generally cannot quantify the likely extent of problems in population statistics.

A key advantage of experiments is that they can demonstrate how much of a difference various alternatives actually make in terms of response distributions, which can range from negligible to profound. One challenge is that there may be no clear measure of "truth" by which to gauge alternative response distributions. Nevertheless, researchers may be guided by a priori assumptions about expected responses. For example, a dangling qualifier may include important information designed to prompt respondents to include additional information when responding. A researcher could logically hypothesize that moving the qualifier earlier in the question, when it is more likely to be heard, could increase reports of a particular behavior. If the alternative version actually produced higher reports, the findings would support that hypothesis.

Our experiments also produced data other than response distributions, including measures of key interviewer and respondent behaviors, and time required to obtain acceptable responses. One advantage of collecting these data is that they improve our ability to evaluate tradeoffs of various design decisions. For example, we might find that one version of a question produces responses that are slightly more consistent with expectations than an alternative version, but also produces significantly higher evidence of response difficulty – possibly including longer response times, more frequent requests for clarification, and a greater need for probing to produce acceptable answers. When there are conflicting metrics regarding which version is "best," researchers can use the available evidence to maximize whichever criterion is deemed of greatest importance.

Although this sort of experimentation can be used to evaluate specific variants of questions, we think it also has great value for producing cumulative evidence that will lead to improved questionnaire design principles. For example, we conducted a number of experiments comparing questions with dangling qualifiers to alternatives, in which material was presented prior to the question mark. Generally, these experiments suggested that overall complexity was a larger problem than any particular question structure, and varying that structure did not usually make a large difference in response distributions or behaviors. However, we identified at least one example in which a definition trailing the question mark was apparently ignored some of the time, and which also produced undesirable response behaviors. Thus on the whole, there was some evidence for a case against the use of dangling qualifiers. Similarly, we conducted several experiments which pointed to undesirable consequences of overly complex single questions. In one case, words highlighting relatively minor details obscured the overall meaning of the question; in others, combining multiple concepts in the same question seemed to produce a combination of recall and judgment errors. Some of these problems could be reduced through the use of multiple questions. Finally, other experiments expanded our understanding of context effects by showing how reports of relatively innocuous behaviors may vary when framed as apparently positive or negative by prior questions. These experiments highlight particular interests of ours, but undoubtedly experimentation can illuminate numerous other aspects of questionnaire design and cognitive aspects of survey measurement.

All of the experiments reported in this chapter were based upon probability samples of household telephone numbers in the United States. Probability sampling was a key component of our research design because we wanted the data to accurately reflect the range of responses (and response problems) that would be present in the population. That being said, we accepted significant deviation from optimal probability samples of the U.S. population. Even at the time these data were collected (mostly in 2005–2006), concerns were mounting about whether RDD samples truly represented the overall household population, and we made no effort to include households without landline telephones. Furthermore, we accepted any adult member of participating households, made limited efforts at callbacks and refusal conversions, and did not weight the data to population parameters. These decisions were made in order to use limited resources to obtain as many completed cases as possible from our initial sample, and we do not believe that any of them fundamentally affected the conclusions of our experiments. In our view, random assignment of question versions across both respondents and interviewers was ultimately more important for ensuring the validity of results, and deviations from representativeness were also spread across treatments. Probability samples were meant to provide a reasonable starting point for the study, balanced by decisions to maximize cost-effectiveness as the data collection progressed.

The decision to conduct the experiments through telephone surveys was also driven by practical constraints, but was consequential. Many of the draft questions that we used were originally intended for use on face-to-face surveys. Administering such questions over the telephone, which removes visual cues and reduces interviewer–respondent contact, generally increases their difficulty. Although face-to-face interviewing would have been optimal, doing so was beyond our budgetary resources. Telephone data collection was the next-best option, still allowing for interviewer-administered questions within a national probability sample.

For study purposes, we recognized that administering the questions in an overly challenging mode might amplify certain problems. This could be useful for purposes of exploring performance differences, although the overall prevalence of some problems might be lower through face-to-face interviewing. Again, we considered randomization of treatments to be the most important attribute of the study.

Finally, we note that experiments of this type can be informative even with limited scopes and budgets. Our sample sizes were determined by available funding rather than statistical power calculations. Although in some cases larger samples might have produced more striking differences, even a few hundred cases were often sufficient to produce differences significant at the 0.05 level. We suggest that even small experiments conducted within such limits can still be very informative if driven by clear hypotheses, administered with randomized treatment, and carried out in samples that generally reflect response distributions and response difficulties of the larger population.

References

Beatty, P. (2010). Considerations regarding the use of global survey questions. Paper presented at the Bureau of Labor Statistics Consumer Expenditures Survey Methods Workshop (December 2010). http://www.bls.gov/cex/methwrkshp_pap_beatty.pdf

Beatty, P. (2013). I'm also a client: cognitive interviewing from user and client perspectives. Paper presented at the Question Evaluation Standards (QUEST) Workshop, Washington, DC (9–11 April 2013).

Beatty, P. and Maitland, A. (2008). The accuracy of decomposed vs. global behavioral frequency questions. Paper presented at the American Association for Public Opinion Research Conference held in New Orleans, LA (14–18 May 2008).

Beatty, P. and Willis, G.B. (2007). The practice of cognitive interviewing. *Public Opinion Quarterly* 71: 287–311.

Beatty, P., Fowler, F.J., and Fitzgerald, G. (1999). Construction strategies for complex survey questions. Paper presented at the 1999 AAPOR Conference held in St. Pete Beach, FL (13–16 May 1999). Proceedings of the Section on Survey Research Methods, American Statistical Association, 1999.

Belli, R.F., Schwarz, N., Singer, E., and Talarico, J. (2000). Decomposition can harm the accuracy of behavioural frequency reports. *Applied Cognitive Psychology* 14: 295–308.

Bradburn, N., Sudman, S., and Wansink, B. (2004). *Asking Questions: The Definitive Guide to Questionnaire Design—For Market Research, Political Polls, and Social and Health Questionnaires*. San Francisco, CA: Jossey-Bass.

Converse, J.M. and Presser, S. (1986). *Survey Questions: Handcrafting the Standardized Survey Questionnaire*. Newbury Park, CA: Sage.

Fowler, F.J. (1995). *Improving Survey Questions*. Thousand Oaks, CA: Sage.

Fowler, F.J. (2004). The case for more split-ballot experiments in developing survey instruments. In: *Methods for Testing and Evaluating Survey Questionnaires* (ed. S. Presser, J.M. Rothgeb, M.P. Couper, et al.). Hoboken, NJ: Wiley-Interscience.

Fowler, F.J. and Cannell, C.F. (1996). Using behavioral coding to identify cognitive problems. In: *Answering Questions: Methodology for Determining Cognitive Processes in Survey Research* (ed. N. Schwarz and S. Sudman). San Francisco, CA: Jossey-Bass.

Holbrook, A.L., Krosnick, J.A., Moore, D., and Tourangeau, R. (2007). Response order effects in dichotomous categorical questions presented orally: the impact of question and respondent attributes. *Public Opinion Quarterly* 71: 325–348.

Just, M.A. and Carpenter, P.A. (1992). A capacity theory of comprehension. *Psychological Review* 99: 122–149.
Krosnick, J.A. (2011). Experiments for evaluating survey questions. In: *Question Evaluation Methods: Contributing to the Science of Data Quality* (ed. J. Madans, K. Miller, A. Maitland and G.B. Willis). Hoboken, NJ: Wiley.
Narayan, S. and Krosnick, J.A. (1996). Education moderates some response effects in attitude measurement. *Public Opinion Quarterly* 60: 58–88.
Oksenberg, L., Cannell, C.F., and Kalton, G. (1991). New strategies for pretesting survey questions. *Journal of Official Statistics* 7: 349–365.
Payne, S.L. (1951). *The Art of Asking Questions*. Princeton, NJ: Princeton University Press.
Schuman, H. and Presser, S. (1981). *Questions and Answers in Attitude Surveys: Experiments in Question Form, Wording, and Context*. San Diego, CA: Academic Press.
Sudman, S., Bradburn, N., and Schwarz, N. (1996). *Thinking About Answers: The Application of Cognitive Processes to Survey Methodology*. San Francisco, CA: Jossey-Bass.
Tourangeau, R., Rips, L., and Rasinski, K. (2000). *The Psychology of Survey Response*. Cambridge: Cambridge University Press.
Willis, G.B. (2015). *Analysis of the Cognitive Interview in Questionnaire Design*. New York: Oxford University Press.

7

Impact of Response Scale Features on Survey Responses to Behavioral Questions

Florian Keusch[1] and Ting Yan[2]

[1] *Department of Sociology, School of Social Sciences, University of Mannheim, Mannheim, Germany*
[2] *Westat, Rockville, MD, USA*

7.1 Introduction

When designing a response scale, question writers have to make design decisions pertaining to the number of scale points (Krosnick and Fabrigar 1997), the use of numeric labels (Schwarz et al. 1991; O'Muircheartaigh et al. 1995; Schwarz and Hippler 1995; Schwarz et al. 1998), the assignment of verbal labels to all or some of the scale points (Krosnick and Berent 1993; Krosnick and Presser 2010), the spacing of response options (Daamen and de Bie 1992; Tourangeau et al. 2004), the shading of response options (Tourangeau et al. 2007), and the alignment of the scale or the decision to present scales horizontally or vertically on a screen or paper (Christian et al. 2009). Survey literature has demonstrated that many of these design features of response scales affect how survey respondents process the scale and how they use these features to construct their responses. Features of a response scale can potentially contribute to measurement error in surveys.

This chapter takes advantage of an experiment on a probability online panel to examine both the main effects and the joint effects of three scale features on answers to behavioral questions using a frequency scale. The first design feature is the direction of a frequency scale. A frequency scale could descend from the highest to the lowest point (e.g. from "all of the time" to "never"). Hofmans and colleagues refer to this format as decremental scales (Hofmans et al. 2007). The same scale could also ascend from the lowest to the highest point (e.g. "never" to "all of the time"); this format is designated as incremental scales (Hofmans et al. 2007). The first research question to be answered by our experiment is whether or not the direction of a frequency scale affects survey responses, holding other features of the scale constant. The experimental setting of the study also allows us to examine measurement error in the answers to these behavioral questions, validity of the answers in particular, under different scale directions.

A second feature of particular relevance to web surveys is scale alignment – the presentation of scales on a computer screen. Scales can be shown either horizontally or vertically. While horizontally presented scales were very common in web surveys for a long time, especially to present several items with the same scale in a grid, vertical rating scales are now more often used on smaller, upright screens of smartphones to avoid horizontal scrolling (Fuchs 2008; Peytchev and Hill 2010). The second and the third research questions to be answered in this chapter are:

Experimental Methods in Survey Research: Techniques that Combine Random Sampling with Random Assignment, First Edition.
Edited by Paul J. Lavrakas, Michael W. Traugott, Courtney Kennedy, Allyson L. Holbrook, Edith D. de Leeuw, and Brady T. West.

Companion Website: www.wiley.com/go/Lavrakas/survey-research

(2) whether or not scale alignment affects answers to survey questions and (3) whether or not scale alignment moderates the effects of scale direction.

A third scale feature is the vagueness of scale labels. Survey questions measuring frequency of behaviors either ask respondents to report an exact frequency in terms of numbers (e.g. the number of times something happened in a specific time period) or to choose from a list of vague quantifiers (such as often, rarely, etc.). The fourth and fifth research questions to be answered are: (4) whether or not the vagueness of scale labels affects answers to survey questions and (5) whether or not the vagueness of scale labels moderates the effects of scale direction.

We also look at several respondent characteristics (age and education) and paradata (response times) that are highly indicative of respondents' likelihood to satisfice. This will help us to answer the sixth research questions: (6) whether or not the impact of these scale features is stronger for those with a higher likelihood to satisfice.

7.2 Previous Work on Scale Design Features

7.2.1 Scale Direction

Research on the influence of scale direction on survey responses dates back to the 1960s when Belson (1966) found empirically that survey responses tended to shift toward the starting point of a rating scale regardless of whether that starting point is the low/negative end or the high/positive end of the scale. Later studies on scale direction effects turned up mixed evidence – scale direction effects are observed in some self-administered paper-pencil surveys (e.g. Belson 1966; Chan 1991; Friedman et al. 1993, 1988; Israel 2006; Krebs and Hoffmeyer-Zlotnik 2010; Sheluga et al. 1978) but not in others (Dickson and Albaum 1975; Israel and Taylor 1990; Powers et al. 1977; Weng and Cheng 2000). For web surveys, four studies show a significant impact of scale direction on answers (Hofmans et al. 2007; Liu and Keusch 2017; Toepoel et al. 2009; Stapleton 2013), and one study fails to demonstrate this effect (Ramstedt and Krebs 2007). Results are mixed in Malhotra (2009), Christian et al. (2009), Höhne and Krebs (2017), and Krebs (2012), who found a significant impact of scale direction on some questions, but not others.

All but one study examines scale direction effects in attitudinal items. Carp (1974) is the only study that looks into the impact of scale direction on behavioral questions. The study examines 10 questions on frequency of trips using an 8-point fully labeled frequency scale but fails to find evidence indicating that scale direction affects answers to these behavioral questions. More research is needed before one can conclude that scale direction has no impact on answers to behavior items.

Scale direction effects tend to be considered as a special case of response order effects attributed to satisficing (Krosnick 1999; Krosnick and Presser 2010; Krebs and Hoffmeyer-Zlotnik 2010). Empirically, there is little evidence that scale direction effects are stronger under conditions that are conducive to satisficing (e.g. fast interview pace in Mingay and Greenwell (1989), survey questions placed toward the end of the questionnaire as in Carp (1974) and Yan and Keusch (2015)). One exception is Malhotra (2008), who reports that respondents with low-formal education who speeded through the questionnaire showed the strongest scale direction effects on items employing unipolar rating scales. In the current research, we extend Malhotra's work to examine whether scale alignment and labels in behavioral questions vary as a function of variables (e.g. education or speeding) that are associated with satisficing.

7.2.2 Scale Alignment

Research on scale alignment shows only minor differences between vertically and horizontally presented rating scales in a web survey. Funke et al. (2011), for instance, study the alignment of a visual analog scale and a 7-point rating scale and find no difference in break-off rate, mean ratings, response distribution, and tendency to choose the middle scale point by scale alignment. However, ratings on horizontal scales take significantly less time to complete than on vertical scales. Two studies show that respondents are more likely to select response options at the bottom of a vertical scale than on the right side of a horizontal scale (Scott and Huskisson 1979; Toepoel et al. 2009). This effect is stronger among respondents aged 65 and older (Toepoel et al. 2009). Peytchev and Hill (2010) demonstrate that scale alignment affects answers provided on a smartphone only when the scale is partially visible and respondents have to scroll to see the entire scale.

Furthermore, Tourangeau et al. (2004, 2007) indicate that respondents apply the "up means good" heuristic when using a vertical scale. The heuristic assumes that respondents expect the upmost response option in a vertically presented scale to be the most positive one. Christian et al. (2009) provide empirical support to the use of this heuristic by demonstrating that respondents take longer to cognitively process a scale when it is presented in a format inconsistent with their expectations (i.e. the negative end of the vertical rating scale is shown on top). They also show that scale alignment moderates the scale direction effect – scale direction has a significant impact on answers when scales are presented vertically rather than horizontally. Höhne and Lenzner (2015) confirmed this interaction effect in an eye-tracking study with attitudinal questions.

7.2.3 Verbal Scale Labels

Survey questions asking about relative frequencies employ vague quantifiers such as very often, pretty often, seldom, rarely, and so on. Studies examining vague quantifiers reveal three issues with the use of vague quantifiers in frequency scales (e.g. Bradburn and Miles 1979; Schaeffer 1991). First, respondents can assign different absolute values to the same quantifiers. In other words, the same vague quantifier (e.g. very often) might mean different frequencies to different people. Second, the distance between adjacent quantifiers might not be perceived as being equal (that is, the distance between "very often" and "pretty often" might not be the same as that between "pretty often" and "seldom"). Third, the meaning of quantifiers may change in relation to the overall frequency of the event in question. For instance, drinking coffee "very often" implies a different absolute frequency (e.g. twice a day) than attending college football games "very often" (e.g. once a week during the football season). Consequently, many survey researchers argue against the use of vague quantifiers to measure frequency (Tourangeau et al. 2000). However, the latest work on frequency scales using vague quantifiers demonstrates that they can perform very well in terms of accuracy and predictive validity. Lu et al. (2008) show that a 6-point vague quantifier scale performs as well as an 11-point numeric percentage scale with regard to mean differences between reported and actual results. Al Baghal (2014a) compares the use of a 6-point frequency scale with vague quantifiers to an open-ended numeric response format in a recall test asking respondents to rate how often a word appears in a list. He finds that the vague quantifiers improve accuracy relative to the numeric open-ended response format. In a second study with data from the National Survey of Student Engagement, Al Baghal (2014b) shows that responses to frequency questions on active-collaborative learning and student–faculty interaction have higher levels of predictive validity for grades and satisfaction measures when vague quantifiers are used than when numeric open-ended responses are used.

To the best of our knowledge, there is no study yet examining the moderating effect of vagueness of scale labels on scale direction effects. However, survey literature has shown that cooperative respondents tend to use other scale features to interpret vague and ambiguous scale labels such as numerical scale labels in Schwarz et al. (1991), spacing and order in Tourangeau et al. (2004), and shading in Tourangeau et al. (2007). Scale direction is potentially another cue that can be used to interpret vague labels and to construct answers. The anchoring-and-adjustment heuristic (Tversky and Kahneman 1974; Yan and Keusch 2015), for instance, suggests that the start of a response scale is used by respondents as an anchor to form an answer. Respondents might be more likely to use the start of the scale to anchor their answers to scales employing vague quantifiers. If that were the case, we would expect stronger scale direction effects in scales using vague quantifiers than scales with precise scale labels.

7.3 Methods

Data for this study come from an experiment conducted between 16 November 2013 and 27 November 2013 on GfK's KnowledgePanel (KP). KP is a large-scale online panel based on a representative sample of the U.S. population. Panel members are recruited through telephone random digit dialing (RDD) sampling and postal mail via address-based sampling (ABS) methodologies. Households are provided with access to the Internet and a netbook computer if they do not have Internet access or a computer at the time of recruitment. Only persons sampled through these probability-based techniques are eligible to join KP (GfK 2013).

A sample of 2664 panel members was randomly drawn from KP. The recruitment rate for this study was 14.2% and the profile rate – the rate of panel members answering the GfK "core profile survey" – was 65.6%. A total of 1729 invited panel members started and completed the web questionnaire, yielding a completion rate of 64.9%. The cumulative response rate for this study was 6.1% (=14.2% × 65.6% × 64.9%).

Panel members already having own computers and Internet access when joining KP received the standard cash-equivalent postsurvey incentive, an amount equivalent of $1–1.50 depending on when they joined the panel. Those who did not have computers and Internet access prior to joining KP received laptop computers and Internet access upon joining the panel and were not provided additional monetary incentives for completing this survey.

Our experiment was implemented on two sets of survey item: (i) six items modeled after the K6 inventory on nonspecific psychological distress (Kessler et al. 2002) used in the National Health Interview Survey (NHIS) and (ii) four health related behavioral items (question wordings are displayed in Appendix 7.A). Both sets of questions used a 5-point fully labeled unipolar frequency scale with radio buttons for the respondents to select a response. Each question was presented on an individual screen. Responses to the experimental items were coded so that lower values represent lower frequencies and higher values represent higher frequencies (1 = "never"/"zero days"/"never [zero days]"; 5 = "always"/"seven days"/"all of the time [seven days]"). In addition, we created three indices by summing up the responses to the six items on nonspecific psychological distress, the two items on negative health behaviors ("eat foods that are high in fat and/or calories" and "eat fast food"), and the two items on positive health behaviors ("eat a variety of fresh fruits and vegetables" and "do physical exercise"). For all three indexes, higher values indicate higher reporting of nonspecific psychological distress, negative health behaviors, and positive health behaviors.

A full factorial experiment crossed three scale features for both sets of items. The first scale feature is the direction of the frequency scales. For a random half of the sample, the scale descended from high to low (e.g. "all of the time" – "most of the time" – "some of the time" – "a little of the time" – "never"). For the other half, the scale ascended from low to high (e.g. "never" – "a little of the time" – "some of the time" – "most of the time" – "all of the time"). The second feature varied in our experiment is the alignment of the frequency scale on the computer screen. The frequency scale was presented either horizontally or vertically. The third factor varies the vagueness of scale verbal labels, creating three versions. One version employed only vague quantifiers ("all of the time" – "most of the time" – "some of the time" – "a little of the time" – "never"). The second version showed precise frequency labels ("seven days" – "five or six days" – "three or four days" – "one or two days" – "zero days"). In the third version, a combination of both labels was used ("all of the time [seven days]" – "most of the time [five or six days]" – "some of the time [three or four days]" – "a little of the time [one or two days]" – "never [zero days]"). Crossing the three experimental factors led to 12 different versions of the questionnaire. Respondents were randomly assigned to 1 of the 12 versions and answered all questions of the questionnaire in this condition. The key outcome variable for this experiment are responses (i.e. distributions and means) to the 10 target questions on nonspecific psychological distress and health behaviors under different experimental conditions as well as the three indices.

After the experimental questions, a set of 15 filler items was presented to the respondents. These questions were displayed in a grid on two consecutive screens. In order to not confuse the respondents, the direction of the response scales used for the filler questions was consistent with the experimental conditions assigned to the experimental items that came before. That is, if the scale ascended from low to high in the experimental questions, the response scale for these filler items also ascended from low to high, and vice versa. Filler questions were, however, not subject to the alignment and label experiment.

To directly assess the influence of scale features on measurement error, we use different validation measures. Ten open-ended frequency questions asked respondents to report a numeric value ("During an average week, on how many days do you …?") to the same domains of nonspecific psychological distress and health behaviors of the experimental questions.[1] These questions are intended for validation purpose since the responses to the open-ended numeric questions are free from the impact of scale features manipulated in this experiment. We, therefore, use the correlations between responses to the experimental questions and the open-ended frequency questions as indicators of validity with higher correlations indicating more valid responses to the experimental questions. This set of correlational analyses allows us to tap into the quality of the answers obtained under various scale features when true values are not available.

Two variables from the GfK profile survey that every panel member has to fill out upon enrollment are also used to examine the validity of answers to the experimental questions. Health behaviors have proven to be highly correlated with self-rated health (e.g. Sargent-Cox et al. 2014). In the GfK profile survey, self-rated health was measured on a 5-point fully labeled scale ("excellent" – "very good" – "good" – "fair" – "poor"). We calculate the correlation between the two health behavior indices and the self-rated health question from the GfK profile to further examine the effect of scale design features on the validity of the survey responses. The K6 items on nonspecific psychological distress were developed to identify cases of serious mental illness, such as anxiety and depression (Kessler et al. 2002). Diagnosis for depression had been measured as a yes/no-question in the GfK profile survey. We use the correlation between the index of nonspecific psychological distress from our experiment and the diagnosis of depression

1 Since the validation questions were asked after the experimental manipulation of the scale features, we tested whether the manipulation had an influence on responses to the validation questions. We ran ANOVAs with the responses to the validation questions as dependent variables and dummy variables for the scale features as independent variables and found that the manipulations were independent of the validation questions ($p > 0.05$).

reported in the GfK profile survey as another indicator of validity for survey responses under different scale design conditions.

Finally, we select age, education, and time spent answering the questionnaire as indicators of satisficing. Previous research demonstrates that respondents who speed through a questionnaire (speeders) are more likely to exhibit primacy effects in unordered response options and show more straightlining behavior in multi-item scales (Malhotra 2008; Calegarro et al. 2009; Kaminska et al. 2010; Zhang and Conrad 2014). In this chapter, we define speeders as respondents who spend less than 300 ms/word to complete the survey, as suggested by Zhang and Conrad (2014). Our questionnaire consisted of 523 words, setting the threshold for speeding to 157 seconds (=300 ms/word times 523 words). About 11% of respondents answered the questionnaire faster than the threshold and are flagged as speeders. Previous research finds that older respondents (Knäuper 1999; Krosnick 1991, 1999) and respondents with low education (Krosnick 1991, 1999; Mingay and Greenwell 1989) are more likely to satisfice than younger respondents and respondents with high education, respectively. For our analysis, we dichotomized on age and divided respondents into two groups – respondents aged 65 or older and respondents less than 65 years old. Similarly, we dichotomized on education and divided respondents into two groups – respondents with at least some college and respondents who had a high school degree or less. Nineteen percent of respondents were over 65 years of age and 39% of respondents had a high school degree or less.

All analyses were conducted using R version 3.2.2 (R Core Team 2015). We calculated weighted estimates in our analysis applying the weights provided by KP to correct for biases in sampling and nonresponse in several stages of panel recruitment and survey participation. The weights consist of a base weight adjusting for known sources of deviation from an equal probability of selection design, a panel demographic poststratification weight based on demographic distributions from the most recent data from the Current Population Survey that reduces the effects of nonresponse and noncoverage bias in the overall panel membership before the study sample is drawn, and a set of study-specific poststratification weights to adjusting for the study's sample design and study-specific nonresponse (GfK 2013). When applying KP survey weights, we used the survey package (Lumley 2014) in R, unless otherwise specified.

7.4 Results

Given the nature of the experiment, we first conducted three-way ANOVAs on the weighted answers to each of the 10 items and the 3 indexes to examine the main effect of the 3 experimental variables and their interaction effects. As can be seen in the summary of Table 7.1, the only scale feature that consistently has an influence on responses is the vagueness of scale labels. The vagueness of scale labels has a significant main effect on responses to two (out of the six) items on nonspecific psychological distress, three (out of the four) health behavior items, and all three indices. The main effect of scale direction, main effect of scale alignment, and interaction effects between the three scale features are not consistently statistically significant at the $p = 0.05$ level (see Appendix 7.B for detailed ANOVA results).

To further understand how the vagueness of scale labels affects answers, we next examined response distributions to all 10 items by scale label manipulation. Visual inspection of Figures 7.1 and 7.2 showing the weighted distribution of responses reveals that precise scale labels lead to more selection of the extreme scale points than vague or vague + precise scale labels. On the other hand, vague scale labels attract more selection of the middle category than the other two labels. Vague labels consistently yield higher mean values on individual items and indexes compared to precise labels and, to some extent, also vague + precise labels, as shown in Table 7.2.

Table 7.1 Number of significant effects in 10 ANOVAs.

Main effects	
Scale direction	0
Scale alignment	0
Vagueness of scale labels	8
Two-way interaction	
Scale direction × scale alignment	1
Scale direction × vagueness of scale labels	1
Scale alignment × vagueness of scale labels	0
Three-way interaction	
Scale direction × scale alignment × vagueness of scale labels	0

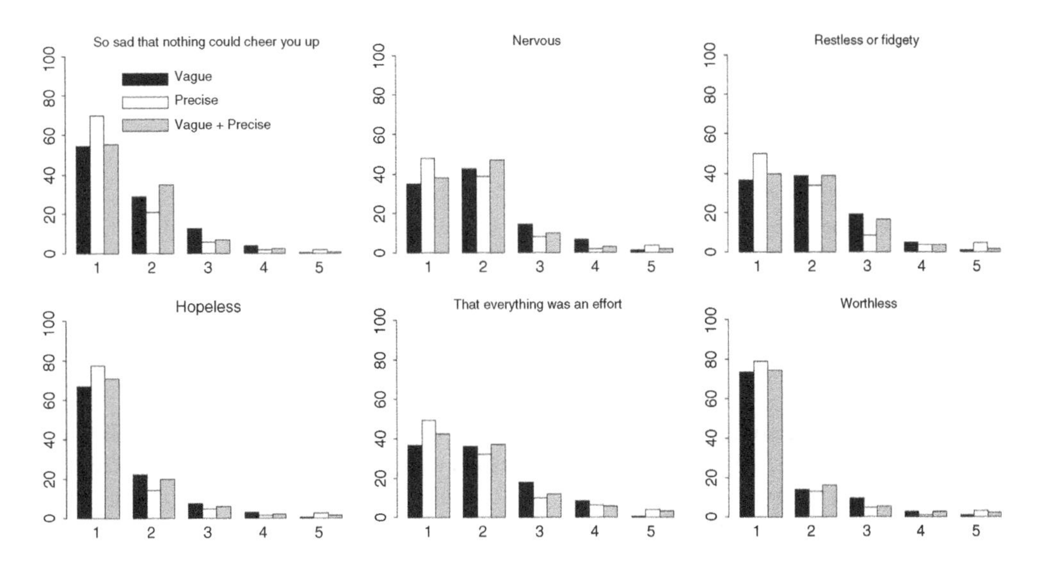

Figure 7.1 Weighted responses to nonspecific psychological distress questions by scale labels.

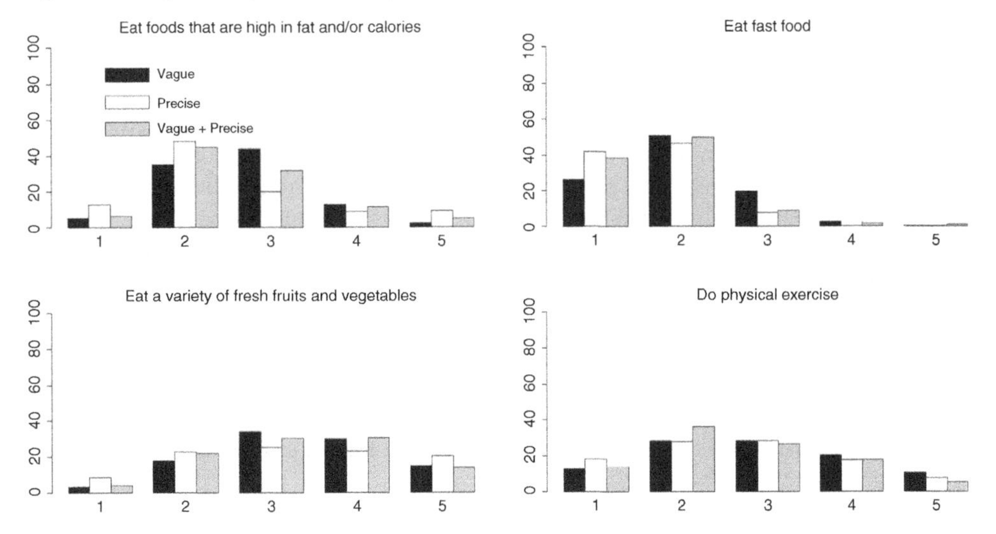

Figure 7.2 Weighted responses to health behavior questions by scale labels.

Table 7.2 Weighted mean responses by vagueness of scale labels.

	Vagueness of verbal labels			Design-based *t*-tests[a)]
	Vague	Precise	Vague + precise	
During an average week, how often do you feel …				
… so sad that nothing could cheer you up?	1.66	1.44	1.58	a, c
… nervous?	1.96	1.74	1.83	a
… restless or fidgety?	1.95	1.78	1.89	
… hopeless?	1.48	1.38	1.43	
… feel that everything was an effort?	1.99	1.84	1.90	
… worthless?	1.43	1.37	1.43	
Nonspecific psychological distress index	10.46	9.54	10.01	a
During an average week, how often do you …				
… eat foods that are high in fat and/or calories?	2.72	2.54	2.64	a
… eat fast food?	2.00	1.72	1.76	a, b
… eat a variety of fresh fruits and vegetables?	3.35	3.24	3.28	
… do physical exercise?	2.88	2.69	2.65	b
Negative health behavior index	4.73	4.26	4.40	a, b
Positive health behavior index	6.24	5.92	5.92	b

Notes: Unweighted $N = 1729$. 1 = lowest value ("never"/"0 d"/"never [0 d]"); 5 = highest value ("all of the time"/"7 d"/"all of the time [7 d]").

a) … Results of pairwise design-based *t*-test with Bonferroni correction for multiple comparisons: (a) …"vague" vs. "precise" ($p < 0.05$); (b) …"vague" vs. "vague + precise" ($p < 0.05$); (c) …"precise" vs. "vague + precise" ($p < 0.05$).

To examine whether or not the three scale features have a stronger impact among people who are more likely to satisfice, we fit two MANOVA models with the unweighted responses to the six items on nonspecific psychological distress and the four health behavior items as outcome variables. We specify both main effects models including indicators of the three experimental variables plus indicators for speeding, age, and education as well as full models that also include two-way interaction terms of the experimental variables with each other and with age, education, and speeding, and a three-way interaction term of the experimental variables.[2]

Tables 7.3 and 7.4 confirm that vagueness of scale labels has a significant main effect on responses after controlling for experimental variation of scale direction and scale alignment as well as for indicators of satisficing. While the indicators of satisficing (age and education) have significant main effects on the survey responses, only education is found to moderate the effect of scale direction on nonspecific psychological distress items and to moderate the effect of scale alignment on health behavior items. Interestingly, for the nonspecific psychological distress items, people with at least some college tended to have higher means when answering the scale descending from high to low than when answering the scale ascending from low to high (one of the seven mean comparisons were statistically significant at $p < 0.1$), indicating that they

2 Malhotra (2008) found that that low-formal education in combination with speeding leads to strong scale direction effects. We specified models that included a two-way interaction term for education by speeding and a three-way interaction term for education by speeding by scale direction. Since neither of the two additional interaction terms are statistically significant and results on other main effects and interaction effects are the same, we presented here models without these interaction terms.

selected response options closer to the start of the scale regardless of the direction of the scale. For low-educated people this effect was reversed with generally lower means on the descending scale compared to the ascending scale (three of the seven mean comparisons were statistically significant at $p < 0.1$). For the health behavior items, people with high school or less had higher means if the scale was presented vertically compared to if the scale was presented horizontally (two of the six mean comparisons were statistically significant at $p < 0.1$). For people with at least some college, no differences in mean responses by scale alignment were observed ($p > 0.1$).

Finally, we examined the validity of the answers to the 10 experimental questions through correlations. We first calculated correlations between the 10 experimental items and their equivalent open-ended frequency items for each of the 12 experimental conditions using the wtd.cors function from the weights package in R (Pasek et al. 2014), yielding a total of 120 weighted correlation coefficients. We then regressed the weighted correlation coefficients on the three

Table 7.3 MANOVA results for responses to nonspecific psychological questions.

	Main effects model			Interaction effects model		
	Wilks	Approx. *F*		Wilks	Approx. *F*	
Main effects						
Scale direction	0.998	0.52		0.998	0.52	
Scale alignment	0.999	0.37		0.999	0.38	
Vagueness of scale labels	0.983	2.41	**	0.983	2.41	**
Age: 65+ years	0.983	4.80	***	0.983	4.80	***
Education: HS degree or less	0.987	3.79	***	0.986	3.79	***
Speeder	0.994	1.73		0.994	1.73	
Two-way interactions						
Scale direction × scale alignment				0.997	0.87	
Scale direction × vagueness of scale labels				0.997	0.47	
Scale alignment × vagueness of scale labels				0.994	0.77	
Scale direction × age				0.999	0.31	
Scale direction × education				0.989	3.10	**
Scale direction × speeding				0.995	1.33	
Scale alignment × age				0.997	0.87	
Scale alignment × education				0.996	1.21	
Scale alignment × speeding				0.998	0.45	
Vagueness of scale labels × age				0.993	0.95	
Vagueness of scale labels × education				0.991	1.18	
Vagueness of scale labels × speeding				0.996	0.54	
Three-way interaction						
Scale direction × scale alignment × vagueness of scale labels				0.993	0.99	

Notes: $N = 1665$; * ...$p < 0.05$; ** ...$p < 0.01$; *** ...$p < 0.001$.

Table 7.4 MANOVA results for responses to health behavior questions.

	Main effects model			Interaction effects model		
	Wilks	Approx. *F*		Wilks	Approx. *F*	
Main effects						
Scale direction	0.999	0.54		0.999	0.54	
Scale alignment	0.999	0.56		0.999	0.56	
Vagueness of scale labels	0.962	8.23	***	0.962	8.22	***
Age: 65+ years	0.964	15.59	***	0.964	15.58	***
Education: HS degree or less	0.966	14.83	***	0.966	14.83	***
Speeder	0.994	2.34		0.994	2.33	
Two-way interactions						
Scale direction × scale alignment				0.995	1.92	
Scale direction × vagueness of scale labels				0.998	0.51	
Scale alignment × vagueness of scale labels				0.991	1.72	
Scale direction × age				0.998	0.63	
Scale direction × education				0.999	0.32	
Scale direction × speeding				0.998	1.00	
Scale alignment × age				0.999	0.13	
Scale alignment × education				0.994	2.48	*
Scale alignment × speeding				0.997	1.33	
Vagueness of scale labels × age				0.998	0.31	
Vagueness of scale labels × education				0.994	1.18	
Vagueness of scale labels × speeding				0.997	0.72	
Three-way interaction						
Scale direction × scale alignment × vagueness of scale labels				0.997	0.54	

Notes: $N = 1691$. * ...$p < 0.05$; ** ...$p < 0.01$; *** ...$p < 0.001$.

scale features (i.e. scale direction, scale alignment, and vagueness of verbal scale labels). Model results are shown in the left pane of Table 7.5. Similarly, we calculated correlations between the index of nonspecific psychological distress and the depression indicator from the profile survey, and between the two health behavior indices and the self-rated health measure[3] from the profile survey under each of the 12 experimental conditions, producing a total of 36 correlations. We again regressed the weighted correlation coefficients on the three scale features and the regression results are displayed on the right pane of Table 7.5.

The left pane of Table 7.5 shows that the vagueness of verbal scale labels has a significant influence on the strength of the correlations between the experimental items and the open-ended frequency items. Using precise and vague + precise verbal labels in the rating scale leads to significantly stronger correlations with responses from the open-ended frequency questions. This

3 Responses to the two questions that form the index of negative health behavior ("eat foods that are high in fat and/or calories" and "eat foods that are high in fat and/or calories") were recoded so that all correlation coefficients would go in the same direction.

Table 7.5 Results from linear regression of weighted correlation coefficients between experimental items and open-ended frequency items (left) and indexes and GfK profile survey items (right).

	DV: Weighted correlation coefficients between experimental items and open-ended frequency items		DV: Weighted correlation coefficients between indexes and GfK profile survey items
	β (s.e.)		β (s.e.)
Intercept	0.77 (0.02)	***	0.14 (0.10)
Scale direction			
High to low	—		—
Low to high	−0.03 (0.01)		0.02 (0.09)
Scale alignment			
Horizontal	—		—
Vertical	−0.02 (0.01)		0.01 (0.09)
Vagueness of scale labels			
Vague	—		—
Precise	0.12 (0.02)	***	0.01 (0.11)
Vague + precise	0.08 (0.02)	***	0.06 (0.11)
R^2	0.31		0.01
Adjusted R^2	0.29		−0.11
N	120		36

Notes: * ...$p < 0.05$; ** ...$p < 0.01$; *** ...$p < 0.001$.

is not surprising because the open-ended frequency items ask for the number of days in the past week respondents felt a certain way or did a certain thing, which corresponds to what is shown on the precise scale labels and vague + precise labels. Scale direction and scale alignment have no significant effect on the strength of correlations between responses to the experimental frequency scale questions and the open-ended frequency questions.

As the right pane of Table 7.5 shows, the experimental scale features did not predict correlation coefficients between indices and the GfK profile variables. An explanation for this finding might be that, in general, the correlations between the indexes and the GfK profile survey items were relatively low.

7.5 Discussion

The experiment presented in this chapter manipulates three scale features – scale direction, scale alignment, and vagueness of verbal scale labels. We found that responses to six questions on the frequency of experiencing a certain type of nonspecific psychological distress and four questions on the frequency of health behaviors are neither affected by the direction of the frequency scale (i.e. whether the scale ascends from the lowest to the highest point or descends from the highest to the lowest point) nor by the alignment of the scale (i.e. whether the scale is presented vertically or horizontally on the screen). Contradictory to the findings in earlier research on attitudinal questions (Christian et al. 2009; Höhne and Lenzner 2015), we did not find an interaction between these two scale features. Our results indicate that neither a specific direction of the scale nor one alignment contribute to measurement error in questions about behaviors.

However, responses to the frequency scale questions are affected by the verbal labels used for the scale. Precise scale labels, such as "seven days" – "five or six days" – "three or four days" – "one or two days" – "zero days," lead to more selection of the extreme scale points than vague quantifiers ("never" – "a little of the time" – "some of the time" – "most of the time" – "all of the time") or a combination of vague and precise scale labels. Vague scale labels produce, by contrast, more selection of the middle category compared to the two other types of labels. In addition, there is some evidence that education moderates the effects of scale features on responses to behavioral survey items; interestingly, higher educated people showed stronger scale direction effects than people with lower education.

We used correlations between answers to the 10 target items with answers to 10 open-ended frequency items and correlations between three indices and profile variables collected earlier in the online panel as indicators of validity. Again, scale direction and scale alignment are found to have only very minor influence on the validity of the collected data. Vague quantifiers produce lower correlations between the experimental questions and open-ended frequency questions on the same domains, indicating more variation in absolute values assigned to vague quantifiers.

Three points warrant additional discussion. First, our experiment is conducted via GfK's KP, a large-scale online panel based on a representative probability sample of the U.S. population. What makes KP special and different from most online panels is that KP is not a volunteer panel, and it includes people with and without Internet access and ownership of a computer. Mullinix et al. (2015) compare the results of experimental studies conducted in KP to those from experiments conducted on several convenience samples (e.g. from Amazon's Mechanical Turk). They find that results from the convenience samples generally provide estimates of causal effects comparable to those found on the population-based samples from KP. However, population-based survey experiments, such as the ones conducted in KP, have a number of advantages over experiments using convenience samples. They allow researchers to generalize the experimental findings to a larger population, to account for heterogeneous treatment effects that might exist in different subpopulations, and to ensure stability of the sample over time. Our experiment is such a population-based survey experiment that simultaneously manipulates three scale features. It enables us to empirically examine the main effect of scale direction, scale alignment, and vagueness of verbal scale labels on survey responses to behavioral survey questions as well as the interaction between the three scale design features. Because of the nature of KP, the findings reported here are high on both external validity and internal validity. They benefit not only survey researchers but also researchers and scientists in various disciplines that rely on survey data collection, such as psychology, political science, public health, and education.

Second, the findings from our experiment contribute new knowledge to the survey literature and have important practical implications for the survey field. We added to the literature of scale direction effects one more experimental study looking into the effects of scale direction on answers to behavioral items. Survey literature has shown that scale direction seems to play a crucial role in the formation of answers to attitudinal item. Our experiment provides more evidence indicating that answers to behavioral items are not affected by scale direction, consistent with Carp (1974). Of course, we did not experimentally manipulate the type of survey items, and we could not conclude that behavioral items are in fact less prone to scale direction than attitudinal items. We can also only speculate about why attitudinal items and behavioral questions might be affected differently by scale direction. One potential explanation is that respondents do not have existing attitudes that are stored in memory and they tend to construct attitudes on the fly (Schwarz 2007; Zaller 1992). By contrast, behavioral questions ask respondents to recall behaviors that already happened. As a result, self-report of attitudes are believed to be more prone to context effects than reports of behaviors.

We further demonstrate that scale alignment does not affect survey answers, consistent with Funke et al. (2011) and Peytchev and Hill (2010). Different from recent work on vague quantifiers (Lu et al. 2008; Al Baghal 2014a, b), we find that vague quantifiers produce data of lower validity than precise verbal scale labels.

Third, we could not find any indication that satisficing contributes to differences in responses to frequency questions using different scale features. While researchers on scale direction effects have cited satisficing as the cause, we did not find stronger scale direction effects on behavioral questions for respondents most prone to satisficing. Answers to frequency scales do differ depending on the verbal labels attached to the scale points, but we do not find evidence indicating that these differences are caused by the use of cognitive shortcuts when responding (i.e. satisficing). Age and speeding do not moderate the effect of the vagueness of verbal scale labels, and the moderating effect of education is the opposite of what would be expected as evidence for satisficing. One potential explanation why we did not find much evidence of satisficing in our study could be that satisficing in general tends to be lower on the Internet than, for example, on the telephone (Chang and Krosnick 2009). More research is needed to further understand how respondents process and use response scale features to arrive at their answers in web surveys.

Even though the set-up of our study in the probability-based KP allows us to generalize the results of our experiment to the larger U.S. adult population, we cannot generalize beyond the items used in this study. In particular, our findings are limited to the six items measuring nonspecific psychological distress modeled after NHIS questions and the four health related behavioral items used in our study. Future research needs to replicate our findings for other behavioral domains frequently measured in social surveys, such as consumer expenditure or leisure activities. In addition, we encourage researchers to experimentally vary the use of frequency and attitudinal scales for questions on the same domain to validate whether frequency items are less prone to variation of design features of a scale compared with attitudinal items.

Acknowledgment

Data collection for this study was supported by a grant through TESS's Special Competition for Young Investigators in 2013 to the first author. TESS (Time-sharing Experiments for the Social Sciences) is funded by the National Science Foundation (SES-0818839).

7.A Question Wording

7.A.1 Experimental Questions (One Question Per Screen)

Q1a. Now we are going to ask you some questions about feelings you may have experienced during an average week. During an average week, how often do you feel so sad that nothing could cheer you up?

Q1b. During an average week, how often do you feel nervous?

Q1c. During an average week, how often do you feel restless or fidgety?

Q1d. During an average week, how often do you feel hopeless?

Q1e. During an average week, how often do you feel that everything was an effort?

Q1f. During an average week, how often do you feel worthless?

Q2a. During an average week, how often do you eat foods that are high in fat and/or calories?

Q2b. During an average week, how often do you eat fast food?

Q2c. During an average week, how often do you eat a variety of fresh fruits and vegetables?

Q2d. During an average week, how often do you do physical exercise?

Table 7.A.1 Experimental variation of response categories.

Scale direction	Scale alignment	Vagueness of verbal labels
High to low For example, "all of the time" – "most of the time" – "some of the time" – "a little of the time" – "never"	*Vertical*	*Vague quantifiers* "all of the time" – "most of the time" – "some of the time" – "a little of the time" – "never" *Precise frequency* "7 d" – "5 or 6 d" – "3 or 4 d" – "1 or 2 d" – "0 d"
Low to high For example, "never" – "a little of the time" – "some of the time" – "most of the time" – "all of the time"	*Horizontal*	*Vague quantifiers + precise frequency* "All of the time (7 d)" – "most of the time (5 or 6 d)" – "some of the time (3 or 4 d)" – "a little of the time (1 or 2 d)" – "never (0 d)"

7.A.2 Validation Questions (One Per Screen)

Q5a. During an average week, on how many days do you feel so sad that nothing could cheer you up? ___ days

Q5b. During an average week, on how many days do you feel nervous? ___ days

Q5c. During an average week, on how many days do you feel restless or fidgety? ___ days

Q5d. During an average week, on how many days do you feel hopeless? ___ days

Q5e. During an average week, on how many days do you feel that everything was an effort? ___ days

Q5f. During an average week, on how many days do you feel worthless? ___ days

Q6a. During an average week, on how many days do you eat foods that are high in fat and/or calories? ___ days

Q6b. During an average week, on how many days do you eat fast food? ___ days

Q6c. During an average week, on how many days do you eat a variety of fresh fruits and vegetables? ___ days

Q6d. During an average week, on how many days do you do physical exercise? ___ days

7.A.3 GfK Profile Questions (Not Part of the Questionnaire)

In general, would you say your physical health is ...

Excellent – Very good – Good – Fair – Poor
Have you been diagnosed with any of the following medical conditions? Depression?
No – Yes

7.B Test of Interaction Effects

Table 7.B.1 Weighted ANOVA results for responses to nonspecific psychological distress questions.

		During an average week, how often do you feel...									Nonspecific psychological distress Index	
		So sad that nothing could cheer you up		Nervous		Restless or fidgety	Hopeless	That everything was an effort	Worthless			
	df	Rao-Scott LRT		Rao-Scott LRT		Rao-Scott LRT	Rao-Scott LRT	Rao-Scott LRT	Rao-Scott LRT		Rao-Scott LRT	
Main effects												
Scale direction	1	1.62		0.59		1.82	<0.01	<0.01	1.43		12.99	
Scale alignment	1	0.02		0.04		0.58	0.06	<0.01	<0.01		0.78	
Vagueness of scale labels	2	14.20	***	14.05	**	8.02	2.65	6.28	1.20		241.80	*
Two-way interaction												
Scale direction × scale alignment	1	2.19		0.51		0.15	1.04	0.01	0.30		7.50	
Scale direction × vagueness of scale labels	2	0.92		8.38	*	2.91	0.49	0.09	0.41		14.59	
Scale alignment × vagueness of scale labels	2	0.45		4.84		3.90	0.42	3.81	0.69		38.35	
Three-way interaction												
Scale direction × scale align-ment × vagueness of scale labels	2	1.91		1.17		1.00	1.22	0.39	0.55		3.56	
Unweighted N		1720		1716		1705	1708	1708	1708		1666	

Notes: * ...$p < 0.05$; ** ...$p < 0.01$; *** ...$p < 0.001$.

Table 7.B.2 Weighted ANOVA results for responses to health behavior questions.

		During an average week, how often do you...								Negative health behavior index		Positive health behavior index	
		Eat foods that are high in fat and/or calories		Eat fast food		Eat a variety of fresh fruits and vegetables		Do physical exercise					
	df	Rao-Scott LRT		Rao-Scott LRT		Rao-Scott LRT		Rao-Scott LRT		Rao-Scott LRT		Rao-Scott LRT	
Main effects													
Scale direction	1	0.54		0.57		2.61		0.15		2.32		3.03	
Scale alignment	1	1.34		0.03		1.39		0.02		1.58		1.01	
Vagueness of scale labels	2	9.90	*	26.83	***	3.64		17.35	*	66.07	***	38.03	*
Two-way interaction													
Scale direction* Scale alignment	1	3.33		0.13		0.19		8.71	*	2.23		12.38	
Scale direction × vagueness of scale labels	2	0.02		1.91		0.53		1.95		1.74		3.78	
Scale alignment × vagueness of scale labels	2	3.79		0.05		9.18		11.01		4.09		30.53	
Three-way interaction													
Scale direction × scale alignment × vagueness of scale labels	2	0.38		1.06		3.61		5.42		3.12		17.34	
Unweighted N		1713		1712		1710		1710		1708		1702	

Notes: * ...$p < 0.05$; ** ...$p < 0.01$; *** ...$p < 0.001$.

References

Al Baghal, T. (2014a). Numeric estimation and response options: An examination of the accuracy of numeric and vague quantifier responses. *Journal of Methods and Measurement in the Social Sciences* 5: 58–75.

Al Baghal, T. (2014b). Is vague valid? The comparative predictive validity of vague quantifiers and numeric response options. *Survey Research Methods* 8: 169–179.

Belson, W.A. (1966). The effect of reversing the presentation order of verbal rating scales. *Journal of Advertising Research* 6: 30–37.

Bradburn, N.M. and Miles, C. (1979). Vague quantifiers. *Public Opinion Quarterly* 43: 92–101.

Calegarro, M., Yang, Y., Bhola, D.S. et al. (2009). Response latency as an indicator of optimizing in online questionnaires. *Bulletin of Sociological Methodology* 113: 5–25.

Carp, F.M. (1974). Position effects on interview response. *Journal of Gerontology* 29: 581–587.

Chan, J.C. (1991). Response-order effects in Likert-type scales. *Educational and Psychological Measurement* 51: 530–541.

Chang, L. and Krosnick, J.A. (2009). National surveys via RDD telephone interviewing versus the Internet. Comparing sample representativeness and response quality. *Public Opinion Quarterly* 73: 641–678.

Christian, L.M., Parsons, N.L., and Dillman, D.A. (2009). Designing scalar questions for web surveys. *Sociological Methods & Research* 37: 393–425.

Daamen, D.D.L. and de Bie, S.E. (1992). Serial context effects in survey items. In: *Context Effects in Social and Psychological Research* (ed. N. Schwarz and S. Sudman), 91–114. New York: Springer.

Dickson, J. and Albaum, G. (1975). Effects of polarity on semantic differential scales in consumer research. *Advances in Consumer Research* 2: 507–514.

Friedman, H.H., Herskovitz, P.J., and Pollack, S. (1993). The biasing effects of scale-checking styles on response to a Likert scale. In: *Proceedings of the Survey Research Methods Section,* 792–795.

Friedman, H.H., Weiser Friedmann, L., and Gluck, B. (1988). The effects of scale-checking styles on responses on a semantic differential scale. *Journal of the Market Research Society* 30: 477–481.

Fuchs, M. (2008). Mobile web survey: a preliminary discussion of methodological implications. In: *Envisioning the Survey Interview of the Future* (ed. F.G. Conrad and M.F. Schober), 77–94. New York, NY: Wiley.

Funke, F., Reips, U.-D., and Thomas, R.K. (2011). Sliders for the smart: type of rating scale on the web interacts with educational level. *Social Science Computer Review* 29: 221–231.

GfK (2013). KnowledgePanel design summary. http://images.politico.com/global/2014/09/12/knowledgepanelr-design-summary-description.pdf.

Hofmans, J., Theuns, P., Baekelandt, S. et al. (2007). Bias and change in perceived intensity of verbal qualifiers effected by scale orientation. *Survey Research Methods* 1: 97–108.

Höhne, J.K. and Krebs, D. (2017). Scale direction effects in agree/disagree and item-specific questions: A comparison of question formats. *International Journal of Social Research Methodology* 2017. https://doi.org/10.1080/13645579.2017.1325566.

Höhne, J.K. and Lenzner, T. (2015). Investigating response order effects in web surveys using eye tracking. *Psiholgija* 48: 361–377.

Israel, G.D. (2006). Visual cues and response format effects in mail surveys. Revised version of the paper presented at the Annual Meeting of the Southern Rural Sociological Association, Orlando, FL (7 February).

Israel, G.D. and Taylor, C.L. (1990). Can response order bias evaluations? *Evaluation and Program Planning* 13: 365–371.

Kaminska, O., McCutcheon, A.L., and Billiet, J. (2010). Satisficing among reluctant respondents in a cross-national context. *Public Opinion Quarterly* 74: 956–984.

Kessler, R.C., Andrews, G., Colpe, L.J. et al. (2002). Short screening scales to monitor population prevalances and trends in nonspecific psychological distress. *Psychological Medicine* 32: 959–976.

Knäuper, B. (1999). The impact of age and education on response order effects in attitude measurement. *Public Opinion Research* 63: 347–370.

Krebs, D. (2012). The impact of response format on attitude measurement. In: *Methods, Theories, and Applications in the Social Sciences* (ed. S. Salzborn, E. Davidov and J. Reinecke), 105–103. Springer, VS: Wiesbaden.

Krebs, D. and Hoffmeyer-Zlotnik, J.H.P. (2010). Positive first or negative first? Effects of the order of answering categories on response behavior. *Methodology: European Journal of Research Methods for the Behavioral and Social Sciences* 6: 118–127.

Krosnick, J. (1991). Response strategies for coping with the cognitive demands of attitude measures in surveys. *Applied Cognitive Psychology* 5: 213–236.

Krosnick, J. (1999). Survey research. *Annual Review of Psychology* 50: 537–567.

Krosnick, J.A. and Berent, M.K. (1993). Comparisons of party identification and policy preferences: the impact of survey question format. *American Journal of Political Science* 37: 941–964.

Krosnick, J.A. and Fabrigar, L.R. (1997). Designing rating scales for effective measurement in surveys. In: *Survey Measurement and Process Quality* (ed. L. Lyberg, P. Biemer, M. Collins, et al.), 141–164. New York: Wiley-Interscience.

Krosnick, J.A. and Presser, S. (2010). Questionnaire design. In: *Handbook of Survey Research* (ed. P.V. Marsden and J.D. Wright), 263. West Yorkshire, England, 313: Emerald Group.

Liu, M. and Keusch, F. (2017). Effects of scale direction on response style of ordinal rating scales. *Journal of Official Statistics* 33: 137–154.

Lu, M., Safren, S.A., Skolnik, P.R. et al. (2008). Optimal recall period and response task for self-reported HIV medication adherence. *AIDS and Behavior* 12: 86–94.

Lumley, T. (2014). Survey: analysis of complex survey samples. R package version 3.30.

Malhotra, N. (2008). Completion time and response order effects in web surveys. *Public Opinion Quarterly* 72: 914–934.

Malhotra, N. (2009). Order effects in complex and simple tasks. *Public Opinion Quarterly* 73: 180–198.

Mingay, D.J. and Greenwell, M.T. (1989). Memory bias and response-order effects. *Journal of Official Statistics* 5: 253–263.

Mullinix, K.J., Leeper, T.J., Druckman, J.N., and Freese, J. (2015). The generalizability of survey experiments. *Journal of Experimental Political Science* 2: 109–138.

Pasek, J. with some assistance from Alex Tahk, some code modified from R-core; Additional contributions by Gene Culter and Marcus Schwemmle (2014). Weights: weighting and weighted statistics. R package version 0.80. http://CRAN.R-project.org/package=weights.

Peytchev, A. and Hill, C.A. (2010). Experiments in mobile web survey design. Similarities to other modes and unique considerations. *Social Science Computer Review* 28: 319–335.

Powers, E.A., Morrow, P., Goudy, W.J., and Keith, P.M. (1977). Serial order preference in survey research. *Public Opinion Quarterly* 41: 80–85.

R Core Team (2015). *A language and environment for statistical computing*. Vienna, Austria: R Foundation for Statistical Computing. http://www.R-project.org.

Ramstedt, B. and Krebs, D. (2007). Does response scale format affect the answering of personality scales? Assessing the Big Five dimensions of personality with different response scales in a dependent sample. *European Journal of Psychological Assessment* 23: 32–38.

Sargent-Cox, K., Cherbuin, N., Morris, L. et al. (2014). The effect of health behavior change on self-rated health across the adult life course: a longitudinal cohort study. *Preventive Medicine* 58: 75–80.

Schaeffer, N.C. (1991). Hardly ever or constantly? Group comparisons using vague quantifiers. *Public Opinion Quarterly* 55: 395–423.

Schwarz, N. (2007). Attitude construction: evaluation in context. *Social Cognition* 25: 638–656.

Schwarz, N., Grayson, C.E., and Knäuper, B. (1998). Formal features of rating scales and the interpretation of question meaning. *International Journal of Public Opinion Research* 10: 177–184.

Schwarz, N. and Hippler, H.J. (1995). The numeric values of rating scales: a comparison of their impact in mail surveys and telephone interviews. *International Journal of Public Opinion Research* 7: 72–74.

Schwarz, N., Knäuper, B., Hippler, H.J. et al. (1991). Rating scales: numeric values may change the meaning of scale labels. *Public Opinion Quarterly* 55: 570–582.

Sheluga, D., Jacoby, J., and Major, B. (1978). Whether to agree-disagree or disagree-agree: the effect of anchor order on item response. *Advances in Consumer Research* 5: 109–113.

Scott, J. and Huskisson, E.C. (1979). Vertical or horizontal visual analogue scales. *Annals of the Rheumatic Diseases* 38: 560.

Stapleton, C.E. (2013). The smartphone way to collect survey data. *Survey Practice* 6: http://www.surveypractice.org/index.php/SurveyPractice/article/view/75/html.

Toepoel, V., Das, M., and van Soest, A. (2009). Design of web questionnaires: the effect of layout in rating scales. *Journal of Official Statistics* 25: 509–528.

Tourangeau, R., Couper, M.P., and Conrad, F. (2004). Spacing, positioning, and order. Interpretive heuristics for visual features of survey questions. *Public Opinion Quarterly* 68: 368–393.

Tourangeau, R., Couper, M.P., and Conrad, F. (2007). Colors, labels, and interpretative heuristics for response scales. *Public Opinion Quarterly* 71: 368–393.

Tourangeau, R., Rips, L.J., and Rasinski, K. (2000). *The Psychology of Survey Responses*. Cambridge University Press.

Tversky, A. and Kahneman, D. (1974). Judgment under uncertainty: heuristics and biases. *Science* 185: 1124–1131.

Weng, L.-J. and Cheng, C.-P. (2000). Effects of response order on Likert-type scales. *Educational and Psychological Measurement* 60: 91–112.

Yan, T. and Keusch, F. (2015). The effects of the direction of rating scales on survey responses in a telephone survey. *Public Opinion Quarterly* 79: 145–165.

Zaller, J. (1992). *The Nature and Origins of Mass Opinion*. Cambridge, UK: Cambridge University Press.

Zhang, C. and Conrad, F.G. (2014). Speeding in web surveys: the tendency to answer very fast and its association with straightlining. *Survey Research Methods* 8: 127–135.

8

Mode Effects Versus Question Format Effects: An Experimental Investigation of Measurement Error Implemented in a Probability-Based Online Panel

Edith D. de Leeuw[1], *Joop Hox*[1], *and Annette Scherpenzeel*[2]

[1] *Department of Methodology and Statistics, University of Utrecht, Utrecht, the Netherlands*
[2] *Chair for Economics of Aging, School of Management, Technical University of Munich, Munich, Germany*

8.1 Introduction

8.1.1 Online Surveys Advantages and Challenges

Online surveys have rapidly become a major mode of data collection (Baker et al. 2010). From a total survey quality perspective (Biemer and Lyberg 2003; Biemer 2010, p. 818) online surveys have many advantages; they are cost effective and can provide fast and timely data (de Leeuw 2012; Smyth and Pearson 2011; Callegaro et al. 2015). Furthermore, thanks to advanced technology, complex questionnaires and skip patterns may be used, paradata (e.g. response times) can be routinely collected, and online surveys can enhance measurement (e.g. advanced graphical presentation, less social desirable answer, and no interviewer bias). When we further focus on Total Survey Error (i.e. coverage, sampling, nonresponse, and measurement), online surveys have to deal with three challenges. Although Internet coverage is high in Northern Europe (e.g. 95.5% for the Netherlands in 2017), it is lower in other countries (for the most recent statistics see http://www.internetworldstats.com/stats4.htm#europe). Furthermore, even with a relative high Internet penetration, the digital divide is still substantial and the off-line population differs from the online on important characteristics, such as, sex, age, education, race, household composition, and life satisfaction (Mohorko et al. 2013; Leenheer and Scherpenzeel 2013; Sterret et al. 2017). Besides undercoverage, nonresponse poses another threat; compared to other data collection methods Internet surveys do have lower response rates (e.g. Lozar Manfreda et al. 2008; Wengrzik et al. 2016). As a solution to the coverage and nonresponse problems that are facing a single online mode approach, several researchers have advocated mixed-mode surveys. In a mixed-mode survey, survey designers try to combine the best of all possible worlds by exploiting the advantages of different modes to compensate for individual weaknesses (de Leeuw 2005), thereby achieving a good coverage of the intended population and a high response rate at still affordable costs (for an overview, see de Leeuw and Berzelak 2016; Dillman 2017; Tourangeau 2017).

Experimental Methods in Survey Research: Techniques that Combine Random Sampling with Random Assignment, First Edition.
Edited by Paul J. Lavrakas, Michael W. Traugott, Courtney Kennedy, Allyson L. Holbrook, Edith D. de Leeuw, and Brady T. West.

Companion Website: www.wiley.com/go/Lavrakas/survey-research

8.1.2 Probability-Based Online Panels

A third challenge for online surveys concerns sampling procedures, as there is no general sampling frame available of Internet users. At least not for the general population; there may be adequate sampling frames for special population, such as, students, employees, or customers. A potential solution is the establishment of probability-based Internet panels, where a state of the art probability sample is drawn using off-line sampling frames (e.g. a registry-based sample or an address-based sample). The resulting sample, which will include people with and without Internet, is than approached via off-line methods, such as face-to-face and telephone interviews, and asked if they want to participate in a panel. This approach is the only way to cover the complete probability sample of the general population. There are two ways to engage willing respondents without Internet (DiSogra and Callegaro 2016). The first is to provide non-Internet users with the necessary equipment; this has been done by the Longitudinal Internet Studies for the Social Sciences (LISS) Panel in the Netherlands (Scherpenzeel 2011; Scherpenzeel and Das 2011). Similar procedures were followed by the German Internet Panel (GIP), the French Longitudinal Study by Internet for the Social Sciences (ELIPSS) Panel in Europe, Knowledge (networks) Panel (GFK), and the Face-to-Face Recruited Internet Survey Platform (FFRISP) in the United States. The second is establishing a mixed-mode panel and use more traditional survey methods for those without an Internet connection, such as a postal mail survey (e.g. the GESIS-panel in Germany, the Gallup panel in the United States), or telephone interviews (e.g. EKOS in Canada and Pew ATP in the United States).

In sum, mixed-mode designs with online surveys or panels as part of the mix may offer many advantages, such as timeliness, affordable costs, and a reduction of coverage and nonresponse error. However, there is the danger of increased measurement error due to specific mode-measurement effects (Tourangeau 2017; de Leeuw and Berzelak 2016).

8.1.3 Mode Effects and Measurement

When looking at data from mixed-mode surveys, it is important to distinguish between mode (self) selection effects and mode-measurement effects (Tourangeau et al. 2013). The selection effects may be beneficial if each mode brings in different types of respondents, and doing so reduces nonresponse and coverage error. If in a mixed-mode design, *no* mode-selection effects occur, this implies that the same type of respondents react to all different modes, and researchers may as well stay with the least expensive single web mode.

Conversely, mode-measurement effects are unwanted effects. When different subgroups of respondents are surveyed with different modes of data collection, mode-measurement effects can influence the resulting data and differential measurement error may threaten the validity of the results. This danger is far greater when disparate modes, such as telephone and online (e.g. EKOS panel) are combined than when more comparable modes, such as paper-mail and online (e.g. GESIS panel) are mixed. This has been corroborated by a long series of mode comparison studies and meta-analyses (e.g. Groves 1989; de Leeuw 1992; de Leeuw and Hox 2011; Tourangeau et al. 2013), showing that there is a clear dichotomy between modes with an interviewer and modes without an interviewer. A consistent finding is that both paper and online questionnaires give rise to lesser social desirability than interviews.

A main difference between self-administered and interview modes is the channel of communication used to present questions and provide answers. In telephone surveys and to a slightly lesser degree in face-to-face interviews, the auditory channel is used to convey information, while in mail and online surveys the visual channel is mainly used. This influences the cognitive burden for the respondent. To accommodate the limitations of the auditory channel, researchers have designed special question formats for telephone interviews. However, changing the question format in one mode may exacerbate mode-measurement effects, result in differential measurement error between modes, and threaten the comparability of the resulting data.

In a series of online experiments, we investigated this effect of different question formats on responses both within and between modes. While most question format experiments are done with selected or convenience samples, we had the opportunity to implement these experiments using a probability-based online panel in the Netherlands (LISS). The sample for this panel was randomly drawn by Statistics Netherlands from the general population. This enabled us to combine the internal validity of a true experimental design with the external validity of a probability-based sample.

In the following sections, we first describe and summarize the potential of probability-based online panels for experimentation, and then we focus more specifically on question format experiments, and present new experimental research.

8.2 Experiments and Probability-Based Online Panels

8.2.1 Online Experiments in Probability-Based Panels: Advantages

Online experiments are the modern equivalent of split-ballot telephone surveys and laboratory experiments. However, online experiments are often implemented in so-called convenience samples (e.g. students, opt-in online surveys, and Mechanical Turk participants), and although the random assignment of experimental conditions provides a high internal validity, the external validity or generalizability may be lacking. In contrast, probability-based online panels provide a database of respondents that are representative of the general population, thereby ensuring both internal and external validity in online experiments conducted with panel members.

A second advantage of this approach is that online panels collect additional information about their respondents. This information can consist of standard demographic variables, but can also include psycho- and sociographic information that has been collected in different studies conducted in the panels, or in special "core" modules. Often, this extensive auxiliary information about each panel member can be merged with the data of a specific experiment, thus enlarging possibilities for analyses and reducing the number of questions that have to be asked in the experiment itself. This is especially true for panels built for scientific research, since data collected in market research panels for commercial customers is usually not publicly available for researchers. Moreover, the available information offers possibilities of selecting specific subgroups of respondents for targeted experiments (e.g. targeted on panel members with children, panel members who own stocks, or panel members with high scores on a specific scale, such as social desirability), as well as matching of experimental and control group members on relevant variables. An important and noteworthy advantage is that the additional information was collected prior to and completely independent of the experiment.

A third reason why online panels offer an excellent setting to conduct experiments is that researchers and panel managers have complete control over fieldwork parameters that facilitate the experimental manipulation. For instance, full control is exerted regarding the version of the questionnaire or question format presented to each panel member and online researchers do not have to rely on interviewers' compliance with experimental instructions. Furthermore, researchers have full control over the assignment to experimental and control groups, the sending of tailored invitations to different groups through the online panel management system, and can even check which panel members have clicked on and reacted to the invitation, in order to investigate selective nonresponse. In addition, the response in all experimental groups can be monitored in real time and paradata, such as key stroke data and response time measures, can be used as experimental outcomes or as supplemental control information.

A fourth advantage of experimenting in probability-based online panels is that response rates for the experiments are transparent and can be calculated following standard definitions for cumulative response rates. In opt-in and other forms of nonprobability panels, total cumulative response rates are not clearly determined as the initial response rates are unknown and the target population is not well defined (Disogra and Callegaro 2016).

8.2.2 Examples of Experiments in the LISS Panel

The Dutch LISS panel is based on a probability sample of households. As a typical example of a probability-based online panel designed for scientific research, the LISS panel has been a true social science laboratory and incorporates numerous different experiments. Many experiments had a methodological focus, while others were of substantive interest for specific scientific disciplines. An example of the latter is a series of economic experiments with real payoffs for the respondents. Respondents were asked to make several choices between alternatives with different risks, costs, and possible benefits. At the end of these experiments, respondents got paid the amount of money that they gained by the choices they made (e.g. Noussair et al. 2013).

A major objective of the LISS panel as laboratory was the innovation of survey tools and measurement techniques. As a result, many experiments were of a survey methodological nature. A prime example is the series experiments using life-tracking technology to collect more accurate data and supplemental measures. A first experiment concerned the continuous measurement of weight and fat percentage over time, using Internet connected weighing scales. A second series of experiments used smartphone diary apps and GPS technology to continuously collect time use data, experience sampling data, and mobility data. Furthermore, the physical activity of respondents was measured using accelerometers. A detailed description of all LISS panel experiments with life-tracking techniques can be found in Scherpenzeel (2016).

Improving response and retention rates, minimizing biases, and improving data quality were essential for the maintenance of the LISS panel and many experiments were hence conducted to optimize the methodology. For example, Scherpenzeel and Toepoel (2012) incorporated an experimental design into the regular LISS panel data collection to evaluate the effect of communication materials send to respondents. Das and Couper (2014) implemented experiments aimed at finding the optimal wording for asking respondents to consent to survey data being linked to administrative records. Other experiments include the implementation of do-not-know options, mixed-mode designs, and the emulation of virtual interviewers for probing (see de Leeuw et al. 2016).

The most recent experiments in the LISS panel focus on online data collection using mixed devices, that is, designing for and comparing surveys on personal computers and laptops vs. mobile devices, such as smartphones and tablets (e.g. Antoun 2015; de Bruyne 2015; Lugtig and Toepoel 2015; for an overview, see Couper et al. 2017).

8.3 Mixed-Mode Question Format Experiments

8.3.1 Theoretical Background

Survey modes differ on many aspects, which can be grouped in three main dimensions: mode inherent factors, context specific factors, and implementation specific factors (e.g. de Leeuw and Berzelak 2016). Mode inherent factors are given and do not change across respondents or situations. Examples are interviewer involvement, such as absence of interviewers in self-administered modes, presentation of stimulus and response, such as aural presentation of questions and responses in telephone interviews, and use of computer technology in online surveys. Context specific characteristics depend on the social and cultural aspects of the (sub)populations for which the survey is implemented. Familiarity with the medium used for surveying (e.g. web vs. telephone) and its associated expectations play an important role. Contrary to mode inherent factors, context specific characteristics may change over time. Finally, implementation specific characteristics are fully under the control of the researcher and may be exploited to achieve an optimal survey design (e.g. well-programmed help functions and visual design in online surveys to convey extra information and compensate for the absence of interviewer assistance). Related to implementation specific characteristics is questionnaire design. One should be aware that the way a questionnaire is designed and implemented often differs over modes. In a single-mode design, specific traditions of how questions are structured and presented exist. For example, in interviews, do-not-know is often not explicitly offered, but recorded when spontaneously given by the respondent, while in online surveys, do-not-know is more often explicitly presented as a separate response category. Depending on the design (e.g. explicit do-not-know, mandatory answers, probing) the amount of missing data may differ considerably between modes, but clever design may achieve common measurement across modes (de Leeuw et al. 2016).

Scalar questions, that is, closed-end questions where the response categories form a scale, are often used in survey research. Such questions use response scales like agree/disagree, satisfied/dissatisfied, or best/worst (Fowler 1995). These questions have been adapted frequently to the specific mode used to accommodate mode differences. For an overview on mode adaptations for scalar questions, see Christian et al. (2008), Dillman and Edwards (2016), and Dillman and Christian (2005).

A main difference between web and telephone is the channel of communication that is applied to present questions and provide answers (de Leeuw 1992; Schwarz et al. 1991). In telephone surveys, the auditory channel is used to convey information, while in online, web surveys it is mainly the visual channel; this influences the cognitive burden for the respondent (Tourangeau et al. 2000). Aural-only transmission of information places higher demands on memory capacity than visual transmission: with aural-only the respondents have to remember all information instead of being able to repeatedly refer to the visual information. Especially when longer lists of response alternatives are presented, the telephone survey is at a disadvantage as all response alternatives have to be processed in the working memory of the respondents. However, in online survey's respondents can refer to response alternatives on their screen and in face-to-face interviews visual show cards may be used to relieve cognitive burden (e.g. Jaeckle et al. 2010).

To accommodate the limitations of the auditory channel, researchers have designed special question formats for telephone interviews. Response scales are often simplified in telephone interviews, by providing only the polar endpoint labels to ease the cognitive and memory burden placed on the respondents. In contrast, in web surveys the visual channel is used and response scales are often presented with all scale points fully labeled. A second method to reduce cognitive burden in telephone interviews is "unfolding" in which scalar questions are branched into two steps, instead of presenting all response categories directly in one step. The advantage of a two-step procedure is that respondents are asked to process fewer categories at one time, thereby reducing the burden on the limited short-term memory available when only aural stimuli are processed.

However, from past research we know that question wording and response format affect responses even within one mode; for an overview, see Sudman and Bradburn (1974). Especially, differential labeling of response categories and presentation of the response scale may influence the responses given (e.g. Christian et al. 2008; Krosnick and Fabrigar 1997; Schwarz et al. 1985; Tourangeau et al. 2013). Therefore, changing the question format when changing mode may add to the net effect of mode and so exacerbate mode-measurement effects in mixed-mode surveys (Dillman and Edwards 2016; Jaeckle et al. 2010).

In a series of experiments, we tested the effects of two specific question formats that are commonly used in mixed-mode mail/web-telephone surveys to accommodate the aural only channel in telephone surveys: (i) the effect of "endpoint" vs. "'full" labeling of response categories and (ii) the effect of "unfolding" vs. "one step direct question" both within telephone and web modes and between telephone and web modes. We compared both similar scales across modes and different scales within modes. This enabled us to investigate mode and question format effects independently and study potential interactions between mode and question format. The latter is especially interesting, as to date there is hardly any literature on interaction between mode and question format. Besides differences in response distributions, we also examined differences in reliability of answers, and two well-known response effects: acquiescence and extremeness.

8.3.2 Method

8.3.2.1 Respondents

In this study, members of the Dutch LISS panel were investigated. The LISS panel was established in autumn 2007 and consists of almost 8000 individuals that complete online questionnaires every month. It is based on a probability sample of households drawn from the Dutch population register by Statistics Netherlands. The sample includes people without an Internet connection, and households that could not otherwise participate because they had no Internet access were equipped with a computer and broadband Internet connection. All people in the sample were approached in traditional ways (by letter, followed by telephone call, and house visit) with an invitation to participate in the panel. A detailed description of the LISS panel design, recruitment, and response can be found in Scherpenzeel and Das (2011).

Every month LISS panel members complete online questionnaires; these can be several short questionnaires or one or two longer ones. It usually takes between 30 and 40 minutes to complete all questionnaires; the incentive is €7.50 per half hour. Half of the interview time available in the panel is reserved for the LISS core study; this is the longitudinal part of the study repeated every year. It follows changes in life course and living conditions of panel members and monitors trends in household composition. The other half of the available time is reserved for proposals from academic researchers. In spring 2009, a series of methodological experiments was conducted.

The mode experiment described here involved two modes: computer assisted telephone interviews (CATI) and online. The online fieldwork was done by the regular staff of the LISS panel. The CATI fieldwork was done by the Dutch TNS-NIPO (Kantar) organization, but during the CATI data collection a member of the LISS-research team was involved as extra quality controller. For this experiment all respondents received a bonus incentive of €6.50, in addition to their regular incentive payment.

8.3.2.2 Experimental Design

We performed two separate experimental question format manipulations. In the first manipulation, we investigated the effect of differential labeling of response categories: (i) all response categories fully labeled vs. (ii) only endpoints labeled verbally. In the second manipulation, we investigated the effect of two-step branching (unfolding) technique: (i) all response categories offered directly in one-step vs. (ii) a two-step unfolding procedure. These manipulations were fully crossed in a 2 × 2 design and embedded in the survey-mode experiment.

First, LISS-panel members were randomly assigned to a data collection mode: CATI or online survey. Second, within each mode respondents were again randomly assigned to the experimental question format conditions in a two by two design (fully labeled vs. endpoint labeled and one-step vs. two-step unfolding). For a description of the experimental design and examples of the questions, see the appendix (see also the online materials at https://www.dataarchive.lissdata.nl/study_units/view/63).

8.3.2.3 Questionnaire

Eight bipolar opinion questions were manipulated. Each question had a response scale with five response categories, ranging from (1) "Strongly disagree" to (5) "Strongly agree." "Do-not-Know/No Opinion" was not offered explicitly in either the web-survey or the telephone survey, but was accepted when given. In the web version, respondent could skip a question by using the "next" button; this approach was taken to make the web-survey equivalent to the telephone survey, in which it is always possible to spontaneously say "do-not-know" to an interviewer.

The topic of the survey was the acceptability of usage of advanced medical technology. The questionnaire was balanced, in the sense that four questions were statements on acceptability and four on unacceptability. Examples are: "If you can save human lives, everything is permitted to achieve that goal," and "It is not desirable to do all that is medically possible." The questions were part of a well-tested questionnaire that was used previously to investigate the acceptability in the Dutch population of advanced medical technology (Steegers et al. 2008).

8.3.3 Analyses and Results

8.3.3.1 Response and Weighting

Two thousand LISS-panel members were assigned to the CATI-condition and 6134 were assigned to the online condition. In the CATI-condition 1207 members responded (60.4%), and in the online-condition 4003 members responded (65.3%). As the response rates between the two modes differed slightly, we first checked if differential nonresponse influenced the comparability of the two experimental mode conditions. For all the LISS panel members, biographical as well as psychographical information is available. When we compare the respondents in the CATI vs. the online condition on this, we find some small but significant differences. CATI and online respondents differ in age (47.8 vs. 46.2 years, $p = 0.02$), household size (2.9 vs. 2.8, $p = 0.00$), (non)urbanicity (3.1 vs. 3.0, $p = 0.05$), and house-ownership (78.5% vs. 75.0%, $p = 0.00$). In a multivariate analysis using logistic regression, only age and household size remained significant. Although the differences are very small, a propensity score weight was constructed based on a logistic regression using age and household size as predictors. The mean weight is 1.00, with a standard deviation of 0.04 and a range from 0.91 to 1.27. All analyses have been carried out both weighted and unweighted. The differences were negligible; hence the unweighted results are reported here.

8.3.3.2 Reliability

The precision with which respondents answer questions, may be affected by factors that vary across modes. One mode inherent factor is the pace of the interview: in a self-administered online survey, the respondent controls the survey situation, sets the pace, decides on taking time to think about a question, while in a telephone conversation the interviewer, who is the initiator of the conversation, controls the pace. Furthermore, in a telephone conversation pauses may feel awkward, and both interviewer and respondent will try to avoid silences and speed up the interview (de Leeuw 1992; Tourangeau et al. 2000; Couper 2011). A second factor has to do with mode of communication; due to the aural presentation in telephone interviews a respondent has to keep the response categories in working memory, while in the visual online mode respondents can read and reread the question and response categories when necessary. These resulting differences in cognitive burden between interview and self-administered modes may produce differences in the question–answer process and the accuracy of the resulting responses (de Leeuw 1992; Krosnick and Alwin 1987; Tourangeau et al. 2000; Couper 2011). Empirical mode comparisons focusing on reliability are scarce. de Leeuw (1992) reports less consistent responses and more random error in a telephone survey compared to a self-administered mail survey. Chang and Krosnick (2009) report more random error in telephone data than in web survey data.

We investigated the reliability for both the CATI and online mode and calculated coefficient alpha of the eight-item scale, after recoding for negatively formulated items. Tise reliability coefficient was calculated in both the CATI and online condition, for fully labeled vs. endpoint labeling, and for one-step direct question vs. two-step unfolding. Coefficient alpha reflects the consistency with which the eight questions were answered in the experimental conditions; a high coefficient alpha implies the absence of random measurement error, but not necessarily a high validity.

The results are summarized in Table 8.1. The difference between the reliability in the CATI (0.26) vs. online (0.44) condition is large and significant ($p < 0.01$). The differences between the reliability in the fully vs. endpoint labeling conditions and the direct one-step question vs. two-step unfolding conditions are small and insignificant (smallest p-value is 0.28). There is also no statistically significant interaction between the mode and the question wording conditions: online respondents answer with less random error than telephone phone and this applies to all question formats. This result is in line with the findings of Chang and Krosnick (2009).

8.3.3.3 Response Styles

We found that the reliability in the Internet condition was higher than in the telephone condition. In other words, the data collected online showed less random error than the data collected via the telephone. However, reliability reflects the absence of random error and a high reliability does not necessarily imply high validity, the latter also reflects the absence of systematic error. In the next section, we investigate if the data collection modes and the question formats influence systematic measurement error.

The aural transmission of information in telephone interviews taxes memory capacity more than online visual presentation; furthermore specific response scale formats also place different burdens on memory capacity (Tourangeau et al. 2000, 2013). An increased burden on memory may heighten the task difficulty and lower the respondent's ability and motivation to give an optimal answer (Krosnick 1991). In other words, due to the mode the question is asked in, and due to question characteristics, respondents will fall back on a simpler answer strategy (Krosnick and Alwin 1987) and satisfice (Groves 1989). This will result in clear response styles in

Table 8.1 Reliabilities (coefficient alpha) across modes for fully vs. endpoint labeling and one-step direct question vs. two-step unfolding.

	Labeling		Question unfolding		
	Fully	Endpoint	One step	Unfolding	Overall
CATI	0.27	0.25	0.22	0.29	0.26
Online	0.44	0.43	0.46	0.42	0.44

rating scales (Roberts 2016). Furthermore, the greater difficulty to communicate cues in telephone interviews compared to face-to-face interviews (Conrad et al. 2008) may result in lesser opportunities for interviewers to help, motivate, and correct respondents when they fall back on response styles triggered by specific question characteristics, and mode and question format may interact in producing response styles.

Below we focus on two well-known response effects; acquiescence (Holbrook et al. 2003; Krosnick 1991), and extremeness (Baumgartner and Steenkamp 2001).

Acquiescence Acquiescence is defined as the tendency to agree with any assertion regardless of its content (Holbrook 2008; Roberts 2016), and can be seen as a form of satisficing (Krosnick 1991, 1999). In aural only presentation, different communication rules, faster pace, and increased response burden over the telephone (e.g. de Leeuw 1992; Schwarz et al. 1991; Tourangeau, et al. 2000), may lead to more satisficing and previous research did show indications of more acquiescence when reporting over the phone (Jordan et al. 1980; Holbrook et al. 2003). Furthermore, specific response scale formats place different burdens on memory capacity (Tourangeau et al. 2013). Although endpoint labeled scales and unfolding scales are easier to hold in memory, fully labeled one-step rating scales help clarify the meaning of the concepts measured, and numerous empirical studies have found that both the reliability and the validity of the data is higher when fully labeled scales are used (for an overview see Krosnick and Fabrigar 1997; Saris and Gallhofer 2007).

In our data, eight balanced questions on the acceptability of advanced medical technology were available: four questions were formulated as statements on acceptability and four as statements on unacceptability. This enabled us to compute an acquiescence score, defined as the proportion "strongly agree" on the original eight questions.

All main effects were significant: for unfolding $F(1, 5199) = 70.5, p = 0.00$, for endpoint labeling $F(1, 5199) = 22.7, p = 0.00$, and for mode $F(1, 5199) = 40.4, p = 0.00$. The interaction between unfolding and mode (F 15 199) = 5.1, $p = 0.02$ was significant, while the interaction between endpoint labeling and mode (F 15 199) = 0.8, $p = 0.38$ was clearly not significant. It should be noted that the effect sizes are small and the explained variance is low: partial eta-square for unfolding is 0.013, for labeling 0.004, for mode 0.008, and for the interaction 0.001.

Table 8.2 presents the expected marginal means in the various conditions. For consistency with the earlier table, we again report means for both interaction effects, but note that the interaction of labeling with mode is not significant.

Extremeness Another response style is extremeness, the tendency to endorse the most extreme response categories (Baumgartner and Steenkamp (2001). Extremeness can also be seen as a form of satisficing (Belli et al. 1999; Kaminskaya et al. 2010). Previous research indeed showed indications of more extremeness when reporting over the telephone (Christian et al. 2008; Groves and Kahn 1979; de Leeuw 1992; Dillman et al. 2009; Ye et al. 2011) and also more

Table 8.2 Mean of acquiescence across modes for fully vs. endpoint labeling and one-step direct question vs. two-step unfolding.

	Labeling		Question unfolding		
	Full	Endpoint	One step	Unfolding	Overall
CATI	0.24	0.26	0.23	0.27	0.25
Online	0.19	0.23	0.18	0.24	0.21
Overall	0.22	0.25	0.20	0.26	0.22

extremeness was reported with certain question formats (e.g. Christian 2007; Christian et al. 2008; Kieruj 2012; Moors et al. 2014).

We defined extremeness as either answering 1 or 5 on a 5-point response scale. We then calculated the proportion extreme answers (either 1 or 5) over the eight acceptability questions. Analysis of variance was used to investigate the effects of mode and of question-format.

We found a small, but significant main effect of mode on extremeness of answers ($F(1, 5199) = 5.2$, $p = 0.02$, partial $\eta^2 = 0.001$). Respondents in the telephone mode had a slightly higher proportion of extreme answers (mean proportion = 0.31) than those in the web condition (mean proportion = 0.29). There was a significant effect of the way answer categories were labeled ($F(1, 5199) = 49.7$, $p = 0.00$, partial $\eta^2 = 0.009$). Respondents who were presented with a full verbally labeled response scale gave less extreme answers (mean proportion of extreme answers = 0.27), than respondents who were presented with endpoint only labeled response scales (mean proportion of extreme answers = 0.33). There was also a significant effect of "unfolding" ($F(1, 5199) = 210.3$, $p = 0.00$, partial $\eta^2 = 0.039$). Respondents to the two-step unfolding format had clearly a higher proportion of extreme answers (mean proportion = 0.35) than those answering to the one-step direct question (mean proportion = 0.25). Interestingly, there were significant interaction effects between mode and question formats: interaction of mode and labeling ($F(1, 5199) = 8.5$, $p = 0.00$, partial $\eta^2 = 0.002$), and interaction of mode and unfolding ($F(1, 5199) = 33.7$, $p = 0.00$, partial $\eta^2 = 0.006$). Fully labeled and direct one-step questions resulted in less extremeness and this effect was somewhat stronger for the visual online mode.

Table 8.3 presents the expected marginal means in the various conditions.

Differences in Distribution of Acceptability of Medical Technology Both mode and question format do influence systematic measurement error, as shown in the analyses above. The effects found were statistically significant but small. How does this influence survey practice? Survey practitioners are often mainly interested in mean scores and response distributions.

To investigate this, we first recoded all responses in such a way that a higher number always indicated more acceptability of modern medical technology. We then calculated an average score of acceptability of medical innovations (ranging from 1 to 5). A two-way analysis of variance showed a small, but significant main effect of mode on acceptance of new medical technology ($F(1, 5185) = 13.3$, $p = 0.00$, partial $\eta^2 = 0.003$). Respondents reported a slightly lower acceptance of medical technology over the telephone (mean acceptance = 2.93) than when answering through the Internet (mean acceptance = 2.99). We found a nonsignificant effect of response category labeling: endpoint labeled (mean acceptance 2.95) vs. fully labeled (mean acceptance 2.96); $F(1, 5185) = 1.0$, $p = 0.33$). But, there was also a small, but significant effect of "unfolding" ($F(1, 5185) = 11.5$, $p = 0.00$, partial $\eta^2 = 0.002$). Respondents in the two-step unfolding condition reported a lower acceptance of medical technology (mean acceptance = 2.93) than respondents in the one-step direct question condition (mean acceptance = 2.98).

Table 8.3 Mean of extremeness across modes for fully vs. endpoint labeling and one-step direct question vs. two-step unfolding.

	Labeling		Question unfolding		
	Fully	**Endpoint**	**Direct**	**Unfolding**	**Overall**
CATI	0.29	0.32	0.28	0.34	0.31
Online	0.25	0.33	0.22	0.37	0.29
Overall	0.27	0.33	0.25	0.35	0.30

Table 8.4 Score on acceptability of medical technology.

	Mean	Standard deviation	*N* of cases
Mode	$p = 0.00$		
CATI	2.93	0.44	1207
Online	2.99	0.49	3986
Labeling	n.s		
Fully labeled	2.96	0.46	2601
Endpoint only	2.95	0.50	2592
Unfolding	$p = 0.00$		
No: one-step	2.98	0.44	2605
Yes: two-step	2.93	0.51	2588

Minimum is 1 (totally unacceptable, maximum is 5 (totally acceptable) mean and standard deviation and N of cases for experimental conditions.

No significant interaction effects between mode and question format were found. Both the interaction of mode and labeling ($F(1, 5185) = 0.6, p = 0.453$) and of mode and unfolding were nonsignificant ($F(1, 5185) = 0.04, p = 0.88$). Table 8.4 contains the means and standard deviations of the acceptability scores.

8.4 Summary and Discussion

Mixed-mode surveys are increasingly popular and have many advantages, such as, lower costs and higher response rate, but mode-measurement effects may threaten data integrity. In a large study of the Dutch population, we found small but consistent mode effects *and* question format effects, but only limited interaction effects between data collection mode and question format. These results cannot be attributed to self-selection of respondents in the different modes, since weighting for selective nonresponse did not change the results.

The largest effects were found for reliability. Respondents in the telephone mode produced less consistent responses and more random error; this is consistent with earlier findings by de Leeuw (1992) and Chang and Krosnick (2009). Furthermore, independent of scale format, telephone respondents provided lower mean ratings than online respondents. However, the differences were small, the average difference between telephone and web was only 0.06 on a 5-point scale. Telephone respondents showed a greater tendency to agree independent of the content of questions (acquiescence) then online respondents. We also found that telephone respondents more often chose the extreme response categories, a finding earlier reported by Christian et al. (2008) for a student population in the United States.

Regardless of mode, question format changed the responses. Although there were no question format effects on the reliability of measures, respondents to the two-step unfolding format provided lower mean ratings than respondents to the one-step format. These differences were also small: the average difference was 0.05 on a 5-point scale. In addition, respondents to the two-step unfolding format and to the endpoint labeling condition showed more acquiescence and also more often chose the extreme response categories.

These findings from the Dutch general population confirm earlier findings from special samples (e.g. students) in the United States (e.g. Christian 2007, Christian et al. 2008) and the

Netherlands (Moors et al. 2014). Therefore, for survey practice it is advised to fully label all scale points if possible, as previous research (Krosnick and Fabrigar 1997; Saris and Gallhofer 2007) shows that fully labeled scales have better psychometric properties.

We should emphasize that only a few interaction effects were found for mode and question formats regarding acquiescence and extreme response tendencies; the majority of the interaction effects are not significant. These results are encouraging for mixed-mode surveys; both mode and question format effects were small with question format effects larger than mode effects. Question format choices are under the control of the researcher and thus can in theory be controlled for. However, in survey practice different modes often use different question formats (Dillman and Christian 2005; Dillman and Edwards 2016); therefore we should be extremely careful and be aware that when different question formats are used within different modes, the small measurement effects add up to potential large effects. For instance, we found a slightly lower acceptance of medical technology score in telephone surveys and also when using the two-step unfolding format. However, in practice, two-step unfolding is often used in the aural telephone mode, while the one-step direct question is used in visual modes, such as mail or web. The two small effects then sum up, and the total effect between the two *mode systems* is then clearly larger. We, therefore, advise not using differential question formats in mixed-mode studies, but to adhere to a unified mode design (Dillman and Edwards 2016) and use equivalent questions whenever possible.

Acknowledgments

The authors gratefully acknowledge the review comments from our editors, Allyson Holbrook and Mike Traugott. We also sincerely thank Jon Krosnick (Stanford University) for his suggestion to investigate acquiescence and Stephanie Stam (CentERdata, Tilburg University) and Robert Zandvliet (TNS-NIPO) for their valuable assistance during the data collection phase. The LISS panel data were collected by CentERdata (Tilburg University, The Netherlands) through its MESS project funded by the Netherlands Organization for Scientific Research (www.lissdata.nl).

References

Antoun, C. (2015). Mobile web surveys: a first look at measurement, nonresponse, and coverage errors. PhD dissertation. University of Michigan.

Baker, R., Blumberg, S.J., Brick, J.M. et al. (2010). AAPOR report on online panels. *Public Opinion Quarterly* 74 (4): 711–781.

Baumgartner, H. and Steenkamp, J.-B.E.M. (2001). Response styles in marketing research: a cross-national investigation. *Journal of Marketing Research* 38: 143–156.

Belli, R.F., Herzog, A.R., and Van Hoewyk, J. (1999). Scale simplification of expectations for survival: cognitive ability and the quality of survey responses. *Cognitive Technology* 4 (2): 29–38.

Biemer, P.P. (2010). Total survey error: design, implementation, and evaluation. *Public Opinion Quarterly* 24 (5): 817–848.

Biemer, P. and Lyberg, L. (2003). *Introduction to Survey Quality*. New York: Wiley.

de Bruyne, M. (2015). Designing web surveys for the multi-device Internet. PhD dissertation. Tilburg University.

Callegaro, M., Lozar Manfreda, K., and Vehovar, V. (2015). *Web Survey Methodology*. London: Sage.

Chang, L. and Krosnick, J.A. (2009). National surveys via RDD telephone interviewing versus the Internet: comparing sample representativeness and response quality. *Public Opinion Quarterly* 73 (4): 641–678.

Christian, L.M. (2007). How mixed-mode surveys are transforming social research: the influence of survey mode on measurement in web and telephone surveys. PhD dissertation. Washington State University.

Christian, L.M., Dillman, D.A., and Smyth, J.D. (2008). The effect of mode and format on answers to scalar questions in telephone and web surveys. In: *Advances in Telephone Survey Methodology* (ed. J.M. Lepkowski, C. Tucker, J.M. Brick, et al.), 250–275. New York: Wiley.

Conrad, F.G., Schober, M., and Dijkstra, W. (2008). Cues of communication difficulty in telephone interviews. In: *Advances in Telephone Survey Methodology* (ed. J.M. Lepkowski, C. Tucker, J.M. Brick, et al.), 212–230. Hoboken, NJ: Wiley.

Couper, M.P. (2011). The future of modes of data collection. *Public Opinion Quarterly* 75 (5): 889–908.

Couper, M.P., Antoun, C., and Mavletova, A. (2017). Mobile web surveys. In: *Total Survey Error in Practice* (ed. P. Biemer, E. de Leeuw, S. Eckman, et al.), 133–154. New York: Wiley.

Das, M. and Couper, M.P. (2014). Optimizing opt-out consent for record linkage. *Journal of Official Statistics* 30 (3): 479–497.

Dillman, D.A. (2017). The promise and challenges of pushing respondents to the web in mixed-mode surveys. *Survey Methodology* 43 (1): 3–30.

Dillman, D.A. and Christian, L.M. (2005). Survey mode as a source of instability across surveys. *Field Methods* 17: 30–52.

Dillman, D.A. and Edwards, M.L. (2016). Designing a mixed-mode survey. In: *The Sage Handbook of Survey Methodology* (ed. C. Wolf, D. Joye, T.W. Smith and Y.-C. Fu), 255–268. London: Sage.

Dillman, D.A., Phelphs, G., Tortorra, R. et al. (2009). Response rate and measurement differences in mixed mode surveys using mail, telephone, interactive voice response, and the Internet. *Social Science Research* 38: 1–18.

DiSogra, C. and Callegaro, M. (2016). Metrics and design tool for building and evaluating probability-based online panels. *Social Science Computer Review* 34 (1): 26–40.

Fowler, F.J. (1995). *Improving Survey Questions: Design and Evaluation.* Thousand Oaks: Sage.

Groves, R.M. (1989). *Survey Errors and Survey Costs.* New York: Wiley.

Groves, R.M. and Kahn, R. (1979). *Surveys by Telephone; A National Comparison with Personal Interviews.* New York, NY: Academic Press.

Holbrook, A. (2008). Acquiescence response bias. In: *Encyclopedia of Survey Research Methods,* vol. 1, 3–4 (ed. P.J. Lavrakas). Thousand Oakes, CA: Sage.

Holbrook, A.L., Green, M.C., and Krosnick, J.A. (2003). Telephone versus face-to-face interviewing of national probability samples with long questionnaires: comparisons of respondent satisficing and social desirability response bias. *Public Opinion Quarterly* 67: 79–125.

Jaeckle, A., Roberts, C., and Lynn, P. (2010). Assessing the effect of data collection mode on measurement. *International Statistical Review* 78 (1): 3–20.

Jordan, L.A., Marcus, A.C., and Reeder, L.G. (1980). Response styles in telephone and household interviewing: a field experiment. *Public Opinion Quarterly* 44: 210–222.

Kaminskaya, O., McCutcheon, A.L., and Billiet, J. (2010). Satisficing among reluctant respondents in a cross-national context. *Public Opinion Quarterly* 74 (5): 956–984.

Kieruj, N.D. (2012). Question format and response style behavior in attitude research. PhD thesis. Tilburg University.

Krosnick, J.A. (1991). Response strategies for coping with the cognitive demands of attitude measures in surveys. *Applied Cognitive Psychology* 5: 213–236.

Krosnick, J.A. (1999). Survey research. *Annual Review Psychology* 50: 537–567.

Krosnick, J.A. and Alwin, D.F. (1987). An evaluation of a cognitive theory of response order effects in survey measurement. *Public Opinion Quarterly* 51: 201–219.

Krosnick, J.A. and Fabrigar, L.R. (1997). Designing rating scales for effective measurement in surveys. In: *Survey Measurement and Process Quality* (ed. L. Lyberg, P. Biemer, M. Collins, et al.), 141–164. New York: Wiley.

Leenheer, J. and Scherpenzeel, A.C. (2013). Does it pay off to include non-Internet households in an Internet panel? *International Journal of Internet Science* 8 (1): 17–29.

de Leeuw, E.D. (1992). *Data Quality in Mail, Telephone, and Face-to-Face Surveys*. Amsterdam: TT-publikaties https://edithl.home.xs4all.nl/pubs/disseddl.pdf.

de Leeuw, E.D. (2005). To mix or not to mix data collection modes in surveys. *Journal of Official Statistics* 21 (2): 233–255.

de Leeuw, E.D. (2012). The quality of Internet surveys. *Bulletin de Méthodologie Sociologique* 114: 68–78.

de Leeuw, E.D. and Berzelak, N. (2016). Survey mode or survey modes? In: *The Sage Handbook of Survey Methodology* (ed. C. Wolf, D. Joye, T.W. Smith and Y.-C. Fu), 142–156. London: Sage.

de Leeuw, E.D. and Hox, J.J. (2011). Internet surveys as part of a mixed mode design. In: *Social and Behavioral Research and the Internet: Advances in Applied Methods and Research Strategies* (ed. M. Das, P. Ester and L. Kaczmirek), 45–76. New York: Routledge.

de Leeuw, E.D., Hox, J.J., and Boeve, A. (2016). Handling do-not-know answers: exploring new approaches in online and mixed-mode surveys. *Social Science Computer Review* 34 (1): 116–132.

Lozar Manfreda, K., Bosjnak, M., Berzelak, J. et al. (2008). Web surveys versus other survey modes: a meta-analysis comparing response rates. *International Journal of Market Research* 50 (1): 79–104.

Lugtig, P. and Toepoel, V. (2015). The use of PCs, smartphones, and tablets in a probability-based panel survey: effects on survey measurement error. *Social Science Computer Review* 34: 78–94.

Mohorko, A., de Leeuw, E., and Hox, J. (2013). Internet coverage and coverage bias in Europe: developments across countries and over time. *Journal of Official Statistics* 29 (4): 609–622.

Moors, G., Kieruj, N.D., and Vermunt, J. (2014). The effect of labelling and numbering of response scales on the likelihood of response bias. *Sociological Methodology* 44 (1): 369–399.

Noussair, C.N., Trautmann, S.T., and van de Kuilen, G. (2013). Higher order risk attitudes, demographics, and financial decisions. *Review of Economic Studies* 81 (1): 325–355.

Roberts, C. (2016). Response styles in surveys. In: *The Sage Handbook of Survey Methodology* (ed. C. Wolf, D. Joye, T.W. Smith and Y.-C. Fu), 570–596. London: Sage.

Saris, W.E. and Gallhofer, I.N. (2007). *Design, Evaluation, and Analysis of Questionnaires for Survey Research*. New York: Wiley.

Scherpenzeel, A. (2011). Data collection in a probability based Internet panel: how the LISS panel was built and how it can be used. *Bulletin de Méthodologie Sociologique* 109 (1): 56–61.

Scherpenzeel, A. (2016). Mixing online panel data collection with innovative methods. In: *Proceedings of the ASI-Conference Mixed-Mode-Befragungen, Cologne, 2015*, 25–47. Heidelberg: Springer.

Scherpenzeel, A. and Das, M. (2011). True longitudinal and probability-based Internet panels: evidence from the Netherlands. In: *Social and Behavioral Research and the Internet: Advances in Applied Methods and Research Strategies* (ed. M. Das, P. Ester and L. Kaczmirek), 77–104. New York: Routledge.

Scherpenzeel, A. and Toepoel, V. (2012). Recruiting a probability sample for an online panel: effects of contact mode, incentives and information. *Public Opinion Quarterly* 76 (3): 470–490.

Schwarz, N., Hippler, H.-J., Deutsch, B., and Strack, F. (1985). Response categories: effects on behavioural reports and comparative judgements. *Public Opinion Quarterly* 49: 388–395.

Schwarz, N., Hippler, H.-J., Strack, F., and Bishop, G. (1991). The impact of administration mode on response effects in survey measurement. *Applied Cognitive Psychology* 5: 193–212.

Smyth, J.D. and Pearson, J.E. (2011). Internet survey methods: a review of strengths, weaknesses, and Innovations. In: *Social and Behavioral Research and the Internet: Advances in Applied Methods and Research Strategies* (ed. M. Das, P. Ester and L. Kaczmirek), 11–44. New York: Routledge.

Steegers, C., Dijstelbloem, H., and Brom, F.W.A. (2008). *Meer dan status alleen. Burgerperspectieven op embryo-onderzoek.* [In Dutch: Assessment and points of view of citizens on technology and embryo-research]. The Hague: Rathenau-Instituut, TA rapport 0801.

Sterret, D., Malato, D., Benz, J. et al. (2017). Assessing changes in coverage bias in web surveys in the United States. *Public Opinion Quarterly* 81 (1): 338–356.

Sudman, S. and Bradburn, N.M. (1974). *Response Effects in Surveys.* Chicago: Aldine.

Tourangeau, R. (2017). Mixing modes: tradeoffs among coverage, nonresponse, and measurement error. In: *Total Survey Error in Practice* (ed. P. Biemer, E. de Leeuw, S. Eckman, et al.), 115–132. New York: Wiley.

Tourangeau, R., Rips, L.J., and Rasinski, K. (2000). *The Psychology of Survey Response.* Cambridge: Cambridge University Press.

Tourangeau, R., Conrad, F.G., and Couper, M.P. (2013). *The Science of Web Surveys.* New York: Oxford University Press.

Wengrzik, J., Bosnjak, M., and Lozar Manfreda, K. (2016). Are web surveys still inferior? An updated and extended meta-analysis comparing response rates. Presentation at RC33 Conference, Leicester, UK (September 2016). https://doi.org/10.13140/RG.2.2.34679.21923.

Ye, C., Fulton, J., and Tourangeau, R. (2011). More positive or more extreme? A meta-analysis of mode differences in response choice. *Public Opinion Quarterly* 75 (2): 349–365.

9

Conflicting Cues: Item Nonresponse and Experimental Mortality

David J. Ciuk[1] *and Berwood A. Yost*[2]

[1] *Department of Government, Franklin & Marshall College, Lancaster, PA, USA*
[2] *Franklin & Marshall College, Floyd Institute for Public Policy and Center for Opinion Research, Lancaster, PA, USA*

9.1 Introduction

One of the primary threats to the validity of any experimental design is experimental mortality, which in a survey experiment is best described as the noncompliance of research participants (Campbell and Stanley 1966). Broadly, noncompliance refers to an experimental participant's refusal to follow prescribed experimental protocols. In medical research testing drug efficacy, for example, noncompliant participants might not take their prescribed medication on time, in the correct dosages, or for the full duration of the trial. In survey research experiments, noncompliance appears when a participant refuses to answer or provides a "don't know" response that is not explicitly offered. Noncompliance in surveys is commonly referred to as item nonresponse. Noncompliance is of particular concern when participants drop out of one experimental condition at a higher rate than another condition, suggesting that the treatment itself may be the cause (McDermott 2012). Specifically, this chapter examines the implications of survey-based experiments that by their design may increase item nonresponse. As with any form of noncompliance, item nonresponse produces less information, less efficient estimates, and statistical tests with less statistical power (de Leeuw et al. 2003). From a total survey error perspective, understanding the conditions that increase item nonresponse can help survey researchers identify the causes of error and either eliminate or better predict and model them (Groves and Lyberg 2010).

9.2 Survey Experiments and Item Nonresponse

The use of random assignment experiments in political science journals has increased significantly since the end of the last century, and many of those experimental designs are survey based (Druckman et al. 2012). Population-based experiments are also being used more often by sociologists, psychologists, and economists (Mutz 2011). The growth of survey-based experiments likely stems at least in part from the fact that they provide greater external validity than the convenience samples offered by on-campus research (Sniderman 2012), although they do not eliminate these concerns altogether. Perhaps the greatest benefit of survey-based experimentation

Experimental Methods in Survey Research: Techniques that Combine Random Sampling with Random Assignment, First Edition.
Edited by Paul J. Lavrakas, Michael W. Traugott, Courtney Kennedy, Allyson L. Holbrook, Edith D. de Leeuw, and Brady T. West.

Companion Website: www.wiley.com/go/Lavrakas/survey-research

is its strong internal validity; but the validity and underlying assumptions about experimental manipulation are questionable in circumstances where experimental outcomes are affected by treatment assignment (Gaines and Kuklinski 2012). A fundamental assumption of experimental designs is that those who comply and those who do not comply with an experimental treatment are roughly proportional across treatment conditions. This assumption is violated when compliance is a function of group assignment.

Some experimental treatment conditions have the potential to produce higher noncompliance rates by producing much higher proportions of "don't know" and nonresponse to survey questions. For example, treatment conditions that offer conflicting cues, where a participant is asked to respond to a condition that ostensibly offers conflicting information produce greater rates of nonresponse. This is illustrated in Ciuk and Yost's (2016) paper that created conflicting cues by linking a specific policy cue to a specific party cue in a way that could appear counterintuitive; for example, Democrats who support "fracking" over environmental protection. Boudreau and MacKenzie (2014) also identified similar treatment conditions, which they termed counter-stereotypical conditions, as a source of conflicting research findings. Boudreau (2013) noted that 12–13% more respondents within an experimental condition that received conflicting information did not answer the question. Experiments in political communications research that manipulate party and policy cues are particularly likely to have these kinds of conditions. Why are these conditions likely to increase item nonresponse?

Being asked to respond to a survey question invokes a specific set of cognitive tasks. One model suggests that there are four distinct cognitive tasks involved in responding to a survey question: comprehension, retrieval, judgment, and response formation (Tourangeau et al. 2000). The first cognitive task begins with efforts to comprehend the request for a response. The respondent will listen to the question and accompanying instructions, attempt to interpret the form of the question, attempt to identify the type of information being sought, and then attempt to link the ideas in the question to key concepts that the individual understands or holds relevant. At this stage of the process, questions that have unclear or ambiguous terms or cause confusion could prevent respondents from answering (Krosnick 2002). Ambivalence and a lack of understanding could also decrease the likelihood of responding.

The next cognitive task, retrieval, requires respondents to recall pertinent information from memory. For attitudinal questions, respondents must recall and retrieve relevant information that could include existing evaluations, general impressions, general values, and any feelings or beliefs associated with the topic. Those with less strongly held feelings about an issue are likely to have a more difficult path to retrieval and receiving a conflicting cue could make that task even more difficult for these respondents.

The judgment task requires a respondent to process, combine, supplement, and otherwise make sense of the information they have retrieved. Attitude questions tend to be heavily context dependent. It may be possible for some respondents who offered a specific response to a prior question to respond differently to a later question; these are called conditional context effects. The upshot of these context effects is primarily that they affect the accessibility of information used to make a judgment about the question being asked. Retrieval and judgment tasks often rely on stereotypes and schema to help respondents formulate a response. It is plausible that questions prior to an experiment that invoke political responses could influence the information that is available for use during the judgment task. If partisan attachments have been called to mind by previous questions, offering conflicting cues could invoke confusion and even changes in judgment (Boudreau 2013).

The final task involves offering a response to the question. The decision to offer a response involves decisions about fitting the response onto the offered response categories and it also involves deciding whether to answer. The decision about whether to answer a question can be a product of whether a respondent believes a question is sensitive; social desirability, concerns about an invasion of privacy, or concerns about some risk of disclosure could all contribute to perceptions of a sensitive question. If context effects during the judgment stage have encouraged strong partisan thinking, it is reasonable to expect that strong partisans might choose not to answer a question that offers a position in conflict with their preferences but seemingly aligned with their party, or vice versa.

This cognitive model of survey response suggests how conflicting cues can increase nonresponse. In part, it seems likely that offering conflicting cues will cause confusion, particularly among those who are most likely to recognize the conflicts. It is possible that conflicting cues complicate the retrieval task, perhaps by calling into play more potential information and by encouraging respondents to square the question with their existing knowledge and beliefs about the topic – those who employ the cognitive effort might assemble the information they need to answer while others might be discouraged and cease the retrieval process which could encourage them to fall back on shortcuts and cues or to not respond. Further, when the question itself provides information that is counter-stereotypical it is easy to imagine this could interfere with the construction of a response, the judgment task. These judgments could be further influenced by where the experiment appears in the survey and what other topics have been addressed during the interview. For researchers interested in identifying the effects of party and policy cues on preferences, the sorting of the electorate into increasingly homogenous groups makes it more likely that respondents will recognize conflicting cue conditions and react to them.

One could also think about nonresponse as a product of motivated reasoning. If citizens are not "Bayesian updaters," and the process of updating prior evaluations based on new information is "subject to a range of unconscious biases designed to support prior preferences, rather than to rationally update them" (Redlawsk 2002, p. 1022), then one might expect respondents encountering conflicting cues to have an especially difficult time resolving the cognitive dissonance in time to respond to the relevant survey item. Instead, these respondents may leave the item blank and be able to go on with their day without having to change closely held existing beliefs. In other words, item nonresponse offers these motivated reasoners (Kunda 1987, 1990; Lodge and Taber 2000) a way out of a situation that would otherwise require them to deal with information that challenges the preexisting beliefs.

One concern about communications research that relies on the use of cues is the similarities in how people process information about politics and how they process requests for a survey response. A primary interest in politics is how deeply partisans process policy and party relevant information. Certain conditions, such as issue salience and strong partisanship and ideology (Arceneaux 2008; Ciuk and Yost 2016) have been shown to encourage systematic processing, i.e. deeper thinking, about political topics, which suggests that partisans are able to go beyond partisan cues in forming opinions. But, the use of systematic processing is highly conditional and partisans will often use such cues to engage in heuristic processing. This dual-processing model can also be applied to models of survey response. Satisficing, where respondents choose not to answer because it is too much cognitive work to optimize a response, is an example of this (Krosnick 2002). Some questions encourage deeper thinking than others, and some respondents are also more likely to think deeply about their answers. Respondents, for example, appear less susceptible to issue-specific party cues when they have a deep understanding of an issue that does not comport with a specific cue (Slothuus 2010). More generally, respondents who are more politically aware (Kam 2005), respondents who have a greater need for cognition (Bullock 2011), and respondents evaluating an issue that is more personally salient (Ciuk and Yost 2016) are more likely to use systematic processing. It seems that respondents switch between systematic and heuristic processes as the characteristics of issues motivate them to do so and it seems likely that survey questions evoke the same types of response. Much more work needs to be done to understand how the cognitive processes involved in survey-based experiments influence what we think we understand as a result of such experiments.

9.3 Case Study: Conflicting Cues and Item Nonresponse

In most observational studies about party cues, respondents are asked to place themselves and the parties on a series of issues. Those who answer the questions about party stances are assumed to have received cues (e.g. information they encounter on an everyday basis) about those stances, and correlations between self and party placement are then taken as evidence that party cues affect individuals' attitudes. While this approach has been used in important research, it relies on several potentially problematic assumptions. First, it assumes that people actually do receive party cues – that they are not expressing their views on issues and policies about which they are uninformed. Second, it assumes that the information contained in the party cues dictates respondents' perceptions of party stances. That is, it assumes that people are not projecting their own attitudes onto the parties, that there is no endogeneity. Third (and this is a problem with many observational studies), it assumes that the regression model used to estimate the effects of party cues on policy attitudes is properly specified and that there are no omitted variables correlated with the reception of cues and the dependent variable (i.e. there is no omitted variable bias). Finally, in a real-world setting, party cues are often tightly correlated with other politically relevant variables (e.g. policy information), so estimating the relative effects of both of those variables is exceedingly difficult to do. In this case, random assignment might be necessary to effectively pick apart the effects of party cues and policy information on individuals' attitudes.

9.4 Methods

To illustrate some potential problems associated with noncompliance, we focus on a study that relied on a survey experiment where some respondents were exposed to treatment conditions with conflicting cues. Ciuk and Yost (2016), in an attempt to clarify circumstances when voters might engage in systematic versus heuristic processing, embedded an experiment in a poll of 622 Pennsylvania registered voters that was in the field from 29 January 2013 to 3 February 2013.[1] In terms of sex, education, and income, the sample was representative of the population of registered voters in the state of Pennsylvania. Participants were randomly assigned to one of two salience conditions, either a fracking (higher salience) condition or a storm-water management (lower salience) condition. Participants in both groups were then read a policy statement that included one of the three party cue conditions (no cues/Democrats support/Republicans support) and one of the two policy conditions (liberal/conservative).[2]

1 The live-interviewer telephone survey was conducted at the Center for Opinion Research at Franklin and Marshall College. The final sample included 313 Democrats, 232 Republicans, and 77 respondents registered as Independent/Other. The sample of registered voters was obtained from Voter Contact Services. The survey had a cooperation rate of 86% (AAPOR CR3), a response rate of 24% (AAPOR RR3), and a refusal rate of 4% (AAPOR Ref R2). Survey results were weighted by region, gender, and party using an iterative weighting algorithm to reflect the known distribution of those characteristics as reported by the Pennsylvania Department of State.

2 Consistent with other research in this field, the publication resulting from the experiment was focused on the influence of party cues, policy cues, and issue salience on partisans, meaning that those who did not identify as Republicans or Democrats were excluded from our analyses.

9.5 Issue Selection

An essential element of our experimental design was the choice of issues to represent the high salience and low salience treatment conditions. We considered three primary criteria when selecting the issues we would manipulate. First, we thought both issues should be from the same issue domain so respondents in each group would have the opportunity to tap the same subset of attitudes when forming a response. Both fracking and storm-water management tap into issues related to the economy and the environment. Respondents taking policy information into account should draw from the same basic subset of attitudes.

Second, we thought it was essential that the claims of office holders taking counter-stereotypical stances were believable (e.g. Republicans endorsing a liberal policy, Democrats endorsing a conservative policy). Though storm-water runoff management is an issue that the Pennsylvania state legislature has discussed, it is neither engrained in political rhetoric nor in public opinion. The credibility of counter-stereotypical stances for storm-water management, if there are stereotypical stances in the first place, is of little concern. Fracking, although more prominent and politicized in Pennsylvania politics than storm-water management, has been one issue where political figures do not always adhere to a standard party line. Although it is generally the case that Pennsylvania Republicans are "profracking" and Pennsylvania democrats are "antifracking," support and opposition to fracking is not uniformly partisan – there are profracking Democrats and antifracking Republicans that have made their stances publicly known. While it may not be normal for a Republican to take an antifracking stance, or for a Democrat to take a profracking stance, these combinations are not unknown. We thought it likely that our conflicting cues would be perceived as believable.

Our third consideration in selecting issues for this experiment was that media coverage about the issues, and the personal importance respondents attributed to the issues, must differ. There is little doubt that fracking was a "hotter" issue than storm-water management at the time of the experiment. Governor Tom Corbett, during his 2014 reelection campaign, and all four major Democratic opponents in the 2014 Pennsylvania gubernatorial race, cited various aspects of fracking (e.g. regulation, taxation, land availability, etc.) as central issues of their campaigns. Survey data collected by the Center for Local, State, and Urban Policy at the University of Michigan suggested that, as of November 2012, 86% of Pennsylvanians had at least some information on fracking (Brown et al. 2013). An item in our survey asked respondents to indicate how often they read newspaper articles about their assigned issue, and on average, respondents encountered significantly more stories about fracking ($M = 1.89$) than they did storm-water management ($M = 2.47$), $t\ (601) = 7.189, p < 0.001$ (lower scores indicate more exposure). Last, a LexisNexis keyword search for fracking and hydraulic fracturing in Pennsylvania's two most widely read newspapers, the *Philadelphia Inquirer* and the *Pittsburgh Post-Gazette*, yielded 323 results while a search for storm-water runoff and storm-water management yielded only 46 results during the same time periods. All this suggested that people heard more about fracking than they did about storm-water management at the time of the experiment. In addition to being a hotter issue, data suggests that fracking is more salient than storm-water runoff management at the individual level as well. According to the 2012 survey data from the Center for Local, State, and Urban Policy, 84% of Pennsylvania residents see fracking as a very or somewhat important issue for the state's economy, and 82% of the Pennsylvania respondents could name at least one environmental risk associated with fracking (Brown et al. 2013). A second item in our survey asked respondents to rate the importance of the issue to which they were assigned. On average, respondents in the fracking group ($M = 2.06$) attributed more importance to their issue than did respondents in the storm-water management group ($M = 2.38$), $t\ (595) = 4.411$, $p < 0.001$ (lower scores indicate greater importance).

9.6 Experimental Conditions and Measures

Subjects in the fracking group were read a policy statement that included one of three party cues conditions (no cues/Democrats support/Republicans support) and one of two policy conditions (liberal/conservative). The experimental fracking conditions are as follows:

Fracking: Liberal Policy Content

Pennsylvania [lawmakers/Democrats/Republicans] are close to winning a key policy battle as the state legislature may consider two bills concerning hydraulic fracturing or fracking: the first prohibits fracking in state parks, and the second allows municipalities to limit where fracking can take place within their boundaries. [Supporters/Democratic supporters/Republican supporters] of the bills said they are needed to protect residents from water contamination and air pollution, as well as to preserve state parks for future generations. [Opponents/Republican opponents/Democratic opponents] say that they may lead to higher energy prices, a greater dependence on foreign oil, and a decrease in energy security.

Fracking: Conservative Policy Content

Pennsylvania (lawmakers/Democrats/Republicans) are close to winning a key policy battle as the state legislature may consider two bills concerning hydraulic fracturing or fracking: the first allows fracking in state parks, and the second prohibits municipalities from limiting where fracking can take place within their boundaries. (Supporters/Democratic supporters/Republican supporters) of the bills said they are needed to lower energy prices, decrease our dependence on foreign oil, and increase energy security. (Opponents/Republican opponents/Democratic opponents) say they may lead to water contamination, air pollution, and the eventual degeneration of state parks.

Subjects in the storm-water management group followed the same basic process. The text of the storm-water management policy cues are as follows:

Storm-Water Management: Liberal Policy Content

Pennsylvania (lawmakers/Democrats/Republicans) are close to winning a key budget battle as the state legislature may consider a bill that increases funds for municipalities that have been affected by storm-water runoff. (Supporters/Democratic supporters/Republican supporters) of the bill say the funds will allow municipalities to create trenches and storm-water repositories in an effort to prevent flooding, improve water quality, and maintain ecosystem health. (Opponents/Republican opponents/Democratic opponents) say the bill will lead to reduced school funding, higher taxes, and a possible budget deficit.

Storm-Water Management: Conservative Policy Content

Pennsylvania (lawmakers/Democrats/Republicans) are close to winning a key budget battle as the state legislature may consider a bill that decreases funds for municipalities that have been affected by storm-water runoff. (Supporters/Democratic supporters/Republican supporters) of the bill say the cuts are needed to balance a budget that increases school funding without raising taxes or cutting other social welfare programs. (Opponents/Republican opponents/Democratic opponents) say the cuts may prohibit municipalities from creating trenches and storm-water repositories designed to prevent flooding, improve water quality, and maintain ecosystem health.

After being read the stimulus, participants were asked, “Taking everything you have heard into consideration, do you approve or disapprove of the changes to [fracking/storm water] policy?" Responses were coded on a 4-point scale such that 4 = strongly approve, 3 = approve, 2 = disapprove, and 1 = strongly disapprove. We also asked several additional questions to use as covariates that identified the personal importance of each issue,[3] the amount of exposure each respondent had to the issue,[4] and each respondent's expressed need for cognition.[5]

9.7 Results

Subjects who received a party cue and a policy cue that conflicted (e.g. Republicans supporting a liberal policy solution, Democrats offering a conservative policy solution) were classified as receiving a conflicting cue. Nearly one in four (27%) participants who received a conflicting cue offered a "don't know" response while only one in seven (14%) of those who received consistent or neutral cues offered a "don't know" response. Table 9.1 shows that nonresponse differed across the treatment conditions. There was no major effect between treatment condition and item nonresponse, but there is a significant interaction between party cue and policy cue ($F(6, 592) = 3.286$, $p = 0.003$). Table 9.1 shows that item nonresponse is highest for the low salience issue when it is presented with a conflicting cue (i.e. Democrats supporting a conservative policy position and Republicans supporting a liberal policy position).

Besides the loss of sample power, the primary concern about losing cases in an experimental design of this type is that the stable unit treatment value assumption is violated. Simply put, experimental outcomes should not be affected by which treatment a participant is assigned to. If there is no discernible pattern of nonresponse among participants assigned to the conflicting cue condition then there is less concern that final estimates produced by the experiment will be biased. The pattern of nonresponse for the conflicting cues suggests that our estimates are not maximally efficient, and calls into question whether or not they are unbiased. A logistic regression analysis reveals several patterns within the data. First, it reconfirms that offering a conflicting cue nearly doubles the odds that a respondent will offer a "don't know" response. Second, the analysis shows that variables we might expect to be associated with our outcomes are significantly associated with nonresponse. Specifically, self-described conservatives have about twice the odds of offering a "don't know" response and those who say the issue they were asked about is "very important" to them are less likely to do so (see Table 9.2). There is clearly

Table 9.1 Item nonresponse by treatment group.

	Party cue					
	Neutral intro.		Democrat intro.		Republican intro.	
Policy cue	**Mean**	**Count**	**Mean**	**Count**	**Mean**	**Count**
High salience liberal policy	0.16	51	0.14	59	0.23	42
High salience conservative policy	0.13	35	0.15	44	0.15	66
Low salience liberal policy	0.14	50	0.03	53	0.33	54
Low salience conservative policy	0.31	58	0.37	54	0.16	57

3 How important an issue is (fracking/storm water management) to you, personally? Is it extremely important, somewhat important, not very important or not at all important?

4 How often have you seen (fracking/storm water management) stories in the news in the past year or so: frequently, sometimes, seldom, or never?

5 Some people prefer to solve simple problems instead of complex ones. Other people prefer to solve complex problems instead of simple ones. Do you prefer to solve simple or complex problems? Do you greatly prefer, somewhat prefer, or only slightly prefer solving (simple/complex) problems compared to (complex/simple) problems?

Table 9.2 Logistic regression analysis predicting noncompliance.

Coefficients	Estimate	Std. error
(Intercept)	−2.21***	(.680)
Republican	0.24	(.495)
Democrat	0.35	(.500)
Conflict cue	0.61**	(.241)
Age 35–54	0.18	(.454)
Age 55 and older	0.19	(.432)
ED some college	−0.31	(.300)
ED college or more	−0.49*	(.283)
Nonwhite	0.32	(.451)
Conservative	0.63**	(.289)
Liberal	−0.14	(.350)
Issue importance	−1.14***	(.407)
Issue exposure	−0.18	(.288)
Need for cognition	0.04	(.089)
Female	0.47*	(.245)
High salience issue	−0.30	(.241)

Note: $^{*}p < 0.10$; $^{**}p < 0.05$; $^{***}p < 0.01$.
n: 560.
AIC: 506.31.
Cell entries are logit coefficients with standard errors in parentheses.

a pattern whereby conservatives, those less invested in an issue, and those offered a conflicting cue are generally less likely to respond.

This pattern of nonresponse raises some concern that the stable unit treatment assumption has been violated and that our final estimates could be biased. When we reviewed the characteristics of the total sample according to assignment to treatment condition, we found that the groups were statistically equivalent in terms of gender, education, political ideology, and political party affiliation. If the initial randomization process created statistically equivalent groups, what effect did the identified nonresponse have on treatment group assignment? The short answer appears to be none: comparing treatment assignment only for those who provided a response once again shows no differences in terms of gender, education, political ideology, or political party affiliation.

Although a cursory review of the data suggests the stable treatment assumption has not been violated, conducting a more thorough exploration of the potential effects of nonresponse on our final estimates seems a reasonable next step.

9.8 Addressing Item Nonresponse in Survey Experiments

The power of a randomized experiment is rooted in *ignorability*. That is, the assignment mechanism (i.e. random assignment) creates a situation in which respondents' assignment to a treatment condition is statistically independent from potential outcomes, and the only cause of differences between treatment groups on the outcome variable is the treatment itself. When nonresponse occurs, it is likely *nonignorable* (i.e. caused by factors that may affect potential outcomes). If, in this situation, researchers employ listwise deletion and omit nonrespondents from their analysis as seems common, the potential for biased estimates of treatment effects exists (Little et al. 2009).

Thinking about how respondents process survey questions suggests how the use of conflicting cues in experiments creates conditions that are likely to increase item nonresponse, but what strategies might decrease item nonresponse? The first and obvious remedy to item nonresponse is to write the best possible questions, i.e. questions that are clear, understandable, and not burdensome. Knowing that they are using conflicting cues should encourage researchers to ensure that the concepts and issues included in the experiment are conceptually clear and understandable. Issue selection may also be important, as less salient issues may complicate the recall task and increase nonresponse, although including highly salient issues may also raise concerns about pretreatment effects (Druckman and Leeper 2012). Another strategy might be to use probes to encourage respondents to think again about their response and offer a substantive response, although this runs the risk of encouraging people who do not truly hold an opinion to form one (Beatty and Herrmann 2002). Researchers should remember that part of the decision to respond is motivational, and that concerns about burden and sensitivity can suppress response.

It might seem wise to try to avoid offering such cues. One way of avoiding the use of conflicting cues and the confounds that might arise is to offer designs that eliminate conflicting information (Boudreau and MacKenzie 2014). This solution is viable when assessing the effects of cues over multiple issues, but it requires a sufficient number of issues so that respondents can receive both reinforcing and nonreinforcing cues. The other limitation of this approach is that it does not allow researchers to account for circumstances where some partisans have positions outside the mainstream of their party.

Besides selecting the right issues and writing good questions, another remedy for dealing with item nonresponse could reside in more sophisticated treatment of missing data. More sophisticated data analytic strategies may be one corrective in circumstances where nonresponse is present, but these techniques don't seem common in experimental circumstances. For instance, it seems to be common practice to either eliminate nonrespondents from all analyses or substitute average scores when conducting analyses of the experimental outcomes (see, for example, Arceneaux 2008; Boudreau 2013; Boudreau and MacKenzie 2014).

There are imputation methods available to reduce the hazardous effects of nonresponse, but most traditional methods of imputation (e.g. mean imputation, regression-based imputation, or any other deterministic method of imputation) are known to produce statistically undesirable results. More specifically, these methods of imputation routinely underestimate variances and produce biased parameter estimates that are based on variances (e.g. regression coefficients). In a similar vein, these imputation methods underestimate standard errors, which lead to inflated test statistics and p-values. Briefly, conventional statistical software has no way to distinguish "real" observations (where respondents complied with the protocol) from imputed observations. As such, the sample n is artificially inflated and estimates of variance and standard errors are underestimated. This problem becomes more troublesome the larger the fraction of missing data. Allison (2009) recommends that researchers avoid conventional methods of imputation and that results based on these methods are read with a healthy dose of skepticism.

In many instances, multiple imputation provides a viable alternative when nonresponse is a problem. The main ideas of multiple imputation are straightforward: identify a variable (or variables) with problematic missing data then build a model to predict the underlying values of the missing data. Since the imputation model is a predictive model (rather than an analytical model) one is encouraged to add all possible relevant variables into the model to improve predictive power. Variables that are highly correlated with the variable being modeled, including those that might introduce endogeneity, should be added to the model (King et al. 2001; Allison 2002; Little and Rubin 2002).[6] The process is called *multiple* imputation because it pulls multiple random draws (for example, m draws) from the necessary posterior distributions for each missing value. The imputation program produces m complete datasets on which the researcher can use complete-data analysis methods. The researcher obtains "final" results by averaging over the m analyses on the imputed datasets (Rubin 1987; Schafer 1997; King et al. 2001; Allison 2002).

Under what conditions can multiple imputation be used to impute values on the dependent variable? If the data are missing at random (MAR) and there are several auxiliary variables (i.e. variables, in addition to treatment assignment, that can be used to model missing data on the dependent variable), then multiple imputation is a viable strategy that increases statistical efficiency and, in some cases, reduces bias. If, however, the data are MAR and there are no auxiliary variables, then multiple imputation is not a viable strategy (Allison 2009). Additionally, it may be the case that "don't know" is an accurate response, not an unobserved response (King et al. 2001) and that some type of imputation produces data unrepresentative of respondents' thought processes.

Tables 9.3 and 9.4 provide the results of models in which the results presented in the original Ciuk and Yost (2016) paper are replicated using different techniques to deal with missing data. In both tables, column 1 includes the estimates from the original paper based on listwise deletion of cases with missing data, column 2 provides estimates based on mean imputation, column 3 provides estimates based on regression-based imputation, and column 4 provides estimates based on multiple imputation. Note that while substantive results are all very similar, there are important differences between models. Most notably, in Table 9.4 (the model fit to Republican respondents) the smaller standard errors in columns 2 and 3 cause the coefficient on *liberal policy* × *fracking* to reach statistical significance where it fails to do so in columns 1 and 4.[7]

Does multiple imputation work when the problematic missing data is only on the dependent variable? The answer is not straightforward, and it is up for debate. While some argue that it is a viable alternative when there are a sufficient number of auxiliary variables (e.g. King et al. 2001), other scholars argue in the opposite direction. Even when there are sufficient auxiliary variables, von Hippel (2007) argues that imputed values on the dependent variable may add noise and unnecessarily inflate standard errors, and this problem is exacerbated when there is a large amount of missing data on the dependent variable or when only a small number of data sets are imputed. In fact, in a data set that contains missing data on both the dependent and independent variables, von Hippel recommends that the multiple imputation routine be run, but then that observations in the original data set with nonresponse on Y be omitted from ensuing analyses. In other words, multiple imputation provides a way forward when there is problematic missing data in independent variables, but when the problematic missing data is on the dependent variable alone, the answer is less clear.

A conceptually similar method to deal with noncompliance (and nonresponse) is to calculate propensity scores to model noncompliance. A propensity score is "the conditional probability of assignment to a particular treatment given a vector of observed covariates" (Rosenbaum and Rubin 1983, p. 41). The strength of propensity scores as a basis for matching has to do

6 This process is not problematic here because the multiple imputation routine is meant to calculate the joint distribution of all variables under consideration. Variables added to the model that introduce endogeneity or have no effect on the imputation variable do not affect the calculation of the joint distribution; thus, they do not produce serious problems in the model (King et al. 2001, pp. 56–57).

7 For interested readers, R code is available in the supplementary material.

Table 9.3 OLS models: Democrats.

	Model 1	**Model 2**	**Model 3**	**Model 4**
Intercept	2.16***	2.17***	2.15***	2.15***
	(.196)	(.156)	(.154)	(.197)
Fracking	−0.41	−0.33	−0.40	−0.33
	(.306)	(.251)	(.247)	(.331)
Dem. sup. cues	0.91***	0.76***	0.91***	0.85***
	(.232)	(.193)	(.190)	(.259)
× Fracking	−0.80**	−0.61**	−0.78***	−0.65*
	(.355)	(.293)	(.288)	(.338)
Rep. sup. cues	0.22	0.19	0.22	0.30
	(.245)	(.199)	(.196)	(.232)
× Fracking	−0.36	−0.30	−0.36	−0.45
	(.365)	(.300)	(.295)	(.377)
Lib. pol.	−0.12	−0.05	−0.12	−0.05
	(.196)	(.161)	(.158)	(.188)
× Fracking	1.15***	0.88***	1.13***	0.894***
	(.293)	(.240)	(.236)	(.302)
R^2	0.17	0.14	0.20	0.14
N	231	284	284	284

Notes: *$p < 0.10$; **$p < 0.05$; ***$p < 0.01$.
Model 1: Listwise deletion; Model 2: Mean imputation; Model 3: Model-based imputation; Model 4: Multiple imputation.
Cell entries are OLS coefficients with standard errors in parentheses.

with reducing the information of several variables down into a one-dimensional score. And, respondents need only to be matched on that score (Rosenbaum and Rubin 1983).

Follmann (2000) describes two propensity score methods to model noncompliance. The first involves estimating a logit model on respondents in the treatment group in which the compliance information (e.g. did the respondent answer the question or not) is the dependent variable, and the independent variables are baseline covariates. Coefficients from the model are then used to create propensity scores for participants in both the treatment and control groups. Last, the outcome variable is regressed on treatment condition, propensity scores, and the condition by propensity score interaction. The second method is similar to the first insofar as it starts with estimating propensity scores with compliance information for those in the treatment group and baseline covariates for both groups, but it differs in that its focus is on compliers. The next steps are to estimate the average outcome for compliers in the treatment group, which is straightforward (i.e. the mean outcome for respondents in the treatment group), and to estimate the average outcome for the compliers in the control group. To do so requires estimating the average outcome for all those in the control group that would have complied with the treatment if assigned to do so. It should be noted that the comparison one is making when using this method is between compliers in the treatment group and compliers (or, more accurately, those that would have complied if they were in the treatment group) in the control group. In other words, results can be generalized to compliers – not to the whole population. Follman notes that the first method is preferable to the second in most situations.

While it is certainly the case that advances in multiple imputation and propensity scores help researchers deal with noncompliance and nonresponse, to use the methods properly one

Table 9.4 OLS models: Republicans.

	Model 1	Model 2	Model 3	Model 4
Intercept	2.88***	2.80***	2.88***	2.74***
	(.239)	(.174)	(.173)	(.243)
Fracking	−0.54	−0.41	−0.53**	−0.41
	(.326)	(.251)	(.249)	(.310)
Dem. sup. cues	−0.64**	−0.50**	−0.64***	−0.58**
	(.273)	(.213)	(.212)	(.290)
× Fracking	0.38	0.30	0.41	0.30
	(.381)	(.303)	(.301)	(.418)
Rep. sup. cues	0.08	0.09	0.08	0.17
	(.276)	(.211)	(.210)	(.259)
× Fracking	0.20	0.14	0.20	0.10
	(.376)	(.296)	(.294)	(.368)
Lib. pol.	−0.17	−0.16	−0.17	−0.12
	(.222)	(.171)	(.169)	(.233)
× Fracking	0.49	0.42*	0.48*	0.43
	(.310)	(.244)	(.242)	(.281)
R^2	0.07	0.06	0.09	0.08
N	212	268	268	268

Notes: *$p < 0.10$; **$p < 0.05$; ***$p < 0.01$.
Model 1: Listwise deletion; Model 2: Mean imputation; Model 3: Model-based imputation; Model 4: Multiple imputation.
Cell entries are OLS coefficients with standard errors in parentheses.

must be able to accurately model the noncompliance/nonresponse. That is, the data set must have an adequate number of auxiliary variables to satisfy ignorability. Often, data gathered specifically to test a set of hypotheses does not have an adequate number of auxiliary covariates to model noncompliance, and thus, the previously discussed methods may not work. Methods that require fewer assumptions and less data may be more appropriate for researchers operating with less latitude.

9.9 Summary

Researchers engaged in population-based, random assignment survey experiments should be concerned about participant noncompliance, which in the survey experiment manifests itself as item nonresponse. The primary concern is that some experimental conditions may discourage providing a response, potentially violating the stable unit treatment assumption. We have identified a specific circumstance where nonresponse varied based on treatment assignment when participants were presented with conditions where the experimentally manipulated information they received may have been in conflict with their expectations. Other research has exhibited a similar pattern.

Our findings fit with two expectations that arise from common thinking about the cognitive aspects of survey response. First, those with less strongly held feelings about an issue, because they have a more difficult path to retrieval and because the conflicting cue makes that task even more difficult, were less likely to offer a substantive response. Second, we found that conservatives were more likely not to answer a question when it offered a position in conflict with their preferences even though it was aligned with their party, which lends some support to the idea that strong partisans might not respond when cues conflict.

The use of data analytic correctives is called for in circumstances where nonresponse leads to differences in the characteristics of the experimental treatment groups, that is, when the groups are no longer statistically equivalent. Identifying the patterns of nonresponse, as we have done here, and determining whether nonresponse affected treatment assignment is the first step in understanding whether additional data analysis to correct for the biases created by nonresponse are required. We were fortunate that in this experiment, such correctives were not needed.

References

Allison, P.D. (2002). *Missing Data*. Thousand Oaks, CA: Sage.

Allison, P.D. (2009). Missing Data. In: *The SAGE Handbook of Quantitative Methods in Psychology* (ed. R.E. Millsap and A. Maydeu-Olivares). Washington, DC: Sage.

Arceneaux, K. (2008). Can partisan cues diminish democratic accountability? *Political Behavior* 30 (2): 139–160.

Beatty, P. and Herrmann, D. (2002). To answer or not to answer: decision processes related to survey item nonresponse. In: *Survey Nonresponse* (ed. R. Groves, D. Dillman, J. Eltinge and R. Little), 71–85. New York: Wiley.

Boudreau, C. (2013). Gresham's Law of Political Communication: How citizens respond to conflicting information. *Political Communication* 30: 193–212.

Boudreau, C. and MacKenzie, S. (2014). Informing the electorate? How party cues and policy information affect public opinion about initiatives. *American Journal of Political Science* 58 (1): 48–62.

Brown, E., Hartman, K., Borik, C. et al. (2013). *Public Opinion on Fracking: Perspectives from Michigan and Pennsylvania*. Ann Arbor, MI: Center for Local, State, and Urban Policy at the University of Michigan. http://closup.umich.edu/files/nsee-fracking-Zfall-2012.pdf.

Bullock, J.G. (2011). Elite influence on public opinion in an informed electorate. *American Political Science Review* 105 (3): 496–515.

Campbell, D.T. and Stanley, J.C. (1966). *Experimental and Quasi-Experimental Designs for Research*. Chicago: Rand McNally.

Ciuk, D. and Yost, B. (2016). The effects of issue salience, elite influence, and policy content on public opinion. *Political Communication* 33 (2): 328–345.

Druckman, J. and Leeper, T. (2012). Learning more from political communication experiments: pretreatment and its effects. *American Journal of Political Science* 56 (4): 875–896.

Druckman, J.N., Green, D.P., Kulkinski, J.H., and Lupia, A. (2012). *Cambridge Handbook of Experimental Political Science*, 2e. New York: Cambridge University Press.

Follmann, D.A. (2000). On the effect of treatment among would-be treatment compliers: an analysis of the multiple risk factor intervention trial. *Journal of the American Statistical Association* 95: 1101–1109.

Gaines, B.J. and Kuklinski, J.H. (2012). Treatment effects. In: *Cambridge Handbook of Experimental Political Science*, 2e (ed. J.N. Druckman, D.P. Green, J.H. Kuklinski and A. Lupia), 445–458. New York, NY: Cambridge University Press.

Groves, R. and Lyberg, L. (2010). Total survey error past, present, and future. *Public Opinion Quarterly* 74 (5): 849–879.

von Hippel, P.T. (2007). Regression with missing y's: an improved strategy for analyzing multiply imputed data. *Sociological Methodology* 37: 83–117.

Kam, C.D. (2005). Who toes the party line? Cues, values, and individual differences. *Political Behavior* 27: 163–182.

King, G., Honaker, J., Joseph, A., and Scheve, K. (2001). Analyzing incomplete political science data: an alternative algorithm for multiple imputation. *American Political Science Review* 95: 49–69.

Krosnick, J. (2002). The causes of no-opinion responses to attitude measures in surveys: they are rarely what they appear to be. In: *Survey Nonresponse* (ed. R. Groves, D. Dillman, J. Eltinge and R. Little), 87–100. New York: Wiley.

Kunda, Z. (1987). Motivated inference: self-serving generation and evaluation of causal theories. *Journal of Personality and Social Psychology* 53: 636–647.

Kunda, Z. (1990). The case for motivated reasoning. *Psychological Bulletin* 108: 480–498.

de Leeuw, E., Hox, J., and Husman, M. (2003). Prevention and Treatment of Item Nonresponse. *Journal of Official Statistics* 19 (2): 153–176.

Little, R.D. and Rubin, D.B. (2002). *Statistical Analysis with Missing Data*. Hoboken, NJ: Wiley.

Little, R.D., Long, Q., and Lin, X. (2009). A comparison of methods for estimating the causal effect of a treatment in randomized clinical trials subject to noncompliance. *Biometrics* 65: 640–649.

Lodge, M. and Taber, C.S. (2000). Three steps toward a theory of motivated political reasoning. In: *Elements of Reason: Cognition, Choice, and the Bounds of Rationality* (ed. A. Lupia, M.L. McCubbins and S.L. Popkin), 183–213. New York, NY: Cambridge University Press.

McDermott, R. (2012). Internal and external validity. In: *Cambridge Handbook of Experimental Political Science*, 2e (ed. J.N. Druckman, D.P. Green, J.H. Kulkinski and A. Lupia), 27–40. New York: Cambridge University Press.

Mutz, D. (2011). *Population-Based Survey Experiments*. Princeton, NJ: Princeton University Press.

Redlawsk, D.P. (2002). Hot cognition or cool consideration? testing the effects of motivated reasoning on political decision making. *Journal of Politics* 64: 1021–1044.

Rosenbaum, P. and Rubin, D. (1983). The central role of the propensity score in observational studies for causal effects. *Biometrika* 70: 41–55.

Rubin, D.B. (1987). *Multiple Imputation for Nonresponse in Surveys*. New York: Wiley.

Schafer, J.L. (1997). *Analysis of incomplete multivariate data*. New York: Chapman & Hall.

Slothuus, E. (2010). When can political parties lead public opinion? Evidence from a natural experiment. *Political Communication* 27: 158–177.

Sniderman, P. (2012). The logic and design of the survey experiment. In: *Cambridge Handbook of Experimental Political Science*, 2e (ed. J.N. Druckman, D.P. Green, J.H. Kulkinski and A. Lupia), 102–114. New York: Cambridge University Press.

Tourangeau, R., Rips, L., and Rasinski, K. (2000). *The Psychology of Survey Response*. New York: Cambridge University Press.

10

Application of a List Experiment at the Population Level: The Case of Opposition to Immigration in the Netherlands

Mathew J. Creighton[1], *Philip S. Brenner*[2], *Peter Schmidt*[3], *and Diana Zavala-Rojas*[4]

[1] *School of Sociology, Geary Institute for Public Policy, University College Dublin, Stillorgan Road, Dublin 4, Ireland*
[2] *Department of Sociology and Center for Survey Research, University of Massachusetts, 100 Morrissey Blvd., Boston, MA 02125, USA*
[3] *Department of Political Science and Centre for Environment and Development (ZEU), University of Giessen, Karl-Glöcknerstrasse 21 E, 35394 Giessen, Germany*
[4] *Research and Expertise Centre for Survey Methodology and European Social Survey ERIC, Department of Political and Social Sciences, Universitat Pompeu Fabra, C/de Ramon Trias Fargas, 25-27, 08005 Barcelona, Spain*

This chapter briefly describes the design and application of the item count technique (ICT), also known as the list experiment, in population-level survey research. We review recent applications of the technique and connect it to the use of experiments in social research in general. After this a detailed application of this technique on attitudes toward immigrants is presented.[1]

While survey experiments have been the focus of a good deal of recent research (Gaines et al. 2007; Barabas and Jerit 2010; Schlueter and Schmidt 2010; Mutz 2011; Sniderman 2011), they are nearly as old as the sample survey itself. Along with audit studies (LaPiere 1934), split ballots are one of the original forms of survey experiment. This technique is very flexible and varies widely in its application but at its most basic offers two versions of a question (or different question orderings, sets of response categories, instructions, or the like) to two different subsamples of respondents. Statistics computed from these variations are then compared between groups to estimate the effect of the change to the questionnaire. For example, Rugg (1941) compared two versions of a question about freedom of antigovernment speech testing the effect of ostensible antonyms "forbid" and "allow." Hyman and Sheatsley (1950) manipulated the order of questions about journalistic freedom for Soviet and American reporters.

A newer form of survey experiment based on the general structure of the split ballot is used to give the respondent "cover" to admit something that they may feel embarrassed or even threatened to state outright. These techniques allow the respondent to admit unpopular opinions or counter-normative behavior without directly confessing them to an interviewer. This set of related approaches, including the randomized response (the subject of chapter 9 in this volume) and ICTs present direct questions to the respondent in a way that she/he can answer honestly with the knowledge that his or her response to the question about the sensitive attribute or behavior remains hidden from the interviewer. Respondents' hesitance to directly admit counter-normative behavior and attitudes has long been discussed as a source of nonrandom

1 The item count technique (ICT) question experiment used as an example in this chapter was conducted in the LISS (Longitudinal Internet Studies for the Social Sciences) panel (a representative sample of Dutch individuals who participate in monthly internet surveys) administered by CentERdata (Tilburg University, The Netherlands) in September 2014. The experiment was fielded in Dutch language and is back-translated into English in this chapter. See Scherpenzeel and Das (2011) for a more detailed description of the panel.

Experimental Methods in Survey Research: Techniques that Combine Random Sampling with Random Assignment, First Edition.
Edited by Paul J. Lavrakas, Michael W. Traugott, Courtney Kennedy, Allyson L. Holbrook, Edith D. de Leeuw, and Brady T. West.

Companion Website: www.wiley.com/go/Lavrakas/survey-research

measurement error and conceptualized in theories of response behavior in surveys (Saris and Gallhofer 2007; Tourangeau and Yan 2007; Krumpal 2013).[2]

This chapter focuses specifically on the ICT, which is essentially a specialized version of the split ballot technique used to estimate the prevalence of a sensitive behavior, attitude, or trait. In its most basic form, ICT presents two versions of a single survey question to two random subsamples of respondents. Both versions of the ICT item present the same question stem followed by a limited number of categorical response options similar to a mark-all-that-apply list. However, unlike mark-all-that-apply, the respondent is not asked to openly endorse any particular option but rather to count the number of items she/he would endorse and report only the sum. The two versions of the ICT item differ only on the number of response options presented to the respondent: the control version of the question having j options, the manipulation having $j+1$. The extra item in the list is the behavior or attribute of interest, typically being potentially sensitive for the respondent to report. The difference between the means of the two groups estimates the prevalence of the additional, typically sensitive, item from the list.

Starting from this basic 2-item design, ICT has been expanded, increasing both its utility and its complexity. For example, the design can include more than one manipulation, a separate control group with a standard direct question measuring the attribute of interest, or may offer all respondents a direct question on the attribute of interest (typically randomizing whether this question is asked before or after the ICT). These alterations, and others, can be included in the design of ICT to amplify its potential to reduce bias in estimates of the focal concept or to estimate bias in conventional survey self-reports.

Regardless of these more complex design features, a number of concerns common to all variations of ICT must be addressed in the construction of an item. The list of nonsensitive items must be long and varied enough in their ease or difficulty of endorsement to allow respondents "cover." If the list of nonsensitive items is too short or comprised of items that are easy to endorse, the design of the ICT will fail to hide endorsement of the sensitive item. This phenomenon, called the "ceiling effect," has a counterpart called a "floor effect." If items are too difficult to endorse, they will not provide cover to those respondents endorsing the sensitive item. Together, these effects suggest a best practice of designing a list of nonsensitive items of which very few or no respondents endorse all or none.

Like other methods using self-reports from respondents, ICT can be validated against administrative and record data, when available, at the aggregate level (e.g. voting turnout statistics) to estimate bias in survey self-reports, although few studies have taken this approach (Rosenfeld et al. 2015). Comparing the estimate of the focal behavior or attitude produced by ICT to that produced by a conventional direct survey question allows for the estimation of the relative reduction in bias attributable to the indirect measurement of ICT (cf. Holbrook and Krosnick 2010). Typically, however, higher reports of counter-normative behaviors, attitudes, and traits are assumed to be more valid and lower reports of normative behaviors, attitudes, and traits are assumed to be more valid. As this suggests, measurement methods that yield these higher reports (for counter-normative) or lower reports (for normative) are assumed to be better measures of their focal construct.

More commonly, ICT has been used to estimate the prevalence of typically counter-normative and difficult-to-measure attitudes, behaviors, and traits, although with somewhat mixed results. One series of papers uses the ICT to achieve more valid reports of counter-normative behaviors like corruption, cheating, and criminal behavior. Coutts et al. (2011) assessed the efficacy of ICT in measuring plagiarism in undergraduate students' papers. They found ICT

2 As an alternative statistical approach to control for social desirability, a line of research has used confirmatory factor analysis and bifactor models to control for social desirability tendencies (Podsakoff et al. 2003).

failed to increase validity of reports (assuming that more reports of plagiarism are more valid). Kiewiet De Jonge (2015) used ICT to measure the rate of vote buying – candidates or campaigns bribing individual voters to vote in a particular way – in 10 surveys in Latin American countries. When compared to estimates from a direct survey question, the ICT resulted in higher estimates of vote buying in six of these surveys. Similarly, Ahlquist et al. (2014) used ICT to estimate the rate of voter fraud – ineligible individuals voting or eligible voters voting more than once – humorously comparing it to the (negligible) rate of Americans claiming alien abduction.

Malesky et al. (2015) used ICT to examine the situations that increase the probability of engaging in bribery and Kuha and Jackson (2014) used reports of criminal behavior (specifically, purchasing stolen goods) from ICT to examine key design features of the technique.

Other work has examined counter-normative attitudes. Kuklinski et al. (1997a) found that ICT uncovered very high rates of racist sentiment in the South, and higher rates than in other regions. Similar research using ICT found much higher rates of anger over affirmative action in the South, where anger was nearly universal, compared to other regions (Kuklinski et al. 1997b). Knoll (2013) measured nativism using the technique. His finding that the rate of nativism when directly measured (64%) exceeds that from ICT (45%) suggests that many people overreport their antiimmigrant or antiforeigner sentiment, in line with a decades-old finding by LaPiere (1934). Finally, Kane et al. (2004) found little evidence for implicit antisemitic attitudes that would prevent voting for vice presidential candidate Joe Lieberman or an unnamed Jewish candidate.

In comparison, relatively fewer studies have used ICT to examine the overreporting of normative behaviors or attitudes and does not consistently yield less biased estimates than direct questions. Holbrook and Krosnick (2010) compared turnout estimates from ICT and a direct survey question in a phone interview. They found that ICT yielded about a 20 percentage point reduction in reported voting in bias compared to the direct survey question, using turnout statistics as a criterion. Comşa and Postelnicu (2013) conduct a similar study of vote overreporting using direct questions and ICT in the Romanian Presidential Elections Study and find a similar level of bias in the direct reports. However, Tsuchiya et al. (2007) examine direct reports and ICT estimates of blood donation, but find that ICT does not yield improved measurement of this socially desirable behavior.

10.1 Fielding the Item Count Technique (ICT)

The purpose of ICT is to measure what might otherwise be masked. The intuition of the technique is rooted in the manipulation of the amount of anonymity guaranteed to respondents. As with any experimental design, the basic set-up involves independent samples designated as treatment(s) and control (see Figure 10.1). The control sample is presented with a list of items. For the example used in this chapter, translated from Dutch, the following question (C1) was posed to all respondents in the control sample:

> (C1) *Of the following three statements, HOW MANY of them do you AGREE with? We don't want to know which statements, just HOW MANY?*
>
> *The Netherlands should increase assistance to the poor*
> *The Netherlands should decrease the tax on petrol and diesel*
> *The Netherlands should allow large corporations to pollute the environment*

Respondents can respond with a number between zero and three. These items are not of analytic interest, but serve to establish a distribution. Responses to this question allow for the

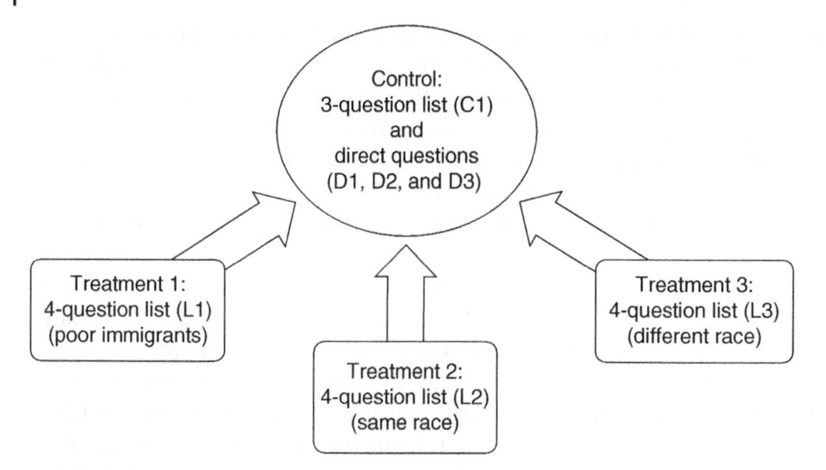

Figure 10.1 Relationship between control and treatment list questions in an application of the item count technique (ICT).

calculation of any measure of central tendency, particularly the mean and standard deviation. An independent treatment group is presented a similar list question, but with an additional item included, referred to as the focal item. For the example used in this chapter, the following question (L1) was posed to the first of three treatment groups:

(L1) *Of the following four statements, HOW MANY of them do you AGREE with? We don't want to know which statements, just HOW MANY?*

The Netherlands should increase assistance to the poor
The Netherlands should decrease the tax on petrol and diesel
The Netherlands should allow large corporations to pollute the environment
The Netherlands should allow people from poorer countries outside Europe to come and live here

The two remaining treatment questions, designated L2 and L2 below, received a similar list question, but distinct focal items. In these cases, the focal item was replaced with one of the following two:

(L2) *The Netherlands should allow people of the same race or ethnic group as most Dutch people to come and live here*
(L3) *The Netherlands should allow people of a different race or ethnic group from most Dutch people to come and live here*

To the treatment list question, respondents can answer with a number that ranges from zero to four. Identical to the control list, the aggregate of all responses allow for the estimation of measures of central tendency. There are somewhat distinct designs that have been used that vary the number of baseline list items in the control, but the underlying manipulation, which must include a control list with j number of items and a treatment list with $j + 1$ items, remains the same. Regardless of the design employed, the key information extracted from the two types of list questions is the mean response pattern and standard deviation, which is used to calculate the implicit measure of agreement with the focal item. The difference between these two means gives the proportion that select the focal item (i.e. L2 and L3).

In some cases, it is desirable/required to extract a measure of the size of the difference between the implicit measure revealed via the ICT and an explicit measure that is measured via a direct question akin to what one might find in a standard survey. This is one of the novel aspects of the ICT in that it can help address social desirability bias (SDB) and it can offer insight into the extent that this bias poses a problem. To do this requires one additional dichotomous question. It can be asked of the same control group or, to avoid contamination due to asking repeated related questions, of an additional random sample. For the example given below, the direct question (D1) was asked of a subsample of the control sample that received the three-question list (see Figure 10.1). This design trades sample size for independence of observations as no respondent is asked more than one direct question, but the direct question is only asked of a subsample of the larger control sample.[3] The wording for the direct question, which is directly comparable to the focal item in the first treatment list (L1), is as follows:

> (D1) *Do you think the Netherlands should allow people from poorer countries outside Europe to come and live here?*
> *Allow to come and live here*
> *Do not allow to come and live here*

As shown in Figure 10.1, this direct question (D1) is comparable to the response implicitly derived from the comparison of the first treatment list question (L1) to the control list question (C1). Similarly, direct questions (D2 and D3) are comparable to the focal item in treatment second (L2) and third treatment (L3) lists and take the following form:

> (D2) *Do you think the Netherlands should allow people of the same race or ethnic group as most Dutch people to come and live here?*

> (D3) *Do you think the Netherlands should allow people of a different race or ethnic group from most Dutch people to come and live here?*

In the end, the list (C1, L1, L2, and L3) and direct (D1, D2, and D3) questions allow for the estimation of three aspects of the respondents' sentiment: (i) implicit sentiment, (ii) explicit sentiment, and (iii) the difference between the two (i.e. SDB).

10.2 Analyzing the Item Count Technique (ICT)

Differencing the mean response to the treatment list question from the mean response to the control list derives the proportion offers a direct measure of implicit agreement with and of the focal items included in the treatment lists (L1, L2, and L3). This is formalized by Eq. (10.1):

$$Z = \overline{X}_L - \overline{X}_C \tag{10.1}$$

where Z is the proportion of the sample that select the additional list item in the treatment, which is derived from the difference between the mean response to the treatment list (e.g. L1),

---

3 In the design deployed, one comparison offers some utility in assessing whether any specific direct question changed the response pattern (i.e. mean) of the control list relative to any other specific direct question. The mean of the control list derived from the subsample asked one of the three direct question (e.g. mean of L1 in a sample of respondents also asked D1) can be compared to the mean of the control list for the subsample in not asked the direct question of interest (e.g. mean of L1 in a sample of respondents also asked D2 or D3, but not D1). This results in three t-tests – one for each direct question. In no case does the test approach significance at the $p < 0.10$ level.

defined by the indicator L, and the mean response to the control list, defined by the indicator C. For this example, the relevant n in this calculation is the total sample of the control group as all respondents were asked the control list question (C1). A standard t-test offers a formal assessment of whether the proportion selecting the additional item in the treatment is significantly greater than zero. Given that the treatment and control samples are independent, the upper and lower bound of the estimate are derived from the combined observed standard error of the treatment and control for the list question. The null hypothesis in this case is no difference, assuming no design effects, as the mean for the treatment list should never be less than that of the control list.

The result of Eq. (10.1) is interpretable as the proportion of the sample that has selected the additional list item in the treatment group, which provides an implicit estimate of the outcome of interest – *positive* attitudes toward immigrants in this case. For the first treatment group in the example above (T1), it is the proportion of the population agreeing that "*The Netherlands should allow people from poorer countries outside Europe to come and live here.*" When the resulting implicit estimate, determined using Eq. (10.1) is subtracted from the explicit estimate, derived from questions D1–D3 posed to the control group, the resulting value is interpretable as the level of SDB, which is formalized by Eq. (10.2):

$$B = \overline{X}_D - Z \tag{10.2}$$

where B is direct measure of SDB that, when converted to a percentage scale, is typically interpreted as the number of percentage points difference between the explicit, derived from the control sample, and the implicit estimate (Z), derived from Eq. (10.1). The explicit response is calculated from the sum of responses to the binary direct question (D1, D2, or D3), which is defined by the indicator D, indicating that the calculation is limited to the subsample of the control group that was posed the direct question of interest, over the relevant total subsample of control. This example assumes that explicit agreement, derived from D1–D3, may be over reported so Eq. (10.2) will result in a positive value. If theory suggests that SDB may result in under reporting, it may be preferable to subtract the explicit measure ($\overline{X}_D$) from the implicit estimate (Z) to avoid interpreting a negative value.

10.3 An Application of ICT: Attitudes Toward Immigrants in the Netherlands

In the following application, a representative sample of the adult population of the Netherlands expressed implicit and explicit opinions about immigrants defined by racial (D2, D3, L2, and L3) and economic characteristics (D1 and L1). The following example is illustrative and will cover the estimation of implicit selection of the focal item, using Eq. (10.1), and SDB, using Eq. (10.2).

10.3.1 Data – The Longitudinal Internet Studies for the Social Sciences Panel

The Longitudinal Internet Studies for the Social Sciences (LISS) panel, initiated in 2007 as part of the Measurement and Experimentation in the Social Sciences (MESS) project in the Netherlands, offers a representative sample of the general population of the Netherlands. The mode of collection is via web survey and, aside from initial contact, involved no direct interaction with a respondent. The full LISS panel, drawn from the population register held by Statistics Netherlands, is a true probability sample and includes approximately 8000 individuals residing in 5000 households. The LISS panel is available to outside researchers conditional on approval

Table 10.1 Descriptive statistics.

	n	Mean (std. dev.) or proportion
(C1): Control list question	2114	1.43 (0.01)
(L1): Treatment 1 list question	699	1.62 (0.03)
(L2): Treatment 2 list question	701	1.62 (0.03)
(L3): Treatment 3 list question	698	1.83 (0.03)
(D1): Control direct question 1	709	0.38
(D2): Control direct question 2	699	0.62
(D3): Control direct question 3	704	0.62

Note: The direct questions D1, D2, and D3 are asked of subsamples of the sample asked C1. The difference between the size of the C1 sample ($n = 2114$) and the n derived from the sum of D1, D2, and D3 ($n = 709 + 699 + 704 = 2113$) is attributable to two respondents who did not answer the direct question, but did answer the control list question. In addition, the relevant independent sample to compare ICT results to the direct estimates includes only respondents who received the control list, but not the direct question that is the focal question in the ICT estimated. To compare with D2, ICT estimate for L2 is calculated using a sample that includes no respondents who received D2.

of a submitted research proposal and all collected data is publically available to registered users via the LISS data archive (Scherpenzeel and Das 2011).[4]

The data used in this chapter were collected in September of 2014 as part of a larger experiment comprised of eight distinct experimental groups. The total number of usable cases for the entire data collection, of which only a subset are used in this chapter, was 5615. This total sample reflects the number of nonmissing cases out of a total of 6558 selected individuals from the full LISS panel with nonresponse of 14.3% and three incompletes. For the example presented in this chapter, we consider only four of the groups constituting one control and three treatments (see Figure 10.1). The control group ($n = 2114$), all of which received the control list question (C1), was further divided into three random subgroups of approximately 700 respondents (see Table 10.1 for the exact sample sizes, which vary slightly) to which each subsample was presented one of the three direct question (D1, D2, and D3). This design generates a relatively large sample for the control list question (C1), but avoids multiple direct questions from being posed to the same control-group participant. Each treatment group received a distinct list question (L1, L2, and L3) each of which consisted of about 700 respondents.

10.3.2 Implicit and Explicit Agreement

Using Eq. (10.1), implicit measures of agreement can be derived by subtracting the mean response to the control list question (C1) from each of the three list questions (L1, L2, and L3). For the list item about agreement with allowing poor immigrants to settle in the Netherlands, the implicit estimate (0.19) and related 95% confidence interval (± 0.07) are derived differencing the mean for the control list (1.43) from the first treatment list (1.62). Because the two samples are independent, the confidence interval (± 0.07) from the difference

4 http://www.lissdata.nl/dataarchive.

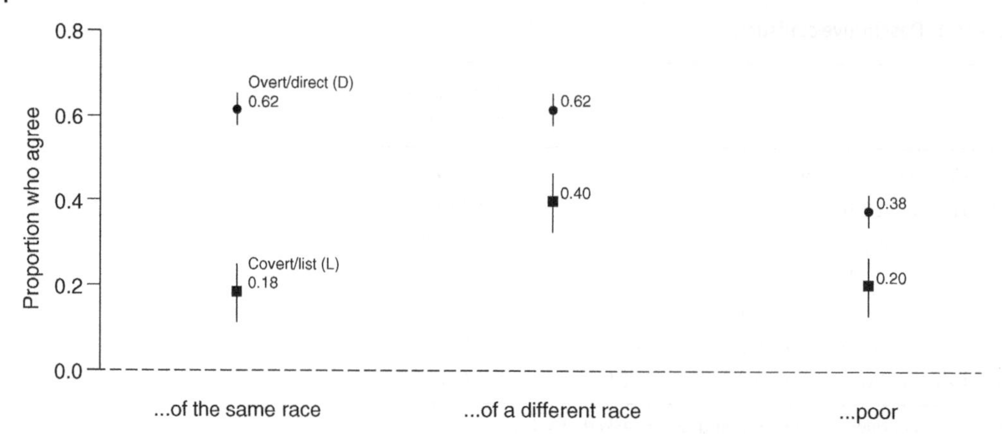

Figure 10.2 Overt/direct (D) and covert/list (L) estimates: the Netherlands should allow immigrants who are....

between the two is the sum of the two standard deviations (see Figure 10.2). Given that the LISS panel is representative of the national population, the interpretation in percentage terms is that 19% of the Netherlands' adult population implicitly agree that poor immigrants should be allowed to settle.

A similar exercise can be done for the two remaining focal items that ask about immigrants of the same (L2) or different race (L3). The estimate is nearly identical to that for poor immigrants (L1) with an implicit estimate for the focal item about the same race (L2) of 0.19 (±0.07). In other words, in percentage terms, 19% of the Netherlands adult population implicitly agrees that immigrants of the same race should be allowed to settle (see Figure 10.2). The third implicit measure, derived from the third treatment group (L3), is significantly higher at 0.40 (±0.07). In this case, again using percentage terms, 40% of the Netherlands implicitly agrees that immigrants of a different race should be allowed to settle (see Figure 10.2).

Explicit measures are straightforward to ascertain, being derived from direct questions (D1, D2, and D3) posed to subsets of the control sample. In this example, about 700 respondents from the total control sample of 2114 were assigned one of three direct questions (D1, D2, and D3), which corresponded to a list item (L1, L2, or L3), which are shown in Table 10.1. Figure 10.2 reports the explicit estimates and associated 95% confidence interval. The confidence interval is calculated in the standard way by multiplying the proportion that agree by the proportion who do not, dividing by the relevant sample size and taking the square root of the result. For each of the direct questions, the relevant sample is only the subsample of the control group that received the direct question of interest. For the questions about race, same and different, the estimated proportion is the same at 0.62 (±0.03). In percentage terms, the formal interpretation is that 62% of the Netherlands explicitly agrees that immigrants of the same/different race should be allowed to settle (see Figure 10.2). Poor immigrants are confronted with significantly less accommodation in that only 38% of the population explicitly agree that poor immigrants should be allowed to settle in the Netherlands.

10.3.3 Social Desirability Bias (SDB)

The estimate of SDB, which is reported for each ICT application in Figure 10.3, is derived using Eq. (10.2). It is clear from the outset that masking is observed (i.e. SDB) for all types of sentiment expressed toward immigrants covered by this application of the ICT. The interpretation of SDB is not the percentage change. Instead it is the percentage *point* difference between the explicit

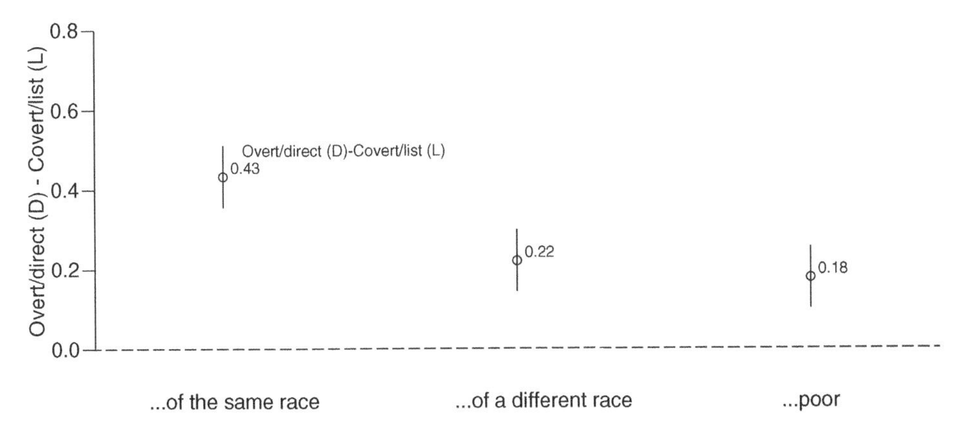

Figure 10.3 Social desirability bias: the Netherlands should allow immigrants who are....

measure and implicit measure. If one would like to ascertain the percentage difference one would need to designate a baseline (e.g. explicit) and divide the SDB derived from Eq. (10.2) by that value. In this case, in percentage point terms, there is a significant difference between the explicitly stated agreement and the implicitly stated agreement to allow immigrants of different economic and racial characteristics into the Netherlands.

To ascertain whether the difference between the two proportions is significant, a standard z-test can be used. For the explicit measure, the relevant sample in this example is the subsample of the control to which the direct question was posed (e.g. $n = 709$ in the case of D1). The sample size for the implicit measure, derived from Eq. (10.2), is the sum of the relevant treatment group (e.g. $n = 699$ for L1) and the control group to which the control list (C1) was presented, but who did not receive a direct question (e.g. $n = 1405$ when excluding D1). Excluding respondents who received a direct question used to compare with the ICT estimates preserves independence of the two samples (i.e. no respondent contributes to both the ICT and direct estimates). This process results in three comparative samples for the implicit measures – 2104 (C1 – D1 + L1), 2116 (C1 – D2 + L2), and 2108 (C1 – D3 + L3) – and three comparative samples for the direct measures – 709 (D1), 699 (D2), and 704 (D3).

The interpretation of SDB, often referred to as masking, is straightforward. In this example, there is a 42-percentage-point difference between explicit and implicit agreement that immigrants of the same race should be allowed to settle in the Netherlands, which is significant ($z = -20.31$; $p < 0.01$). Put it in another way, these results suggest that 42% of respondents masked their lack of agreement with permitting same-race immigrants to settle in the Netherlands when asked directly. Attitudes toward immigrants of a different race and poor immigrants also show significant SDB, albeit less. In the case of poor immigrants, 18% of respondents in the Netherlands mask their lack of agreement with permitting settlement, which is significant ($z = -10.13$; $p < 0.01$). This level of masking is echoed in the case immigrants of a different race where 22% of respondents in the Netherlands mask their lack of agreement, which is also significantly different than zero ($z = 8.38$; $p < 0.01$).

The direct measurement of SDB is an appealing aspect of any application of the ICT where an explicit measure of the focal list item is available whether internal to the design, as this example shows, or derived from a secondary source (see Moseson et al. 2015, for an example that compares implicit measures, derived from the ICT, to the Demographic Health Survey). Rather than simply correcting for SDB, the ICT allows for the direct estimation of the bias. Therefore, depending on the research question, the pattern and degree of masking can be of analytic interest in of itself.

10.4 Limitations of ICT

The ICT offers a way forward when it is likely that masking will make direct measures problematic. However, it is not a perfect solution and a number of limitations are worth considering before selecting the ICT as a preferred method. This section will underline five of the main concerns/limitations/assumptions of the ICT.

The first limitation stems from the fact that the ICT requires a binary response, which may or may not be appropriate for the outcome of interest. In the example presented here, the outcome of interest is agreement with permitting certain immigrant groups to settle in the Netherlands. The ICT can easily be adapted, and has been, to sentiment such as opposition, agreement, being upset or angry. Regardless of the type of response being expressed, the technique will only reveal the proportion opposed/in agreement/upset/angry and offers no other scale (e.g. somewhat opposed, very opposed, neither agree nor disagree). In other words, the true opinion of those that do not select the focal item can be somewhat more heterogeneous than a binary response permits. For some outcomes that require ordinal measures or 3+ categories, it can be difficult to plausibly convert such questions to a binary and, therefore, may be reason enough to consider an approach other than the ICT.

A second concern is that the anonymity normally afforded by list question is violated when a respondent has no variation in his or her response to all of the items in the list. This can be problematic in the treatment and control. If the response is equal to the total number of list items, which would reveal the respondents preferences for any one of the items in the list, it is called a ceiling effect. A floor effect is when the response is zero, which also reveals the respondent's preference for any given list item. In either case anonymity is not plausibly maintained and, therefore, the ICT does not offer any advantage over a direct question. Table 10.2 shows the percentage of respondent who select all (ceiling effects) or none (floor effects) of the list items.

A third concern emerges from the design of the question. The control list and the treatment list, by definition, have different numbers of items, j in the former and $j+1$ in the latter. If responses are determined, at least in part, by the number of items, the estimate of the proportion counting the additional focal item would conflate indirectly expressed sentiment with the effect of the design of the question itself. Randomization of the list order is a good practice to reduce the likelihood of introducing a design effect based on list item ordering, but does not alleviate the effect of the total number of items in the list. There are recent efforts to test for the presence of a design effect and, in some cases, correct for it (see Blair and Imai

Table 10.2 Floor and ceiling effects.

	n	Floor effect (%)	Ceiling effects
(C1): Control list question	2114	5.77	2.13
(L1): Treatment 1 list question	699	4.72	1.14
(L2): Treatment 2 list question	701	6.28	0.57
(L3): Treatment 3 list question	698	3.87	1.00

Note: The column ceiling effect reports the proportion of respondents in the specified condition for which the response is equal to the total number of list items. The column floor effect reports when the response is zero. n is the total number of observations in each condition.

2012). That said, Kiewiet De Jonge and Nickerson (2014) highlight the problems that arise in the presence of ceiling and/or floor effects, which are a known concern in any application of the ICT.

A fourth consideration is the interpretation of causality. A clear attraction of experimental design is that causality, being derived from the manipulation itself, is generally straightforward to interpret (Pearl et al. 2016). The ICT does involve a manipulation, but the causal interpretation is limited to the difference between the treatment and control list question. In all designs of the ICT, if the difference in the mean response to the treatment and control list is significant, the proportion that selects the additional focal list item in the treatment list is significantly different than zero. This difference is caused by the manipulation but, in the absence of a direct measure either derived from a separate experimental group or via a secondary source, this is not usually of particular interest to the researcher beyond a measurement correction.

In the design where a direct question is given to a control group, the difference between explicit measure and the implicit measure derived from ICT is interpretable as causal in a formal sense. It follows that the causal interpretation of differences between subpopulations (e.g. female, politically conservative, etc.) are similarly restricted. In addition to limiting causal interpretation to differences directly linked to the manipulation, there is an interpretative aspect. The attribution of causality normally assumes that the treatment condition is relatively more valid. The difference between the two estimates is due to bias when posing a direct question and the mechanism underlying any observed bias is assumed, albeit not directly tested in the manipulation, to be social desirability pressure.

A fifth aspect that should be considered is the inability of the ICT to offer estimates of the outcome of interest, opposition to immigration in this case, or bias attributable to social desirability at the individual level. This issue arises from the way in which anonymity is protected in the presentation of the list of items. In other words, the fact that a given individual's response can never be known is what creates a credible reduction in social desirability pressure, but, at the same time, it precludes the assessment of individual-level attitudes. It is, therefore, not feasible to use the ICT to correct for bias for a given individual's responses. This has implications for the implementation of multivariate techniques (see Imai 2011, for a discussion of the issue) and can mask within-group heterogeneity in response patterns and response bias.

More recent efforts have developed techniques that use the predicted responses from ICT experiments as explanatory variables in regression models (Imai et al. 2015). These approaches do assume that the predicted response is equal for different levels of the latent explanatory variable and require a large sample, which has implications in terms of the cost of the data collection. That said, the ICT approach in general, relative to traditional approaches, require somewhat large samples, particularly when direct measures are of interest.

To reiterate, the intention of this chapter is to provide an introduction to the ICT in population-level survey experiments. Although strengths and weaknesses of the approach are outlined, this approach has, is and will be continuously refined and critiqued. Although the applications of the ICT are wide, ranging from sociology to political science to public health and beyond, it is far from a magic bullet or a truth serum. There are plausible interpretations of any ICT application in terms of reducing SDB and causality, which we have described in some detail, but there are a number of assumptions that should always be at the forefront of any application. There is little doubt that skepticism about direct measurement of controversial topics in surveys will continue. We suggest that the ICT offers a reasonable way forward for some questions about sensitive topics level like criminal behavior or prejudice on the aggregate level. Moreover, its ability to be rapidly and consistently fielded at the population level guarantees its relevance in the near, medium, and long term.

References

Ahlquist, J.S., Mayer, K.R., and Jackman, S. (2014). Alien abduction and voter impersonation in the 2012 US general election: evidence from a survey list experiment. *Election Law Journal* 13: 460–475.

Barabas, J. and Jerit, J. (2010). Are survey experiments externally valid? *American Political Science Review* 104 (2): 226–242.

Blair, G. and Imai, K. (2012). Statistical analysis of list experiments. *Political Analysis* 20 (1): 47–77.

Comşa, M. and Postelnicu, C. (2013). Measuring social desirability effects on self-reported turnout using the item count technique. *International Journal of Public Opinion Research* 25: 153–172.

Coutts, E., Jann, B., Krumpal, I., and Näher, A.-F. (2011). Plagiarism in student papers: prevalence estimates using special techniques for sensitive questions. *Journal of Economics and Statistics* 231 (5/6): 749–760.

Gaines, B.J., Kuklinski, J.H., and Quirk, P.J. (2007). The logic of the survey experiment reexamined. *Political Analysis* 15 (1): 1–20.

Holbrook, A. and Krosnick, J.A. (2010). Social desirability bias in voter turnout reports: tests using the item count technique. *Public Opinion Quarterly* 74 (1): 37–67.

Hyman, H.H. and Sheatsley, P.B. (1950). The current status of American public opinion. In: *The Teaching of Contemporary Affairs* (ed. J.C. Payne), 11–34. National Council of Social Studies.

Imai, K. (2011). Multivariate regression analysis for the item count technique. *Journal of the American Statistical Association* 106 (494): 407–416.

Imai, K., Park, B., and Greene, K.F. (2015). Using the predicted responses from list experiments as explanatory variables in regression models. *Political Analysis* 23 (2): 180–196.

Kane, J.G., Craig, S.C., and Wald, K.D. (2004). Religion and presidential politics in Florida: a list experiment. *Social Science Quarterly* 85 (2): 281–293.

Kiewiet De Jonge, C.P. (2015). Who lies about electoral gifts? *Public Opinion Quarterly* 79 (3): 710–739.

Knoll, B.R. (2013). Assessing the effect of social desirability on nativism attitude responses. *Social Science Research* 42 (6): 1587–1598.

Krumpal, I. (2013). Determinants of social desirability bias in sensitive surveys: a literature review. *Quality and Quantity* 47 (4): 2025–2047.

Kuha, J. and Jackson, J. (2014). The item count method for sensitive survey questions: modelling criminal behaviour. *Journal of the Royal Statistical Society, Series C - Applied Statistics* 63: 321.

Kuklinski, J.H., Cobb, M.D., and Gilens, M. (1997a). Racial attitudes and the 'New South'. *Journal of Politics* 59 (2): 323–349.

Kuklinski, J.H., Sniderman, P.M., Knight, K. et al. (1997b). Racial prejudice and attitudes toward affirmative action. *American Journal of Political Science* 41 (2): 402–419.

LaPiere, R.T. (1934). Attitudes vs. Actions. *Social Forces* 13 (2): 230–237.

Malesky, E.J., Gueorguiev, D.D., and Jensen, N.M. (2015). Monopoly money: foreign investment and bribery in Vietnam, a survey experiment. *American Journal of Political Science* 59 (2): 419–439.

Moseson, H., Massaquoi, M., Dehlwendorf, C. et al. (2015). Reducing under-reporting of stigmatized health events using the list experiment: results from a randomized, population-based study of abortion in Liberia. *International Journal of Epidemiology* 44 (6): 1951–1958.

Mutz, D.C. (2011). *Population-Based Survey Experiments*. Princeton, NJ: Princeton University Press.

Pearl, J., Glymour, M., and Jewell, N.P. (2016). *Causal Inference in Statistics: A Primer*. Chichester: Wiley.

Podsakoff, P.M., MacKenzie, S.B., Lee, J.-Y., and Podsakoff, N.P. (2003). Common method biases in behavioral research: a critical review of the literature and recommended remedies. *Journal of Applied Psychology* 88 (5): 879–903.

Rosenfeld, B., Imai, K., and Shapiro, J.N. (2015). An empirical validation study of popular survey methodologies for sensitive questions. *American Journal of Political Science* https://doi.org/10.1111/ajps.12205.

Rugg, D. (1941). Experiments in wording questions: II. *Public Opinion Quarterly* 5 (1): 91–92.

Saris, W.E. and Gallhofer, I.N. (2007). *Design, Evaluation, and Analysis of Questionnaires for Survey Research*. New York: Wiley.

Scherpenzeel, A.C. and Das, M. (2011). "True" longitudinal and probability-based internet panels: evidence from the Netherlands. In: *Social and Behavioral Research and the Internet: Advances in Applied Methods and Research Strategies* (ed. M. Das, P. Ester and L. Kaczmirek). Routledge.

Schlueter, E. and Schmidt, P. (2010). Special issue: survey experiments. *Methodology* 6 (3): 93–95.

Sniderman, P.M. (2011). The logic and design of the survey experiment: an autobiography of a methodological innovation. In: *Cambridge Handbook of Experimental Political Science* (ed. J.N. Druckman, D.P. Green, J.H. Kuklinski and A. Lupia), 102–114. New York: Cambridge University Press.

Tourangeau, R. and Yan, T. (2007). Sensitive questions in surveys. *Psychological Bulletin* 113 (5): 859–883.

Tsuchiya, T., Hirai, Y., and Ono, S. (2007). A study of the properties of the item count technique. *Public Opinion Quarterly* 71: 253–272.

Part IV

Introduction to Section on Interviewers

Brady T. West[1] *and Edith D. de Leeuw*[2]

[1] *Survey Research Center, Institute for Social Research, University of Michigan, Ann Arbor, MI, USA*
[2] *Department of Methodology & Statistics, Utrecht University, Utrecht, the Netherlands*

In face-to-face and telephone surveys, interviewers play an essential role in securing cooperation, instructing and motivating respondents, collecting and recording survey measures, collecting paradata (e.g. interviewer observations), administrative tasks, and correctly processing the data collected. High-quality performance from interviewers is thus essential for a face-to-face or telephone survey research project to succeed. Unfortunately, interviewers can also have differential effects on a variety of survey outcomes when considering the total survey error paradigm (Groves 2004; West and Blom 2017). When considering the possibility of embedding experiments in surveys using these two modes of data collection, the experimental designs need to carefully consider the roles of interviewers, and potential effects that the interviewers may have on the implementation of the experimental design. This section of the book discusses important perspectives on the roles of interviewers when experiments are embedded in interviewer-administered surveys.

The section contains two chapters. Chapter 11 by Holbrook, Johnson, and Krysan considers the effects of the race/ethnicity of interviewers on survey participation and survey responses, focusing on Black, White, and Latino/a interviewers and randomly assigning cases in a telephone survey to each group of interviewers. The authors first review the literature in this area and then report a variety of interesting outcomes from their experimental study, presenting new findings like reduced accuracy of respondent perceptions of interviewer race/ethnicity over the telephone.

Chapter 12 by Lavrakas, Kelly, and McClain focuses more on how interviewers should be accounted for when embedding experimental studies in surveys. The authors first describe how the effects of interviewer characteristics can be analyzed like the effects of experimental factors when cases are randomly assigned to different interviewers, and then provide strong empirical evidence in support of interviewers only administering one type of experimental treatment in an experimental study (rather than having to administer multiple treatments). These authors also reiterate the importance of more consistent efforts to control for interviewer effects when analyzing the effects of embedded experimental factors in interviewer-administered surveys, which is important advice for all survey practitioners. Recent computational advances in multilevel modeling (Hox et al. 2018) make it very easy for researchers to account for these effects,

Experimental Methods in Survey Research: Techniques that Combine Random Sampling with Random Assignment, First Edition.
Edited by Paul J. Lavrakas, Michael W. Traugott, Courtney Kennedy, Allyson L. Holbrook, Edith D. de Leeuw, and Brady T. West.

Companion Website: www.wiley.com/go/Lavrakas/survey-research

provided that data producers supply interviewer ID codes for respondents along with the experimental data.

There are many interesting directions for future work in this area. We view the use of advanced analytic techniques such as multilevel modeling and prediction of random interviewer effects for "live" interviewer monitoring during data collection as essential for ensuring that interviewer performance remains steady and outliers do not emerge. This is especially important during the implementation of embedded experiments, so that survey managers can ensure that interviewers are not adversely affecting the experimental administration for whatever reason. We hope that the studies described will inspire this type of data-driven monitoring using new technologies and motivate additional research in this area, given that interviewers will continue to play an important role in survey administration and survey experiments for years to come.

References

Groves, R.M. (2004). The interviewer as a source of survey measurement error. In: *Survey Errors and Survey Costs*, Chapter 8, 2ee. New York: Wiley-Interscience.

Hox, J.J., Moerbeek, M., and van de Schoot, R. (2018). *Multilevel Analysis: Techniques and Applications*, Quantitative Methodology Series, 3ee. New York: Routledge.

West, B.T. and Blom, A.G. (2017). Explaining interviewer effects: a research synthesis. *Journal of Survey Statistics and Methodology* 5: 175–211.

11

Race- and Ethnicity-of-Interviewer Effects*

Allyson L. Holbrook[1], *Timothy P. Johnson*[2], *and Maria Krysan*[3]

[1] *Departments of Public Administration and Psychology and Survey Research Laboratory, University of Illinois at Chicago, Chicago, IL, USA*
[2] *Department of Public Administration and the Survey Research Laboratory, University of Illinois at Chicago, Chicago, IL 60607, USA*
[3] *Department of Sociology and the Institute of Government & Public Affairs, University of Illinois at Chicago, Chicago, IL, USA*

11.1 Introduction

11.1.1 The Problem

Surveys administered by interviewers, either by telephone or face-to-face, offer a number of advantages relative to surveys employing other modes of data collection. These include the opportunity for the interviewer to observe and address respondents' confusion or difficulties. However, interviewers also have the potential to introduce error into survey data. Interviewer variance is the variance in survey data that is due to the interviewer (e.g. Kish 1962). Although researchers might hope that data collected by different interviewers is interchangeable (and therefore there is no interviewer variance), empirical evidence suggests that interviewers may have a substantial effect on survey data (see Davis et al. 2009 for a review). The effect of an interviewer is a function of the extent to which they influence the survey variance and the interviewer workload (Groves 2004; see West and Blom 2017 for a review). That is to say that even a small interviewer effect can substantially affect survey data if the interviewer conducts a large number or proportion of interviews. Interviewer variance may be the result of the effect of interviewers on nonresponse (the characteristics of respondents from whom different interviewers obtain interviews vary; e.g. West and Olson 2010) as well as the effect of interviewers on responses to individual survey questions (e.g. Cannell et al. 1981). Interviewers may influence participation or survey responses either actively through their behavior or even their own perceptions or ratings (e.g. Hill 2002), or interviewers may influence survey data more passively as a result of their observable characteristics (e.g. race or gender; Schuman and Converse 1971; Groves and Fultz 1985).

This chapter focuses on the effect of interviewers' race or ethnicity – one usually observable characteristic that may influence respondents' decisions about whether or not to participate and their answers to survey questions. In the following sections, we describe the

*Note: Portions of the results presented in this chapter were previously reported in two presentations given at the 2009 annual meeting of the American Association for Public Opinion Research. This chapter is dedicated in memory of Ingrid Graf who served as the project coordinator for the 2008 Chicago Area Study.

Experimental Methods in Survey Research: Techniques that Combine Random Sampling with Random Assignment, First Edition.
Edited by Paul J. Lavrakas, Michael W. Traugott, Courtney Kennedy, Allyson L. Holbrook, Edith D. de Leeuw, and Brady T. West.

Companion Website: www.wiley.com/go/Lavrakas/survey-research

theoretical models that would predict race- and ethnicity-of-interviewer effects on survey participation and responses, and describe the predictions of each. Next, we review the literature on race- and ethnicity-of-interviewer effects and describe how our survey extends those findings. We then describe a telephone survey that we conducted to test the impact of race- and ethnicity-of-interviewers on survey participation and responses and report the results of our analyses of those data. Finally, we describe the implications of our results for survey research, the limitations of our findings, and our suggestions for future research.

11.1.2 Theoretical Explanations

11.1.2.1 Survey Participation

Several factors may underlie race-of-interviewer effects on survey participation. When viewed from the perspective of the *social distance model,* some potential respondents may feel less comfortable being asked to interact with persons they perceive as being less like themselves. Given the segregated nature of social life in the United States, this discomfort may stem from limited experience interacting with persons of different racial or ethnic backgrounds (Dovidio et al. 2011). Lack of direct experience with persons of varying backgrounds may lead to distrust, suspicion, or even fear. It may also lead to reliance on negative stereotypes. Negative personal experiences with persons of different race/ethnic backgrounds may, of course, also lead to negative stereotypes and increased social distance. Greater perceived social distance between respondents and interviewers may in turn be associated with greater levels of survey nonparticipation, both directly because of refusals, as well as indirectly by making oneself unavailable for contact and/or screening. Conversely, interviewers of similar race/ethnicity to the potential respondent are more likely to engender greater trust and less suspicion, leading to greater willingness to accept a survey request (Dotinga et al. 2005). Statistical analyses employing interactions between interviewer and respondent race/ethnicity can be used to test these ideas based on social distance.

Interviewer race/ethnicity may also influence a respondent's willingness to participate in a survey interview based on racialized perceptions of the legitimacy of the request. Specifically, white interviewers (in the US racial system) may be more likely to be perceived as representing "mainstream" society, and the request to participate in a survey may therefore appear to be more legitimate coming from a white interviewer. Unlike the *social distance model,* this perspective can be tested by examining the main effects of interviewer race/ethnicity on survey response indicators. Significant effects of interviewer race indicating higher levels of contact, cooperation and/or response to white interviewers, regardless of respondent race/ethnicity, can be interpreted as evidence supportive of a *mainstream conformity model* of race-of-interviewer effects.

A contrasting perspective is the *mainstream distrust model,* in which prospective survey respondents, particularly those who are members of minority race and ethnic groups, are less willing to participate when invited to do so by white interviewers. Unlike the *social distance model,* which associates respondent distrust, suspicion and unwillingness to participate with any race/ethnicity differences between respondent and interviewer, the *mainstream distrust model* focuses more narrowly on opposition to majority culture by members of one or more minority groups. This opposition may be the consequence of negative personal experiences of discrimination, awareness of broader patterns of discrimination against one's race/ethnic group currently or in the past (e.g. the Tuskegee Syphilis Study; Jones 1993), or of a combination of these factors.

11.1.2.2 Survey Responses

The available literature suggests that race- and ethnicity-of-interviewer effects are most likely in response to questions concerned with relevant (e.g. race- or ethnicity-related) attitudes (e.g. Davis et al. 2009). Examples of racial/ethnic attitude questions might include those asking about opinions toward affirmative action or immigration policies. As with survey participation, interviewer effects on survey responses may assume different forms. To date, the majority of studies examining interviewer race/ethnicity and survey responses have focused on White and Black interviewers and respondents. Much less research has been conducted with interviewers from other racial groups (e.g. Asian) or with Latino interviewers.

The *social distance model* is useful for explaining possible race/ethnicity of interviewer effects on the actual responses reported in surveys. Specifically, differences in the attitudes and opinions regarding racial/ethnic topics that are reported to interviewers of different racial/ethnic backgrounds would be attributed to social distance. The presumption is that respondents would provide more candid answers to those interviewers who share their racial/ethnic status and with whom they have less social distance. All else equal, one would expect less social distance between interviewer and respondent dyads to be associated with greater trust, less suspicion of interviewers and their intentions and therefore more candid responses.

Another explanation for race- and ethnicity-of-interviewer effects is the *deference model*. This model interprets these effects as reflecting respondent efforts to maintain positive social interactions with their interviewer by providing responses to race-relevant questions that they imagine are deferential to their interviewer's actual or perceived racial/ethnic identity (Campbell 1981; Anderson et al. 1988b). The *deference model* is closely related to social desirability explanations for respondent behavior that emphasize the presentation of a positive self-image (Davis and Silver 2003; Finkel et al. 1991). This model is also closely related to the *social attribution model*, which suggests that respondents take cues from interviewer social characteristics, including race/ethnicity, to make inferences regarding interviewer norms, which in turn condition respondent answers in order to conform to perceived interviewer norms and expectations (Johnson et al. 2000). These models would each predict main effects of interviewer race/ethnicity on respondent answers.

It is also possible that the discrimination and oppression that members of minority populations in any society inevitably feel and experience may lead to greater levels of interviewer effects among members of minority groups, relative to those in the majority. The ramifications of social distance and the need for social conformity may both be more salient to minority populations, resulting in stronger race of interviewer effects among respondents from these subgroups. For example, in the United States there is widespread perception and evidence of differential treatment of offenders, by race/ethnicity, within the criminal justice system (Johnson and Bowman 2003). Concern with more severe treatment may render minority groups more sensitive to interviewer race/ethnicity when they are asked to answer questions concerned with criminal justice issues or experiences in social surveys.

11.1.3 Existing Evidence

11.1.3.1 Survey Participation

Before an interview begins, interviewers must successfully contact and gain cooperation from respondents. Interviewers' characteristics (including their race or ethnicity) may influence this process. Relatively little research has examined the impact of interviewer race/ethnicity on survey participation (in contrast to the larger body of evidence regarding the impact of race/ethnicity on responses to survey questions). The social distance model suggests that respondents will be more cooperative with interviewers who are similar to them. In other words, respondents should be more likely to participate when approached by a same race/ethnicity interviewer than when approached by a different race/ethnicity interviewer. Consistent with this model, Moorman et al. (1999) found higher cooperation rates in a telephone survey when respondent and interviewer race/ethnicity was concordant. Similarly, in a mall intercept study of White and Latino[1] interviewers and respondents, Webster (1996) found that matching respondents and interviewers resulted in higher cooperation with the survey request (and more answers to open-ended questions). However, in a study of arrestees, Lord et al. (2005) found no differences in likelihood of participation in a face-to-face survey, or in willingness to provide a urine sample when interviewed by a same-race versus different-race.

Other researchers have found evidence that the effect of interviewer race/ethnicity on survey participation may depend upon the characteristics and environment of the selected household and respondent. In an examination of the 1999, 2000, and 2001 National Surveys on Drug Use and Health (NSDUH), a nationally representative face-to-face survey that included White, Black, and Latino interviewers, Chromy et al. (2005) found that although the race/ethnicity of the interviewer had few main effects on contact or cooperation, significant interactions were found when one examined the composition of the neighborhood in which a selected household was located and the race/ethnicity of the selected respondent. Specifically, both Black and Latino interviewers were less likely to make contact with White respondents (no effects of interviewer race/ethnicity were observed for Black or Latino respondents). In contrast, Black and Latino respondents were more likely to cooperate with White interviewers than with Black or Latino interviewers (there was no effect of interviewer race/ethnicity on cooperation among White respondents). A follow-up analysis of NSDUH data collected between 2002 and 2007 examined interactions between the composition of the area in which a household was located, the race- and ethnicity-of-interviewers, and the race/ethnicity of the selected respondent in affecting cooperation with the survey request (Wang et al. 2013). Their findings did not show that there was an advantage to matching interviewers to respondents of similar race or ethnicity and did not support the *social distance model*. The authors concluded that "[t]here is little evidence that interview cooperation rates were maximized when the [interviewer's] race/ethnicity matched that of the screener report of the race/ethnicity of the respondent" (Wang et al. 2013, p. 21). So evidence regarding social distance theory and survey participation is mixed.

11.1.3.2 Survey Responses in Face-to-Face Interviews

In contrast to research on participation, a great many studies have examined the effects of interviewers' race or ethnicity on survey responses in face-to-face surveys where visual information relevant to interviewers' race/ethnicity is available to respondents, and quite a few of these have found that interviewers' race or ethnicity does influence survey responses, particularly for questions that are related to race or ethnicity. Much of the early work on race-of-interviewer effects focused on differences between survey reports from Black respondents when interviewed face-to-face by White versus Black interviewers. Evidence from a 1968 race-of-interviewer experiment suggested that Black respondents interviewed by White interviewers reported being less favorable towards Black militant protest and less hostile

1 For simplicity and consistency, we use the term "Latino" rather than "Hispanic" to refer to people of Latino, Spanish or Hispanic ethnicity throughout the chapter even when referring to other research in which different terminology is used.

toward Whites than did those interviewed by Black interviewers, but no race-of-interviewer effects were found for questions about other race-related policies (Schuman and Converse 1971). The authors conclude that this pattern of effects is due to deference toward White interviewers. Consistent with this, Williams (1964) found that race-of-interviewer effects among Black respondents were greater when social distance was high (high-rank interviewers and low-rank respondents) and for more threatening questions (i.e. the greater the potential reward or cost of answering in a particular way). Also consistent with the deference explanation, Black respondents interviewed by a White interviewer were more likely to report warmth and closeness to Whites when interviewed by a White interviewer than when interviewed by a Black interviewer (Anderson et al. 1988b; also see Campbell 1981). Black respondents were more likely to report closeness (but not warmth) toward Blacks when interviewed by a Black interviewer than when interviewed by a White interviewer and interviewer race was inconsistently associated with policy positions on a variety of race-related policies (Anderson et al. 1988b; also see Schaeffer 1980). Race-of-interviewer effects have also been observed for the behavioral self-reports of Black respondents in face-to-face interviews. Black respondents interviewed by Black interviewers were more likely to falsely report that they voted (when they did not) than were Black respondents interviewed by White interviewers (Anderson et al. 1988a), suggesting higher social desirability pressure (and less honest responses). Similarly, Fendrich et al. (1996) found that Black juvenile arrestees interviewed by Black interviewers were less likely to report that they had ever used cocaine in their life than those interviewed by White interviewers. However, Cody et al. (2010) found that race-of-interviewer effects observed in the National Black Election Survey (NBES) were eliminated when the workload of interviewers was accounted for. They argue that race-of-interviewer effects may be weaker than previously observed and the appearance of effects may be increased by uneven workload across interviewers.

A smaller body of evidence has examined race-of-interviewer effects among White respondents in face-to-face surveys (e.g. Athey et al. 1960; Campbell 1981; Hatchet and Schuman 1975). Fendrich et al. (1996) did not find a race-of-interviewer effect on reports of lifetime cocaine use among White juvenile arrestees. However, there is some evidence that White respondents give more pro-Black survey responses (e.g. reports that they would not mind if a relative married a Black person and more positive ratings of Blacks) when interviewed by a Black interviewer than when interviewed by a White interviewer (Campbell 1981; Hatchet and Schuman 1975). Campbell (1981) found similar effects for White and Black respondents (both were more positive toward the other group when interviewed by members of that racial group) showing deference. However, Hatchet and Schuman (1975) argue that evidence that this effect is somewhat stronger among more educated White respondents suggests that the mechanism for White respondents may be motivation to be or to appear tolerant. Consistent with this, Schaeffer (1980) found that race-of-interviewer effects among White respondents were primarily evident on items where the respondent reported more liberal race-related policies and opinions when interviewed by a Black interviewer than when interviewed by a White interviewer. Consistent with this, Athey et al. (1960) found that White respondents were less likely to report that Blacks moving into their neighborhoods would negatively affect property values when being interviewed by a Black interviewer than when being interviewed by a White interviewer.

Some research examining interviewer race/ethnicity has found similar effects across respondent race/ethnic groups. For example, Johnson and Parsons (1994) found that the homeless were more likely to report at least some forms of drug use to White interviewers than to Black interviewers and this effect did not differ for White and Black respondents. Similarly, Loewenstein and Varma (1970) found higher reports of health care utilization to White interviewers than to Black interviewers across all race/ethnic categories of respondents.[2]

Although most of this research exclusively focuses on Black–White interviewer differences, there is a small body of research considering additional racial and ethnic groups (e.g. Lord et al. 2005; Reese et al. 1986; Weeks and Moore 1981; Worcester and Kaur-Ballagan 2002). For example, Welch et al. (1973) examined in a face-to-face survey whether the responses of Mexican-American respondents to questions about health and politics were affected by whether the respondent was interviewed by a Mexican-American interviewer, an Anglo interviewer, or a pair of interviewers (one Mexican-American and one Anglo). They found few significant differences and those that they did find were almost all eliminated when respondent demographics (education and age in particular) were included as control variables. In another small study of non-English-speaking children from four different ethnic groups (Cubans, Chicanos, Native Americans, and Chinese), Weeks and Moore (1981) found that interviewer ethnicity did not affect answers to nonsensitive questions. In a larger study limited to White and Latino respondents, White respondents gave more deferential responses to questions about Mexican-American culture when interviewed by a Latino interviewer than when interviewed by a White interviewer (Latino respondents did not show race-of-interviewer effects for these questions, but might have shown similar deference if they had been asked questions about Whites; Reese et al. 1986). Similarly, White respondents reported more acceptance of Asians when interviewed by an Asian interviewer than when interviewed by a White interviewer (Athey et al. 1960). However, consistent with the *social distance model*, another study found that Asian respondents were more comfortable giving responses that went against popular opinion when interviewed by an Asian interviewer than when interviewed by a White interviewer (Worcester and Kaur-Ballagan 2002).

11.1.3.3 Survey Responses in Telephone Interviews

Much of this early work on race-of-interviewer effects came from household surveys in which respondents were interviewed face-to-face. Under these circumstances, it is likely that a respondent (or a potential respondent) can accurately identify their interviewer's race/ethnicity. However, in telephone interviews, respondents (or potential respondents) do not have visual cues and their perceptions of interviewers' race/ethnicity are based on what interviewers say and how they say it (e.g. interviewers' vocal characteristics, pronunciation, or dialect).

Despite the possible ambiguity of interviewers' race or ethnicity over the telephone, a number of studies have nonetheless found that interviewer race influences telephone survey responses (e.g. Cotter et al. 1982), although these comparisons have been confined to Black and White interviewers. In a telephone survey of Virginia voters, Finkel et al. (1991) found that White respondents were more likely to say that they would vote for a Black gubernatorial candidate when interviewed by a Black interviewer than when interviewed by a White interviewer. Similarly, in a telephone survey conducted with adult residents of Alabama, White respondents were more likely to give pro-Black responses to 9 of the 13 racial questions (e.g. Whether or not school busing was justified) when interviewed by a Black interviewer than when interviewed by a White interviewer (Morin 1989). Responses among Black respondents were not affected by interviewers' race, but the sample size of Black respondents was much smaller ($n = 103$) than the sample of White respondents ($n = 439$). However, several reports of telephone surveys conducted by media organizations have suggested that Black respondents may give more

2 This main effect was qualified by some evidence that interviewer race/ethnicity and gender interacted to influence reports of health care utilization among some categories of respondents, but the findings are reported as inconsistent (findings were not reported in detail).

pro-White responses when interviewed by a White interviewer than when interviewed by a Black interviewer (e.g. Meislin 1987; Morin 1989). For example, Black respondents were more likely to say that a lawyer representing a Black victim of a racial attack was acting irresponsibly when interviewed by a White interviewer than when interviewed by a Black interviewer (Meislin 1987). Davis (1997a) also found that Black respondents were more likely to acquiesce (or agree with) contradictory statements when they are made by White interviewers than when they are made by Black interviewers and that Black respondents reported less political influence and power when interviewed by a White interviewer than when interviewed by a Black interviewer.

There is also some evidence that interviewer's race may influence behavioral reports in telephone surveys – specifically, respondents may give more honest (and less socially desirable) responses to interviewers of the same race. For example, Livert et al. (1998) found that both Black and Latino respondents were less likely to report substance use when interviewed by a White interviewer than when interviewed by a Black interviewer (there were no Latino interviewers included in that study). Similarly, Johnson et al. (2000) combined racial similarity in an overall index of demographic similarity (along with gender, age category, and education) and found that lower social distance (more demographic similarity) was associated with higher levels of reported lifetime and recent drug use.

The studies reviewed above used the interviewer's self-reported race to explore race-of-interviewer effects. This assumes, however, that respondents can accurately assess interviewers' race (or ethnicity) over the telephone. A number of researchers have explored the accuracy of respondents' perceptions of interviewers' race. The accuracy expected by chance alone varies depending upon the number of interviewer race or ethnicities included. A study of Black respondents with all Black interviewers reported an overall accuracy rate of 73.1% (Wolford et al. 1995). Additionally, perceived interviewer race had a consistent effect on questions about Whites and race-related policies such that responses showed greater deference to the perceived interviewer's racial group (Wolford et al. 1995). In a similar study, Harms (1995) reported that 75% of Black respondents believed they were talking to a Black interviewer and 14% believed they were talking to a White interviewer. Furthermore, perceived interviewer race influenced survey responses such that respondents who believed that they were being interviewed by a White interviewer reported less pro-Black opinions and beliefs (Harms 1995). Of course, since all the interviewers in these studies were Black, it is impossible to compare the effect of actual and perceived interviewer race.

Studies using interviewers from multiple races or ethnicities allow one to assess the association between perceived and actual interviewer race as well as comparing their effects on survey responses. The studies that have done so have focused on comparing responses to White and Black interviewers. In a study of Black respondents where both White and Black interviewers were used, 79% of respondent perceptions of interviewer race were correct and actual and perceived race had similar effects on the responses (Davis 1997b). Although the exact proportions vary, other similar telephone surveys suggest that respondents correctly identify interviewers' race between 60% and 90% of the time (e.g. Meislin 1987; Tate 1998; Wilson and Olesen 2002). Although these are relatively high accuracy rates, it is important to note that White interviewers tend to be perceived accurately a greater percentage of the time and that Black respondents are more accurate in their perceptions of White interviewers than White respondents are of Black interviewers. For example, in a telephone survey conducted by the Gallup Organization, 82.2% of White interviewers were correctly perceived as White, but only 73.8% of Black interviewers were correctly perceived as Black (Wilson and Olesen 2002). Additionally, Black respondents correctly perceived 89.5% of White interviewers as White while White respondents only perceived 75.8% of Black interviewers as Black. These findings are confirmed by laboratory studies showing that people can accurately and quickly perceive race based on vocal characteristics and dialect (e.g. Lass et al. 1979; Purnell et al. 1999). This literature suggests that respondent perceptions of race in telephone surveys are reasonably accurate (roughly three quarters of the time), but also that there is substantial error in these estimates. However, the research has largely been limited to Black and White respondents and Black and White interviewers. Accuracy rates may drop if interviewers from more racial or ethnic groups are included and these studies do not directly address how accurate respondents might be at identifying Latino interviewers in telephone surveys.

Despite this evidence of the accuracy of perceptions, actual and perceived interviewer race do not always have consistent effects. Davis and Silver (2003) found that 50% of Black respondents reported that they were interviewed by a Black interviewer (when the true proportion was 67%), and 25% of White respondents reported they were interviewed by a Black interviewer (when the true proportion was 55%). They found, consistent with the notion of stereotype threat (i.e. that respondents who are aware of a culture stereotype may be distracted from a task by trying to avoid the stereotype), that Black respondents gave fewer correct responses to a political knowledge quiz when they believed that they were being interviewed by a White interviewer than when they believed that they were being interviewed by a Black interviewer (Davis and Silver 2003). No such effects were observed for White respondents. However, these effects were observed for perceived interviewer race only and not actual interviewer race.

11.1.3.4 Survey Responses in Surveys that Are Not Interviewer-Administered

Although typically one might think that race- and ethnicity-of-interviewer effects would primarily occur when a survey is administered by an interviewer, there is some evidence that the race of an interviewer may affect responses even if the survey is not interviewer-administered, and work examining virtual interviewers suggests that the race of an interviewer influences responses even if s/he is not physically present. For example, Liu and Wang (2016) found that interviewers' race/ethnicity influenced survey responses (from White, Black, and Latino respondents) such that respondents gave answers more favorable toward the interviewers' race/ethnic group when being interviewed by a member of that group, even when the questions were part of a self-administered portion of the questionnaire.

In a study of live versus "virtual" interviewers, Black respondents reported less liberal views to White interviewers than to Black interviewers regardless of whether they were virtual or live, suggesting that an interviewer did not need to be present (or even real) to affect survey responses (Krysan and Couper 2003). However, White respondents gave more racially conservative responses to some questions when interviewed by a virtual Black interviewer than when interviewed by a virtual White interviewer. This suggests the possibility that in the virtual interviewer condition, White respondents might have been affected by stereotype activation rather than deference to the interviewer (since more liberal responses would be predicted by deference). A follow-up experiment in a web survey provided only weak support for this hypothesis, however (Krysan and Couper 2006).

11.1.3.5 Other Race/Ethnicity of Interviewer Research

The work that we've reviewed here has focused on research done in the US context. Researchers in other countries have also examined the effects of interviewer race/ethnicity on survey responses (e.g. Kappelhof and De Leeuw 2019) and the challenges that present themselves in interviewing racial minorities (e.g. Kappelhof 2017), although little of this work has involved experimentally assigning respondents to interviewers of different races/ethnicities. Researchers have also considered the impact of interviewer race/ethnicity on indicators of data quality (rather than the distribution of responses, per se) such as acquiescence response bias (e.g.

Olson and Bilgen 2011) and size of response effects (e.g. Sudman and Bradburn 1974) but the evidence for race- and ethnicity-of-interviewer effects here is weak and inconsistent (and may be affected by whether the questions used to measure data quality are related to race or not).

11.2 The Current Research

The current research is designed to build on the existing evidence regarding race- and ethnicity-of-interviewer effects and to address some of the limitations of past research. Much of the evidence to date has focused on Black and White interviewers and respondents, and even the work that has considered other ethnicities has done so in a narrow way (e.g. only including White and Latino interviewers). Although researchers are beginning to consider interviewers' ethnicity – specifically whether or not they are Latino (e.g. Reese et al. 1986) – very little research has examined more than two racial/ethnic groups simultaneously. Furthermore, in most of the research that has been conducted, the distribution of race/ethnicity among both respondents and interviewers has been uneven – with greater percentages of non-Latino Whites in both categories (with the exception of research that targets one racial or ethnic group; e.g. Davis 1997b). Almost no work has used substantial numbers of White, Black, and Latino interviewers to interview a population with sizeable numbers of White, Black, and Latino respondents. In addition, only a few researchers have examined the impact of race or ethnicity on survey participation or the role of social context on this relationship. Finally, although experiments in which respondents or households are randomly assigned to interviewers have been conducted (e.g. Cotter et al. 1982; Hatchet and Schuman 1975; Schuman and Converse 1971), most of the research examining interviewers' race or ethnicity has not involved random assignment.

The telephone survey that we conducted was designed to address some of these limitations. It was conducted in the City of Chicago, where there are large White, Black, and Latino populations. Neighborhoods in Chicago also vary in their racial/ethnic composition, so we were able to explore the impact of social context on potential respondents' reactions to interviewers of different races and ethnicities. Finally, respondents were also asked about their perceptions of their interviewers' race/ethnicity at the end of the interview to assess the accuracy of their perceptions.

One of the weaknesses of much of the evidence regarding the effects of interviewer race/ethnicity is lack of experimental evidence. When cases are not randomly assigned to interviewer by race/ethnicity, it is difficult to draw conclusions about the causes of interviewer race/ethnicity effects on survey responses because other variables are confounded with interviewer race/ethnicity. Random assignment of cases to specific interviewers (i.e. a given case is always worked by a single interviewer) might seem the best approach, but this makes it difficult because different interviewers might make contact attempts on different days (and at different times of day). Random assignment to a specific interviewer may therefore confound interviewer race/ethnicity with some other aspect of administration. We dealt with this issue by randomly assigning telephone numbers sampled for the survey to different interviewer race/ethnicity categories, but not to a specific interviewer. That is to say that each number was assigned to be called by only members of one race/ethnic group, but it might be called by one interviewer within that race/ethnic group and another interviewer within that race/ethnic group during a different shift. A number assigned to the Black interviewer condition might be called on different days/times of day by different Black interviewers, but it would not be called by a White or Latino interviewer (unless it was identified as Spanish-speaking). We attempted to have at least one interviewer from each racial and ethnic group work each scheduled shift so that numbers were equivalently worked. This means that all contact attempts were made by interviewers from the same race/ethnicity, although not all were made by the same interviewer. Approximately equal proportions of the sample were worked by interviewers of each racial/ethnic group (non-Latino White, non-Latino Black, and Latino).

The current research addresses a number of questions with regard to the impact of interviewers' race/ethnicity on survey data. Specifically, it was designed to address the following seven research questions in three broad areas related to race- and ethnicity-of interviewer effects. After each research question, we have listed some hypotheses that follow from each. These hypotheses are based on the theoretical perspectives and literature described above, with the caveat that most of the evidence regarding interviewer effects examines only white and black interviewers and respondents and the theoretical perspectives sometimes make for competing hypotheses.

Survey Participation

(1) How does the race or ethnicity of the interviewer influence the decision to participate in a telephone interview?
   a. *Hypothesis 1*: Response rates overall will be highest for White interviewers and lower for Black and Latino interviewers.

(2) How does neighborhood racial/ethnic composition influence the decision to participate in a telephone interview?
   a. *Hypothesis 2*: Response rates will be higher in geographic areas with higher percentages of White residents than in areas with higher levels of non-White residents.

(3) Does the effect of interviewers' race/ethnicity differ as a function of neighborhood racial/ethnic composition?
   a. *Hypothesis 3*: Response rates will be higher among interviewers of a given race/ethnicity in geographic areas with high percentages of residents of the same race/ethnicity and lower in areas with low percentages of residents of the same race/ethnicity.

Respondents' Perceptions of Interviewers' Race/ethnicity

(4) How successful are respondents at identifying the race or ethnicity of their interviewer?
   a. *Hypothesis 4*: Respondents will be less successful at identifying interviewers' race/ethnicity than in past work that has involved fewer race/ethnic groups and fewer non-White interviewers.

(5) Who is correctly identified (i.e. which interviewers) and who correctly identifies (i.e. which respondents)?
   a. *Hypothesis 5a*: White interviewers will be identified more successfully than Black or Latino interviewers.
   b. *Hypothesis 5b*: Respondents will be more successful at identifying interviewers' race/ethnicity when interviewers and respondents race/ethnicity are the same than when they are different.

Survey Responses

(6) What are the effects of actual interviewer race/ethnicity on racial/ethnic-related attitudes?
   a. *Hypothesis 6a*: Respondents will report more positive attitudes toward policies that benefit a particular group when interviewed by an interviewer of that race/ethnicity.
   b. *Hypothesis 6b*: Respondents will report more positive attitudes toward a racial/ethnic group (or a member of that groups) when being interviewed by an interviewer of that race/ethnicity.

(7) What are the effects of perceived interviewer race/ethnicity on racial/ethnic-related attitudes?
   a. *Hypothesis 7b*: Effects of perceived interviewer race/ethnicity will follow a similar pattern, but be stronger than effects of actual interviewer race/ethnicity.

11.3 Respondents and Procedures

This chapter utilized data from the 2008 Chicago Area Study (CAS; Holbrook and Johnson 2008) survey. This survey was conducted between 19 April 2008 and 16 August 2008 by the Survey Research Laboratory (SRL) of the University of Illinois at Chicago (UIC) in collaboration with graduate students enrolled in a practicum-based survey research methods course. Using a design similar to the Detroit Area Study (Clemens et al. 2002), interviews were conducted both by students enrolled in the practicum course and by professional SRL interviewers. The survey was conducted via telephone and used a random digit dialing (RDD) design (including only households with landline phones) to sample adults age 18 and older in the City of Chicago. In households with more than one eligible respondent, the Troldahl–Carter–Bryant selection method (Bryant 1975) was used to select a respondent to interview within the household. Interviews were conducted in both English and Spanish. On average, the interviews took 24 minutes to complete. The AAPOR Response Rate 3 for the survey was 20.3%. All research procedures were approved by the UIC Institutional Review Board.

The collection of survey data from one metropolitan region is both an advantage and a potential liability. A more representative national sample would enhance the external validity of the results. Data from a nationally representative sample would also allow for broader inferences about the effects of interviewer race/ethnicity. However, limiting the survey to the City of Chicago also has advantages. The sample for a national survey would have smaller proportions of Black and Latino respondents whereas the City of Chicago has substantial numbers of both groups. This allows us to explore the impact of interviewer race on participation in geographic areas with different characteristics (e.g. mostly White, mostly Black, a heterogeneous mixture of races) and to obtain a relatively large sample of respondents of each racial and ethnic group to assess the impact of interviewer race/ethnicity on respondents' answers.

11.4 Measures

Below we describe the measures and coding for our key variables. Continuous variables were consistently coded to range from 0 to 1 where 0 was given to the lowest value on the continuum and 1 was given to the highest value on the continuum. Equally spaced values between 0 and 1 were given to intermediate values between the high and low. This means that in an ordinary least squares regression the size of the coefficient reflects the amount of change in the dependent variable from the lowest possible value of the continuous variable to the highest possible value of the continuous variable. This coding allows one to compare the relative sizes of the effects of continuous variables and is a common approach to coding continuous variables, particularly in political science (e.g. Alvarez and Brehm 1997; Branton et al. 2011; Campbell et al. 2011).

11.4.1 Interviewer Race/Ethnicity

Sampled telephone numbers were randomly assigned to four groups. The numbers assigned to one condition were called only by non-Latino White interviewers. The numbers assigned to a second condition were called only by non-Latino Black interviewers. The numbers assigned to a third condition were called only by Latino interviewers. Finally, because data collection was done in collaboration with the CAS practicum course and there were class members who did not fit into these three interviewer categories, a fourth condition (excluded from our analyses here) were numbers assigned to be called by interviewers of other races and ethnicities. This latter group of selected numbers was only worked during the student portion of the interviewing and was not as thoroughly called as the other three conditions. In addition, households identified as Spanish speaking from all four conditions were separated and these interviews were conducted by bilingual interviewers.

11.4.2 Sample Frame Variables

11.4.2.1 Disposition Codes

As in most telephone surveys, each telephone number was identified with a final disposition code. The disposition codes used were consistent with recommendations from the American Association for Public Opinion Research (AAPOR 2015) and were used to create dichotomous variables indicating whether or not the number was coded as a complete, whether or not the number was successfully contacted, and whether or not the number was coded as a refusal. The completion or participation variable was coded 1 if the number resulted in a completed interview, missing if the case was known to be ineligible (e.g. a business) and 0 otherwise. The contact variable was coded 1 if the number resulted in contact (e.g. a completed interview, a refusal, a household where the individual respondent was never available) and 0 if the household was not successfully contacted (e.g. only automated communications or cases where eligibility was unknown). Known ineligible cases were coded as missing. The refusal variable was coded 1 if the final disposition code was a refusal (either household or respondent-level) and 0 otherwise. Again, known ineligible cases were coded as zero.

11.4.2.2 Neighborhood Level Variables

Telephone numbers were matched to ZIP codes (using respondent reports for responding households and a matching procedure for nonresponding households; see Johnson et al. 2006). Census data about ZIP-code level racial/ethnic composition was appended to the sample frame. Four types of ZIP codes were identified: (i) primarily African-American (those with 80% or more African-American residents); (ii) primarily White (those with 80% or more non-Latino White); (iii) mixed (those with more than 20% from two or more of the three racial/ethnic groups we were focusing on); and (iv) other (those ZIP codes that did not fit into any of the other three categories including one ZIP code that was majority Latino).

11.4.3 Respondent Variables

11.4.3.1 Race/Ethnicity

The race/ethnicity of respondents was measured using two questions. First, respondents were asked: "Which of the following racial groups best describes you: White, Black, Asian, Pacific Islander, Native American, or something else?" Respondents were also asked: "Are you Mexican, Mexican American, or Chicano, Puerto Rican, or Cuban, or some other Spanish origin?" Respondents who said they were White and of Latino origin were coded as Latino. Respondents who said they were White and not of Latino origin were coded as White. Respondents who said they were Black and not of Latino origin were coded as Black. All other respondents were excluded from our analyses.[3]

3 One hundred sixty eight (168) respondents were dropped from our analyses either because they were interviewed by interviewers in the "other" race/ethnicity group, because their interviews were conducted in Spanish, or because they did not identify as White, Black, or Latino (or some combination of these three factors). Respondents who said they were "something else" were asked to specify their race. Respondents who answered this specific question with a clear race or ethnicity were coded appropriately.

11.4.3.2 Perceived Race/Ethnicity

At the end of the interview, respondents were told "This last question is just for research purposes. You may not have thought about this, but I'd like to ask you to guess my race or ethnicity. Would you guess that I am White, Black, Latino, Asian, or some other race?" A variable indicating accuracy was coded 1 if perceived race/ethnicity matched interviewers' actual race/ethnicity and 0 if it did not.

11.4.3.3 Confidence

Respondents were also asked: "How confident are you in your guess? Would you say extremely confident, very confident, somewhat confident, not very confident, or not at all confident?" Confidence was coded to range from 0 to 1 where higher values indicated greater confidence.

11.4.3.4 Male

Gender was coded by the interviewer (male = 1; female = 0).

11.4.3.5 Age

Respondents were asked: "Now, we'd like to find out a little bit about you. In what year were you born?" This variable was coded to range from 0 (18 years old) to 1 (91 years old) with ages between these two extremes given equally spaced intermediate values (between 0 and 1). So, for example, a 19 year-old respondent received a value of $1/(91 - 18)$, a 20 year-old respondent receiving a value of $2/(91 - 18)$, and so on.

11.4.3.6 Years of Education

Respondents were asked: "What is the highest grade of school or year of college you have completed?" This variable was coded to range from 0 (zero years of education reported) to 1 (17 or more of education). Intermediate values were given to years between (i.e. 2–16). For example, a value of 1/17 was given to respondents who reported one year of education, a value of 2/17 was given to respondents who reported two years of education, and so on.

11.4.3.7 Household Income

Respondents were asked: "Was your total household income for the year 2007, from all sources, before taxes, more or less than \$60 000?" Respondents who said their income was less than 60k were asked: "Was it less than \$40 000?" Those who said it was less than 40k were asked: "Was it less than \$20 000?" Respondents who said their income was more than 60k were asked: "Was it more than \$80 000?" Respondents who said their income was more than 80k were asked: "Was it more than \$100 000?" Income was coded 0 for less than 20k, 0.20 for 20–40k, 0.40 for 40–60k, 0.60 for 60–80k, 0.80 for 80–100k, and 1 for more than 100k.

11.4.3.8 Political Ideology

Respondents were asked: "In general, would you describe your political views as very conservative, conservative, moderate, liberal, or very liberal?" Ideology was coded 1 for respondents who said "very liberal" 0.75 for respondents who said liberal, 0.5 for respondents who said "moderate," 0.25 for respondents who said "conservative," and 0 for respondents who said "very conservative."

11.4.3.9 Affirmative Action

Two questions in the survey asked respondents their opinion about affirmative action. First, respondents were told: "Now let me read you two brief statements on affirmative action programs for Blacks and other minorities. I'm going to ask which one comes closer to your own point of view. Affirmative action programs are still needed to make up for the effects of discrimination against minorities, and help reduce racial inequality, or affirmative action programs have gone too far in favoring minorities, and should be phased out because they are unfair to whites." This variable was coded 1 if respondents said that affirmative action programs were still needed and 0 if they said they thought such programs had gone too far.

Second, respondents were asked: "Do you feel the use of affirmative action programs are generally going in the right direction, or do you feel they are going in the wrong direction?" This dichotomous variable was coded 1 if respondents said "right direction" and 0 if they said "wrong direction."

11.4.3.10 Immigration Policy

Respondents were asked two questions about their views on immigration policy. First, they were asked: "Next, I want to ask you some questions about immigration in the United States. Which comes closest to your view about what government policy should be toward illegal immigrants currently residing in the United States. Should the government: one, deport all illegal immigrants from the United States; two, allow illegal immigrants to remain for a limited amount of time in the United States in order to work, or three, allow illegal immigrants to remain in the United States and become US citizens, but only if they meet certain requirements?" This variable was coded so that values were ordered from least supportive of immigration to most supportive of immigration. Specifically, the recoded variable was coded 0 if respondents said all illegal immigrants should be deported (a position which is not at all supportive of immigration), 0.5 if respondents said immigrants should be allowed for a limited time to work (a moderately supportive position), and 1 if respondents said immigrants should be allowed to stay with conditions (the most supportive position). We chose to code the variable in this way because these responses reflect degrees of support for immigration, a continuous dimension.

Second, respondents were asked: "Do you feel government policies about immigration are generally going in the right direction, or do you feel they are going in the wrong direction?" Responses to this dichotomous variable were coded 1 if a respondent said "right direction" and 0 if s/he said "wrong direction."

11.4.3.11 Attitudes Toward Obama

Finally, respondents were asked their opinion about then-Presidential candidate Barack Obama. Respondents were told: "Now, I'm going to read you the names of several public figures and I'd like you to rate your feelings toward each one as either very positive, somewhat positive, neutral, somewhat negative, or very negative. If you don't know the name, please just say so. Barack Obama." Responses were coded 0 for very negative, 0.25 for somewhat negative, 0.5 for neutral, 0.75 for somewhat positive, and 1 for very positive.

11.5 Analysis

11.5.1 Sample Frame Analyses

Our analyses of the effects of interviewer race/ethnicity on participation used the sample frame including data about which interviewer race/ethnicity condition the telephone number had been assigned to, appended census data regarding the racial/ethnic composition of the corresponding ZIP code (as a measure of neighborhood characteristic), and disposition code data. We used multilevel logit models to predict dichotomous indicators of response, refusal, and contact as the dependent variables, including random ZIP-code effects to control for clustering at the ZIP-code level (level 2). Independent variables in these analyses are dummy variables representing interviewer condition (White interviewers were the comparison group) and ZIP-code composition. (Predominantly White was the comparison group.) All analyses used the `melogit` command in Stata 14.0 with robust standard errors (because we were not able to control for clustering at the interviewer level). We use unweighted data for this set of analyses because none of the information used to construct weights (e.g. probability of selection and demographic characteristics) is available for nonrespondents. We did not use interviewer as a level in this analysis for several reasons. First, it is not a simple analytic task because telephone numbers were not nested within interviewers – a telephone number could be called by multiple interviewers within a race/ethnic group so any such analysis would have to control for each interviewer who made a contact attempt on that case. Furthermore, the sample frame does not include interviewer IDs for all contact attempts. One could imagine accounting for interviewer variance by, for example, assigning each case to the interviewer who contacted that number last (see West and Olson 2010 for a discussion of approaches for controlling for interviewer variance in these cases). However, this approach raises concerns as well because the last contact attempt is not always the attempt that determines the disposition code (e.g. refusals, always busy). Our sample-frame analyses are restricted to the 6274 telephone numbers that were assigned to non-Latino White interviewers, non-Latino Black interviewers, and Latino interviewers.

11.5.2 Survey Sample Analyses

Our analyses of the survey sample were weighted to adjust for the probability of selection and to match census demographic characteristics (age, race/ethnicity, and gender), and are restricted to English language non-Latino White, non-Latino Black, and Latino respondents who were assigned to the non-Latino White, non-Latino Black, and Latino interviewer conditions (weights were not scaled within interviewers). Cross-tabulations and multilevel logit models (including random effects for interviewer ID to control for interviewer variance) were used to assess the effects of interviewer and respondent race/ethnicity on the accuracy of respondents' perceptions of interviewers' race/ethnicity and dichotomous survey responses (analyses were conducted using the `svy: melogit` command in Stata 14.0 with the assumption that interviewers had a weight of 1). Multilevel general linear modeling was used to assess the impact of interviewers' and respondents race/ethnicity and interviewers' perceived race/ethnicity on continuous survey responses. These models included random interviewer effects to control for interviewer variance and were fitted using the `svy: meglm` command in Stata.

11.6 Results

11.6.1 Sample Frame

Our first set of analyses involved the full set of sampled telephone numbers that were assigned to either the White interviewer condition, the Black interviewer condition, or the Latino interviewer condition. We began by estimating indicators of survey participation using formulae suggested by the American Association for Public Opinion Research (AAPOR). The survey achieved a response rate (AAPOR response rate #3) of 20.3%, a refusal rate (AAPOR refusal rate #2) of 39.9%, and a contact rate (AAPOR contact rate #2) of 38.1%. We next examined the value of each of the three survey outcome variables estimated from final disposition codes for each sampled number. Table 11.1 shows the means of these survey outcome variables by race- and ethnicity-of-interviewer and ZIP-code composition.

Table 11.1 Survey outcomes by race- and ethnicity-of-interviewer and ZIP-code composition.

	Race- and Ethnicity-of-Interviewer			
ZIP code type	**White**	**Black**	**Latino**	**All**
Response (n = 3165)				
Predominantly White (%)	14.9	20.0	16.8	17.4
Predominantly Black (%)	24.9	19.5	25.2	23.0
Mixed (%)	22.4	17.9	25.6	21.8
Other (%)	13.0	12.2	15.4	13.5
All (%)	19.1	16.6	21.0	18.9
Refusal (n = 2210)				
Predominantly White (%)	57.8	59.1	50.0	55.8
Predominantly Black (%)	46.3	51.5	49.1	49.1
Mixed (%)	50.7	57.9	45.3	51.4
Other (%)	52.6	58.8	47.6	53.1
All (%)	50.9	56.7	47.4	51.8
Contact (n = 3181)				
Predominantly White (%)	70.3	73.3	60.5	67.9
Predominantly Black (%)	83.3	80.5	83.3	82.3
Mixed (%)	71.9	71.4	73.6	72.3
Other (%)	57.3	60.7	57.7	58.6
All (%)	69.3	70.2	68.6	69.4

11.6.1.1 Survey Participation or Response

For survey participation or response (shown in the top panel of Table 11.1), the response rate is relatively similar across interviewer conditions, although Black interviewers have the lowest response rate. There is greater variation by ZIP-code composition with response rates being higher in the Mostly Black and Mixed ZIP-codes than in the Mostly White ZIP-codes. Households in the Other ZIP-codes also had substantially lower response rates than the other three ZIP-code types.

The results of analyses predicting survey response outcomes (response, refusal, and contact) are shown in Table 11.2. Interviewer race/ethnicity did not have a significant effect on response (see column 1), but response rates were higher in the Mostly Black and Mixed Race ZIP-codes than in the comparison group (Mostly White) ZIP-codes. We next added to the model the interactions between interviewer race and ZIP-code composition to test whether the impact of interviewer race/ethnicity was moderated by social environment (see column 2 of Table 11.2). The two marginally significant interactions in the bottom part of column 2 suggest that although response was higher in Mostly-Black and Mixed-Race ZIP-codes when the interviewer was White, these effects were weaker when the interviewer was Black.

11.6.1.2 Refusals

We next examined survey refusals using a similar approach (see the middle panel of Table 11.1 and columns 3 and 4 in Table 11.2). Refusals were more likely when the interviewer was Black than when s/he was White and somewhat less likely in ZIP-codes that were mostly Black as

Table 11.2 Multilevel logit models (with robust standard errors) predicting survey outcomes with interviewer race and ZIP-code composition (standard errors in parentheses).

Predictor	Dependent variable					
	Response		Refusal		Contact	
Interviewer race/ethnicity condition[a)]						
Black	−0.17 (0.12)	0.38 (0.33)	0.23* (0.10)	0.02 (0.29)	−0.0003 (0.10)	0.11 (0.26)
Latino	0.14 (0.11)	0.19 (0.36)	−0.13 (0.11)	−0.44 (0.31)	−0.03 (0.12)	−0.49* (0.25)
ZIP-code composition[b)]						
Mostly Black	0.56** (0.22)	0.86** (0.33)	−0.32+ (0.17)	−0.57* (0.28)	0.99** (0.38)	0.93* (0.44)
Mixed Race	0.50** (0.20)	0.73* (0.32)	−0.22 (0.16)	−0.39 (0.26)	0.54 (0.35)	0.39 (0.40)
Other	−0.04 (0.22)	0.13 (0.34)	−0.18 (0.18)	−0.38 (0.28)	0.06 (0.36)	−0.14 (0.40)
Interviewer race condition × ZIP-code composition						
Black interviewer × mostly Black		−0.70+ (0.40)		0.20 (0.36)		−0.33 (0.36)
Black interviewer × mixed race		−0.67+ (0.38)		0.27 (0.34)		−0.15 (0.30)
Black interviewer × other		−0.51 (0.41)		0.27 (0.36)		−0.02 (0.31)
Latino interviewer × mostly Black		−0.18 (0.41)		0.56 (0.38)		0.49 (0.36)
Latino interviewer × mixed race		−0.01 (0.38)		0.21 (0.35)		0.56+ (0.30)
Latino interviewer × other		0.01 (0.41)		0.31 (0.37)		0.58+ (0.30)
n (Household)	3165	3165	2210	2210	3181	3181
n (ZIP-code)	40	40	40	40	40	40
ZIP-code variance (95% confidence interval)	0.21 (0.10–0.39)	0.21 (0.10–0.39)	0.18 (0.08–0.44)	0.19 (0.08–0.45)	0.58 (0.42–0.80)	0.58 (0.42–0.80)

$+p < 0.10$, $^{*}p < 0.05$, $^{**}p < 0.01$.

a) White interviewer condition is the comparison group.

b) Mostly White is the comparison group.

compared to ZIP codes that were mostly White. There were no significant interactions between interviewer race/ethnicity condition and ZIP-code racial/ethnic composition.

11.6.1.3 Contact

We next examined successful contact as a dependent variable (see the bottom panel of Table 11.1 and columns 5 and 6 in Table 11.2). Interviewer race/ethnicity did not have a main effect on contact, but contact was higher in Mostly Black ZIP-codes than in Mostly White ZIP-codes. The model including interactions (shown in column 6 of Table 11.2) suggests that

Latino interviewers had significantly lower contact rates in Mostly White ZIP-codes (the main effect for the Latino interviewer variable in this model), but that this effect was significantly weaker for Mixed Race and Other ZIP-codes. Although not significant, the coefficient for the interaction between the Latino interviewer variable and the Mostly Black ZIP-code variable suggests that Latino interviewers likely do not have lower contact rates than White interviewers in these ZIP-codes as well.

11.6.2 Perceptions of Interviewer Race

We next examined perceptions of interviewer race among respondents who participated in the survey. Table 11.3 shows the number of English-language interviews by the race/ethnicity of the interviewer (condition) and the race/ethnicity of the respondent. Our analyses are limited to White, Black, and Latino respondents who were interviewed in English by either a White, Black, or Latino interviewer ($N = 504$) – in other words, only those respondents in the highlighted cells in Table 11.3 were included.

Table 11.4 shows the percentage of perceived race- and ethnicity-of-interviewer by actual race- and ethnicity-of-interviewer. Respondents correctly "guessed" interviewer's race/ethnicity 71% of the time when the interviewer was White. However, correct guesses dropped to 53% for Black interviewers and only 21% for Latino interviewers. This is consistent with the literature that suggests that respondents more often correctly perceive the race of White interviewers than interviewers of other races and ethnicities.

Table 11.3 Race- and ethnicity-of-interviewer by race- and ethnicity-of-respondent.

	Interviewer race/ethnicity				
Respondent race/ethnicity	**White**	**Black**	**Latino**	**Other**	**Total**
White	75	82	71	20	248
Black	67	72	76	14	229
Latino	16	18	27	10	71
Other race	10	6	6	1	23
Total	168	178	180	45	571

Table 11.4 Cross tabulation of perceived race- and ethnicity-of-interviewer by actual race- and ethnicity-of-interviewer.

	Actual race- and ethnicity-of-interviewer		
Perceived race- and ethnicity-of-interviewer	**White**	**Black**	**Latino**
White (%)	71	29	64
Black	6	53	7
Latino	12	5	21
Other	4	6	5
DK/refused	8	7	4
Total (%)	100	100	100
n	158	172	174

Table 11.5 Percentage "correct guesses" on interviewer race/ethnicity by interviewer race/ethnicity and respondent race/ethnicity.

	Respondent race/ethnicity		
	White	Black	Latino
Interviewer race/ethnicity			
White (%)	79	66	59
Black (%)	51	59	29
Latino (%)	17	24	26
n	228	215	61

Note: Cell entries are the percentage of respondents within each group that guessed their interviewer's race/ethnicity correctly. For example, 79% of white respondents correctly identified their interviewer's race as white.

Table 11.5 shows the percent correct responses on interviewer race/ethnicity by respondent race/ethnicity. All respondents were most accurate at correctly identifying the race/ethnicity of their interviewer if s/he was White and least accurate if s/he was Latino.

We first examined the predictors of the accuracy of respondents' perceptions of their interviewers' race/ethnicity. The dependent variable for these analyses is accuracy (1 = correct, 0 = not correct) of respondents' perception of their interviewers' race/ethnicity. The results are shown in Table 11.6. Model 1 includes dummy variables for Black and Latino respondents and Black and Latino interviewers. In both cases the comparison group is Whites. Model 1 also includes demographic variables and two other predictors: respondent's confidence that their guess about their interviewer's race/ethnicity is correct, and political ideology. The results examining respondent and interviewer race/ethnicity confirm what was observed from Tables 11.4 and 11.5. Accuracy is lower when the interviewer is Black or Latino than when s/he is White. Respondents who reported they were more confident and those who reported more liberal ideology were also more likely to be accurate in their assessments of interviewer race/ethnicity. Gender, age, education, and income did not significantly predict accuracy.

Model 2 includes interactions between interviewer race and respondent race. These analyses suggest that although respondents are less accurate when guessing the race/ethnicity of Latino interviewers (than when guessing the race/ethnicity of Black or White interviewers), this effect is qualified by interaction such that Black and Latino respondents are not as inaccurate. To interpret this interaction, analyses to examine the effect of interviewer race/ethnicity were conducted separately with respondents from each racial and ethnic group. Among White respondents, Latino interviewers were associated with less accurate perception of interviewer race/ethnicity (coefficient = −5.76, SE = 1.72, $p = 0.002$). This effect was weaker but still significant among Black respondents (coefficient = −2.66, SE = 1.07, $p = 0.02$) and nonsignificant among Latino respondents (coefficient = −1.01, SE = 0.75, $p = 0.19$).

11.6.3 Survey Responses

We next examined the impact of interviewer race/ethnicity on race- and ethnicity-related survey responses. Because our results suggest that the concordance between perceived and actual interviewer race is not high, we examined the impact of both actual and perceived race on these survey responses in separate analyses. The results of these analyses are shown in Table 11.7.

Table 11.6 Weighted multilevel logit models predicting accuracy of perceived interviewer race/ethnicity (standard errors in parentheses).

Predictor	Model 1	Model 2
Respondent race[a]		
Black	−0.30 (0.41)	−1.18* (0.57)
Latino	−0.65 (0.90)	−1.85** (0.59)
Interviewer race condition[b]		
Black	−1.29* (0.61)	−1.72** (0.75)
Latino	−3.13** (0.85)	−4.63** (1.02)
Respondent race × interviewer race		
Respondent Black × interviewer Black		0.77 (0.77)
Respondent Black × interviewer Latino		1.98* (0.96)
Respondent Latino × interviewer Black		0.19 (1.09)
Respondent Latino × interviewer Latino		3.20+ (1.69)
Male	0.29 (0.71)	0.40 (0.36)
Age	0.29 (0.32)	0.33 (0.68)
Years of education	1.26 (1.53)	1.11 (1.75)
Household income	0.03 (0.51)	0.09 (0.53)
Confidence	2.07** (0.65)	2.14** (0.72)
Political ideology	1.84** (0.55)	1.97** (0.54)
n (Respondents)	413	413
n (Interviewers)	32	32
Interviewer variance (95% confidence interval)	1.39 (0.49–3.94)	1.41 (0.49–4.05)

+$p < 0.10$, *$p < 0.05$, **$p < 0.01$.
a) White respondents are the comparison group.
b) White interviewers are the comparison condition.

Models with main effects are shown for all dependent variables. When there were significant interactions between interviewer race/ethnicity (either actual or perceived) and respondent race/ethnicity, we also presented estimates from a second model reporting the interactions.[4]

One consistent finding is that the actual race of the interviewer had few significant main effects on survey responses when respondent demographic characteristics are controlled. Respondents were more likely to report that immigration was headed in the right direction when interviewers were Latino (see Model 9), but this was the only main effect of actual interviewer race/ethnicity. However, in a number of cases, there were significant interactions between respondent race/ethnicity and actual interviewer race/ethnicity (see Models 4, 7, and 10 in Table 11.7). Model 4 shows that the effect of interviewer race/ethnicity was not significant among White respondents (main effects of interviewer race/ethnicity in this model). However, among Latino respondents, these effects were both larger such that Latino respondents were

4 The analyses including interactions would not converge for several of the models shown in Table 11.7 (specifically Models 1 and 11). This may have been due to the relatively small number of Latino respondents in the sample we analyzed ($n = 40$). These analyses were rerun with only Black and White respondents. The results of these analyses are not shown in Table 11.7 because no significant interactions between respondent and interviewer race/ethnicity were found.

Table 11.7 Impact of interviewer race/ethnicity on responses to race/ethnicity related survey questions (standard errors shown in parentheses).

	Affirmative action still needed		Affirmative action right direction			Support for pro-immigrant policies			Immigration right direction			Positive evaluation of barack Obama	
Model	1	2	3	4	5	6	7	8	9	10	11	12	13
I race measure	Actual	Perceived	Actual		Perceived	Actual		Perceived	Actual		Perceived	Actual	Perceived
Respondent race													
Black	3.44** (0.42)	3.63** (0.42)	0.24 (0.19)	0.33 (0.37)	0.33 (0.17)	0.02 (0.05)	0.10 (0.07)	0.04 (0.05)	0.04 (0.35)	0.99* (0.47)	0.09 (0.33)	0.15** (0.03)	0.16** (0.03)
Latino	4.49** (1.09)	4.84** (1.07)	0.46 (0.47)	−0.84 (0.33)	0.58 (0.45)	0.19** (0.05)	0.30** (0.08)	0.21** (0.05)	−0.19 (0.35)	0.29 (0.67)	−0.15 (0.44)	0.12** (0.04)	0.14** (0.04)
Interviewer race													
Black	−0.31 (0.33)	1.21* (0.42)	0.36 (0.25)	0.39 (0.34)	0.09 (0.48)	0.02 (0.04)	0.09 (0.07)	−0.03 (0.05)	0.20 (0.43)	1.10+ (0.56)	−0.06 (0.29)	0.02 (0.04)	0.04 (0.03)
Latino	−0.08 (0.25)	0.35 (0.71)	0.26 (0.22)	−0.05 (0.16)	−1.19* (0.44)	0.01 (0.05)	0.08 (0.06)	0.08 (0.05)	0.49* (0.50)	1.06* (0.70)	−1.00* (0.38)	0.02 (0.04)	0.04 (0.03)
Other	NA	0.69 (0.53)	NA	NA	0.38 (0.38)	NA	NA	0.12 (0.06)	NA	NA	−0.31 (0.59)	NA	0.05 (0.03)
Male	−0.95** (0.23)	−1.02* (0.28)	0.32 (0.24)3	0.41+ (0.24)	0.33 (0.24)	−0.03 (0.02)	−0.04 (0.03)	−0.02 (0.03)	−0.20 (0.31)	−0.23 (0.29)	−0.19 (0.31)	−0.04 (0.02)	−0.04 (0.02)
Age	1.27+ (0.69)	1.58* (0.72)	−0.11 (0.74)	0.09 (0.67)	−0.24 (0.72)	−0.05 (0.08)	−0.05 (0.08)	−0.03 (0.09)	1.20* (0.53)	1.36* (0.52)	1.03+ (0.52)	0.05 (0.05)	−0.03 (0.05)
Years of education	4.89** (1.04)	4.11* (1.08)	1.08 (0.90)	0.44 (0.98)	1.23 (0.86)	0.16 (0.21)	0.17 (0.22)	0.20 (0.20)	0.05 (0.84)	−0.19 (0.97)	0.07 (0.81)	0.12 (0.11)	0.16 (0.12)
Household income	−0.64 (0.40)	−0.37 (0.41)	−0.99* (0.40)	−0.73+ (0.43)	−1.02** (0.39)	−0.13+ (0.07)	−0.13+ (0.08)	−0.13+ (0.07)	−1.18* (0.49)	−1.15* (0.50)	−0.1.14* (0.50)	−0.06 (0.04)	−0.06 (0.04)
Political ideology	4.04** (0.67)	4.47** (0.66)	1.29** (0.52)	1.37** (0.52)	1.11* (0.53)	0.18** (0.08)	0.18* (0.08)	0.22** (0.08)	−1.34* (0.49)	−1.37** (0.48)	−1.56** (0.52)	0.24** (0.06)	0.26** (0.06)

(Continued)

Table 11.7 (Continued)

	Affirmative action still needed		Affirmative action right direction			Support for pro-immigrant policies			Immigration right direction			Positive evaluation of barack Obama	
Model	1	2	3	4	5	6	7	8	9	10	11	12	13
I race measure	Actual	Perceived	Actual		Perceived	Actual		Perceived	Actual		Perceived	Actual	Perceived
R × I race													
R Black × I Black				–.46 (0.53)			–.11 (0.10)			–1.63** (0.59)			
R Black × I Latino				0.29 (0.45)			–0.12+ (0.07)			–1.08 (0.93)			
R Latino × I Black				2.15+ (1.21)			–0.20* (0.09)			–1.60 (1.17)			
R Latino × I Latino				1.63* (0.68)			–0.13+ (0.07)			–0.21 (0.69)			
N (Respondents)	408	406	373	373	371	419	419	416	375	375	372	435	417
N (Interviewers)	32	32	32	32	32	32	32	32	32	32	32	32	32
Interviewer variance (95% confidence interval)	0.00 (0.00–0.00)	0.00 (0.00–0.00)	0.00 (0.00–0.00)	0.00 (0.00–0.00)	0.00 (0.00–0.00)	0.001 (0.00–0.17)	0.001 (0.00–5.66)	0.003 (0.00–0.04)	0.36 (0.07–1.68)	0.33 (0.06–1.90)	0.32 (0.09–1.09)	0.002 (0.00–0.02)	0.001 (0.00–0.02)

Note: The `melogit` command in Stata was used to analyze all dependent variables except "Immigrants Stay with Conditions" and "Evaluation of Obama" which used `meglm`. All analyses were weighted using the `svyset` and svy: commands in Stata and all analyses specified random intercepts by interviewer ID to account for interviewer variance. $+p < 0.10$, $^{*}p < 0.05$, $^{**}p < 0.01$.

more likely to say affirmative action was moving in the right direction when interviewed by either a Black or Latino interviewer (see I Black × R Latino and I Latino × R Latino interactions in Model 4).

Although interviewer race/ethnicity was unassociated with support for pro-immigrant policies among White respondents (see main effects of interviewer race/ethnicity in Model 7), Latino respondents were less supportive of these policies when interviewed by a Black interviewer (see I Black × R Latino). Somewhat surprisingly, both Black (see I Latino × R Black interaction) and Latino (see I Latino × R Latino interaction) respondents were also somewhat less likely to report support for pro-immigrant policies when interviewed by a Latino interviewer.

Finally, there were interactions between respondent race/ethnicity and actual interviewer race/ethnicity for beliefs about whether immigration policies are headed in the right direction or the wrong direction. Specifically, among White respondents being interviewed by either a Black or Latino interviewer, there were more reports that immigration policies were headed in the right direction (see the main effects of interviewer race/ethnicity in Model 10). However, the effect of being interviewed by a Black interviewer on these reports was not significant among Black respondents (see I Black × R Black interaction in Model 10).

Perceptions of interviewer race had somewhat more consistent and theoretically sensible effects. Respondents who believed they were being interviewed by a Black interviewer were more likely to report that affirmative action was still needed (see Model 2). Respondents who thought that they were being interviewed by a Latino interviewer were less likely to report that affirmative action policies were headed in the right direction (see Model 5). Finally, respondents who thought they were being interviewed by a Latino interviewer were less likely to say that immigration was headed in the right direction than were those who thought they were being interviewed by a White interviewer (see Model 10). This effect is in the expected direction because this survey came shortly after a period of large protests about immigration (one of the largest of which was in Chicago) and increased Latino opposition to legislative efforts to make immigration laws more restrictive. None of these results were qualified by significant interactions with respondent race/ethnicity. In fact, there were no significant interactions between perceived interviewer race/ethnicity and respondent race/ethnicity for models predicting any of the survey responses we examined.

11.7 Discussion and Conclusion

This research tests the effects of interviewer race/ethnicity on survey data – including these effects on the decision to participate and responses to survey questions. We found a relatively small number of effects of interviewer race/ethnicity. Overall, Black interviewers had higher refusal rates than White or Latino interviewers (providing limited support for H1). Somewhat unexpectedly, response (and contact) rates were higher in areas with higher levels of non-White residents and refusal rates were lower (contradicting H2). The only evidence consistent with social distance theory was that Latino interviewers had lower contact rates in Mostly White ZIP codes than did Black or Latino interviewers (somewhat consistent with H3).

Our hypotheses about respondents' identification of interviewers had the strongest empirical support. Respondents did accurately identify interviewer race/ethnicity less frequently than has been found in past work with more limited race/ethnic groups and non-White interviewers (H4). White interviewers were the most successfully identified (H5a), and there was some evidence that respondents were more successful at identifying interviewers who shared their race/ethnic identity (H5b).

When considering interviewers' actual race/ethnicity, there was little evidence that respondents reported more favorable opinions about policies likely to benefit the interviewer's race/ethnic group or to rate members of the interviewer's race/ethnic group more positively (H6a and H6b; with the exception that all respondents were more likely to say immigration policies were going in the right direction when interviewed by a Latino interviewer than when interviewed by a White or Black interviewer). Perceived interviewer race/ethnicity had more consistent and theoretically sensible effects, supporting our hypothesis that the effects of race/ethnicity would be stronger for perceived rather than actual race/ethnicity (H7).

Perhaps the most interesting and surprising evidence from this study comes from our investigation of the accuracy of perceptions of interviewer race/ethnicity. Overall, we found lower accuracy than past research, perhaps because the question we used to measure perceived race/ethnicity provided several racial and ethnic groups as examples. We also asked about race/ethnicity whereas some previous researchers have only asked about perceived race (e.g. Davis and Silver 2003). Black and Latino respondents were less accurate at identifying White interviewers than were White respondents. All categories of respondents were less accurate at identifying Black interviewers than White interviewers. Finally, White respondents were less accurate at identifying Latino interviewers than they were at identifying White interviewers, but Black and Latino respondents were not.

11.7.1 Limitations

This work has a number of limitations. Unfortunately, the sample of Latino respondents in our survey was smaller than anticipated and so it was more difficult to detect effects among this subgroup. In addition, there is a language confound among Latino respondents because those interviewed in Spanish were (out of necessity) interviewed by Latino interviewers and were therefore excluded from our analyses. So our analyses likely excluded the least acculturated Latino respondents.

Our analyses of the sample frame did not account for interviewer variance in part because of our experimental design in which cases were assigned to an interviewer group rather than a specific interviewer. In our case, it would be ideal to account for the effect of each interviewer who contacted a case, but this presents analysis challenges. Future research on interviewer effects may want to focus on the benefits of different approaches to designing race-of-interviewer experiments and analysis strategies for addressing this issue.

In addition, our analysis of survey responses did not consider contextual factors such as ZIP-code composition. Because our analyses of participation involved a larger set of cases, we were able to consider these variables for the analysis of sample frame data. However, for the survey data itself, we did not have sufficient power to consider interviewer race/ethnicity, respondent race/ethnicity, and ZIP-code composition simultaneously (particularly because the latter two are highly correlated).

A final limitation is that all of the research we reviewed was conducted in the United States, although there is research on interviewer race/ethnicity being conducted in other geographic areas such as Africa (e.g. Adida et al. 2015) and the Netherlands (e.g. Dotinga et al. 2005; Heelsum 2013). Although the theoretical framework for explaining race- and ethnicity- of interviewer effects would be similar in other countries (e.g. the social distance model, deference when answering questions about the interviewers' racial or ethnic group), the diversity of racial and ethnic groups and the complexity of the social and political environments make studying interviewer race/ethnicity in many other countries exponentially more complex.

11.7.2 Future Work

In addition, the mechanisms underlying race- and ethnicity-of-interviewer effects require further exploration. Another future direction will be to examine how these effects change over time in the United States. As the US population becomes more diverse, it will become more complex to study the role of interviewer race/ethnicity. It also seems likely that the proportion of the US population who identify with multiple racial groups (2.4% of Americans in the 2000 census) will grow over time which will also contribute to the complexity.

In addition, the mechanisms underlying race- and ethnicity-of-interviewer effects require further exploration. Although a number of explanations have been inferred from the patterns of data that have been observed, there is little direct evidence for these proposed mechanisms. More sophisticated survey experiments, including measures of proposed mechanisms in survey experiments designed to test race- and ethnicity-of-interviewer effects, and multimethod investigations that integrate findings from laboratory and survey experiments are needed to test the proposed mechanisms directly. Additional research that includes both random assignment and probability sampling to examine race- and ethnicity-of-interviewer effects will be most useful to contributing to our understanding of the effects interviewer race/ethnicity is likely to have in large scale public opinion surveys.

Finally, race/ethnicity effects occur in the context of other respondent and interviewer characteristics (e.g. Johnson et al. 2000; Callegaro et al. 2005) and social environments. Future research should continue to explore the extent to which the effects of race/ethnicity (both interviewer and respondent) are moderated by other characteristics and environmental factors.

References

Adida, C.L., Ferree, K.E., Posner, D.N., and Robinson, A.L. (2015). Who's Asking? Interviewer Coethnicity Effects in African Survey Data. Afrobarometer Working Paper No. 158. http://afrobarometer.org/publications/wp158-whos-asking-interviewer-coethnicity-effects-african-survey-data (accessed 25 February 2019).

Alvarez, R.M. and Brehm, J. (1997). Are Americans ambivalent towards racial policies? *American Journal of Political Science* 41 (2): 345–374.

American Association for Public Opinion Research (2015). Standard definitions: final dispositions of case codes and outcome rates for surveys. http://www.aapor.org/AAPORKentico/AAPOR_Main/media/publications/Standard-Definitions2015_8theditionwithchanges_April2015_logo.pdf (accessed 25 February 2019).

Anderson, B.A., Silver, B.D., and Abramson, P.R. (1988a). The effects of race of the interviewer on measures of electoral participation by blacks in SRC national election studies. *Public Opinion Quarterly* 52 (1): 53–83.

Anderson, B.A., Silver, B.D., and Abramson, P.R. (1988b). The effects of race of the interviewer on race-related attitudes of black respondents in SRC/CPS national election studies. *Public Opinion Quarterly* 52 (3): 289–324.

Athey, K.R., Coleman, J.E., Reitman, A.P., and Tang, J. (1960). Two experiments showing the effect of the interviewer's racial background on responses to questionnaires concerning racial issues. *Journal of Applied Psychology* 44 (4): 244–246.

Branton, R., Cassese, E.C., Jones, B.S., and Westerland, C. (2011). All along the watchtower: acculturation fear, anti-latino affect, and immigration. *The Journal of Politics* 73 (3): 664–679.

Bryant, B.E. (1975). Respondent selection in a time of changing household composition. *Journal of Marketing Research* 12 (2): 129–135.

Callegaro, M., De Keulenaer, F., Krosnick, J.A., and Daves, R.P. (2005). Interviewer effects in an RDD telephone pre-election poll in Minneapolis 2001: an analysis of the effects of interviewer race and gender. Paper presented at the Annual Meeting of the American Association for Public Opinion Research. Miami, Florida.

Campbell, B.A. (1981). Race-of-interviewer effects among southern adolescents. *Public Opinion Quarterly* 45 (2): 231–244.

Campbell, D.E., Green, J.C., and Layman, G.C. (2011). The party faithful: partisan images, candidate religion, and the electoral impact of party identification. *American Journal of Political Science* 55 (1): 42–58.

Cannell, C.F., Miller, P.V., and Oksenberg, L. (1981). Research on interviewing techniques. *Sociological Methodology* 12: 389–437.

Chromy, J.R., Eyerman, J., Odom, D.M. et al. (2005). Association between interviewer experience and substance use prevalence rates in NSDUH. In: *Evaluating and Improving Methods Used in the National Survey on Drug Use and Health* (ed. J. Kennet and J. Gfroerer). Rockville, MD: Substance Abuse and Mental Health Services Admin., Office of Applied Studies.

Clemens, J., Couper, M.P., and Powers, K. (2002). *Detroit Area Study 1952–2001: Celebrating 50 Years*. Ann Arbor, MI: University of Michigan.

Cody, J., Davis, D., and Wilson, D.C. (2010). Race of interviewer effects and interviewer clustering. Paper presented at the Annual Meeting of the American Political Science Association, Washington, D.C.

Cotter, P.R., Cohen, J., and Coulter, P.B. (1982). Race-of-interviewer effects in telephone interviews. *Public Opinion Quarterly* 46 (2): 278–284.

Davis, D.W. (1997a). The direction of race of interviewer effects among African-Americans: donning the black mask. *American Journal of Political Science* 41 (1): 309–322.

Davis, D.W. (1997b). Nonrandom measurement error and race of interviewer effects among African Americans. *Public Opinion Quarterly* 61 (1): 183–207.

Davis, D.W. and Silver, B.D. (2003). Stereotype threat and race of interviewer effects in a survey on political knowledge. *American Journal of Political Science* 47 (1): 33–45.

Davis, R.E., Couper, M.P., Janz, N.K. et al. (2009). Interviewer effects in public health surveys. *Health Education Research* 25: 14–26.

Dotinga, A., van den Eijnden, R.J.J.M., Bosveld, W., and Garretsen, H.F.L. (2005). The effect of data collection mode and ethnicity of interviewer on response rates and self-reported alcohol use among Turks and Moroccans in the Netherlands: an experimental study. *Alcohol and Alcoholism* 40 (3): 242–248.

Dovidio, J.F., Eller, A., and Hewstone, M. (2011). Improving intergroup relations through direct, extended and other forms of indirect contact. *Group Processes & Intergroup Relations* 14 (2): 147–160.

Fendrich, M., Johnson, T.P., Wislar, J.S., and Shaligram, C. (1996). The impact of interviewer characteristics on cocaine use underreporting by Male Juvenile Arrestees. In: *Proceedings of the American Statistical Association Section on Survey Research Methods*, 1014–1019. Alexandria, VA: American Statistical Association.

Finkel, S.E., Guterbock, T.M., and Borg, M.J. (1991). Race-of-interviewer effects in a preelection poll: Virginia 1989. *Public Opinion Quarterly* 55 (3): 313–330.

Groves, R.M. (2004). *Survey Errors and Survey Costs*. Hoboken, NJ: Wiley.

Groves, R.M. and Fultz, N.H. (1985). Gender effects among telephone interviewers in a survey of economic attitudes. *Sociological Methods and Research* 14 (1): 31–52.

Harms, W. (1995). Sanders: perceptions of race affect survey responses. *The University of Chicago Chronicle* 15 (4). Available at: http://chronicle.uchicago.edu/951026/sanders.shtml (accessed 25 February 2019).

Hatchet, S. and Schuman, H. (1975–1976). White respondents and race-of-interviewer effects. *Public Opinion Quarterly* 39 (4): 523–528.

Heelsum, A.v. (2013). The influence of the ethnic background of the interviewer in a survey among Surinamese in the Netherlands. In: *Surveying Immigrants/Ethnic Minorities: Methodological Challenges and Research Strategies*, IMISCOE research series (ed. J. Font and M. Méndez), 111–130. Amsterdam: AUP.

Hill, M.E. (2002). Race of the interviewer and perception of skin color: evidence from the multi-city study of urban inequality. *American Sociological Review* 67 (1): 99–108.

Holbrook, A. and Johnson, T. (2008). *Chicago Area Study: Race, Ethnicity, and Political Participation*. Chicago, IL: University of Illinois at Chicago.

Johnson, T.P. and Bowman, P.J. (2003). Cross-cultural sources of measurement error in substance use surveys. *Substance Use & Misuse* 38 (10): 1447–1490.

Johnson, T.P. and Parsons, J.A. (1994). Interviewer effects on self-reported substance use among homeless persons. *Addictive Behaviors* 19: 83–93.

Johnson, T.P., Fendrich, M., Shaligram, C. et al. (2000). An evaluation of the effects of interviewer characteristics in an RDD telephone survey of drug use. *Journal of Drug Issues* 30: 77–101.

Johnson, T.P., Cho, Y.I., Campbell, R.T., and Holbrook, A.L. (2006). Using community-level correlates to evaluate nonresponse in a telephone survey. *Public Opinion Quarterly* 70: 704–719.

Jones, J.H. (1993). *Bad Blood: The Tuskegee Syphilis Experiment*. New York: Free Press.

Kappelhof, J. (2017). Survey research and the quality of survey data among ethnic minorities. In: *Total Survey Error in Practice* (ed. P.P. Biemer, E. de Leeuw, S. Eckman, et al.). New York: Wiley.

Kappelhof, J.W.S. and de Leeuw, E.D. (2019). Estimating the impact of measurement differences introducing by efforts to reach a balanced response among non-western minorities. *Sociological Methods & Research* 48 (1): 116–155.

Kish, L. (1962). Studies of interviewer variance for attitudinal variables. *Journal of the American Statistical Association* 57: 92–115.

Krysan, M. and Couper, M.P. (2003). Race in the live and the virtual interview: racial deference, social desirability, and activation effects in attitude surveys. *Social Psychology Quarterly* 66 (4): 364–383.

Krysan, M. and Couper, M.P. (2006). Race of interviewer effects: what happens on the web. *International Journal of Internet Science* 1 (1): 17–28.

Lass, N.J., Tecca, J.E., Mancuso, R.A., and Black, W.I. (1979). The effect of phonetic complexity on speaker race and sex identification. *Journal of Phonetics* 5: 105–118.

Liu, M. and Wang, Y. (2016). Race-of-interviewer effects in the computer-assisted self-interview module in a face-to-face survey. *International Journal of Public Opinion Research* 28 (2): 292–305.

Livert, D., Kadushin, C., and Schulman, M. (1998). Do interviewer-respondent race effects impact the measurement of illicit substance use and related attitudes? In: *Proceedings of the American Statistical Association Section on Survey Research Methods*, 894–899. Alexandria, VA: American Statistical Association.

Loewenstein, R. and Varma, A.A.O. (1970). Effect of interaction of interviewers and respondents in health surveys. *Public Opinion Quarterly* 34: 472.

Lord, V.B., Friday, P.C., and Brennan, P.K. (2005). The effects of interviewer characteristics on arrestees' responses to drug-related questions. *Applied Psychology in Criminal Justice* 1 (1): 36–54.

Meislin, R. (1987). Racial divisions seen in poll on howard beach attack. *New York Times* (8 January).
Moorman, P.G., Newman, B., Millikan, R.C. et al. (1999). Participation rates in a case-control study: the impact of age, race, and race of interviewer. *Annals of Epidemiology* 9 (3): 188–195.
Morin, R. (1989). Polling – in black and white. *Washington Post* (5 November).
Olson, K. and Bilgen, I. (2011). The role of interviewer experience on acquiescence. *Public Opinion Quarterly* 75 (1): 99–114.
Purnell, T., Isardi, W., and Baugh, J. (1999). Perceptual and phonetic experiments on American English dialect identification. *Journal of Language and Social Psychology* 18 (1): 10–30.
Reese, S.D., Danielson, W.A., Shoemaker, P.J. et al. (1986). Ethnicity-of-interviewer effects among Mexican-Americans and Anglos. *Public Opinion Quarterly* 50 (4): 563–572.
Schaeffer, N.C. (1980). Evaluating race-of-interviewer effects in a national survey. *Sociological Methods and Research* 8 (4): 400–419.
Schuman, H. and Converse, J.M. (1971). The effects of black and white interviewers on black responses in 1968. *Public Opinion Quarterly* 35 (1): 44–68.
Sudman, S. and Bradburn, N.M. (1974). *Response Effects in Surveys*. Chicago, IL: Aldine Publishing.
Tate, K. (1998). *National Black Election Study 1996 [Codebook]*. Ann Arbor, MI: Inter-university Consortium for Political and Social Research.
Wang, K., Kott, P., and Moore, A. (2013). Assessing the Relationship Between Interviewer Effects and NSDUH Data Quality. Report prepared by RTI for Substance Abuse and Mental Health Services Administration.
Webster, C. (1996). Hispanic and Anglo interviewer and respondent ethnicity and gender: the impact on survey response quality. *Journal of Marketing Research* 33 (1): 62–72.
Weeks, M.F. and Moore, R.P. (1981). Ethnicity-of-interviewer effects on ethnic respondents. *Public Opinion Quarterly* 45 (2): 245–249.
Welch, S., Comer, J., and Steinman, M. (1973). Interviewing in a Mexican-American community: an investigation of some potential sources of response bias. *Public Opinion Quarterly* 37 (1): 115–126.
West, B.T. and Blom, A.G. (2017). Explaining interviewer effects: a research synthesis. *Journal of Survey Statistics and Methodology* 5 (2): 175–211.
West, B.T. and Olson, K. (2010). How much of interviewer variance is really nonresponse error variance? *Public Opinion Quarterly* 74 (5): 1004–1026.
Williams, J.A. Jr., (1964). Interviewer-respondent interaction: a study of bias in the information interview. *Sociometry* 27 (3): 338–352.
Wilson, D.C. and Olesen, E.P. (2002). Perceived race of interviewer effects in telephone. Paper presented at the Annual Conference of the American Association for Public Opinion Research. St. Pete's Beach, FL.
Wolford, M.L., Brown, R.E., Marsden, A. et al. (1995). Bias in telephone surveys of African Americans: the impact of perceived race of interviewer on responses. In: *Proceedings of the American Statistical Association Section on Survey Research Methods*, 799–804. Alexandria, VA: American Statistical Association.
Worcester, R.M. and Kaur-Ballagan, K. (2002). Who's asking: answers may depend on it. *Public Perspective* 13 (3): 42–43.

12

Investigating Interviewer Effects and Confounds in Survey-Based Experimentation*

Paul J. Lavrakas[1], *Jenny Kelly*[1], *and Colleen McClain*[2]

[1] *NORC, University of Chicago, 55 East Monroe Street, Chicago, IL 60603, USA*
[2] *Program in Survey Methodology at the University of Michigan, Institute for Social Research, 426 Thompson Street, Ann Arbor, MI 48104, USA*

An area of survey research that has received only modest attention to date is the possibility of unwanted interviewer effects on experimental data that are gathered in telephone or in-person surveys. There are two basic subsets of survey-based experimental designs that lend themselves to investigating interviewer-related effects: one involves *post hoc* analysis and the other is an *a priori* design. By "*post hoc*" we refer to those interviewer-based experimental analyses of survey datasets that were gathered for purposes other than the experimental analyses that the researchers are later interested in conducting – thus a form of secondary analysis research. An example of this type would be using interviewer-gathered survey data that were collected for some other purpose, in which cases were randomly assigned to interviewers (as is done in many general population telephone surveys), to investigate whether the gender of the interviewer affected answers to certain questions. In contrast, by "*a priori*" we refer to those survey-based experiments for which a study of interviewer effects was specifically planned and conducted. An example of this subset would be planning a survey experiment to determine whether the gender of an interviewer affected answers to a series of questions about equitable pay for female and male employees by randomly assigning cases to interviewers.

In this chapter, we first present a review of the *post hoc* approach to studying interviewer effects that are associated with the characteristics of interviewers, including an original example of how that approach can easily be carried out. We then present examples of *a priori* approaches including the results of an original experiment that we conducted in 2015 to investigate interviewer effects in the administration of an experiment involving a telephone survey introduction. It is these *a priori* experiments that most often come to mind when one thinks of interviewer-administered survey-based experiments, in part because the researchers have direct control over the variables that are used. In *post hoc* experimental analyses, the researchers must do the best they can with data that were gathered either primarily or exclusively for some other purpose(s).

The survey literature refers to surveys where randomly sampled cases are randomly assigned (i.e. allocated) to survey interviewers as "interpenetrated designs" (Stokes 1988; O'Muircheartaigh and Campanelli 1999; Gillikin 2008). Traditionally, this was accomplished

* Parts of this chapter were adapted from an unpublished chapter for inclusion in a book that was never published. That earlier chapter was first written by the lead author in the late 1990s and it was revised in 2003 before a decision was made by the authors to stop publication of that book.

Experimental Methods in Survey Research: Techniques that Combine Random Sampling with Random Assignment, First Edition. Edited by Paul J. Lavrakas, Michael W. Traugott, Courtney Kennedy, Allyson L. Holbrook, Edith D. de Leeuw, and Brady T. West.

Companion Website: www.wiley.com/go/Lavrakas/survey-research

by performing the random assignment of cases *before* in-person interviewers began working on recruitment and data collection. However, sample randomization of cases to interviewers can (and does) occur each day in the field period of many telephone surveys.[1] Whenever sampled cases are assigned to survey interviewers in an essentially random fashion through this daily field operations process, a "natural experiment" has been implemented. This form of experiment occurs because the random assignment of the cases to interviewers results in the characteristics of interviewers (e.g. gender, race, experience, etc.) being available to serve as independent variables in experimental analyses.[2]

There have been many *a priori* experimental designs in which the researchers controlled the independent variable(s) that interviewers administered within a survey-based experimental design. Many of these independent variables are question wording or ordering experiments where interviewers read whichever version of a question was randomly assigned to a particular respondent (e.g. Schuman and Presser 1981). In these experiments, the interviewer's role is assumed to be a passive one in that they are expected to merely read what they see in the questionnaire in a nondirective unbiased fashion.

However, there is another subset of experiments in which interviewers administer different types of survey treatments for which they must play a very active role. These experiments include those that test different strategies for recruiting or selecting respondents into the survey (e.g. Groves and McGonagle 2001). In these survey-based experiments, the interviewer actively plays a major role in determining exactly what a respondent is exposed to and thus in defining the actual treatment to which the respondent is exposed, on a case-by-case basis. Unlike what happens when an interviewer reads a specific question to a respondent, often in a highly standardized fashion, most of the treatments that are administered by interviewers in these latter types of *a priori* experiments likely will vary by a fair amount from respondent to respondent (cf. West et al. 2017). This occurs both across interviewers and within interviewers. Thus, any time an experiment is carried out in which an interviewer is asked to use a certain approach to recruiting a respondent, the exact exchange between the interviewer and the respondent may vary considerably from interviewer to interviewer and from respondent to respondent.

In a subsequent section of our chapter that addresses the *a priori* approach to interviewer-administered experimental treatments, we present an experiment that we devised to specifically address the issue of what may happen when interviewers are assigned to administer more than one (or as many as all) of the experimental treatments being studied in a survey-based experiment. However, in the following section we first address issues related to interviewer effects in *post hoc* designs so as to illustrate interviewer-related experimental analyses using existing datasets gathered for another purpose than the experimental analyses.

12.1 Studying Interviewer Effects Using a *Post hoc* Experimental Design

Any survey design, including those explicitly planned with an interpenetrated design, that randomly assigns cases (e.g. phone numbers, addresses, or names) to interviewers lends itself

1 Of note, this randomization of cases to interviewers most often does not occur throughout the field periods of surveys that use responsive and adaptive designs, as is becoming more common nowadays.

2 It should be noted that there are often operational complexities that are introduced in the implementation of these studies that a researcher who wants to use the data for these natural experiments will have to deal with (e.g. some of the completions were achieved by supervisors, some interviewers did not work enough to reasonably be included in the analyses, demographic characteristics of some interviewers are not available, some cases were specially assigned to certain types of interviewers, etc.).

to secondary analysis of possible interviewer effects that may have occurred in this "natural experiment" (Dunning 2012).[3] A likely reason that these opportunities have not yielded more experimental research on interviewer effects is that an interviewer ID variable and associated interviewer characteristics often are not included in these datasets. One likely reason for this is that some Human Resource departments are uncomfortable with these types of data being added to survey datasets. As is explained later in this chapter, without such variables the proper analytic models cannot be fitted – models that allow for the accounting of interviewer effects in the analysis of the experiment, as described by Dijkstra (1983), Hox et al. (1991), and a long line of other research taking a multilevel approach to studying interviewer effects (e.g. Raudenbush and Bryk 1986).

This does not mean that a researcher might not explicitly anticipate and thus plan for such analysis of the survey data. Regardless, as long as the allocation of cases to interviewers is random, the resulting datasets often will support investigations with strong internal validity of the causal relationships between effects related to interviewer characteristics and the dependent variables that were gathered in the survey. Although the researcher has not controlled – in the sense of "creating" and thereby "controlling/manipulating" – the independent variables (e.g. the gender of an interviewer) used in such an experimental analysis, she/he nevertheless often can use such interviewer attributes as the independent variables in a "causal" analysis providing the cases were randomly assigned across interviewers.[4]

Survey designs that are most likely to meet the criteria for generating such data are the ones that are conducted via the telephone mode of recruitment with data collection from one centralized location. This is because in-person surveys typically utilize a clustered sampling design whereby interviewers are assigned to one or a limited number of the geographic subareas that the survey is sampling. In that case, any geographic variation in clusters of cases along the variables of interest likely will be confounded with the attributes of the individual interviewers assigned to work specific clusters (e.g. using bilingual interviewers in areas where many non-English speakers live). Thus, even if cases are randomly assigned to different in-person interviewers within a geographic subarea – and here the reader is cautioned to recall that it is random *assignment* that is being discussed, not random *sampling* – it almost never will be true that the cases themselves are so homogenous across the geographic clusters as to yield a truly "equivalent" assignment of cases across all interviewers.[5]

Furthermore, a telephone survey that is conducted from more than one location will not meet the criterion of random assignment of cases to interviewers, unless the pool of interviewers varies randomly across the locations as well as the cases being randomly assigned to them. The reason for this is that groups of telephone interviewers who work at different locations are unlikely to be equivalent in the way that random assignment requires. Of note, if cases are randomly assigned to interviewers at each location, then the researcher could add a "location" variable as a covariate into the analyses to try to control for the nonequivalency of interviewers

---

3 By "natural experiment," we refer to any survey design that results in an experimental feature – as the result of random assignment – being incorporated into the execution of the survey, even if it was not the explicit intent of the researchers to plan the resulting experiment, but nevertheless allows for subsequent analyses of a cause-and-effect nature. For example, when the normal field operational procedures that are used in assigning cases to interviewers happen to be a random process, then the researchers have a ***natural experiment of interviewer effects***, whether or not they intended it.

4 An important corollary to this general rule is addressed later in this chapter and concerns instances in which an unconfounded experimental design requires that interviewers themselves be randomly assigned to one, and only one, experimental condition as well as having cases randomly assigned to the experimental conditions and to interviewers.

5 Granted, the threats to internal validity that are associated with in-person survey data collection in such studies can be addressed by statistical approaches that strive to control for these area-level effects, but there still will remain uncertainty as to the integrity of the experimental analyses that then are carried out.

across the different locations. But this statistical approach does not assure an unconfounded experimental analytic design.

Ultimately, it is the responsibility of the researcher to carefully think through whether cases truly have been randomly assigned to interviewers before pursuing experimental analyses.

12.1.1 An Example of a *Post hoc* Investigation of Interviewer Effects

It has been demonstrated empirically that most humans develop a fairly accurate ability to discern some of the demographic characteristics of a stranger to whom they are speaking on the telephone, including the other person's gender, race, and age (cf. Lavrakas 1993). This is not to say that respondents and interviewers are perfect in this ability, but merely that they do appear to be accurate enough to provide a reliable foundation to justify the construct validity for such empirical investigations of interviewer effects in certain telephone surveys.

The following analysis, which was conducted solely for this chapter, illustrates the use of secondary analysis to investigate interviewer effects, *post hoc,* using a dataset that was gathered wholly for another purpose.[6] We did the analysis simply to show what can be done with many datasets. In our case, we did not have any hypotheses in which we were particularly interested. Rather we chose these analyses merely because they are illustrative of what can be done in a *post hoc* fashion.

The analysis involves the use of a national random-digit dialing (RDD) landline telephone survey dataset that was gathered for Nielsen Media Research in 2001 by the Center for Survey Research at Ohio State University. This survey interviewed 1913 adults aged 18–34 years in either English or Spanish. Cases were assigned essentially randomly across interviewers (and thus across interviewer demographic characteristics) during the 10-week field period. Thus, by definition, it also was a random process that assigned an interviewer to each respondent regardless of the characteristics of the respondent or of the interviewer. As part of the questionnaire, a series of psychographic questions was asked. One of those items used a five-point Likert scale asking respondents the extent to which they "agreed" or "disagreed" with the following statement: "*I would like to see a return to more traditional standards for sexual relationships.*"[7]

This survey and this particular item demonstrate the presence of significant interaction effects when considering respondent gender and interviewer age. To conduct these analyses correctly a generalized linear mixed model (GLMM) approach was utilized. This approach allows the researcher to model (and thus control for) random interviewer effects. In conducting this secondary analysis with this dataset, we decided to drop the data associated with 15 of the 79 interviewers from our analyses as none of them completed more than five interviews; in fact, all 15 interviewers accounted for only 37 of the 1913 completions. This left 1876 cases completed by 64 interviewers for use in the GLMM analysis. Among the 64 interviewers who completed the 1876 interviews, 41% were in the same age cohort as the respondents (18–34 years old), whereas 59% were not. As shown by the descriptive results in Table 12.1, which parallel the results from the GLMM, there is no main effect associated with the age of the interviewers – that is, interviewers in the younger cohort elicited essentially similar answers overall about the appeal of traditional sexual standards as did the cohort of older interviewers; approximately

6 These data come from a survey conducted in 2001 by the Ohio State University survey center for which the lead author had been the founding faculty director from 1996 to 2000 on a project for which the lead author was the principal investigator at the Nielsen company.

7 The reader may find it of interest to know that a variation of this analysis was first conducted at the time the first author's 2003 unpublished chapter on interviewer effects in experiments was written to demonstrate, in part, how readily available RDD data sets are for such post hoc causal investigations of interviewer effects.

Table 12.1 Percentage agreeing[a] with traditional sexual standards item by gender of respondent and age of interviewer ($n = 1876$).

	18–34 years old interviewer (%)	35 years or older interviewer (%)
Female respondent	69	75
Male respondent	68	62
Total	69	69

Note: Data were gathered in a 2001 survey conducted by the Ohio State University Center for Survey Research for Nielsen Media Research.

a) The percentage agreeing was made up of those respondents who answered "strongly agree" or "agree" to the Likert five-point scale used for this item. These simple percentages are displayed to make the interpretation of the GLMM analysis more intuitive to most readers. Later in the chapter, we use GLMM to analyze the original experiment we conducted for this chapter and present details on those analyses.

69% of respondents agreed regardless of the interviewers' ages. However, there was a significant interaction effect ($p < 0.001$), with female respondents being more likely to agree when they were interviewed by an older interviewer, whereas male respondents were more likely to agree when they were interviewed by a younger interviewer.

The ease with which this empirical example was identified for use in this chapter demonstrates how prevalent such opportunities are across databases of interviewer-administered surveys that have cases/respondents randomly assigned to interviewers, and that have an interviewer ID variable for each case in the dataset.[8] All that typically is missing from such databases are the interviewer characteristics needed for the analyses.

Interviewer characteristics such as tenure and past performance (in productivity, quality, or both) are typically produced by the operational departments of survey organizations, who will often provide these data on the basis that they are kept confidential. Demographic characteristics however – particularly those that fall under employment protection laws, such as age, sex, and race – are often not available for use (in the United States at least) due to the federal employment protections. However, depending on the purpose of the study, if such factors are considered to be potentially explanatory of the results, a reasonable proxy can sometimes be constructed by having coders listen to a sample of each interviewer's recordings and code on *perceived* age, sex, and race from audible cues, since these are exactly the perceptions to which respondents would be reacting.[9]

As noted above, researchers who conduct *post hoc* experimental analyses using datasets that were gathered for other purposes than investigating interviewer-related effects must be very careful in deciding the extent to which they can claim that the cases in the dataset were in fact assigned randomly across the interviewers that worked on the survey. If the researchers believe

8 For example, the Behavioral Risk Factor Surveillance Survey (BRFSS) that is carried out each year at the state level in the United States produces datasets that contain an interviewer ID variable (cf. Elliott and West 2015).

9 At the time this chapter was being written a study was released from Pew (http://www.pewresearch.org/fact-tank/2017/09/27/many-poll-respondents-guess-wrong-on-their-interviewers-race-or-ethnicity/) that suggested that humans cannot do this well. However, the methodology used in that study did not utilize an unconfounded design. That is, each of the respondents who made a judgment about the interviewer who had just finished interviewing them via telephone reported their judgments to the interviewer to whom the respondent was speaking. Thus, these data likely suffer from social desirability bias because many respondents would be prone to give answers that they believed the interviewer to whom they were speaking would be pleased to hear.

that they are in fact justified in drawing cause-and-effect conclusions from their findings, then they should feel obligated to fully disclose the reasons that they believe that cases were assigned randomly across interviewers. To the extent the researchers cannot claim with confidence that the entire process was random, they should address the limitations and what they did (if anything) to try to remove their confounding effects on any causal inferences about interviewer effects that they are advancing from their observed results.

12.2 Studying Interviewer Effects Using *A priori* Experimental Designs

It has long been known that interviewer characteristics and respondent characteristics may interact to affect the data that respondents provide in interviewer-administered surveys when the topic of the survey is related to the characteristics in question (cf. West and Blom 2017). These are study designs that were planned *a priori* to test for such effects. These investigations use an analytic design that is founded on the random allocation of cases to a pool of interviewers and vice versa. The best known of these types of studies focus on the race of the interviewer and race of the respondent in in-person preelection polls (e.g. Finkel et al. 1999) whereby many White respondents have been found to provide strikingly different answers about whether or not they plan to vote for a Black candidate depending on whether a Black or White interviewer gathers data from them. The internal validity (cause-and-effect validity) of this type of research is provided by the fact that cases have been randomly assigned to interviewers and thus the natural variation of the interviewers' characteristics and the respondents' characteristics become the independent variables in the analyses.

Other examples include effects associated with the race of the interviewer and the respondent when racial attitudes are being measured (e.g. Schuman and Converse 1971; also see Chapter 11 in this volume on race effects) and effects associated with the gender of the interviewer and respondent when sexual attitudes or behaviors are being measured. For example, Lavrakas (1992) reported an example of large gender-related interviewer/respondent effects that occurred in a telephone survey on the topic of sexual harassment in the workplace. In an RDD landline survey of Chicago, interviewers were assigned a random set of cases each time they worked on the study. By knowing the gender of the interviewer and of the respondent, an analysis with strong internal validity was planned *a priori* to investigate the extent to which a respondent was willing to report a past instance where s/he had (or may have been perceived as having) sexually harassed someone. It was found that male respondents were nearly twice as likely to report having sexually harassed someone when interviewed by a male (31%) vs. when interviewed by a female (17%). Similarly, women were twice as likely to report having sexually harassed someone when interviewed by a female (10%) vs. when interviewed by a male (5%).

12.2.1 Assignment of Experimental Treatments to Interviewers

But there is another type of interviewer effect for which survey experiments need to be planned in an *a priori* fashion and with which this chapter is primarily concerned. This effect concerns the manner in which interviewers are assigned to administered treatments within a survey experiment and how that manner may confound an experiment, possibly rendering it uninterpretable.

In all interviewer-administered surveys that incorporate experimental designs, it is paramount that researchers anticipate whether the experimental design will be confounded if cases are not randomly assigned to interviewers. For example, if a researcher is planning an experiment on question-wording and the cases are not randomly assigned to interviewers, this could render the results uninterpretable if different interviewers disproportionately administered one version of the wording of the question while the other experimental versions of the wording were disproportionately administered by another set of interviewers. In this type of confounded design the researcher would not know if any observed differences across experimental conditions were due to the wording differences or to the fact that the different versions of the wording were administered by different interviewers.

There also are *a priori* survey-based interviewer-administered experiments for which it is not enough merely to assign cases randomly to interviewers. Instead, interviewers also must be randomly assigned to one, and only one, of the different treatment protocols. That is because not doing so may create a confound due to interviewers' differential preferences for, and abilities with, administering multiple treatments. And yet, too often, many otherwise-expert researchers who plan such *a priori* survey-based experiments appear to have not recognized this design necessity. And in turn, many otherwise-expert journal editors and reviewers also appear to have missed this likely design flaw in some of the articles that they have approved for publication.

An example of this possible confound may have occurred in a series of experiments that were reported in prominent journals in the 1980s about various within-household respondent selection methods that were used to select one adult as the designated respondent for a given household (cf. Gaziano 2005). Essentially these experimental studies contrasted the effects of different within-unit respondent selection methods – such as the Kish method (Kish 1949), the Trohdal–Carter–Bryant method (Bryant 1975), and the next-birthday method (Salmon and Nichols 1983) – on cooperation rates by randomly assigning *cases* to only one of the selection methods but allowing interviewers to work each of the methods being compared (e.g. Czaja et al. 1982; Salmon and Nichols 1983; Oldendick et al. 1988).

In these types of experiments, even when cases are randomly assigned to interviewers, the design may be confounded unless the interviewers are assigned randomly to one, and only one, of the experimental conditions. And, ideally, the interviewers should be kept unaware (i.e. blind) to the experimental manipulation of which they are part. The reason for keeping interviewers blind is that simply knowing that they are part of an experimental treatment could unintentionally lead to a diffusion, dilution, and/or corruption of the integrity of the treatment(s) to which they are assigned to administer. This could occur if interviewers assigned to different treatment conditions talk to each other and alter (i.e. distort) the manner in which they administer the treatment to which they were assigned due to these conversations. It also could occur if interviewers learn about other treatments and become less enthusiastic about administering the treatment to which they are assigned.

In the series of experiments that evaluated the effects on cooperation of the different respondent-selection techniques, the interviewers were assigned to administer each of the treatment protocols, even though they often were very familiar with one of the protocols that were being tested and not so familiar with the other protocols.[10] And, being familiar with one of the within-unit selection techniques because they were used to using it in prior surveys could be expected to lead the interviewers to (i) personally prefer that technique over the other selection techniques that were being tested in the survey experiment and/or (ii) be more skilled in delivering the technique that they had past experience with compared to the other selection techniques that they may have been using for the first time. Thus, the observed results of the experiments would be confounded if the interviewers had differential preferences for and/or

10 This familiarity was due to the survey organization for which the interviewers work using a standard selection method for the surveys the organization conducted with which interviewers were very familiar.

experiences with the selection techniques being tested in the experiment. And, this hypothesis is one that we investigate in the original experiment that we report later in this chapter.

Thus, whenever interviewers are not randomly assigned to one and only one of the treatment protocols in an experiment that is testing various interviewer-administered treatments, it is possible for many types of confounding processes to undermine the integrity of the experimental design. Therefore, any differences observed between the outcome measures (e.g. cooperation rates) for the different protocols may have nothing to do with the protocols themselves, but rather may be a result of the differential motivation and/or differential skills of interviewers (including interviewers' varying skills at multitasking) in administering the respective protocols which they were assigned to administer.

Because of this, and since much of the literature on the effects of within-household respondent selection methods on cooperation rates has been based on possibly flawed experimental designs, we believe that this body of research might be subject to alternative explanations in drawing internally valid conclusions about which of the selection methods leads to higher/lower cooperation rates.[11]

12.3 An Original Experiment on the Effects of Interviewers Administering Only One Treatment vs. Interviewers Administrating Multiple Treatments

The issue that we will now focus upon falls within the domain of interviewer-related error. That is, to what extent might the behavior of the interviewers who administer the different treatments in a survey-based experiment bias the results of the experiment, thereby confounding any clean interpretation of the observed findings? To that end, we will report on an experiment that we implemented within a dual-frame random-digit dialing (DFRDD) telephone survey that serves as a formal test of this question:

> Do the response rates that are generated by interviewers who were trained on (and administered) only one treatment within an experiment differ to a reliable extent from the response rates generated by interviewers who were trained on and administered all the treatments in an experiment?

To our knowledge, no one has conducted research in a way that allows them to investigate the possible confounds of having interviewers administer all treatments in an experiment. That is not to suggest that no one has set up an *a priori* survey-based experiment that might allow for testing this issue, but rather that we are unaware that anyone has reported out those results. Furthermore, we acknowledge that in some cases groups of interviewers have been randomly assigned to one and only one condition in an interviewer-implemented survey-based-experiment (e.g. West et al. 2017).

12.3.1 The Design of Our Survey-Based Experiment

The data gathered in our experiment come from a large DFRDD probability survey of the State of California ($n = 122\,185$ phone numbers were processed), which was conducted in February and March of 2015 by NORC at the University of Chicago. Two-hundred and seventy-two

11 Of note, the data gathered in these within-unit selection method studies could be reanalyzed by allowing the effects of the selection method to randomly vary across interviewers, and then reassess the importance of the selection method effects.

NORC interviewers were randomly assigned to one of three experimental groups pertaining to the type(s) of introduction that the group used to gain cooperation from a contacted household in order to complete the screener for the survey.

There were two types of introductions:

1. A *conventional introduction* (CI) was to be read as written by the interviewers trained in this approach to state traditional information about the survey as soon as someone answered the phone. This information included (i) the interviewer's name and (ii) the organization that was conducting the survey, (iii) that a state agency was sponsoring the survey, (iv) a general sense of the survey topic (i.e. health insurance), (v) a statement that nothing was being sold, (vi) a statement that the call would be monitored or recorded for quality purposes, and (vii) a query of whether the person being spoken to was an adult. Interviewers were trained not to deviate from this ordering of information within the five-sentence introduction and to relay all this information without pausing, unless interrupted by the respondent. In other words, unless the respondent interrupted the interviewer, approximately 20 seconds typically passed with only the interviewer speaking. The interviewer then engaged the respondent with the first screener question.
2. A *progressive engagement introduction* (PEI) (cf. Lavrakas et al. 2015) was used by interviewers trained in this approach to convey all the same information as in the CI script, but the pacing of the much more conversational approach that the PEI interviewers were trained to use was quite different than the pacing of the CI approach. In the PEI approach, the CATI script was programmed, and the interviewers were trained to build in short but distinctive pauses after each sentence so as to "allow" and "invite" the respondent to engage the interviewer with a comment or question. Furthermore, as soon as the interviewer introduced herself/himself, stated who they were calling from and on whose behalf, she/he asked the respondent a question along the lines of "How are you doing today?" or "How is your morning/afternoon/evening going?" and then paused for the respondent to reply. In other words, an interviewer using the PEI spoke for less than 10 seconds before the respondent was asked to say something. Once the PEI-trained interviewer covered all the requisite content in the introduction and responded to whatever comments or questions the respondent contributed, the interviewer then engaged the respondent with the first screener question.

As noted above, interviewers were randomly assigned to one of the three training groups. And phone numbers were randomly assigned to the two introductions. One group of interviewers was trained only to administer the CI, another group was trained only to administer the PEI, and the third group was trained to administer both introductions (CI and PEI). Furthermore, the 122k phone numbers were randomly assigned to one and only one type of introduction, either the CI or the PEI. All calls made to a phone number used the same introduction and all the interviewers who processed calls to that number had been trained to use the introduction that was assigned to the case. Interviewers were assigned to the cases that they had been trained to call on a random basis each time they came to work. Interviewers trained to administer both introductions worked only one of those introductions during a given interviewing shift and the introduction they worked during a given shift was varied on every other day that they worked.

12.3.2 Analytic Approach

As Dijkstra (1983) demonstrated in his seminal article, interviewer-related variance can seriously bias the findings of studies using interviewer-administered survey protocols if that variance is not explicitly taken into consideration as part of the analyses. Research on interviewer-administered survey-based experiments therefore needs to take this potential interviewer variance into account in the analyses performed, and Dijkstra (1983) discusses the negative implications of failing to do this. Of note, none of the publications reviewed by Gaziano (2005) in her meta-analysis of research on the experimental testing of interviewer-administered within-unit respondent selection methods used analyses that accounted for interviewer-level variance.

To carry out the correct analytic approach, a random interviewer effect must be incorporated into the analytic model; and to do this the researchers must be able to identify each interviewer who worked on the study with the cases they worked. The required variable will contain unique codes that identify the interviewer associated with each case. Given this type of interviewer ID variable, multilevel regression modeling can be used to analyze the random interviewer effects and estimate the relationships of various experimental protocols with different types of dependent variables (nominal, binary, etc.) when controlling for the random effects. In this way, the variance in the observed data that is caused by the variable effects of each interviewer (which introduces clustering of the data by interviewer) can be parceled out.

In order to account for the clustered assignment of cases to interviewers and the potential for interviewer effects to introduce bias into the analysis of our experimental design, we estimated multilevel logistic regression models that allowed us to model our nested data appropriately. Such an approach has long been recognized as necessary in the analysis of data collected via nested designs generally, and interviewer-nested designs specifically (see, e.g. Hagenaars and Heinen 1982; Dijkstra 1983; Raudenbush and Bryk 1986, 2002; Hox et al. 1991). For analyzing our original experiment, we created two dependent variables to evaluate the effects of our experimental design. One was whether the case was successfully screened for eligibility, and the other was whether the case ended as a completed interview. Both were binary (0, 1) variables.

The first analysis, focusing on successful screening of a case (regardless of the outcome – that is, including those found to be ineligible via a fully completed screener or via some other determination by the interviewer that the case was ineligible), uses all cases that were still active at, or had been finalized by, the end of the experimental period on March 19, 2015. We excluded all cases that had been deemed ineligible from the analysis, in addition to: (i) those touched by interviewers with a total case load of fewer than five cases during the experiment, (ii) those touched by an interviewer who was incorrectly randomized or trained, and (iii) those touched by a senior staff member (e.g. supervisor review). A sample size of 91 732 cases was thus used in the screening analysis. We do not believe that leaving out the abovementioned cases had any discernible impact on our results, as we ran sensitivity analyses which led to this conclusion.

12.3.3 Experimental Design Findings

To provide more context for the experimental results we are about to present, it is informative to note that an eligible respondent was a low-income, uninsured, legal resident of California, who was 18–64 years of age. As such, very few of those contacted and who were successfully screened ($n = 5550$) were found to be eligible ($n = 423$); and therefore, very few of the cases yielded a completed interview ($n = 308$).

Given our primary interest in whether or not interviewers assigned to work one condition or two finalized their cases at different rates, and the fact that cases were often called by more than one interviewer given their random delivery across sites[12] and calling shifts, our first task was to classify the outcomes of interest and determine which interviewer should be given "credit" for a case. We utilized the call outcome codes generated by the NORC CATI system (Voxco) to classify cases in close accordance with the outcome codes utilized in AAPOR's *Standard*

12 At the time of this experiment, NORC had CATI staffing working in Chicago, Las Vegas, and New York.

Table 12.2 Final disposition for screening analysis of cases by interviewer group.

Final outcome as of 19 March, 2015	One intro (%)	Two intros (%)	Total (%)
Eligible completion	0.40	0.25	0.34
Eligible partial	0.14	0.11	0.13
Completed screener but ineligible	6.71	4.13	5.59
Refusal, eligibility unknown	20.93	11.44	16.82
Refusal, known to be eligible	0.04	0.03	0.03
Contact made, never refused, eligibility unknown	9.00	6.98	8.13
Contact made, never refused, known to be eligible	0.03	0.05	0.04
Known residence, answering machine/voice mail outcome but no other human contact, eligibility unknown	54.58	71.70	62.00
Always busy or never answered, eligibility unknown	8.18	5.30	6.93
Total (*n*)	51 958	39 774	91 732

Definitions (AAPOR 2016), focusing on the final disposition of each case at the end of the experiment.[13] Next, we assigned each case to the interviewer who achieved the final disposition for the relevant analysis. In the screening analysis, this would assign a case to the interviewer who completed the screener; or if no screener was completed but a refusal was obtained, to the interviewer who obtained the refusal; or if no refusal was obtained, to the last interviewer who made contact with the respondent/household via an outbound call. This process yielded the distribution of case outcomes, overall and by interviewer group, shown in Table 12.2.[14]

Our second analysis considered only those cases that were screened or otherwise found to be eligible/ineligible and we excluded cases with the same criteria as above, yielding a sample size of 5550 cases. The assignment of interviewers to cases for the completed interview analysis proceeded in the same fashion, except that the interviewer who achieved the completion (or partial completion) was assigned, followed by the interviewer who finished the screener if the interview did not proceed; and so on. Table 12.3 displays the breakdown of final dispositions by experimental introduction group for the completion analysis.

For the analyses[15]:

1. We began by estimating multilevel logistic regression models predicting a completed screener against all other outcomes, and a completed interview among cases that were screened. First, we estimated models that included only the experimental condition as the researcher-controlled independent variable/predictor (i.e. whether the interviewer who achieved the final outcome was assigned to two introductions or one).

13 By "final disposition" we do not mean the outcome of the last contact attempt that was made for each case, but rather the disposition that was assigned as the *final status* of the case in accordance with AAPOR's *Standard Definitions*. For example, if a phone number were called 10 times, and the resulting outcomes of each call were nine ring-no-answers (on the 1st through 4th and 6th through 10th call attempt) and one refusal (on the 5th call attempt), then the final disposition for that case would be a refusal.

14 Of note, the "final disposition" as reported here was only the status of the case when our data collection stopped. Many of these cases had only one or two contact attempts at that point and were yet to be fully worked, thus the response rates for this analysis are much lower than what the final response rates were for the study. Data collection for the experiment was stopped early because it was observed that the PE introductions were yielding better outcomes for the survey sponsor/client (Lavrakas et al. 2015).

15 Interested readers can view the SAS code that was used to conduct our analyses at the Wiley website for supplementary materials for this book at www.wiley.com/go/Lavrakas/survey-research.

Table 12.3 Final outcome for screening analysis, of cases known to be eligible or potentially eligible.

	One intro (%) (n)	Two intros (%) (n)	Total (%) (n)
Eligible completion	5.50 ($n = 207$)	5.65 ($n = 101$)	5.55 ($n = 308$)
Eligible partial	1.89 ($n = 71$)	2.40 ($n = 43$)	2.05 ($n = 114$)
Completed screener but ineligible	92.05 ($n = 3462$)	91.39 ($n = 1635$)	91.84 ($n = 5097$)
Refusal, known to be eligible	0.56 ($n = 21$)	0.56 ($n = 10$)	0.56 ($n = 31$)
Total (n)	3761	1789	5550

2. Next, we added a limited set of interviewer- and case-level covariates that were readily available to the research team, that could be linked on interviewer and/or case ID, and that were deemed theoretically or practically relevant to the analysis. We specifically kept this set small in order to focus on the experimental design as well as by necessity, since very little was known about the unscreened respondents and few interviewer-level covariates were available in this setting (of note, we encourage researchers to obtain as many relevant interviewer covariates as possible in future analyses). We included a variable for the site at which the interviewer achieving the final outcome was located (Chicago, Las Vegas, or New York), as well as case-level predictors for whether or not the case was assigned to a conventional or a progressive introduction (constant throughout the field period) and the log-transformed number of calls that had been made to a respondent before the final outcome was achieved.
3. These models were estimated with random intercepts for each interviewer that called on the project and achieved at least five final outcomes. Importantly, we note that our outcome measures are fairly skewed, owing to the rare events of screening and achieving a completion in a study with very stringent eligibility criteria, and thus we encourage some caution in interpretation of our results as well as replication by future research. We also note the possibility of model misspecification given this skewness and our lack of other covariates.
4. Next, in order to investigate the hypothesis that interviewers may vary in their ability to utilize introductions based on their recruitment status (i.e. new to NORC or experienced with NORC), we estimated the same models on only the cases for which the final outcome was achieved in Chicago (with only that site having interviewers in both groups). In this second set of models, we added an interaction between the assignment to experimental group and the interviewer's tenure at NORC (new vs. old).
5. We estimated all models in SAS 9.4 using PROC GLIMMIX and adaptive quadrature estimation with five integration points (accommodating the skewness of the outcome measures); see Footnote 16 to gain access to the SAS code that was used. For each relevant model, we present logistic regression coefficients and their standard errors, the interviewer variance component along with the significance test for this variance component estimated using a mixture of chi-square distributions (and testing whether or not the interviewer variance component is equal to 0), as well as an estimate of the Bayesian Information Criterion (BIC) as an indicator of model fit (with smaller relative values indicating better fit).

Examining the results presented in Table 12.4, we find that being assigned to one introduction vs. two is associated with an increase in the log-odds of completing a screener; that is, those interviewers using only one introduction were significantly ($p < 0.001$) more successful in screening their contacted respondents. This result holds up upon including covariates in the model, and we note an interviewer variance component significantly ($p < 0.001$) different from 0 in both instances that decreases modestly when including the covariates in the model (along with a very modest decrease in BIC, indicating that we have not estimated a worse-fitting model

Table 12.4 Model results for all interviewers: predicting if eligibility was determined.

	Experimental treatment only β (SE)	With covariates β (SE)
Intercept	−3.134 (0.076)***	−3.167 (0.094)***
Interviewer-level predictors		
One intro	0.604 (0.088)***	0.674 (0.088)***
Site: Chicago vs. Las Vegas		0.141 (0.089)
Site: New York vs. Las Vegas		−0.166 (0.142)
Case-level predictors		
Assignment to conventional intro		0.000 (0.041)
Number of calls (log)		−0.160 (0.024)***
Interviewer variance component	0.239 (0.032)***	0.222 (0.030)***
N	91 732	91 732
BIC	41 150.51	41 120.79

$^{***}p < 0.001$.

with their inclusion). We note that the number of calls is additionally predictive of completing a screener, with its log-transformed predictor inversely related to success ($p < 0.001$). This latter finding intuitively makes sense in that as a case is called more frequently, the odds of success may decrease (if, e.g. previous refusals have been obtained or noncontacts are actually purposefully screening their calls). We find no impact of site on screening, a welcomed result that supports the decision for combining data from different sites in analysis where possible, and also no impact of type of introduction to which the case was assigned on the screening outcome.

We replicated these analytic procedures and present results in Table 12.5 focusing only on cases that achieved their final outcome in Chicago. We further note a significant interaction

Table 12.5 Model results for screener outcome among Chicago interviewers only by experience.

	Experimental treatment only β (SE)	With covariates β (SE)	With interaction β (SE)
Intercept	−3.097 (0.11)***	−2.869 (0.126)***	−2.655 (0.164)***
Interviewer-level predictors			
One intro	0.637 (0.124)***	0.713 (0.121)***	0.411 (0.193)*
New hire (vs. experienced interviewer)		−0.285 (0.107)***	−0.649 (0.212)**
Intro* recruitment: new, one			0.484 (0.244)*
Case-level predictors			
Assignment to conventional intro		−0.083 (0.059)	−0.078 (0.059)
Number of calls (log)		−0.072 (0.032)**	−0.073 (0.032)*
Variance component	0.273 (0.044)***	0.249 (0.041)***	0.240 (0.039)***
N	48 538	48 538	48 538
BIC	23 531.64	23 531.25	23 532.23

$^{*}p < 0.05$, $^{**}p < 0.01$, $^{***}p < 0.001$.

($p < 0.05$) of experimental group and recruitment status, suggesting that assignment to one vs. more than one introduction may differentially (and potentially negatively) impact newly employed interviewers as compared with experienced interviewers. Turning to descriptive percentages to help illustrate this result, we find that screening rates were lower for new hires than experienced interviewers (6.5% vs. 7.9%). Broken down by experimental group, new hires assigned to two introductions had the lowest screening rates (3.6%), followed by experienced interviewers assigned to work two introductions (5.9%), new interviewers assigned to work one introduction (7.5%), and experienced interviewers assigned to work one introduction (8.7%). Although the fixed-effect result matches our expectations, we note that the BIC does not decrease by adding the interaction, and in fact slightly increases. This raises caution in our interpretation of results – though some variance appears to be explained by the inclusion of this interaction. Therefore, we encourage further work in determining which interviewer characteristics may moderate the effectiveness of experimental manipulations and the estimation of random coefficient models, as our study (to our knowledge) is only a first attempt in investigating these issues.

Next, we turn to models predicting completed interviews among those for whom screening (eligible and not eligible) was accomplished; see Table 12.6. We find no impact of assignment to one or two introductions on completing an interview once screening is accomplished, finding only an effect of the number of calls as well as the introduction type (a positive relationship between log-transformed calls to final outcome and completion, and an inverse relationship between assignment to conventional introduction and completion). The introduction type result, while not the focus of the present investigation, is an important result for the design of the experiment itself, as the study also was designed to test whether or not a nonconventional "progressive engagement" introduction could increase response rates over a conventional introduction of the type usually used on similar NORC studies. Mirroring descriptive results presented by Lavrakas et al. (2015), we find that progressive introductions indeed appear to facilitate completed interviews ($p < 0.01$). However, we find that the impact of using *multiple* introductions is mainly restricted to the screening results, which makes intuitive sense. The ability to successfully recruit respondents for long enough to screen them for eligibility requires a different set of behaviors than those required to complete an interview among those

Table 12.6 Model results for all interviewers: predicting completed interview.

	Experimental treatment only β (SE)	**With covariates β (SE)**
Intercept	−2.845 (0.119)***	−2.803 (0.156)***
Interviewer-level predictors		
One intro	−0.033 (0.141)	−0.142 (0.149)
Site: Chicago vs. Las Vegas		0.082 (0.15)
Site: New York vs. Las Vegas		0.127 (0.247)
Case-level predictors		
Assignment to conventional intro		−0.371 (0.126)**
Number of calls (log)		0.256 (0.098)**
Variance component	0.104 (0.074)*	0.108 (0.073)*
N	5550	5550
BIC	2392.61	2398.19

*$p < 0.05$, **$p < 0.01$, ***$p < 0.001$.

Table 12.7 Model results for Chicago interviewers only: predicting completed interview.

	Experimental treatment only	With covariates	With interaction
Intercept	−2.788 (0.171)***	−2.786 (0.192)***	−2.801 (0.224)***
Interviewer-level predictors			
One intro	−0.09 (0.194)	−0.24 (0.192)	−0.212 (0.287)
New hire (vs. experienced interviewer)		0.475 (0.178)**	0.509 (0.311)
Intro* recruitment: new, one			−0.05 (0.379)
Case-level predictors			
Assignment to conventional intro		−0.514 (0.158)**	−0.515 (0.158)**
Number of calls (log)		0.118 (0.126)	0.118 (0.126)
Variance component	0.107 (0.103)	0.030 (0.082)	0.030 (0.082)
N	3284	3284	3284
BIC	1436.47	1432.69	1437.48

$^{*}p < 0.05$, $^{**}p < 0.01$, $^{***}p < 0.001$.

cooperative. Our experimental treatment for this chapter, that is, whether an interviewer was trained and assigned to use one introduction or two, accordingly had little to do with the administration of the questionnaire itself, since the questionnaire and its administration was the same across all interviewers and all respondents.

Finally, we tested the hypothesized interaction effect with completed interviews as the dependent variable. As shown in Table 12.7, the analysis yielded nonsignificant results. In this case, additionally, we do not observe an interviewer variance component significantly different from zero (0).

12.4 Discussion

There is a long history in the survey research literature of researchers reporting the results of experiments in which interviewers and their behaviors are used as the independent variables in "causal" analyses. In some cases, it is the physical characteristics of interviewers (e.g. gender, age, race, etc.) that serve as the independent variable(s) in the analyses in what are often "natural" experiments (West and Blom 2017). In other interviewer-administered experiments, the *a priori* treatments being tested are meant to be passively administered by all interviewers, in a standardized fashion, such as when question-wording or question-ordering is manipulated by the researchers so that different respondents are exposed to different versions (cf. Schuman and Presser 1981). In these cases, the individual interviewers are assumed to be like conduits that simply "pass along" the treatments that have been randomly assigned to respondents and are not thought to actually be part of the experimental treatments.

In other interviewer-administered experiments, the treatments are administered by interviewers in ways that cannot practically be standardized. Thus, not every administration of an experimental treatment condition will be the same. In these experiments, the interviewers are not passive conduits for the treatments that the researchers have created and randomly assigned to respondents. Rather, the treatments that researchers have randomly assigned respondents to receive are actively operationalized by the interviewers who administer them.

Therefore, if a given interviewer is differentially skilled and/or differentially motivated when implementing a given experimental treatment compared to how s/he implements another of the treatments, that differential behavior on the part of the interviewer becomes inextricably linked to the treatment that the researchers created and to which a given respondent is exposed. Thus, the actual manifestation of the treatment to which the respondent is exposed *is* the interviewer's behavior. When this happens, the differential behavior within and across interviewers who are meant to administer the same treatment to all the respondents assigned to that treatment can become a serious confound for the experiment. And this can make it difficult to know how to interpret any observed differences in the dependent variables across experimental conditions: Is it the treatment the researchers created that has caused the observed differences or is it the behavior of the interviewers who administered the treatment, or is it both?

To this end, using statistical procedures such as GLMM provides a valid and manageable approach for sorting out the "true" causes of the observed variation across experimental treatments. By including a random effect that links the interviewer who administered the treatment to each case, the researchers are able to account for interviewer-level variance (see Hagenaars and Heinen 1982; Dijkstra 1983; Raudenbush and Bryk 1986, 2002; Hox et al. 1991). As Dijkstra (1983) demonstrated, conducting cause-and-effect experimental analyses in interviewer-administered experiments without accounting for interviewer-level variation can lead to highly spurious findings. Instead, by studying and quantifying the effects of interviewer variance researchers can more accurately interpret the effects of their experimental factors. Thus, the GLMM approach we have used is an approach that many researchers who test treatments that are interviewer-administered should be using. And it is an analytic approach that journal editors should require be applied in articles that report and interpret the results of many interviewer-administered experiments.

We have reported empirical evidence in our chapter about two basic kinds of interviewer-administered experiments. In one case, we have shown that in many telephone surveys when cases are randomly, or essentially randomly, assigned to interviewers, a natural experiment has been created whereby the characteristics of the interviewers can be used as independent "treatment" variables in the analyses. Our example (results shown in Table 12.1) of doing secondary analyses with such a dataset to investigate the impact of interviewers' age and respondents' gender on respondents' question-answering behavior illustrates a natural experiment approach that we believe is woefully underutilized, in part because too few researchers include a variable in their datasets that identifies the interviewer who is "responsible" for each case.[16] We believe that as long as researchers provide credible evidence that the assignment of cases in the study was essentially random across interviewers, then there is little concern that the results of the experimental analyses have been confounded due to biased assignment of interviewers to conditions.

Although we did not report any original findings in this chapter about interviewer-administered experiments which require interviewers to administer all treatment conditions in a passive (i.e. a neutral conduit-like) fashion, we believe that these types of experiments will not be subject to nonignorable confounds related to interviewer effects, if adequate quality controls are in place to assure that standardized administration of all the treatments takes place. Examples of these interviewer-administered experiments are when a group of interviewers administers differently worded versions of a questionnaire item that the researchers have randomly assigned different respondents to receive.

16 The reader is reminded that this interview ID variable should not make it possible to identify the actual person linked to the ID. And, of course, there is no need to know that for the analytic purposes discussed in this chapter.

However, when a set of "actively administered" experimental treatments results in idiosyncratic administration by interviewers, we believe that it does matter how the researchers chose to allocate interviewers to treatments. As shown in our illustrative experimental testing of two different approaches for introducing a survey, it did matter whether interviewers were trained to and administered only one introduction or whether they were trained to and administered all the introductions. In that original experiment that we conducted for this chapter we found important differences between the groups of interviewers trained to administer one and only one type of introduction, compared to the group of interviewers trained to administer all the introductions that were tested in our experiment. The groups of interviewers who were trained and administered one introduction were significantly more successful ($p < 0.001$) in screening respondents compared to the interviewers who were trained and administered multiple approaches to screening respondents. A plausible explanation for our results is that by having to learn to administer two very different introductory styles, the interviewers in this "all treatments" condition experienced cognitive overload, leading to treatment diffusion, and thereby found it difficult to administer each style as well as they could had they been expected to learn and use only one of the introductory approaches. Furthermore, many of the interviewers who were randomly assigned to both introductions were well experienced with one of the approaches but not the other.

While some previous work leaves open the possibility of diluting treatment effects due to interviewers administering more than one condition, other researchers control for this possibility carefully. For example, the Smyth, Olson, and Matthews chapter in this book (see Chapter 2), which focuses on experiments that compare different within-unit respondent selection methods, reports the results of an original experiment that randomly assigned interviewers to be trained to administer only one of the selection methods that were tested in the experiment. It is our belief that many of the past articles that have been published on this topic may have used confounded experiments because their interviewers could have been differentially skilled and/or differentially motivated in their delivery of the different selection techniques that they administered. Because of this, those articles should be interpreted cautiously due to the designs that were used. Furthermore, to our knowledge, none of the published articles in that literature used a multilevel statistical approach, such as GLMM, to account for random interviewer effects in their analyses.

There are other domains of survey experimentation where interviewers have actively administered survey treatments and where they were assigned to administer more than one or all of the treatments in the experiment. This includes experiments that tested different survey introductions. It also includes experiments that tested other recruitment strategies such as using randomly assigned differential incentives of varying amounts as part of recruitment whereby interviewers actively leveraged the incentives in their efforts to gain cooperation from a given respondent. The same holds for experiments that tested different advance letters that interviewers later leveraged as part of gaining cooperation with respondents who they contacted. And, it includes experiments in which interviewers used different refusal conversion protocols that were randomly assigned to previously refusing respondents. In all these types of experiments where interviewers were used to actively "create" the actual treatment to which a randomly assigned respondent was exposed, if interviewers were required to administer more than one of the experimental treatments being tested then we would argue that the design may have been confounded.

When experimental treatments are administered by interviewers, we also recommend that data be gathered from and about each interviewer so that the researchers can learn how well the interviewers administered the treatments to which they we assigned. Although, due to limited space, we have not reported it in this chapter, that is what we did by gathering data directly from the interviewers about the introduction treatment they used and by gathering data from quality-monitoring staff about how well each interviewer administered the treatment that was randomly assigned to the case the interviewer was working. Information about that part of our research is available on the Wiley website containing Supplemental Materials related to our chapter. Of note, these additional types of data provided support to the conclusions we reported in this chapter.

A final issue that we have addressed previously bears repeating. That is the issue of the variation among interviewers in how they implement (operationalize) the types of experimental treatments that we tested in our experiment that was conducted for this chapter. Those are treatments where the interviewer's own behavior becomes an integral part of the treatment. There is good reason to believe that not only will interviewers vary how they implement the treatment(s) to which they are assigned compared to other interviewers, but that they will vary how they deliver the treatment from one of their respondents to another. Because of this interviewer-level variation, the field of survey research should conduct more research on the "how" and "why" of the variation that occurs within and between interviewers when they administer recruitment and selection protocols within survey-based experiments and in surveys in general.[17] Interviewer-level variation that occurs whenever interviewers are using the protocol(s) that researchers have assigned to them to recruit respondents in any survey that is using interviewer recruitment also needs to be studied more. As part of these empirical investigations it is imperative that researchers use multilevel statistical methods, such as GLMM, that incorporate an interviewer-level random effect within the analytic model (cf. Hox et al. 1991).

In conclusion, we believe that we have identified a potential gap in previous literature that deserves further exploration. We also believe that we have provided evidence that survey researchers in the future who plan experiments for which treatments are administered by an interviewer should be very careful in deciding how to allocate their interviewers to the different treatments that are being tested. If in doubt, the safest course is to train and allocate interviewers to one and only one of the experimental treatments. Interviewers also should be kept blind to the fact that other interviewers are implementing other treatment conditions. At a minimum, this requires that separate interviewer training sessions must be held for each experimental condition.

Finally, we recognize through our past experience, and our experience with the original experiment we conducted for this chapter, that implementing interviewer-administered experiments presents many operational challenges, some of which may be so severe as to threaten or even undermine the integrity of the experiment. As such, researchers must plan *and monitor* such experiments very carefully, especially in light of the complexities that often are faced in implementing and interpreting these designs (cf. West and Olson 2010).

References

AAPOR (2016). *Standard Dispositions: Final Dispositions of Outcome Rates and Case Codes for Surveys*, 9e. Deerfield, IL: American Association for Public Opinion Research http://www.aapor.org/AAPOR_Main/media/publications/Standard-Definitions20169theditionfinal.pdf.

Bryant, B. (1975). Respondent selection in a time of changing household composition. *Journal of Marketing Research* 12: 129–135.

17 We did not do this in our analyses as we did not have the resources to carry this out.

Czaja, R., Blair, J., and Sebestik, J. (1982). Respondent selection in a telephone survey. *Journal of Marketing Research* 19: 381–385.

Dijkstra, W. (1983). How interviewer variance can bias the results of research on interviewer effects. *Quality and Quantity* 17: 179–187.

Dunning, T. (2012). *Natural Experiments in the Social Sciences: A Design-Based Approach*. Cambridge University Press.

Elliott, M.R. and West, B.T. (2015). "Clustering by interviewer": a source of variance that is unaccounted for in single-stage health surveys. *American Journal of Epidemiology* 182 (2): 118–126.

Finkel, S., Guterbock, T., and Borg, M. (1999). Race of interviewer effects in a pre-election poll: Virginia 1989. *Public Opinion Quarterly* 55: 313–330.

Gaziano, C. (2005). Comparative analysis of within-household selection techniques. *Public Opinion Quarterly* 69 (1): 124–157.

Gillikin, J. (2008). Interpenetrated design. In: *Encyclopedia of Survey Research Methods* (ed. P.J. Lavrakas), 359–360. Thousand Oaks, CA: Sage.

Groves, R.M. and McGonagle, K.A. (2001). A theory-guided interviewer training protocol regarding survey participation. *Journal of Official Statistics* 17: 249–265.

Hagenaars, J.A. and Heinen, T.G. (1982). Effects of role-independent interviewer characteristics on responses. In: *Response Behaviour in the Survey-Interview* (ed. W. Dijkstra and J. Van Der Zouwen), 91–130. Academic Press.

Hox, J.J., De Leeuw, E.D., and Kreft, I.G.G. (1991). The effect of interviewer and respondent characteristics on the quality of survey data: a multilevel model. In: *Measurement Errors in Surveys* (ed. P.P. Biemer, R.M. Groves, L.E. Lyberg, et al.), 439–462. New York: Wiley.

Kish, L. (1949). A procedure for objective respondent selection within the household. *Journal of the American Statistical Association* 57: 92–115.

Lavrakas, P.J. (1992). Attitudes towards and experiences with sexual harassment in the workplace. Paper presented at the Midwest Association for Public opinion Research, Chicago IL.

Lavrakas, P.J. (1993). *Telephone Survey Methods: Sampling, Supervision and Selection*, 2e. Newbury Park CA: Sage Publications.

Lavrakas, P.J., Kelly, J., and McClain, C. (2015). Using a progressive engagement introduction to gain cooperation in an interviewer-administered telephone survey. Presented at the 40th Annual Conference of the Midwest Association for Public Opinion, Chicago (20 November 2015).

Oldendick, R., Bishop, G., Sorenson, S., and Tuchfarber, A. (1988). A comparison of the Kish and last-birthday methods of respondent selection in telephone surveys. *Journal of Official Statistics* 4: 307–318.

O'Muircheartaigh, C. and Campanelli, P. (1999). A multilevel exploration of the role of interviewers in survey nonresponse. *Journal of the Royal Statistical Society: Series A (Statistics in Society)* 162 (3): 437–446.

Raudenbush, S. and Bryk, A.S. (1986). A hierarchical model for studying school effects. *Sociology of Education* 59 (1): 1–17.

Raudenbush, S.W. and Bryk, A.S. (2002). *Hierarchical Linear Models: Applications and Data Analysis Methods*, vol. 1. Thousand Oaks, CA: Sage.

Salmon, C. and Nichols, J. (1983). The next birthday method for respondent selection. *Public Opinion Quarterly* 47: 270–276.

Schuman, H. and Converse, J.M. (1971). The effects of black and white interviewers on black responses in 1968. *Public Opinion Quarterly* 35 (1): 44–68.

Schuman, H. and Presser, S. (1981). *Questions and Answers in Attitude Surveys: Experiments on Question Form, Wording and Context*. New York: Academic Press.

Stokes, L. (1988). Estimation of interviewer effects for categorical items in a random-digit dial telephone survey. *Journal of the American Statistical Association* 83 (403): 623–630.

West, B.T. and Blom, A.G. (2017). Explaining interviewer effects: a research synthesis. *Journal of Survey Statistics and Methods* 5 (2): 175–211.

West, B.T. and Olson, K. (2010). How much of interviewer variance is really nonresponse error variance? *Public Opinion Quarterly* 74 (5): 1004–1026.

West, B.T., Conrad, F.G., Kreuter, F., and Mittereder, F. (2017). Can conversational interviewing improve survey response quality without increasing interviewer effects. *Journal of the Royal Statistical Society: Series A (Statistics in Society)* https://doi.org/10.1111/rssa.12255.

Part V

Introduction to Section on Adaptive Design

Courtney Kennedy[1] *and Brady T. West*[2]

[1] *Pew Research Center, Washington, DC, USA*
[2] *Survey Research Center, Institute for Social Research, University of Michigan, Ann Arbor, MI, USA*

Facing increased uncertainty about the performance of a given survey protocol, numerous researchers have turned to adaptive or responsive designs. This approach calls for altering the survey design during the course of data collection with the goal of improving cost efficiency or achieving more precise, less biased estimates (Groves and Heeringa 2006). Often experiments are embedded in an initial phase and their results are used to inform the design of subsequent phases. While embedded experiments are not a requirement of an adaptive or responsive design, they feature prominently in that growing literature (e.g. Kessler et al. 2004; Groves et al. 2005). In addition to the computerization of data collection and the accessibility of paradata, the ability to embed experiments in surveys is one of the foundational tools underpinning the movement toward adaptive designs.

This section contains two chapters. Both provide examples of experimental design and analysis that resemble those in the adaptive design literature, though neither chapter presents a formal adaptive design as defined by Groves and Heeringa (2006). Chapter 13 by Kunz and Fuchs considers dynamic interventions to reduce measurement error in web surveys. The authors implemented several types of feedback instructions addressing either speeding or nondifferentiation in grid questions with the aim of reducing the proportion of satisficing respondents. Such dynamic interventions have a distinctly responsive quality. The interventions use paradata about respondents' behavior captured in real time during the interview to make alterations to the subsequent survey experience. In this case, respondents were shown a prompt with a message like, "Please take some more time for your answers." This general line of research is finding some success in programming surveys to adapt to respondents' behavior and, in turn, eliciting what appear to be more accurate answers.

Chapter 14, by Suzer-Gurtekin, Elkasabi, Lepkowski, Liu, Curtin, and McBee, is an example of using a randomized experimental design to examine a different error source – nonresponse. Their study speaks to how researchers can reduce nonresponse bias by experimentally testing different modes and recruitment features. They evaluated several protocols for recruiting address-based sample members to complete a survey either online or by mail. The research design varied the sequence in which mode options were offered, the number of mailings, and the provision of a preincentive. In the context of a responsive design, the results from such an

Experimental Methods in Survey Research: Techniques that Combine Random Sampling with Random Assignment, First Edition.
Edited by Paul J. Lavrakas, Michael W. Traugott, Courtney Kennedy, Allyson L. Holbrook, Edith D. de Leeuw, and Brady T. West.

Companion Website: www.wiley.com/go/Lavrakas/survey-research

experiment might be used to identify the best performing protocol according to one or more predetermined metrics, and then that protocol would be used for a subsequent data collection phase. Under an adaptive design (as described by Schouten et al. 2013), the results from such an experiment might be used to determine different optimal designs for different subgroups. A future survey would then tailor the data collection protocol to each subgroup based on those past results.

While evaluation studies show that the efficacy of adaptive designs may be marginal (Tourangeau et al. 2017), the central idea of leveraging embedded experiments and other information to improve surveys in a dynamic fashion is likely to endure. This is one of the few strategies that survey designers have at their disposal to shore up probability-based surveys in the face of increasing costs and decreasing participation. The chapters in this section speak to some of the challenges inherent in adaptive designs (e.g. sometimes the expected cost advantages of a given protocol are not realized). These chapters also speak to the promise of adaptive approaches. Suzer-Gurtekin and her colleagues identified a protocol that doubled the response rate (from 17% to 34%) in an address-based sample using web/mail design, while Kunz and Fuchs found significant, positive effects on measurement from using interactive feedback. Together, these chapters provide examples of how randomized experiments can be used to enhance survey quality, in either a static or dynamic way.

References

Groves, R.M. and Heeringa, S.G. (2006). Responsive design for household surveys: tools for actively controlling survey errors and costs. *Journal of the Royal Statistical Society: Series A (Statistics in Society)* 169: 439–457.

Groves, R.M., Benson, G., Mosher, W.D. et al. (2005). *Plan and Operation of Cycle 6 of the National Survey of Family Growth*. Hyattsville, MD: National Center for Health Statistics.

Kessler, R.C., Berglund, P., Chiu, W.T. et al. (2004). The US National Comorbidity Survey Replication (NCS-R): design and field procedures. *International Journal of Methods in Psychiatric Research* 13: 69–92.

Schouten, B., Calinescu, M., and Luiten, A. (2013). Optimizing quality of response through adaptive survey design. *Survey Methodology* 39: 29–58.

Tourangeau, R., Brick, J.M., Lohr, S., and Li, J. (2017). Adaptive and responsive survey designs: a review and assessment. *Journal of the Royal Statistical Society: Series A (Statistics in Society)* 180: 203–223.

13

Using Experiments to Assess Interactive Feedback That Improves Response Quality in Web Surveys

Tanja Kunz[1] *and Marek Fuchs*[2]

[1] *Leibniz-Institute for the Social Sciences, P.O.Box 12 21 55, 68072, Mannheim, Germany*
[2] *Darmstadt University of Technology, Institute of Sociology, Karolinenplatz 5, 64283 Darmstadt, Germany*

13.1 Introduction

Data quality is a major concern in survey research. In particular, measurement error is second only to nonresponse error in terms of the most important threats to data quality in many surveys. Measurement error refers to differences between values reported by the respondents and their true values, with deviations in web surveys potentially stemming from the respondents themselves, from the survey instrument, or from the interaction between both respondent- and instrument-related factors. We begin this chapter with a review of the literature on assessing and preventing measurement error in web surveys.

13.1.1 Response Quality in Web Surveys

One cause of measurement error, known as satisficing, can be attributed to respondents completing the questionnaire with little or no attention to the survey questions. High-quality responses presuppose that respondents thoroughly pass through the four steps of the question–answer process and are carefully interpreting the question meaning, searching their memories for all relevant information, forming judgments based on this information, and accurately reporting these judgments while taking account of the response format provided (Krosnick 1991; Sudman et al. 1996; Tourangeau 1984). Thus, optimal responding to survey questions requires respondents to expend a certain degree of cognitive effort. Satisficing, by contrast, describes a response behavior that aims at keeping the extent of cognitive effort within a limit by applying cognitive shortcuts while passing through the question–answer process, which may lead to less than optimal survey responding. Apart from recording the respondents' eye movements while completing a survey in a laboratory, satisficing cannot be observed directly in a conventional survey setting; this is why various response behaviors are studied as proxies to detect inattentive or otherwise poorly engaged respondents. Various satisficing behaviors can be distinguished, including selecting the "don't know" option, agreeing with assertions, or endorsing the status quo (Krosnick 1991). Although being common satisficing behaviors, "it is [...] important not to assume that these responses *always* constitute satisficing. Rather, they may *sometimes* constitute satisficing and are presumably more likely

Experimental Methods in Survey Research: Techniques that Combine Random Sampling with Random Assignment, First Edition.
Edited by Paul J. Lavrakas, Michael W. Traugott, Courtney Kennedy, Allyson L. Holbrook, Edith D. de Leeuw, and Brady T. West.

Companion Website: www.wiley.com/go/Lavrakas/survey-research

to occur under the conditions that foster satisficing" (Krosnick 1991, p. 220), namely limited respondent ability, low respondent motivation, and high task difficulty.

In this regard, the prevalence of survey satisficing may differ depending on the mode of data collection. Considering the fact that in self-administered surveys such as web surveys, no interviewer is present to render assistance and motivate the respondents to thoroughly process all survey questions, the unsupervised interviewing process can lower data quality (Fricker et al. 2005; Grandjean et al. 2009; Heerwegh and Loosveldt 2008; Lozar Manfreda et al. 2008). Also, web surveys rely solely on visual presentation of questions and response options (David 2005) and thus differ in the cognitive demand. In particular, respondents are supposed to read the questions and response options themselves which is likely to induce satisficing behaviors, particularly among respondents with limited cognitive capacities (Chang and Krosnick 2010).

In addition, the risk of survey satisficing differs depending on the type and format of a survey question. Grid (or matrix) questions are considered convenient when asking a set of rating scale items sharing the same response options because they save space and avoid repetition. However, grid questions are especially prone to various types of satisficing behaviors. They require a high level of respondent ability because it is challenging to combine information from rows with information from columns. At the same time, they strain respondent motivation because answering is repetitive and tiring (Couper et al. 2013; Kunz 2015). This is the reason why respondents are likely to fall back on cognitive shortcuts in order to minimize the effort required to answer the items and to complete them as quickly as possible. Previous studies have shown that completion times are significantly shorter if a set of rating scale items sharing the same response options are presented in a single grid question instead of splitting the set of items into several smaller grid questions or using a single-item-per-screen format (Couper et al. 2001; Toepoel et al. 2009; Tourangeau et al. 2004). Other studies also found shorter completion times in grid questions, but these differences have failed to reach statistical significance (Bradlow and Fitzsimons 2001; Callegaro et al. 2009).

Zhang and Conrad (2013) showed that speeding through questions and "responding too fast to give much thought to answers" (p. 127) can be considered an indicator of survey satisficing because "respondents who are prone to speed are also prone to straight-line regardless of their demographics" (p. 127). However, faster survey completion times per se are not sufficient evidence of respondents actually engaging in satisficing behaviors and providing low-quality responses in grid questions, because "respondents can sometimes answer a question both quickly and accurately depending on factors such as task difficulty and accessibility of an attitude" (Zhang 2013a, p. 8). This is the reason why additional indicators of survey satisficing are used to assess data quality in grid questions. Nondifferentiation (Fricker et al. 2005) and straight-lining (Schonlau and Toepoel 2015) as the most severe form of nondifferentiation are typically considered when assessing grid questions. This is because presenting a set of rating scale items jointly in visual proximity – as is the case in grid questions – implies that these items are considered conceptually more related than if they are presented independently from one another. This in turn tempts respondents to give the same or similar answers to the items of a grid question irrespective of their content (Tourangeau et al. 2004). As a result, the consistency of the respondents' answers to grid questions is likely to be overestimated, while the respondents' true response variation is underestimated (Tourangeau and Rasinski 1988). Previous findings examining the extent of nondifferentiation in grid questions compared to other question formats are rare because nondifferentiation is considered a satisficing behavior that is explicitly inherent to the nature of a grid. While some previous studies actually found a higher level of nondifferentiation in grid questions (Grandmont et al. 2010; Tourangeau et al. 2004), other studies found inconsistent results or no differences between grid and non-grid questions (Couper et al. 2001; Kunz 2015). Despite the limited findings regarding the increased risk of nondifferentiation in grid questions, nondifferentiation is considered a form of strong satisficing because respondents simply answer the first item by selecting a response option that seems reasonable and then adjust subsequent answer choices to this response option (Krosnick 1991).

Respondents are known to engage in satisficing behaviors such as speeding and nondifferentiation in order to reduce the effort needed to complete a grid question. Not investing enough time and effort into the thorough processing of all steps of the question–answer process potentially reduces data quality and ultimately the validity of the results. Subsequently, survey researchers may have to dedicate a nontrivial amount of time and effort to the task of identifying low-quality responses and removing them from analyses (Blasius and Thiessen 2012; Greszki et al. 2015). Web surveys offer a distinct advantage in this regard. Instead of detecting low-quality responses after the field phase has already been completed, survey researchers can implement various kinds of interventions in web surveys during the response process to prevent low-quality responses in real-time.

13.1.2 Dynamic Interventions to Enhance Response Quality in Web Surveys

Past studies have experimented with static motivational statements that are always visible in the questionnaire (e.g. "This question is very important. Please take your time answering it.") in order to elevate response quality (Smyth et al. 2009). Results indicate that respondents do pay attention to such statements and ultimately provide higher quality answers. However, web surveys enable not only static ("always visible") instructions but, due to their interactive nature, also dynamic interventions depending on the respondents' behaviors, in a similar manner to interviewer-administered surveys. Previous research tested dynamic interventions in terms of interactive clarification features, interactive "no opinion" probes, or interactive follow-up probes (Al Baghal and Lynn 2015; Conrad et al. 2007; DeRouvray and Couper 2002; Holland and Christian 2009; Oudejans and Christian 2011). Based on the results of these investigations, the use of interactive feedback can be considered an effective strategy to address measurement error arising from inattentive respondents at an early stage. More specifically, interactive feedback can prevent respondents from processing the survey questions only superficially, and thus promoting high-quality responses. Even though previous research has demonstrated that warnings and prompts pose the risk of more socially desirable responding (Clifford and Jerit 2015), other studies have shown that probes and prompts can be implemented effectively "to help produce higher quality responses, without increasing respondent burden or frustration" (Holland and Christian 2009, p. 199).

Examples in this body of literature demonstrate that interactive feedback is most useful (i) when it is provided immediately after the respondent's relevant action, and (ii) when no additional effort is required from the respondent to obtain the feedback (Conrad et al. 2005, 2006). For example, Conrad et al. (2005) tested the effectiveness of delayed feedback (provided automatically by the system after the answer has been submitted) and concurrent feedback (provided instantly while the question is answered) on the level of response accuracy in a constant sum question (tallies equal to 100%). These authors found that interactive feedback was more effective when it was provided immediately after each action of the respondent. For example, a running tally that adjusted the sum of values after each partial data entry (one of multiple numeric values) was more effective than feedback provided after respondents had submitted their full answer. Concerning the effort that is necessary to obtain the feedback, Conrad et al. (2006) examined several interactive methods of providing clarification features. They showed that additional information was requested more often when respondents could access the clarification features via a mouse rollover, whereas only one mouse click on a hyperlink seemed to require more effort than the respondents were actually willing to expend.

Several studies have examined the effectiveness of delayed feedback addressing speeding or nondifferentiation by using either a pop-up window or a follow-up web page to deliver a prompt to respondents. Conrad et al. (2011) implemented speeding prompts in several frequency questions, with respondents being prompted when their response time fell below the typical reading time for a question defined by the threshold of 350 ms/word. The wording of the prompt message was as follows: "You seem to have responded very quickly. Please be sure you have given the question sufficient thought to provide an accurate answer. Do you want to go back and reconsider your answer?" In the control group (CG), there was no such prompt. The authors found that speeding prompts actually slowed down respondents as indicated by increased response times. Furthermore, a carry-over effect was found: respondents who slowed down in reaction to the speeding prompts in earlier questions exhibited lower levels of straight-lining in a later grid question. Importantly, findings indicated that the prompt messages did not increase the risk of survey breakoff.

Zhang (2012) tested the same kind of delayed feedback (i.e. feedback provided after respondents had submitted their responses by clicking on the "continue" button) in five frequency questions with the speeding prompts occurring either on a subsequent survey screen or in a pop-up window overlapping of the respective frequency question. Again, a threshold of 350 ms/word was used to identify speeding. Zhang (2012) found a significant reduction of speeding in the experimental groups (EGs) with a speeding prompt compared to the control group where no intervention was provided. Limited evidence was found with respect to a reduction of straight-lining in later grid questions. Again, speeding prompts did not seem to cause elevated survey breakoff. Zhang (2013a,b) also examined the effectiveness of speeding prompts occurring with unreasonably short response times below 300 ms/word in four grid questions. The prompts appeared in the form of a pop-up window after respondents had submitted their response, providing them with the opportunity to go back and reconsider their answers. In addition to speeding prompts, nondifferentiation prompts were tested with respondents being prompted for straight-lining (same response for all items in a grid) and near-straight-lining (same responses for all but one item in a grid). The results showed that prompting because of either speeding or nondifferentiation reduces both response behaviors in grid questions. This indicates that each prompt message can be used to address both speeding and nondifferentiation. There was no increase in survey breakoff as a result of the speeding or nondifferentiation prompts.

Hence, previous findings on delayed feedback addressing speeding or nondifferentiation showed that prompts are able to reduce satisficing behaviors in web surveys without increasing the risk of survey breakoff. Moreover, prompting can address several satisficing behaviors, "not just the specific behavior targeted in the prompt" (Zhang 2013a, p. 53). Although these findings on the use of delayed feedback are promising, effect sizes for the prevention of satisficing behaviors are rather small. This may be due to the fact that respondents are supposed to actively return to the grid question affected by satisficing behaviors after they have already submitted their answers which poses additional respondent burden. In fact, the percentage of respondents who went back in reaction to the prompt messages in the four grid questions examined by Zhang (2013a) was rather small; it ranged between 15.5% and 24.6% (with just one exception of 46.0%).

While previous research used delayed feedback that is provided after the respondents have already submitted their responses, in this chapter we assess feedback that is delivered immediately after the respondents have provided their partial response in a grid question. Accordingly, feedback is provided instantly as the relevant respondent actions are detected. This so-called instant feedback that was triggered by the problematic response behavior while the respondents were still answering a particular grid question becomes effective before respondents submit their responses and continue with the next survey page. The underlying idea is that feedback provided instantly, while the items are still being answered, can address satisficing behaviors more efficiently than delayed feedback that is provided after the survey page comprising the grid question is submitted. This is due to the fact that respondents exposed to instant feedback are not required to actively return to a question that they have already submitted in order to revise their initial response.

13.2 Case Studies – Interactive Feedback in Web Surveys

Variations in the visual designs of web surveys and their effects on the prevalence of satisficing behaviors are commonly tested by means of field-experimental studies. The strength of field-experimental designs is that random assignment is combined with random sampling with the aim of enhancing the internal and external validity of the survey results. Internal validity can be achieved by randomly assigning the respondents to different experimental conditions in a split-sample (or split-ballot, or split-half) experiment. By dividing a sample into random groups and assigning them each a different treatment, variations in survey design features and their effects on data quality can be systematically examined. Random assignment of sample members to the experimental (or treatment) group and control group ensures "that, on average, any observed differences between the two groups can be attributed to treatment effects rather than to differences in subsample composition" (Ziniel 2008, p. 384). External validity concerns the process of generalization of survey results obtained from a sample to the target population. External validity premises random sampling, which aims at fielding a sample that is representative of the target population and can be used to make an inference to the target population. Field-experimental studies that feature both random assignment and random sampling support causal inferences that can be generalized to the entire population. We assess and discuss the potential threats to internal and external validity of the experimental design used in our research in the final section of this chapter.

We embedded several randomized between-subjects field experiments in three web surveys to examine the use of interactive feedback enabling direct interaction with the respondents and immediate reaction to their response behavior. More specifically, we implemented two different types of feedback instructions addressing either speeding or nondifferentiation in grid questions with the aim of reducing the proportion of satisficing respondents.

Study 1 draws on data gathered from university freshmen ($n = 1696$) at Darmstadt University of Technology (Germany) in 2014. The response rate was 25.9% (1696 complete cases out of 6540 university freshmen who received an e-mail invitation). The net sample consisted of all respondents who answered at least 50% of the survey questions (according to AAPOR (2016) rule for incomplete cases).

The effectiveness of providing interactive feedback on speeding (Experiment 1.1) and nondifferentiation (Experiment 1.2) was examined by comparing the EGs where feedback was provided with the CG where no feedback was provided. The interactive feedback was provided either in terms of delayed feedback (EG1) or instant feedback (EG2). Aside from the absence or presence of the feedback, the grid questions presented in the control group and experimental groups were completely identical. Details of Experiments 1.1 and 1.2 are depicted in Table 13.1.

Both types of feedback instructions made use of a dynamic responsive design, that is, feedback was provided depending on particular respondent actions either in terms of instant feedback while the respondent was still answering the grid or in terms of delayed feedback triggered

Table 13.1 Description of Experiments 1.1 and 1.2 in Study 1.

	Experiment 1.1 **Speeding prompt**	**Experiment 1.2** **Nondifferentiation prompt**
Feedback instructions		
Delayed feedback (EG1)	"Please take some more time for your answers. If you want to go back to the question, click OK."	"Please try to differentiate more among your answers. If you want to go back to the question, click OK."
Instant feedback (EG2)	"Please take some more time for your answers."	"Please try to differentiate more among your answers."
Experimental questions		
Short grid (8 items)	Choice of field of study	Basic skills
Long grid (13 items)	Leisure activities	Job prospects
Behavior triggering feedback		
Delayed feedback (EG1)	Time span ($click_i - click_{i-1}$): <2000 ms at least once	McCarty and Shrum 2000: nondiff ($grid_{total}$) < 0.50
Instant feedback (EG2)	Time span ($click_i - click_{i-1}$): <2000 ms	McCarty and Shrum 2000: nondiff ($item_1 - item_i$) < 0.50
Dependent variable		
Primary effect	Response time	Nondifferentiation index
Secondary effect	Nondifferentiation index	Response time

after the respondent had submitted the grid. In the instant feedback condition, the threshold of speeding was 2000 ms/item, that is, each time the time span between the clicks on two radio buttons was shorter than this threshold, a speeding prompt was displayed. Considering the average number of words per item, this threshold equals about 250 ms/word. Theoretically, a respondent received as many speed prompts as the number of items comprised in the grid question. In the delayed feedback condition, a speeding prompt occurred only once when clicking on the "continue" button, if the time span between the clicks on two radio buttons fell below the threshold of 2000 ms at least once on the grid question.

The calculation of the extent of nondifferentiation was based on the index used by McCarty and Shrum (2000). Values can range between 0 and 1, with higher values being indicative of lower levels of nondifferentiation. In the present study, the nondifferentiation index was calculated in real-time while the respondents were answering the set of items presented in a grid question. The threshold was set to 0.50, that is, each time the nondifferentiation index fell below this critical value, a nondifferentiation prompt was administered in the instant feedback condition while the respondents were still answering the grid question. In the delayed feedback condition, a nondifferentiation prompt was displayed only once after clicking on the "continue" button if the final nondifferentiation index based on the entire grid question was below the threshold of 0.50 (see Figure 13.1 for an example).

The effectiveness of interactive feedback was assessed by means of the time respondents spent on responding and the extent of nondifferentiation in the grid question. Feedback instructions in response to speeding are primarily expected to increase the average response time (primary effect). Taking more time to answer a grid question may in turn be associated with a decrease in the extent of nondifferentiation (secondary effect). Conversely, feedback instructions in response to nondifferentiation may predominantly decrease the extent of

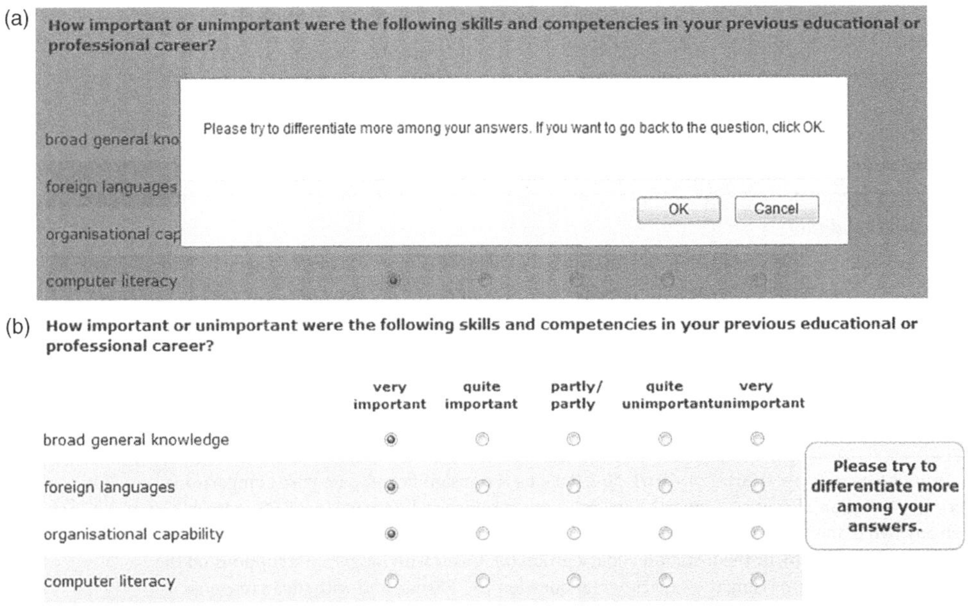

Figure 13.1 Screenshot of examples for feedback instruction in response to "Nondifferentiation" in terms of delayed feedback (a) and instant feedback (b).

nondifferentiation (primary effect) which is also likely to be reflected in longer response times (secondary effect). Thus, we assessed both speeding and nondifferentiation in all experiments (primary and secondary effects), even though the prompts in each experiment were triggered by speeding or nondifferentiation only and contained instructions to either slow down or to differentiate more. Calculations of final response times and final nondifferentiation indexes were based on all items in the respective grid question. To ensure comparability across respondents, only those cases were included in the analyses that actually answered all items in the respective grid question (i.e. no item missing values). Also, respondents with unusually long response times were excluded (i.e. respondents who interrupted and resumed the survey on this particular experimental question and also all respondents who exceeded a response time of two standard deviations above the group mean). The interactive feedback on speeding and nondifferentiation was assessed in both a short grid comprising 8 items and a long grid comprising 13 items. All items were answered on a five-point scale.

For the two continuous dependent variables – response time and nondifferentiation index – analysis of variance (ANOVA) was used to assess the differences in means between the two experimental groups and the control group. In order to account for multiple comparisons of sets of two groups simultaneously, a Bonferroni correction was used. Pair-wise comparisons were tested at a significance level of 0.05.

Results concerning the effectiveness of using interactive feedback on speeding are depicted in Table 13.2. A comparison with the control group showed that both delayed feedback (EG1) and instant feedback (EG2) yielded significantly higher mean response times in both the short grid and the long grid. Differences in response times between the delayed and instant feedback condition were found in the short grid which required significantly longer response times when delayed feedback (EG1) was provided compared to instant feedback (EG2). Concerning the extent of nondifferentiation, a significant decrease in nondifferentiation was found only for the instant feedback (EG2) in the short grid as indicated by higher values on the nondifferentiation

Table 13.2 Effectiveness of feedback instruction in response to "Speeding" in Study 1 (Experiment 1.1).

	Response time (mean s)		Nondifferentiation index (mean, McCarty and Shrum 2000)	
Experimental question	**Short**	**Long**	**Short**	**Long**
Experimental condition				
CG, no feedback	$41.5^{b,c}$	$60.5^{b,c}$	0.505^{c}	0.695
EG1, delayed feedback	$50.9^{a,c}$	69.6^{a}	0.497^{c}	0.693
EG2, instant feedback	$46.3^{a,b}$	66.4^{a}	$0.529^{a,b}$	0.691
F-value	41.11^{***}	23.99^{***}	5.43^{**}	0.25^{ns}
$df_{between, within}$	21 554	21 507	21 554	21 507
n	1557	1510	1557	1510

Notes: Overall *F*-test: $^{***}p < 0.001$, $^{**}p < 0.01$, $^{*}p < 0.05$, ns = nonsignificant; pair-wise comparisons (Bonferroni correction): lowercase superscripts indicate a significant difference ($p < 0.05$ or less) between any two of the three experimental conditions, that is, compared to control group (a), to the delayed feedback group (b), or the instant feedback group (c). Cases with longer interruptions on the respective target page comprising the experimental question (>7200 ms) and with mean response times exceeding two standard deviations above the group mean were excluded.

index. Hence, delayed feedback and instant feedback on speeding seem to be comparably effective in reducing speeding; they both show significant primary effects. By contrast, delayed feedback on speeding has no secondary effect on the extent of nondifferentiation, whereas the effectiveness of instant feedback on speeding in reducing nondifferentiation remains inconclusive.

Results regarding the use of interactive feedback on nondifferentiation are shown in Table 13.3. Again, a comparison between the experimental groups and the control group

Table 13.3 Effectiveness of feedback instruction in response to "Nondifferentiation" in Study 1 (Experiment 1.2).

	Nondifferentiation index (mean, McCarty and Shrum 2000)		Response time (mean s)	
Experimental question	**Short**	**Long**	**Short**	**Long**
Experimental condition				
CG, no feedback	$0.598^{b,c}$	0.610^{c}	$42.2^{b,c}$	52.7^{c}
EG1, delayed feedback	0.623^{a}	0.615^{c}	44.9^{a}	53.4^{c}
EG2, instant feedback	0.629^{a}	$0.631^{a,b}$	47.1^{a}	$59.7^{a,b}$
F-test	10.07^{***}	5.45^{**}	8.66^{***}	20.17^{***}
$df_{between, within}$	21 565	21 491	21 565	21 491
n	1568	1494	1568	1494

Notes: Overall *F*-test: $^{***}p < 0.001$, $^{**}p < 0.01$, $^{*}p < 0.05$, ns = nonsignificant; pair-wise comparisons (Bonferroni correction): lowercase superscripts indicate a significant difference ($p < 0.05$ or less) between any two of the three experimental conditions, that is, compared to the control group (a), to the delayed feedback group (b), or the instant feedback group (c). Cases with longer interruptions on the respective target page comprising the experimental question (>7200 ms) and with mean response times exceeding two standard deviations above the group mean were excluded.

showed that instant feedback (EG2) results in a significantly reduced degree of nondifferentiation in both experimental questions, whereas delayed feedback (EG1) was effective only in the short grid with a significantly lower degree of nondifferentiation compared to the control group. The same pattern occurred with regard to the response times, with respondents taking significantly more time to answer the short grid and the long grid when instant feedback (EG2) was provided compared to the control group where no feedback was provided (CG). Concerning the use of delayed feedback (EG1), a significant increase in the response times occurred only for the short grid. Hence, instant feedback on nondifferentiation yields a significant primary and secondary effect by reliably decreasing both nondifferentiation and speeding in the short- and long-grid questions. By contrast, results concerning the primary and secondary effect of using delayed feedback in response to nondifferentiation are mixed because nondifferentiation and speeding are reduced in the short grid but not in the long grid.

The findings from Study 1 demonstrate the advantages of using instant feedback over delayed feedback because instant feedback on nondifferentiation (Experiment 1.2) reduces both nondifferentiation and speeding more reliably than delayed feedback. Furthermore, delayed feedback on speeding (Experiment 1.1) actually results in longer response times. However, because of the fact that these longer response times are not accompanied by a decrease in nondifferentiation as compared to instant feedback, the additional time spent by the respondents seems to be devoted to tasks other than optimizing response behavior. Additional analyses (not displayed in the tables) suggested that instant feedback on nondifferentiation also reduces straight-lining (i.e. a nondifferentiation index of 0) to a larger extent as compared to delayed feedback. Based on the elevated effectiveness of instant feedback over delayed feedback, we speculate that the timing of interactive feedback is decisive for reducing satisficing behaviors. We assume that instant feedback alters the still ongoing question–answer process, while delayed feedback requires respondents to go back to the submitted question and resume its cognitive processing.

In order to replicate and further test the effectiveness of instant feedback, similar experiments employing a between-subjects design were embedded in two other surveys. Studies 2 and 3 were conducted among university applicants in 2013 and 2014, respectively. The response rate was 40.4% in Study 2 (7395 complete cases out of 18 327 invited university applicants) and 34.5% in Study 3 (5996 complete cases out of 17 357 invited university applicants). The net sample comprised all respondents who answered at least 50% of all survey questions (according to the AAPOR (2016) rule for incomplete cases).

In both studies, a between-subjects design using random assignment was implemented to test the effect of instant feedback on speeding (Experiments 2.1 and 3.1) and nondifferentiation (Experiments 2.2 and 3.2) provided in the EG compared to the CG where no feedback was provided. Details of the four experiments are depicted in Table 13.4.

In Study 2, instant feedback on speeding and nondifferentiation was tested in a short grid consisting of 8 items and a long grid comprising 14 items, respectively. In Study 3, two grid questions with 12 items each were used. In both studies, all items were answered on a five-point scale (see Table 13.4). For each grid question instant feedback on speeding and nondifferentiation was implemented in two independent experimental conditions; thus they are sharing the same control group. The threshold of speeding was 2000 ms/item. A speeding prompt appeared instantly each time the time span between the clicks on two radio buttons was shorter than this threshold. Based on the average number of words per item, the speeding threshold corresponds to about 300 ms/word in Study 2 and 170 ms/word in Study 3. Thus, in both studies, instant feedback was only issued if the respondent clearly fell below 350 ms/word which is typically assumed for serious reading (Conrad et al. 2011). In both studies, the nondifferentiation index was again calculated according to McCarty and Shrum (2000), with a nondifferentiation

Table 13.4 Description of Experiments 2.1/2.2 in Study 2 and Experiments 3.1/3.2 in Study 3.

	Experiments 2.1/3.1 **Speeding prompt**	**Experiments 2.2/3.2** **Nondifferentiation prompt**
Feedback instructions	"Please take some more time for your answers."	"Please try to differentiate more among your answers."
Experimental questions in Study 2		
Short grid (8 items)	Future goals	Basic skills
Long grid (14 items)	Study choices	Job prospects
Experimental questions in Study 3		
Long grid (12 items)	Positive achievement motivation	Positive achievement motivation
Long grid (12 items)	Negative achievement motivation	Negative achievement motivation
Behavior triggering feedback		
Instant feedback	Time span ($\text{click}_i - \text{click}_{i-1}$): <2000 ms	McCarty and Shrum 2000: nondiff (item_1 to item_i) < 0.50
Dependent variable		
Primary effect	Response time	Nondifferentiation index
Secondary effect	Nondifferentiation index	Response time

prompt occurring instantly after each click on a radio button if the index value based on the responses provided up to this point fell below the threshold of 0.50.

Again, the effectiveness of instant feedback concerning speeding or nondifferentiation was assessed by means of response time and the extent of nondifferentiation (primary and secondary effects), with calculations of both indicators being based on the complete grid question. To ensure comparability across respondents, only those cases were included in the analyses where the respondent answered all items (8 items in the short grid, and 14 or 12 items, respectively, in the long grid) in the respective grid question. Also, respondents with unusually long response times were excluded (respondents who interrupted and resumed the survey on this particular question and also all respondents who exceeded a response time of two standard deviations above the group mean).

Comparisons of response time and nondifferentiation index across experimental conditions employed t-tests for independent samples which provide information on whether the experimental group is significantly different from the control group.

Results regarding the use of instant feedback on speeding and nondifferentiation are summarized in Tables 13.5 and 13.6. When instant feedback on speeding was provided (EG) in Experiment 2.1, we found a primary effect in terms of significantly longer response times as well as a secondary effect in terms of a significantly lower degree of nondifferentiation, relative to when no feedback was used (CG). These findings are consistent with the results of Experiment 1.1; however, they were not confirmed in Experiment 3.1, because no significant differences were found between the experimental group and the control group.

Concerning the use of instant feedback on nondifferentiation, significant primary and secondary effects could be found in Experiments 2.2 and 3.2, which support the results of Experiment 1.2. As compared to the control group where no feedback was provided, the use of instant feedback (EG) significantly decreases the degree of nondifferentiation which is also accompanied by significantly longer response times.

Table 13.5 Effectiveness of feedback instruction in response to "Speeding" in Studies 2 and 3.

	Response time (mean s)				Nondifferentiation index (mean, McCarty and Shrum 2000)			
	Experiment 2.1		Experiment 3.1		Experiment 2.1		Experiment 3.1	
Experimental question	Short	Long	Long	Long	Short	Long	Long	Long
Experimental condition								
CG, no feedback	48.9	85.1	150.7	159.0	0.513	0.600	0.613	0.634
EG, instant feedback	56.3	95.7	151.4	154.3	0.531	0.607	0.615	0.632
t-test	7.94***	7.84***	0.20ns	1.48ns	3.76***	2.37*	0.50ns	0.545ns
df	4079	4539	3535	3439	4506	4657	3535	3498
n	4542	4661	3537	3500	4542	4661	3537	3500

Notes: *t*-tests for independent samples (considering Levene's test for variance homogeneity): ***$p < 0.001$, **$p < 0.01$, *$p < 0.05$, ns = nonsignificant. Cases with longer interruptions on the respective target page comprising the experimental question (>7200 ms) and with mean response times exceeding two standard deviations above the group mean were excluded.

Table 13.6 Effectiveness of feedback instruction in response to "Nondifferentiation" in Studies 2 and 3.

	Nondifferentiation index (mean, McCarty and Shrum 2000)				Response time (mean s)			
	Experiment 2.2		Experiment 3.2		Experiment 2.2		Experiment 3.2	
Experimental question	Short	Long	Long	Long	Short	Long	Long	Long
Experimental condition								
CG, no feedback	0.551	0.619	0.613	0.634	45.1	73.3	150.7	159.0
EG, instant feedback	0.596	0.646	0.643	0.653	53.4	83.8	158.1	166.6
t-test	14.05***	11.30***	8.49***	5.78***	9.59***	8.91***	2.10*	2.28*
df	6387	6179	3417	3451	6313	6192	3604	3579
n	6576	6352	3606	3581	6576	6352	3606	3581

Notes: *t*-tests for independent samples (considering Levene's test for variance homogeneity): ***$p < 0.001$, **$p < 0.01$, *$p < 0.05$, ns = nonsignificant. Cases with longer interruptions on the respective target page comprising the experimental question (>7200 ms) and with mean response times exceeding two standard deviations above the group mean were excluded.

Taken together, the substantive findings of Studies 1, 2, and 3 suggest that interactive feedback to respondents is an effective means of increasing data quality in web surveys. Both delayed feedback and instant feedback in response to speeding and nondifferentiation typically have significant primary effects in terms of reduced incidence of speeding and lower extent of non-differentiation, respectively. However, our findings also suggest that instant feedback is somewhat more effective than delayed feedback. We speculate that the higher effectiveness is mostly caused by the fact that the instant feedback intervention in the question–answer process happens at an earlier point in time while respondents are still working on the grid affected by speeding or nondifferentiation and thus, are more willing to modify their response behavior. By contrast, delayed feedback asks respondents to return to a grid question that they have already submitted in order to revise their responses. This is presumably more burdensome and thus compliance with delayed feedback is less pronounced. These findings are consistent with the notion that feedback is most effective when it is provided in close temporal proximity with the respondent's relevant action (Conrad et al. 2005, 2006). In addition, our findings support the notion that feedback is most effective when the respondent's effort that is necessary to comply with the feedback is kept to a minimum.

13.3 Methodological Issues in Experimental Visual Design Studies

In this section, we will discuss methodological implications of the combination of randomized between-subjects field experiments and sample-based survey methods that is being used in the field of visual design research in general and also in the present studies. Given space limitations, we restrict the data reported to support and demonstrate our reasoning to the results of Study 3. Of the three studies conducted, Study 3 is the most recent study that provides many paradata (Kreuter 2013), which we did not collect in the early stages of our research concerning visual design and interactive feedback. This study also provides a rich frame that facilitates a detailed nonresponse bias assessment (Olson 2006).

13.3.1 Between-Subjects Design

In order to make causal inferences about the impact of visual design features on the question–answer process, researchers typically use randomized between-subjects designs (Ziniel 2008) in order to overcome limitations of statistical control of covariates. When survey respondents have been randomly assigned to various experimental conditions that differ only with respect to a particular visual design feature, differences in the response distributions can be used as indicators of the effects of visual design features on the question–answer process. In our case, the experimental conditions resembled the control condition in every respect except for the presence of interactive feedback that appeared if the respondent showed signs of satisficing behavior.

This method has a long tradition in the field of survey methodology in general and in the realm of questionnaire design in particular. Since the early days of this field of research, researchers have aimed at assessing the impact of various question wordings and other features using the split-sample method (Schuman and Presser 1981; Schwarz and Sudman 1995). The split-sample design has proven quite powerful in assessing causal effects of single questionnaire characteristics on the answers provided by respondents. Based on the theoretical framework that was developed as a result of the two Cognitive Aspects of Survey Methodology (CASM) conferences (Jabine et al. 1984; Sirken et al. 1999), an integrated theoretical approach evolved that helped to explain the formation of survey responses and served as a basis for a wide range of hypotheses concerning the impact of various question characteristics on survey responses. More recent years have seen the emergence of factorial survey experiments using vignette techniques (Atzmüller and Steiner 2010; Sauer et al. 2011), which aim to assess the interplay of several causal factors at the same time. However, questionnaire design research has so far made only limited use of this methodology (Auspurg et al. 2015), and the majority of studies still rely on split-sample experiments.

Between-subjects designs are preferred over within-subjects designs because of learning effects and other confounding factors arising from respondents answering the same questions multiple times within the same questionnaire. Because cognitive theory suggests that asking the same questions multiple times typically leads to differential question understanding (Schwarz 1994), it is not safe to assume that the respondents develop the very same question understanding when exposed to the same question wording for a second or third time. Even though visual design features might differ across repetitions of the same question, respondents might not report the very same answer because of the fact that they have already answered this question before. Even when separating iterations of the same question wording by buffer questions, respondents might remember that they were already exposed to the same question before and consequently change their answer not only because of the differences in the visual design, but also because of the fact that they have developed a differential question understanding as a result of the repetition of the question.

13.3.2 Field Experiments vs. Lab Experiments

Over the course of the past 15 years, several studies have focused on visual design features and their impact on the question–answer process in web surveys (Couper 2008; Tourangeau et al. 2013). Generally, it is assumed that visual design has the potential to influence the question–answer process in addition to textual information provided to respondents either as part of the question stem, the response options, or in terms of additional clarification features. However, the effect sizes of visual design factors are typically smaller in comparison to effect sizes of textual information (Smyth et al. 2009; Toepoel and Couper 2011). Accordingly, methodological experimentation relies on large sample sizes in order to generate enough statistical power necessary to demonstrate the impact of visual design features. Hence, field experiments embedded in large ongoing surveys are typically preferred over lab experiments. Indeed, field-experimental studies suffer from limited internal validity because the respondents' exposure to the experimental stimulus while answering survey questions cannot be fully controlled (Krantz 2001; Reips 2002). In our case, respondents had control over the device that they used to answer the survey, and device type can interact with the design features under study. In Study 3, about a quarter of the respondents used a mobile device to answer the questions. Even though many of the main effects showed up in the desktop/notebook group as well as in the group using a mobile device, the proportion of respondents exposed to speeding prompts was much larger in the mobile group (see Table 13.8). In web survey field experiments, the researcher also cedes control over exactly when respondents participate and whether they interrupt answering the questions. Because we were unable to rule out interactions of interruptions and the effects of instant prompts we had to exclude a small fraction of the respondents due to the fact that they terminated and later resumed the survey on the web page with the experimental questions (see notes to Tables 13.2–13.6).

By contrast, in a laboratory setting the choice of the device and aspects such as deviations from standard response behavior, or other conditions distracting respondents that might affect response behavior (e.g. Ansolabehere and Schaffner (2015) noted behaviors such as watching TV, speaking with another person, or answering a phone call) could have been standardized more easily. However, field-experimental studies are nevertheless generally preferred over lab-experimental studies because of the larger number of subjects per experimental condition and greater generalizability of results to typical survey production settings. By comparison, the external validity of lab-experimental studies might be limited because participating respondents are aware of the fact that they are being observed and supervised while answering a survey questionnaire.

13.3.3 Replication of Experiments

Replication of the experiments within the same study is typically advisable because concerns have been raised that the results regarding a particular experimental variation of a question may in fact be tied to the specifics of a particular question and not only to the experimental factor tested in the experiment (Billiet and Matsuo 2012; Holland and Christian 2009; Sniderman and

Grob 1996). Accordingly, we conducted similar experiments using different question topics at different positions throughout the questionnaire. In our case, at least two experiments employing the same experimental stimulus have been embedded in each study. We did not further increase the number of experiments per study, because when introducing new features – in this case, interactive feedback on speeding and nondifferentiation – the extent of respondent burden and potential subsequent survey breakoff cannot be foreseen.

Because most visual design experiments are "piggybacked" on surveys that are conducted primarily for substantive purposes, researchers must, therefore, be mindful that the methodological experimentation does not undermine the substantive estimates. This limits the number of questions that can be used for experimentation in a single survey. Accordingly, replications of the same experimental conditions across multiple studies are necessary. In this chapter, we report results from three field-experimental surveys conducted over a period of two years with two different populations (university freshmen and university applicants). Even though consistency of results is not a proof of the validity of the results, inconsistent results at least raise questions and concerns. In our case, the fact that the instant feedback on nondifferentiation seemed to yield reliable effects while instant feedback on speeding showed inconsistent effects in Study 3 disputes our initial conclusions. In particular, we are no longer convinced that urging respondents to spend more time on a question alone is particularly fruitful in restraining respondents from survey satisficing.

13.3.4 Randomization in the Net Sample

Random assignment of subjects to experimental conditions in the net sample (comprising all cases that started the survey and arrived at the survey page immediately prior to the respective page with the experimental question) is preferable over randomization in the gross sample (all cases invited to the survey) prior to fielding the survey because randomization that is undertaken at the latest possible point avoids the negative impact of accidental but differential unit nonresponse (Rahman and Dewar 2006) on the composition of the experimental group and the control group. Unit nonresponse may be caused by noncontact or refusal but may also be due to technical difficulties that respondents or their computers may have with the programming used to design the experiments. Interactive feedback, for example, is implemented using JavaScript programming. Although nowadays most computers are equipped with JavaScript (Kaczmirek and Thiele 2006), some respondents may have disabled it. Randomization of respondents in the net sample offers the opportunity to actually test the technical requirements for the experiment. By contrast, a random assignment that is undertaken based on the gross sample entries cannot guarantee that a technical difficulty applies to the same proportion of respondents assigned to the control group and the experimental groups. This is because experimental groups are often more demanding in their technical requirements due to the visual design features being tested as compared to control groups.

To conduct the field experiments reported in this chapter, specific online survey software (QuestBack) was employed. The randomization mechanism embedded in this particular software has generally proven reliable. In Study 3 conducted among university applicants in 2014, roughly one third of cases were assigned to either the control group or to one of the two experimental groups addressing either speeding or nondifferentiation (i.e. Experiments 3.1 and 3.2 were embedded in the same two grid questions, they made use of the same random variable for randomization, and they shared the same control group, see also Table 13.4). Nevertheless, the groups of subjects assigned to either the control group or the experimental groups differed with respect to the exact case count. In Study 3, random error in the allocation of cases to the experimental conditions was small but visible (see Table 13.7). In Experiment 3.1, the

Table 13.7 Distribution of background variables (%) in the gross sample, in the net sample, in the subsamples exposed to the experiments and across subsamples in the experimental conditions in Study 3.

			Experiment 3.1			Experiment 3.2		
	Gross sample (1)	Net sample (2)	Total (3)	CG (4)	EG (5)	Total (6)	CG (7)	EG (8)
Case count	17357	5996	3869	1938	1931	3947	1938	2009
Gender								
Male	57	54	53	53	53	54	53	56
Female	43	46	48	48	48	46	48	44
	100	100^{***}	100^{ns}	100^{ns}	100^{ns}	100^{ns}	100^{*}	100^{*}
Region								
Environs	56	59	59	59	58	60	59	61
Germany	42	39	40	39	41	39	39	38
Other	1	1	1	1	1	1	1	1
	100	100^{***}	100^{ns}	100^{ns}	100^{ns}	100^{ns}	100^{ns}	100^{ns}
Subject								
Natural science	16	18	18	17	19	18	17	18
Social sciences/ humanities	51	48	49	50	47	49	50	48
Engineering	31	32	31	31	32	32	31	32
Interdisciplinary	2	2	2	2	2	2	2	2
	100	100^{***}	100^{ns}	100^{ns}	100^{ns}	100^{ns}	100^{ns}	100^{ns}
Change of subject								
Yes	9	6	6	6	6	6	6	6
No	91	94	94	94	94	94	94	94
	100	100^{***}	100^{ns}	100^{ns}	100^{ns}	100^{ns}	100^{ns}	100^{ns}
Freshmen								
Yes	98	99	99	99	99	99	99	99
No	2	1	1	1	1	1	1	1
	100	100^{***}	100^{ns}	100^{ns}	100^{ns}	100^{ns}	100^{ns}	100^{ns}

Notes: The net sample in column (2) represents cases that remained in the sample after exclusion of initial unit nonresponse and incomplete cases (according to the 50% AAPOR (2016) rule). Case count for Experiments 3.1 and 3.2 (Totals in columns (3) and (6)) represents the number of cases exposed to the experiment. The number of cases used for the analysis (see Tables 13.2, 13.3, 13.5, and 13.6) was smaller because outliers and cases with longer periods of inactivity as well as cases with missing values for one or more items in the respective experimental questions were excluded. CG = control group, EG = experimental group; Experiments 3.1 and 3.2 were embedded in the same two grid questions, made use of the same random variable for randomization, and shared the same control group (columns (4) and (7) are identical). Chi-square goodness-of-fit tests were used for the following comparisons: (2) vs. (1), (3) vs. (2), (6) vs. (2). Pearson's chi-square tests for independent samples were used for the following comparisons: (4) vs. (5), (7) vs. (8). $^{***}p < 0.001$, $^{**}p < 0.01$, $^{*}p < 0.05$, ns = nonsignificant.

difference in the case count was only seven across experimental conditions; in Experiment 3.2, misallocation of cases to experimental conditions amounted to 71 cases. This misallocation was due to the fact that the integrated random trigger does not ensure but merely strives for a uniform distribution. Different software products available on the market might differ in the extent of this phenomenon and future improvement in software development may eliminate this problem altogether. In the present study, the uneven distribution of cases across experimental conditions is harmless because group sizes do not differ enough to meaningfully affect the standard errors.

13.3.5 Independent Randomization

Generally, it is advisable to replicate visual design experiments within the same survey using different question topics (see discussion above). In the present case, respondents might in fact be exposed to the same interactive feedback multiple times. In order to avoid an impact of previous experiments in the same survey on later experiments, independent random assignment of subjects to experimental conditions is advisable for each experiment. Instead of randomly assigning subjects to the experimental conditions only once in a survey for all subsequent experiments, randomization is repeated for each experiment immediately prior to the respective experiment. Independent randomization is used in order to minimize the impact of carry-over effects of respondents being exposed to the same experimental variation multiple times. Of course, independent randomization does not avoid a situation where subjects are exposed, for example, to the same visual design stimulus multiple times. In fact, when conducting two experiments with one experimental group and one control group each, a quarter of the respondents are exposed to the experimental group in both experiments. But at the same time those respondents who were exposed to the experimental group in the previous experiment are randomly assigned to the control group and the experimental group in the second experiment. Thus, the effect of prior exposure to a visual design feature is at least randomly spread across all experimental conditions for the second visual design experiment.

To date, little is known regarding the respondents' expectations after they are exposed to an experimental stimulus. For example, it may well be that respondents who were exposed to the instant feedback condition in the first experiment came to expect subsequent grid questions to employ similar features. This may ultimately lead to more inattentiveness and satisficing by respondents, because they assume that they will be reminded to optimize their response behavior when their satisficing exceeds a certain limit. By contrast, one could also assume that respondents from the experimental group in an initial experiment may have changed their behavior for the better, which might then lead to a situation where subjects in the experimental group of a subsequent experiment exhibit less satisficing. In this case, reduced prevalence of satisficing behavior cannot be attributed to the experimental interventions in the subsequent experiment but rather to learning effects due to the initial experiment. In order to minimize potential distortion due to such problems, it is advisable to independently randomize prior to each experiment. This has been undertaken for all experiments that are reported in this chapter.

13.3.6 Effects of Randomization on the Composition of the Experimental Conditions

In order to test whether randomization actually succeeded, we compared the distributions of various background variables in the control group and the experimental groups. For the assessment of the (potentially) differential composition of the control group and the experimental groups, we used variables that were particularly relevant for the presumed effect of the experimental stimulus. General cognitive functioning has been identified as a key variable moderating the impact of visual design features in web surveys (Couper et al. 2004; Redline and Dillman 2002; Tourangeau et al. 2000). However, the surveys used for our experiments were all conducted among university students or university applicants which are highly educated samples providing little variance with respect to these characteristics. Thus, for an assessment of the effects of random assignment on the composition of the various experimental conditions, we assessed gender, region, subject of the study program, change of subject, and application for a higher semester of study as relevant background variables that were available either in the sampling frame or in the dataset. Chi-square goodness-of-fit tests were used for categorical variables in order to compare the distribution of background variables in the net sample (after the application of the 50% AAPOR (2016) rule for incomplete cases) with the subsamples exposed to Experiments 3.1 and 3.2. Pearson's chi-square tests for independent samples were used to compare the distribution of background variables in the experimental group and the control group in Experiments 3.1 and 3.2, respectively.

Table 13.7 shows the results of these comparisons for Study 3. The findings revealed that almost all comparisons yielded no significant differences. The only exception occurred in Experiment 3.2, where the proportion of male respondents was slightly but significantly higher ($p < 0.05$) in the experimental group (56%) than expected based on the joint distribution of CG and EG (54%; see column (6)). Analyses of Studies 1 and 2 yielded similar nonsignificant results.

Randomization seemed to have worked properly; the overwhelming majority of the comparisons of control group and experimental group did not yield any significant differences with respect to the background variables assessed. The only significant result involved a difference of merely two percentage points. Even though we have no definitive proof that randomization yielded an identical or at least similar distribution of cases in all experimental conditions with respect to all relevant background variables, we have only very limited reasons to doubt that randomization has secured internal validity. More generally, studies using experimental designs should perform these types of analyses to ensure that random assignment has been effective.

13.3.7 Differential Unit Nonresponse

The results displayed in Table 13.7 also revealed that initial unit nonresponse led to a significant disturbance of the net sample as compared to the gross sample, potentially hampering the external validity of the experiments. Chi-square goodness-of-fit tests were used in order to compare the distribution of background variables in the net sample (after the application of the 50% AAPOR (2016) rule for incomplete cases) with the gross sample. Comparing the first two columns of Table 13.7, various significant differences of the gross sample and the net sample of Study 3 can be observed. In particular, the proportion of male subjects was moderately but significantly lower in the net sample as compared to the gross sample. While 57% of intended respondents in the gross sample were male, the respective percentage in the net sample was only 54%. Also, the proportion of subjects living in the surroundings of the university location was three percentage points (pp) higher in the net sample (59%) as compared to the gross sample (56%). In addition, small shifts in the composition of the net sample with respect to the subject of the study program were evident: Natural sciences were slightly overrepresented (18% instead of 16%), while social science and humanities were underrepresented (48% instead of 52%). Furthermore, the percentage of participants who changed their subject (−3 pp) and the percentage of university freshmen (−1 pp) was also moderately affected by unit nonresponse.

While these differences were predominately due to initial unit nonresponse, survey breakoff prior to the experiments may have caused additional shifts in the composition of the subsamples exposed to the experiments. However, no further bias in the composition of the subsamples exposed to the various experiments in Study 3 was found (columns (3) and (6) did not differ from column (2) in Table 13.7). Still, it should be noted that incomplete cases according to the 50% AAPOR (2016) rule have been excluded from the sample prior to the analysis. Thus, potential effects of early breakoff are not entirely visible in the table.

13.3.8 Nonresponse Bias and Generalizability

Because certain subpopulations are slightly underrepresented in the net sample as compared to the gross sample, we have to reflect on the question of whether our findings can be generalized to the target population. Accordingly, we tested for potential associations of the background variables assessed for nonresponse bias and the dependent variables in the experiments. For example, gender and subject of the study program showed associations with the time used to complete the experimental grid questions. Female respondents as well as respondents from the social sciences and humanities seemed to answer faster (i.e. fewer seconds were needed to complete the grid question). Similar but less consistent effects were found for nondifferentiation. Even though analyses indicated that gender and subject of the study program had consistent effects on the time needed to complete the experimental questions and (less consistently) on the degree of nondifferentiation, interaction terms such as "gender × experimental condition" or "study program × experimental condition" were typically not significant, indicating that the control group and experimental group were affected by the main effects of gender and study program to about the same extent. Thus, the potential disturbing effect of differential nonresponse on the main effects of the visual design was likely to be limited and can be treated as noise. For the other variables that were prone to nonresponse bias, no associations with time to complete the experimental question and degree of nondifferentiation were found. Future research needs to always consider the potential impact of nonresponse bias on the external validity and generalizability of main effects demonstrated in field-experimental studies concerning visual design features in web surveys.

Even though the impact of nonresponse rates on nonresponse bias is not consistently high and severe (Groves and Peytcheva 2008), we have to accept that response rates in surveys have been steadily declining (Brick and Williams 2013; de Leeuw and de Heer 2002; Stoop 2005). Also, nonresponse may have further unobserved effects on the composition of the net sample which potentially limits generalizability. Thus, future field-experimental research should aim to make use of samples from rich frames providing many relevant background variables in order to assess potential nonresponse bias, and particularly interactions with the main effects of the experiments. In our case, the variables available on the sampling frame were only weakly linked to the question–answer process, and hence other unobserved variables might have caused a stronger nonresponse bias, potentially limiting generalizability.

13.3.9 Breakoff as an Indicator of Motivational Confounding

In addition to initial nonresponse, page-specific breakoff on a web page comprising the experimental question poses a serious threat to the internal validity of the results because of motivational confounding (Reips 2007). If one of the experimental conditions was particularly prone to breakoff on the experimental questions, causal inferences drawn from the comparisons of response distributions of the experimental conditions would be hampered. Accordingly, we examined potential negative side effects of instant feedback by means of analyzing the proportion of respondents who abandoned the survey on the experimental question in each experimental condition. Overall, the extent of breakoff on experimental questions was low (typically about 1%). Most importantly, no significant differences concerning page-specific breakoff rates were found between the EG and the CG in all experiments.

13.3.10 Manipulation Check

In field-experimental studies with dynamic responsive design elements, the exposure of subjects to the treatment in the experimental groups depends on their behavior in the survey instrument. In order to test the effectiveness of interactive feedback, we designed the experiments in a way that makes the occurrence of the interactive feedback rather likely. As Table 13.8 illustrates, 50% to 60% of respondents received instant feedback in Experiment 3.1 (at least one speeding prompt), whereas more than 80% of respondents received instant feedback in Experiment 3.2 (at least one nondifferentiation prompt). For the control groups, we calculated the proportion of respondents who would have received at least one speeding prompt in Experiment 3.1, or at least one nondifferentiation prompt in Experiment 3.2 had they instead been in the experimental group. Pearson's chi-square tests for independent samples were used to compare the proportion of respondents receiving at least one prompt in the experimental group to the simulated proportions in the control group in Experiments 3.1 and 3.2, respectively.

Results indicate that if the control group respondents had been eligible for the instant feedback, they would have received at least one speeding prompt or at least one nondifferentiation prompt at about the same frequency as respondents in the respective experimental group. The fact that the percentage of respondents who received or would have received instant feedback is about the same in the control group and the experimental group again suggests that randomization led to roughly the same composition of control groups and experimental groups with respect to respondents exhibiting satisficing behaviors (see also discussion above).

In order to test whether the occurrence of instant feedback in response to nondifferentiation is actually responsible for the decrease in the extent of nondifferentiation reported earlier in Experiment 3.2, we assessed the mean nondifferentiation index for respondents who received instant feedback on nondifferentiation as compared to respondents who did not receive instant feedback in the experimental group. We compared these two subgroups to their counterparts in the control groups (to those in the control group where we envisaged instant feedback based on the simulation and to those where simulation results did not indicate the occurrence of instant feedback). Means of response time and nondifferentiation index were compared using t-tests for independent samples.

Results in Table 13.8 indicate that the mean difference in the nondifferentiation index was only statistically significant for respondents who received at least one nondifferentiation prompt. This applied to both experimental questions in Experiment 3.2. By contrast, the respondents in the experimental groups who did not receive any nondifferentiation prompts (because of their already optimizing response behavior) showed no difference in the extent of nondifferentiation as compared to respondents in the control groups who would not have received instant feedback based on their response behavior. The fact that respondents who did not receive instant feedback on nondifferentiation as a result of their response behavior in the experimental groups did not differ from their counterparts in the control groups with respect to the extent of nondifferentiation suggests that no other factors might have induced the mean difference in the nondifferentiation index when comparing control groups and experimental groups.

When assessing the primary effects of instant feedback in response to speeding in Experiment 3.1, we found no significant difference in the mean response times for respondents who did and did not receive at least one speeding prompt. Thus, the overall result reported previously, whereby the speeding prompt had no significant impact on response time, holds for the 50–60% of respondents who were actually exposed to instant feedback on speeding. Accordingly, the overall nonsignificant difference of experimental group and control group cannot be attributed to the relatively low percentage of respondents in the experimental groups who were actually exposed to instant feedback on speeding.

Table 13.8 Manipulation check in Study 3.

	Experiment 3.1				Experiment 3.2			
	Speeding prompt				Nondifferentiation prompt			
Experimental question	**Long (pos)**		**Long (neg)**		**Long (pos)**		**Long (neg)**	
Device type (%)	CG	EG	CG	EG	CG	EG	CG	EG
Desktop	73^{ns}	75^{ns}	73^{ns}	75^{ns}	73^{ns}	75^{ns}	73^{ns}	75^{ns}
Mobile	27^{ns}	25^{ns}	27^{ns}	25^{ns}	27^{ns}	25^{ns}	27^{ns}	25^{ns}
1+ prompts (%)								
Desktop	58^{ns}	57^{ns}	46^{ns}	48^{ns}	81^{ns}	81^{ns}	79^{*}	76^{*}
Mobile	68^{ns}	64^{ns}	61^{ns}	60^{ns}	84^{ns}	84^{ns}	82^{ns}	80^{ns}
Total	$61^{ns/***}$	$59^{ns/*}$	$50^{ns/***}$	$51^{ns/***}$	$82^{ns/ns}$	$81^{ns/ns}$	$80^{*/ns}$	$77^{*/*}$
Nondifferentiation index (*mean*)								
Desktop, no prompt	0.626^{ns}	0.625^{ns}	0.641^{ns}	0.635^{ns}	0.681^{ns}	0.682^{ns}	0.693^{ns}	0.687^{ns}
Desktop, 1+ prompts	0.615^{ns}	0.616^{ns}	0.635^{ns}	0.635^{ns}	0.606^{***}	0.643^{***}	0.624^{***}	0.648^{***}
Desktop, total	$0.620^{ns/***}$	$0.620^{ns/**}$	$0.638^{ns/**}$	$0.635^{ns/ns}$	$0.620^{***/***}$	$0.651^{***/**}$	$0.638^{***/**}$	$0.657^{***/**}$
Mobile, no prompt	0.600^{ns}	0.600^{ns}	0.611^{ns}	0.630^{ns}	0.675^{ns}	0.670^{ns}	0.691^{ns}	0.690^{ns}
Mobile, 1+ prompts	0.594^{ns}	0.604^{ns}	0.628^{ns}	0.619^{ns}	0.581^{**}	0.609^{**}	0.606^{**}	0.629^{**}
Mobile, total	$0.596^{ns/***}$	$0.600^{ns/**}$	$0.621^{ns/**}$	$0.624^{ns/ns}$	$0.596^{**/***}$	$0.619^{**/**}$	$0.621^{**/**}$	$0.640^{**/**}$
Response time (*mean s*)								
Desktop, no prompt	158.3^{ns}	162.6^{ns}	156.7^{ns}	158.3^{ns}	145.9^{ns}	137.8^{ns}	155.1^{ns}	163.8^{ns}
Desktop, 1+ prompts	145.0^{ns}	143.7^{ns}	156.1^{ns}	145.8^{ns}	$151.7^{ns/ns}$	158.1^{ns}	156.8^{*}	166.1^{*}
Desktop, total	$150.7^{ns/ns}$	$151.0^{ns/ns}$	$156.4^{ns/ns}$	$152.3^{ns/ns}$	150.7^{ns}	$154.3^{ns/**}$	$156.4^{*/ns}$	$165.6^{*/ns}$
Mobile, no prompt	154.4^{ns}	157.5^{ns}	165.8^{ns}	161.9^{ns}	164.3^{ns}	172.8^{ns}	160.4^{ns}	165.6^{ns}
Mobile, 1+ prompts	149.3^{ns}	149.7^{ns}	166.2^{ns}	159.6^{ns}	148.3^{**}	169.6^{**}	167.3^{ns}	170.8^{ns}
Mobile, total	$150.9^{ns/ns}$	$152.5^{ns/ns}$	$166.0^{ns/ns}$	$160.5^{ns/ns}$	$150.9^{**/ns}$	$169.8^{**/**}$	$166.0^{ns/ns}$	$169.8^{ns/ns}$

Notes: The experimental questions were two long grid questions (12 items, respectively) addressing positive (pos) and negative (neg) achievement motivation. CG = control group, EG = experimental group; Experiments 3.1 and 3.2 were embedded in the same two questions, they made use of the same random variable for randomization and they shared the same control group. Cases with longer interruptions on the respective target page comprising the experimental question (>7200 ms) and with mean response times exceeding two standard deviations above the group mean were excluded. Pearson's chi-square tests for independent samples used for categorical variables (device type, 1+ prompt); *t*-tests for independent samples for continuous variables (nondifferentiation index, response time): $^{***}p < 0.001$, $^{**}p < 0.01$, $^{*}p < 0.05$, ns = nonsignificant. In cells with two indications concerning the results of statistical testing, the first indication (before the slash) refers to the comparison of EG vs. CG (same row); the second indication (after the slash) refers to the comparison of the two cells above this cell (same column).

One key problem for any manipulation check in field-experimental studies concerning visual design in general and interactive feedback in particular is the increasing proportion of respondents who choose to answer the web survey using a mobile device. Mobile devices differ with respect to multiple characteristics as compared to desktop (or notebook) computers (Fuchs 2008; Couper et al. 2017). When it comes to visual design features, the limited screen size of mobile devices is a key aspect. Researchers cannot make sure that respondents using mobile devices are actually able to see the manipulated aspect of the visual design, even though the online survey software may have exposed them to a particular visual design feature. In our case, we were not entirely sure whether respondents using a mobile device could actually see the instant feedback, because respondents may have chosen to zoom in on the web page displaying the experimental grid question. This may have led to a situation where the instant feedback was displayed on a segment of the web page that was actually not visible on the small screen of a mobile device.

We have also taken into account that response behavior differs in many other respects when using a mobile device. For example, mobile respondents might be in a public or crowded place and, accordingly, they might be less focused on the survey. When looking at the results concerning nondifferentiation and response times in grid questions, we often find more nondifferentiation among mobile respondents, even though mobile respondents typically need more time to complete grid questions as compared to desktop respondents (Couper 2016; Guidry 2012; McClain et al. 2012; Revilla et al. 2017; Stern et al. 2016). Similarly, when looking at the results of Experiments 3.1 and 3.2, we find more nondifferentiation in the experimental groups as well as the control groups for mobile respondents as compared to desktop respondents as indicated by significantly lower values on the nondifferentiation index among mobile respondents. We also find longer mean response times in the experimental groups and in the control groups for mobile respondents as compared to desktop respondents (with one exception in Experiment 3.1; see Table 13.8).

Given that the choice of device depends on the respondent's decision and was not randomized, we are unable to determine whether these differences are due to the device or due to other specific characteristics of those respondents who have chosen to answer the survey using a mobile device (composition effects due to self-selection). This key problem cannot be solved in the field-experimental setting we have chosen for our experiments, because the invitation to respondents to our survey did not limit their choice of device. For future field-experimental studies, we might limit the type(s) of devices respondents can use to participate – a decision which would reduce external validity in order to increase internal validity. As previous research has shown, however, restricting devices is difficult to implement in a field-experimental setting because respondents may simply ignore special notification at the beginning of the questionnaire reminding them to complete the survey on a computer or notebook and no other device, or are no longer willing to take part in the survey if they are asked to use a device other than their preferred device (de Bruijne and Wijnant 2013, 2014; Mavletova and Couper 2013; Revilla et al. 2017; Wells et al. 2013). Alternatively, researchers could reduce self-selection bias due to device type selection by statistically controlling for the covariates (respondent characteristics) that predict the likelihood of using or not using a specific device to complete the survey.

When observing the effects of instant feedback for mobile respondents and desktop respondents, our results confirm the overall null effect according to which instant feedback on speeding has no effect in both subsamples exposed to this kind of interactive feedback. This is true for the primary effect of instant feedback in response to speeding on response time as well as for its secondary effect on nondifferentiation. For the instant feedback in response to nondifferentiation, we find significant positive effects on the extent of nondifferentiation for mobile as well as for desktop respondents. Although mobile respondents seem to be more prone to nondifferentiation compared to desktop respondents, the instant feedback has similar positive effects on their response behavior – albeit on a lower level. Because the proportion of respondents who have received at least one nondifferentiation prompt is similar for mobile and desktop respondents, it is likely that mobile respondents in fact saw the instant feedback even on the smaller screen and subsequently reacted to it. When accepting this assumption, the nonsignificant effect of the speeding prompt for mobile respondents hence cannot be attributed to the fact that they might not have seen the instant feedback. A reasonable explanation is still pending; one potential explanation implies that it is just more difficult to slow down respondents when they are using a mobile device.

It is now fairly common for web surveys to have 30–40% of respondents completing the surveys on mobile devices (smartphone or tablet) (Keeter and Weisel 2015; Revilla et al. 2016). This proportion may even be increased to 57%, if a mobile device completion option is explicitly offered in a survey (Toepoel and Lugtig 2014). Presumably, this proportion will increase even further; accordingly, we will have to take precautions to make sure that mobile respondents are actually exposed to the visual design manipulation that is tested in field experiments. Even though our results suggest that mobile respondents react to the visual design manipulation to about the same extent as compared to desktop respondents (which suggests that they have in fact seen the prompt), we should in the future test this assumption in greater detail using a randomized experiment and ruling out potential effects of self-selection concerning the device used to fill in the experimental questions.

13.3.11 A Direction for Future Research: Causal Mechanisms

One over-arching observation about this study is that satisficing and low respondent motivation appear to be a lesser problem in our experiments that were conducted among university students or university applicants than in surveys among the general population. In addition, based on previous studies on nonresponse follow-ups, refusal conversion, and data quality (Olson 2006), we have reasons to assume that nonrespondents are typically less interested in the survey and more prone to satisficing as compared to respondents, resulting in an overestimation of the degree of optimizing behavior based on the net sample. Accordingly, the effect sizes demonstrated in our studies might be underestimated because the proportion of satisficing respondents was presumably much smaller as compared to general population surveys.

Thus, the fact that we were able to demonstrate the effectiveness of instant feedback in this rather motivated group of respondents has two potentially contradictory implications: (i) In the samples used in our experiments, the average degree of nondifferentiation was around 0.60 on a scale from 0 to 1, which is a considerably high value (the maximum possible value in our case was about 0.80, given the number of items and the number of the response options used in the experimental grid questions). The fact that we could increase the value on the nondifferentiation index by means of instant feedback (with higher values indicating decreased nondifferentiation) even though it was already considerably high may lead to the conclusion that in general population samples with larger proportions of satisficing respondents, instant feedback should have an even stronger impact on response behavior. (ii) At the same time, however, we have to acknowledge that the relatively high level of motivation among our respondents may have also fueled the effectiveness of instant feedback, because instant feedback affords a certain level of respondent motivation to be effective. Accordingly, it is plausible to assume that in samples with a higher proportion of satisficing respondents who are presumably less motivated to modify their response behavior the effect of instant feedback might have been smaller.

More generally, it is not clear whether visual design features such as interactive feedback have a stronger impact on respondents who are satisficing or those who are optimizing. Satisficing respondents may be more susceptible to the visual appearance of the questions, and thus easier to influence in their response behavior because they do not activate many cognitive resources while working through the question–answer process. By contrast, optimizing respondents may be in a better situation to generate a self-determined response based on an extensive question–answer process and thus rely to a lesser extent on the visual display. However, they are presumably more motivated to pay attention to visual design features and also more motivated to react to instructions conveyed by interactive feedback. Further research is needed to understand whether visual design in general and interactive feedback in particular is more effective with satisficing respondents or with optimizing respondents. It may well be that both groups show similar effects in their response behavior, however, for different reasons. Thus, future field-experimental research needs to disentangle the underlying causal processes in order to better understand the impact of visual design on response behaviors in web surveys.

References

AAPOR (2016). *Standard Definitions. Final Dispositions of Case Codes and Outcome Rates for Surveys*. Oakbrook Terrace, IL: AAPOR.

Al Baghal, T. and Lynn, P. (2015). Using motivational statements in web-instrument design to reduce item-missing rates in a mixed-mode context. *Public Opinion Quarterly* 79 (2): 568–579.

Ansolabehere, S. and Schaffner, B.F. (2015). Distractions: the incidence and consequences of interruptions for survey respondents. *Journal of Survey and Methodology* 3: 216–239.

Atzmüller, C. and Steiner, P.M. (2010). Experimental vignette studies in survey research. *Methodology* 6 (3): 128–138.

Auspurg, K., Hinz, T., Liebig, S., and Sauer, C. (2015). The factorial survey as a method for measuring sensitive issues. In: *Improving Survey Methods. Lessons from Recent Research* (ed. U. Engel, B. Jann, P. Lynn, et al.), 137–149. New York: Routledge.

Billiet, J. and Matsuo, H. (2012). Non-response and measurement error. In: *Handbook of Survey Methodology for the Social Sciences* (ed. L. Gigeon), 149–178. New York: Springer.

Blasius, J. and Thiessen, V. (2012). *Assessing the Quality of Survey Data*. London: SAGE Publications Ltd.

Bradlow, E.T. and Fitzsimons, G.J. (2001). Subscale distance and item clustering effects in self-administered surveys: a new metric. *Journal of Marketing Research* 38 (2): 254–261.

Brick, J.M. and Williams, D. (2013). Explaining rising nonresponse rates in cross-sectional surveys. *The ANNALS of the American Academy of Political and Social Science* 645 (1): 36–59.

de Bruijne, M. and Wijnant, A. (2013). Comparing survey results obtained via mobile devices and computers: an experiment with a mobile web survey on a heterogeneous group of mobile devices versus a computer-assisted web survey. *Social Science Computer Review* 31 (4): 482–504.

de Bruijne, M. and Wijnant, A. (2014). Mobile response in web panels. *Social Science Computer Review* 32 (6): 728–742.

Callegaro, M., Shand-Lubbers, J., and Dennis, J.M. (2009). Presentation of a single item versus a grid: effects on the vitality and mental health scales of the SF-36v2 health survey. Paper presented at the 64th Annual Conference of the American Association of Public Opinion Research (AAPOR).

Chang, L. and Krosnick, J.A. (2010). Comparing oral interviewing with self-administered computerized questionnaires: an experiment. *Public Opinion Quarterly* 74 (1): 154–167.

Clifford, S. and Jerit, J. (2015). Do attempts to improve respondent attention increase social desirability bias? *Public Opinion Quarterly* 79 (3): 790–802.

Conrad, F.G., Couper, M.P., Tourangeau, R., and Galesic, M. (2005). Interactive feedback can improve the quality of responses in web surveys. Paper presented at the 60th Annual Conference of the American Association of Public Opinion Research (AAPOR).

Conrad, F.G., Couper, M.P., Tourangeau, R., and Peytchev, A. (2006). Use and non-use of clarification features in web surveys. *Journal of Official Statistics* 22 (2): 245–269.

Conrad, F.G., Schober, M.F., and Coiner, T. (2007). Bringing features of human dialogue to web surveys. *Applied Cognitive Psychology* 21 (2): 165–187.

Conrad, F.G., Tourangeau, R., Couper, M.P., and Zhang, C. (2011). Interactive interventions in web surveys can increase response accuracy. Paper presented at the 66th Annual Conference of the American Association of Public Opinion Research (AAPOR).

Couper, M.P. (2008). *Designing Effective Web Surveys*. New York: Cambridge University Press.

Couper, M.P. (2016). Grids versus item-by-item designs on smartphones. Paper presented at the 18th General Online Research Conference (GOR).

Couper, M.P., Traugott, M.W., and Lamias, M.J. (2001). Web survey design and administration. *Public Opinion Quarterly* 65 (2): 230–253.

Couper, M.P., Tourangeau, R., and Kenyon, K. (2004). Picture this! Exploring visual design effects in web surveys. *Public Opinion Quarterly* 68 (2): 255–266.

Couper, M.P., Tourangeau, R., Conrad, F.G., and Zhang, C. (2013). The design of grids in web surveys. *Social Science Computer Review* 31 (3): 322–345.

Couper, M.P., Antoun, C., and Mavletova, A. (2017). Mobile web surveys. In: *Total Survey Error in Practice* (ed. P.P. Biemer, E.D. de Leeuw, S. Eckman, et al.), 133–154. Hoboken, NJ: Wiley.

David, C. (2005). Mode effects. In: *Polling America: An Encyclopedia of Public Opinion*, vol. 2 (ed. S.J. Best and B. Radcliff), 453–457. Westport, CT: Greenwood Press.

DeRouvray, C. and Couper, M.P. (2002). Designing a strategy for reducing "no opinion" responses in web-based surveys. *Social Science Computer Review* 20 (1): 3–9.

Fricker, S.S., Galesic, M., Tourangeau, R., and Yan, T. (2005). An experimental comparison of web and telephone surveys. *Public Opinion Quarterly* 69 (3): 370–392.

Fuchs, M. (2008). Mobile web survey: a preliminary discussion of methodological implications. In: *Envisioning the Survey Interview of the Future* (ed. F.G. Conrad and M.F. Schober), 77–94. New York: Wiley.

Grandjean, B.D., Nelson, N.M., and Taylor, P.A. (2009). Comparing an internet panel survey to mail and phone surveys on willingness to pay for environmental quality: a national mode test. Paper presented at the 64th Annual Conference of the American Association of Public Opinion Research (AAPOR).

Grandmont, J., Graff, B., Goetzinger, L., and Dorbecker, K. (2010). Grappling with grids: how does question format affect data quality and respondent engagement? Annual meeting of the American Association for Public Opinion Research. Retrieved July 2016 from http://www.amstat.org/sections/SRMS/proceedings/y2010/Files/400116.pdf.

Greszki, R., Meyer, M., and Schoen, H. (2015). Exploring the effects of removing "too fast" responses and respondents from web surveys. *Public Opinion Quarterly* 79 (2): 471–503.

Groves, R.M. and Peytcheva, E. (2008). The impact of nonresponse rates on nonresponse bias. *Public Opinion Quarterly* 72 (2): 167–189.

Guidry, K.R. (2012). Response quality and demographic characteristics of respondents using a mobile device on a web-based survey. Paper presented at the 67th Annual Conference of the American Association for Public Opinion Reasearch (AAPOR).

Heerwegh, D. and Loosveldt, G. (2008). Face-to-face versus web surveying in a high-internet-coverage population. *Public Opinion Quarterly* 72 (5): 836–846.

Holland, J.L. and Christian, L.M. (2009). The influence of topic interest and interactive probing on responses to open-ended questions in web surveys. *Social Science Computer Review* 27 (2): 196–212.

Jabine, T.B., Straf, M.L., Tanur, J.M., and Tourangeau, R. (1984). *Cognitive Aspects of Survey Methodology: Building a Bridge Between Disciplines*. Washington, DC: National Academy Press.

Kaczmirek, L. and Thiele, O. (2006). Flash, javascript or PHP? Comparing the availability of technical equipment among university applicants. Paper presented at the 8th General Online Research Conference (GOR).

Keeter, S. and Weisel, R. (2015). *Building Pew Research Center's American Trends Panel*. Pew Research Center. Retrieved July 2016 from http://www.pewresearch.org/methods/2015/04/08/building-pew-research-centers-american-trends-panel/.

Krantz, J.H. (2001). Stimulus delivery on the web: what can be presumed when calibration isn't possible. In: *Dimensions of Internet Science* (ed. U.D. Reips and M. Bosnjak), 113–130. Lengerich, Berlin: Pabst Science Publishers.

Kreuter, F. (2013). *Improving Surveys with Paradata*. Hoboken, NJ: Wiley.

Krosnick, J.A. (1991). Response strategies for coping with the cognitive demands of attitude measures in surveys. *Applied Cognitive Psychology* 5 (3): 213–236.

Kunz, T. (2015). Rating scales in web surveys. A test of new drag-and-drop rating procedures. Doctoral dissertation. Darmstadt University of Technology. Retrieved July 2016 from http://tuprints.ulb.tu-darmstadt.de/id/eprint/5151.

de Leeuw, E.D. and de Heer, W.F. (2002). Trends in household survey nonresponse: a longitudinal and international comparison. In: *Survey Nonresponse* (ed. R.M. Groves, D.A. Dillman, J.L. Eltinge and R.J.A. Little), 41–54. New York: Wiley.

Lozar Manfreda, K., Bosnjak, M., Berzelak, J. et al. (2008). Web surveys versus other survey modes: a meta-analysis comparing response rates. *International Journal of Market Research* 50 (1): 79–104.

McCarty, J.A. and Shrum, L.J. (2000). The measurement of personal values in survey research. A test of alternative rating procedures. *Public Opinion Quarterly* 64 (3): 271–298.

McClain, C., Crawford, S.D., and Dugan, J.P. (2012). Use of mobile devices to access computer-optimized web surveys: implications for respondent behavior and data quality. Paper presented at the 67th Annual Conference of the American Association for Public Opinion Research (AAPOR).

Mavletova, A. and Couper, M.P. (2013). Sensitive topics in PC web and mobile web surveys. Paper presented at the 15th General Online Research Conference (GOR).

Olson, K.M. (2006). Survey participation, nonresponse bias, measurement error bias, and total bias. *Public Opinion Quarterly* 70 (5): 737–758.

Oudejans, M. and Christian, L.M. (2011). Using interactive features to motivate and probe responses to open-ended questions. In: *Social and Behavioral Research and the Internet* (ed. M. Das, P. Ester and L. Kaczmirek), 215–244. New York: Routledge.

Rahman, Z. and Dewar, A. (2006). The impact of mode on the comparability of survey data. In: *Survey Methodology Bulletin*, Special Edition No 58, 3–10. London: Office for National Statistics.

Redline, C.D. and Dillman, D.A. (2002). The influence of alternative visual designs on respondents' performance with branching instructions in self-administered surveys. In: *Survey Nonresponse* (ed. R.M. Groves, D.A. Dillman, J.L. Eltinge and R.J.A. Little), 179–193. New York: Wiley.

Reips, U.-D. (2002). Standards for internet-based experimenting. *Experimental Psychology* 49 (4): 243–256.

Reips, U.-D. (2007). The methodology of internet-based experiments. In: *The Oxford Handbook of Internet Psychology* (ed. A.N. Joinson, K.Y.A. McKenna, T. Postmes and U.-D. Reips), 374–390. Oxford: Oxford University Press.

Revilla, M., Toninelli, D., Ochoa, C., and Loewe, G. (2016). Do online access panels need to adapt surveys for mobile devices? *Internet Research* 26 (5): 1209–1227.

Revilla, M., Toninelli, D., and Ochoa, C. (2017). An experiment comparing grids and item-by-item formats in web surveys completed through PCs and smartphones. *Telematics and Informatics* 34 (1): 30–42.

Sauer, C., Auspurg, K., Hinz, T., and Liebig, S. (2011). The application of factorial surveys in general population samples: the effects of respondent age and education on response times and response consistency. *Survey Research Methods* 5 (3): 89–102.

Schonlau, M. and Toepoel, V. (2015). Straightlining in web survey panels over time. *Survey Research Methods* 9 (2): 125–137.

Schuman, H. and Presser, S. (1981). *Questions and Answers in Attitudes Surveys* (reprint 1996 by Sage ed.). San Diego, CA: Academic Press.

Schwarz, N. (1994). Judgment in a social context: biases, shortcomings, and the logic of conversation. *Advances in Experimental Social Psychology* 26: 123–162.

Schwarz, N. and Sudman, S. (eds.) (1995). *Context Effects in Social and Psychological Research*. New York: Springer.

Sirken, M., Jabine, T.B., Willis, G.B. et al. (1999). *A New Agenda for Interdisciplinary Survey Research Methods: Proceedings of the CASM II Seminar*. Hyattsville, MD: National Center for Health Statistics.

Smyth, J.D., Dillman, D.A., Christian, L.M., and McBride, M. (2009). Open-ended questions in web surveys. Can increasing the size of answer boxes and providing extra verbal instructions improve response quality? *Public Opinion Quarterly* 73 (2): 325–337.

Sniderman, P.M. and Grob, D.B. (1996). Innovations in experimental design in attitude surveys. *Annual Review of Sociology* 22 (1): 377–399.

Stern, M., Sterrett, D., and Bilgen, I. (2016). The effects of grids on web surveys completed with mobile devices. *Social Currents* 3 (3): 217–233.

Stoop, I. (2005). *The Hunt for the Last Respondent*. The Hague: Aksant Academic Publisher.

Sudman, S., Bradburn, N.M., and Schwarz, N. (1996). *Thinking About Answers: The Application of Cognitive Processes to Survey Methodology*, 1ee. San Francisco, CA: Josey-Bass Publishes.

Toepoel, V. and Couper, M.P. (2011). Can verbal instructions counteract visual context effects in web surveys? *Public Opinion Quarterly* 75 (1): 1–18.

Toepoel, V. and Lugtig, P. (2014). What happens if you offer a mobile option to your web panel? Evidence from a probability-based panel of internet users. *Social Science Computer Review* 32 (4): 544–560.

Toepoel, V., Das, M., and van Soest, A. (2009). Design of web questionnaires: the effects of the number of items per screen. *Field Methods* 21 (2): 200–213.

Tourangeau, R. (1984). Cognitive science and survey methods. In: *Cognitive Aspects of Survey Design: Building a Bridge Between Disciplines* (ed. T.B. Jabine, M.L. Straf, J.M. Tanur and R. Tourangeau), 73–100. Washington, DC: National Academy Press.

Tourangeau, R. and Rasinski, K.A. (1988). Cognitive processes underlying context effects in attitude measurement. *Psychological Bulletin* 103 (3): 299–314.

Tourangeau, R., Rips, L., and Rasinski, K. (2000). *The Psychology of Survey Response*. Cambridge: Cambridge University Press.

Tourangeau, R., Couper, M.P., and Conrad, F.G. (2004). Spacing, position, and order. Interpretative heuristics for visual features of survey questions. *Public Opinion Quarterly* 68 (3): 368–393.

Tourangeau, R., Conrad, F.G., and Couper, M.P. (2013). *The Science of Web Surveys*. Oxford: Oxford University Press.

Wells, T., Bailey, J.T., and Link, M.W. (2013). Filling the void: gaining a better understanding of tablet-based surveys. *Survey Practice* 6 (1): 1–9.

Zhang, C. (2012). Designing interactive interventions in web surveys: Humanness, social presence and data quality. Paper presented at the 67th Annual Conference of the American Association for Public Opinion Research (AAPOR).

Zhang, C. (2013a). *Satisficing in Web Surveys: Implications for Data Quality and Strategies for Reduction*. Ann Arbor, MI: University of Michigan.

Zhang, C. (2013b). Speeding and non-differentiation in web surveys: evidence of correlation and strategies for reduction. Paper presented at the 68th Annual Conference of the American Association for Public Opinion Research (AAPOR).

Zhang, C. and Conrad, F. (2013). Speeding in web surveys: the tendency to answer very fast and its association with straightlining. *Survey Research Methods* 8 (2): 127–135.

Ziniel, S. (2008). Split-half. In: *Encyclopedia of Survey Research Methods* (ed. P.J. Lavrakas), 834–835. Thousand Oaks, CA: Sage.

14

Randomized Experiments for Web-Mail Surveys Conducted Using Address-Based Samples of the General Population

Z. Tuba Suzer-Gurtekin[1], Mahmoud Elkasabi[2], James M. Lepkowski[1], Mingnan Liu[3], and Richard Curtin[1]

[1] *Institute for Social Research, University of Michigan, Ann Arbor, MI 48106, USA*
[2] *ICF, The Demographic and Health Surveys Program, Rockville, MD 20850, USA*
[3] *Facebook, Inc. Menlo Park, CA, USA*

14.1 Introduction

Address-based sampling (ABS) with the US Postal Service Delivery Sequence File has the potential to provide virtually complete coverage of the US household population (Iannacchione 2011; Iannacchione et al. 2003; Link et al. 2008). It can be used with a variety of modes of data collection, including face-to-face and mail modes. ABS mail surveys of the household population have been shown to be feasible, provide nearly complete coverage, and have good nonresponse properties (Elkasabi et al. 2014; Link et al. 2006, 2008; Link and Lai 2011; Peytchev et al. 2010).

14.1.1 Potential Advantages of Web-Mail Designs

Empirical findings suggest that ABS mail-based designs may be a suitable alternative to random digit dialing (RDD) telephone designs for some studies. For example, Link et al. (2008) found that due to differences in labor costs, the cost of an RDD landline telephone survey was 12% greater than ABS mail per 1000 completed interviews. Montaquila et al. (2013) showed how a two-phase ABS mail survey yielded better coverage and response rates than an RDD landline telephone survey at the same cost per interview. Thus, relative to RDD approaches, ABS mail surveys can have comparable response rates and some reduction in the cost per completed interview.

Researchers have extended ABS mail surveys to provide the option of responding by the web (Dillman et al. 2009; Messer and Dillman 2011; Messer et al. 2012; Smyth et al. 2010). These web-mail strategies have advantages over mail only, and other modes, in terms of more rapid response (timeliness), measurement flexibility (allowing different ways of presenting questions), lower respondent burden (greater convenience about when and where to respond), and interactive question–answer processes during data collection (Couper 2008, 2011; Messer and Dillman 2011; Smyth et al. 2010; Smyth and Pearson 2011). A web-mail approach also overcomes a severe limitation of web surveys: the lack of good online sampling frames for general household populations (Couper 2000; Fricker and Schonlau 2002; Groves et al. 2009).

Experimental Methods in Survey Research: Techniques that Combine Random Sampling with Random Assignment, First Edition.
Edited by Paul J. Lavrakas, Michael W. Traugott, Courtney Kennedy, Allyson L. Holbrook, Edith D. de Leeuw, and Brady T. West.

Companion Website: www.wiley.com/go/Lavrakas/survey-research

Adding a web completion option to a mail survey may also reduce costs, such as eliminating survey data entry of returned mail questionnaires (Werner 2005; Werner and Forsman 2005). Of course, web data collection does entail additional programming and Internet hosting costs, but overall survey costs compared to mail or other modes are generally reduced.

Additionally, there is a common perception that many find web response more convenient and appealing. Internet accessibility is nearly universal in many populations, and web response may be preferred by subgroups who use email and messaging services routinely (Mohorko et al. 2013). For example, in the United States the percentage of Internet users reached 74.6% in 2015, following a steady increase since 1998 (NTIA 2016). While overall Internet penetration is fairly high in the United States, there are notable gaps in access and use across social and economic subgroups. Those gaps have been shrinking but are still substantial in many instances (Morris 2015; Pew Research Center 2017b). For example, in the United States, adults who are younger, wealthier, more formally educated, Asian or white tend to use the Internet more than older, lower income, less educated, black or Hispanic adults (NTIA 2016). Such disparities illustrate why it can be important for ABS of general populations to still offer mail response in addition to web.

The Internet appears to offer a particularly effective means for reaching younger adults, as access and use rates have consistently been negatively correlated with age (Pew Research Center 2017a). Underrepresentation of younger age groups has been a chronic concern in general population surveys that use telephone, face-to-face, and mail modes (Abraham et al. 2006; Beebe et al. 2012; Schneider et al. 2012). It may be that because younger age groups have higher Internet access and use, are more familiar with the mode, and find the web requires lower cognitive demands to respond, they would prefer the web (Smyth et al. 2014). Therefore, offering a web option could reasonably be expected to improve survey response among young adults.

Consistent with this hypothesis, Smyth et al. (2014) found that younger respondents prefer a web response option over a mail one when administering an ABS web-mail survey in cities in the states of Idaho and Washington. Similarly, in a series of statewide experiments in Washington, Messer and Dillman (2011) found that younger respondents were more likely to respond by web than mail. These patterns in choosing a web option, however, do not guarantee a reduction in differential nonresponse in the aggregated survey results. Messer and Dillman (2011) found that surveys recruiting respondents through the mail and offering a web response option produced samples demographically similar to those of mail-only surveys. That is, while younger respondents chose more often to respond by web than older respondents, younger age groups were still underrepresented in survey results aggregated across chosen modes.

Other research has shown beneficial effects on overall response rates and sample composition when a web response option was offered. The American Community Survey (ACS), a mandatory general population survey of US households, achieved a higher overall response rate and a higher percentage of younger and more educated respondents in a web-mail survey than in a survey offering only a mail response mode (Tancreto 2012).

14.1.2 Design Considerations with Web-Mail Surveys

An important question in research on mixing mail and web modes in ABS sample surveys is how to sequence the mode options in mail recruitment invitations. Is it better to offer mail and web options concurrently throughout, or could higher web completion be achieved if the web were offered before mail? Three general approaches have been considered:

(1) The ***mail-intensive design*** offers a mail response by including a questionnaire and postage paid return mailing envelope in mailings before offering an option to respond via a link to a website in a later mailing.

(2) The ***concurrent web-mail design*** offers a choice between mail completion using an enclosed questionnaire and postage paid return mailing envelope and a web link to an online questionnaire at the first mailing.
(3) The ***web-intensive design*** offers only a web link to an online questionnaire in the first mailings before providing an enclosed questionnaire and postage paid return mailing envelope in addition to the web link in later mailings. This is sometimes called a "push contact strategy."

Previous published studies have compared the characteristics of respondents obtained using these strategies to one another and to the general population from which the sample was selected. Response rates and the web completion rates (the proportion of completed interviews coming from the web) have also been examined (Holmberg et al. 2010; Messer and Dillman 2011; Millar and Dillman 2011; Tancreto 2012). Comparing respondent sample compositions with general population distributions provides an indication of whether differential nonresponse across key demographic subgroups is improved, especially for frequently underrepresented groups. These investigations have often had multiple goals: to increase the overall response rate, to reduce differential nonresponse across key subgroups, and to increase the percentage of web completion. These goals can be achieved if the web option leads to an individual not willing to respond by mail responding by web. Whether this actually occurs is a question explored in this study.

14.1.3 Empirical Findings from General Population Web-Mail Studies

Although the literature has consistently shown that the web-intensive design yields higher percentages of completes by web than the concurrent design, effects on response rates and responding sample composition have not been consistent across studies (Holmberg et al. 2010; Messer and Dillman 2011; Tancreto 2012). Although it is difficult to completely disentangle the effects of mode sequence from the effects of other design features, there are findings indicating that the sequence of response options is related to the responding sample characteristics (Church 1993; Groves and Peytcheva 2008; Heberlein and Baumgartner 1978; Wagner et al. 2014). For example, Holmberg et al. (2010) reported higher response rates for a concurrent web-mail design compared to a web-intensive design, but found no differences in response rates across 10 subgroups defined by gender, age, marital status, and income. The ACS Internet study (Tancreto 2012) did find differences in participation rates between households which were not frequent Internet users and those which were. A concurrent web-mail design yielded higher response rates for the less frequent user group, while a web-intensive design yielded higher response rates for the frequent user group. That is, there was evidence of an interaction between the sequence of mode options and the frequency of Internet use.

In the search for a web-mail design to increase response rates, previous studies found a prepaid cash incentive increases overall response rates in web-mail surveys (Messer and Dillman 2011; Millar and Dillman 2011). Other survey features, such as an advance letter and a replacement questionnaire mailed after nonresponse have increased response rates in mail surveys significantly (Dillman et al. 2009). Although an advance letter and a replacement questionnaire have been adapted for web-mail designs, the size of the effect on web-mail response rates has not been empirically investigated.

Previous web-mail survey research has often used statewide or special population samples (Holmberg et al. 2010; Messer and Dillman 2011; Millar and Dillman 2011; Smyth et al. 2010; Werner and Forsman 2005), or national experiments in mandatory government surveys (Griffin et al. 2001; Schneider et al. 2005; Tancreto 2012). How well those studies generalize to national nongovernment and noncompulsory surveys is unclear.

This chapter reports on a series of experiments designed to examine the effects of these design features in a voluntary national population survey conducted in the University of Michigan's Surveys of Consumers (SOC). The purpose of these experiments was to determine whether the SOC could move from an RDD telephone survey design to an ABS web-mail alternative. In concurrent and web-intensive designs, incentive and mailing protocol alternatives were randomly assigned to sample households. The overall response rate, the share of interviews completed by web, the demographic composition of the responding sample, and the substantive survey estimates were compared across the design features.

In addition, an important design requirement for any SOC data collection alternative is the rate of return of completed surveys over the one-month data collection period. Consumer attitudes change over a one-month period, sometimes in response to major events. Mid-month estimates require a minimum number of interviews after two weeks of data collection. Thus, an even accumulation of interviews over the month assures better measurement of population attitudes month by month. The telephone data collection is managed to provide an even flow of interviews throughout the survey period. An SOC web-mail design must also provide a comparable flow to capture temporal change in attitudes, meet mid-month estimation goals, and decrease the risk of breaking the time series. Thus, this chapter also examines the web-mail rate of completed interviews received over the data collection period.

14.2 Study Design and Methods

The SOC is a monthly, national, two-wave rotating panel survey conducted by telephone by the Survey Research Center at the University of Michigan. The survey estimates change in consumer attitudes and expectations. Since 2015 the SOC sample has been selected from a cell telephone sampling frame. At the time of the ABS web-mail experimental studies reported here (2011), the SOC sample was drawn from a landline telephone frame.

Each month, the SOC conducts approximately 600 telephone interviews with adults living in households in the coterminous United States (48 states plus the District of Columbia). An independent cross-sectional sample of cell telephone numbers is drawn each month. In addition, the respondents interviewed six months previously are also interviewed each month. Approximately one-third of each monthly completed interview sample are these re-interviews.

14.2.1 Experimental Design

In 2011, a series of ABS web-mail experimental studies were conducted simultaneously with the telephone data collection. An ABS sample of residential addresses was selected from the Marketing Systems Group (MSG) US Postal Service Computerized Delivery Sequence File (Iannacchione 2011). Concurrent and web-intensive survey designs were implemented following the Tailored Design Method (Dillman et al. 2009) to collect data using the SOC's monthly questionnaire. Three mailing protocols were used:

(1) **Protocol** I included five mailings: an advance letter, a questionnaire mailing, a reminder postcard, a second questionnaire mailing, and a final reminder postcard. In the concurrent web-mail design, a cover letter in the first questionnaire mailing included a survey Uniform Resource Locator (URL) and a login ID for a web survey alternative. In the web-intensive design, a separate web survey invitation letter followed the advance letter with a survey URL and login ID. The web survey letter also mentioned a paper questionnaire would be sent in a forthcoming mailing. Mailings were seven days apart to provide a flow of completed

Table 14.1 Experimental protocols in the SOC web-mail study.

Sequence	Protocol	Pre-paid $5 cash incentive?	Address sample size
Concurrent	(I) Advance letter, 2 paper q'naires + URL, 2 postcards	Yes	746
Concurrent	(II) Advance letter, 1 paper q'naire + URL, 1 postcard	Yes	750
Concurrent	(III) 2 paper q'naires + URL, 1 postcard	Yes	750
Concurrent	(I) Advance letter, 2 paper q'naires + URL, 2 postcards	No	746
Web-intensive	(I) Advance letter, Web survey invitation letter, 2 paper q'naires + URL, 1 postcard	Yes	745
Web-intensive	(II) Advance letter, Web survey invitation letter, 1 paper q'naire + URL	Yes	750
Web-intensive	(III) Web survey invitation letter, 1 paper q'naire + URL, 1 postcard	Yes	750
Web-intensive	(I) Advance letter, Web survey invitation letter, 2 paper q'naires + URL, 1 postcard	No	746

questionnaires somewhat similar to the flow observed in the monthly RDD landline telephone survey.

(2) **Protocol II** used three mailings. For the concurrent design, the protocol included the first three components of Protocol I (an advance letter, a questionnaire mailing, and a reminder postcard). For the web-intensive design, Protocol II included an advance letter, a web survey invitation letter with URL and login ID, and a questionnaire mailing.

(3) **Protocol III** used a three-mailing design, eliminating the advance letter in both the concurrent and web-intensive designs.

In addition, a prepaid incentive experiment was conducted for Protocol I. For concurrent and web-intensive designs under Protocol I, equal-sized address samples received or did not receive a $5 prepaid incentive in the questionnaire mailing.

Table 14.1 shows eight experimental arms by the type of web-mail design (concurrent or web-intensive), the protocol (I, II, or III), and the use of incentives ($5 prepaid cash or none). In each month, sample addresses were assigned at random to concurrent or web-intensive designs and to protocol-incentive groups within each web-mail design.

This experimental design allows for the investigation of effects for sequence (web-mail designs), protocols, and, for Protocol I, the prepaid incentive. In addition to the main effects for sequence, protocol, and incentive, interactions between sequence and protocol and between sequence and incentive within mailing Protocol I could be examined. For the analysis reported here, the effects examined primarily include sequence for specific protocols as well as incentive within Protocol I.

14.2.2 Questionnaire Design

The ABS web-mail self-administered questionnaires included a subset of SOC questions about contact information, sociodemographic characteristics, and consumer attitudes and expectations. The questions used are presented in the supplementary online website that Wiley has created for this book (www.wiley.com/go/Lavrakas/survey-research).

The paper questionnaire was a 20-page booklet in large type. The paper questionnaire was adapted to the web with four potentially important differences that could affect comparisons. First, multiple questions appear on a page in the paper questionnaire; each question was on a separate screen in the web version. Second, there were no explicit "Don't know" and "Not applicable" options in the paper questionnaire; the web version made the options visible in an automatic prompt presented when a respondent left a question blank. Third, skip routing was by arrows and written instruction in the paper questionnaire, but programmed in the web version. The web version also had data validation questions programmed into the web questionnaire presentation that could not be asked conveniently in the paper questionnaire. Fourth, open-ended numerical responses were used in income and stock ownership questions in the web versions with categorical response options presented to web nonrespondents for these items. The paper questionnaire presented only the categorical response options. Since the questions in the web and the paper versions did not differ in terms of the question wording and formatting, possible differences due to measurement between the paper and the web versions were expected to be small (Couper 2011; De Leeuw 2008; Schwarz et al. 1991; Tourangeau et al. 2000).

14.2.3 Weights

The SOC landline telephone survey system selected a single person aged 18 years or older at random for the telephone household interview. Two units of analysis are then identified: (1) adults 18 years of age and older and (2) householders 18 years of age and older. Both "adults" and "householders" receive weights that first adjust for unequal probabilities of selection of telephone households by accounting for multiple telephone landlines within the household. The "adult" sample requires further adjustment to compensate for unequal probabilities of selection due to within household selection. Finally, raking ratio adjustment weights for the "adult" and the "householder" samples were computed using Current Population Survey (CPS) benchmark data from the prior year (for example, 2010 CPS for 2011 SOC data) on age, sex, income, region, and home ownership for persons 18 years of age or older. The unequal probability adjusted raked weights are trimmed and bracketed (five weight values) to reduce weight variation.

The web-mail paper and web questionnaires included instructions to the household to select only the householder (a person 18 years of age or older in whose name the selected housing unit is owned, being bought, or rented) or his/her partner to complete the questionnaire. The web-mail "householder" sample can thus only be compared to the "householder" sample from the telephone survey. The web-mail "householder" sample does not adjust for unequal probabilities of selection due to multiple telephone lines in the household or selection of only one "adult" from among all eligible adults. The web-mail "householder" weights do employ raking ratio adjustments using CPS benchmark data on age, sex, income, region, and home ownership for "householders." As for the telephone sample weights, web-mail raked weights are trimmed and bracketed (five weight values) to reduce weight variation.

14.2.4 Statistical Methods

The analysis uses two-sample "Z-tests" to compare unweighted response rates, unweighted percent completion by web (PCW), weighted demographic characteristics, and weighted substantive estimates. Comparisons are across alternative web-mail designs, protocols, and, among Protocol I samples, prepaid incentive groups. Unweighted demographic characteristics are also compared between web-mail designs and against 2011 CPS benchmarks. The CPS benchmark computations include only the householder and partners and persons 18 years of age or older. The CPS benchmarks were compared to 95% confidence intervals for survey statistics to determine whether the CPS benchmarks were contained in the interval. The comparisons that did not include CPS benchmarks within 95% confidence intervals were flagged as significant. For the comparisons of the response distributions on demographic items used as raking ratio benchmark data, unweighted results are compared. For weighted estimates, standard errors were computed using a Taylor series approximation in SAS PROC SURVEYFREQ to account for the contribution of weighting to the variation in survey estimates. Standard errors for unweighted estimates are computed using SAS nonsurvey two-sample comparison procedures.

Initial analysis indicated that the demographic and substantive distributions differed across the web-mail designs but not across the mailing protocols. The analysis presented here compares data combined from Protocols I, II, and III for each web-mail design to assess differences between the concurrent and web-intensive designs with greater precision. The prepaid incentive condition effect is examined only within Protocol I.

14.3 Results

Table 14.2 shows that overall response rates for the Protocols I, II, and III neither differ within web-mail designs nor across web-mail designs. Despite evidence of potential interaction effects between mailing protocol and design, interactions are not statistically significant. Thus, there is no web-mail effect within a mailing protocol.

However, there are markedly higher PCWs for the web-intensive design. Table 14.3 shows response rates and PCWs across web-mail designs within Protocol I only, and separately for prepaid and no-incentive samples. The response rates for the concurrent web-mail surveys were indistinguishable when a five-mailing protocol (Protocol I) with prepaid incentive was used, and

Table 14.2 Response rates (AAPOR RR2) and unweighted percent completion by web (PCW) for web-mail designs and protocols.

Web-mail design	Response rates (%)			PCW (%)		
	Protocol I	Protocol II	Protocol III	Protocol I	Protocol II	Protocol III
Concurrent	33.6	31.8	27.9	21.8	17.0	15.2
Web-intensive	31.5	27.1	22.8	52.0	52.7	44.7
Concurrent vs. web-intensive	2.1	4.7	5.1*	−30.2***	−35.7***	−29.5***

p Values for the two sample Z-test: $^{*}p < 0.05$, $^{**}p < 0.01$, $^{***}p < 0.001$

Table 14.3 Response rates (AAPOR RR2) and unweighted percent completion by web (PCW) for Protocol I by web-mail design and incentive condition.

Experimental condition	Response rates (%)			PCW (%)		
	Concurrent (1)	Web-intensive (2)	Concurrent vs. web-intensive [(1)–(2)]	Concurrent (3)	Web-intensive (4)	Concurrent vs. web-intensive [(3)–(4)]
Protocol I without incentive	16.9	17.7	−0.9	21.8	31.1	−9.3
Protocol I with prepaid incentive	33.6	31.5	2.0	21.8	52.0	−30.3***

p Values for the two sample Z-test: $^{*}p < 0.05$, $^{**}p < 0.01$, $^{***}p < 0.001$

when a five-mailing protocol without a prepaid incentive was used. There are clearly important effects of prepaid incentives for both concurrent and web-intensive designs.

In addition, the web-intensive design produced substantially higher PCWs than the concurrent design. This occurred whether a prepaid incentive was provided or not, and furthermore the difference between the concurrent and the web-intensive designs was significant in the prepaid incentive condition.

Figure 14.1a shows the overall cumulative number of resolved cases for the two web-mail designs throughout the survey period in November 2011, an experimental month where mailing

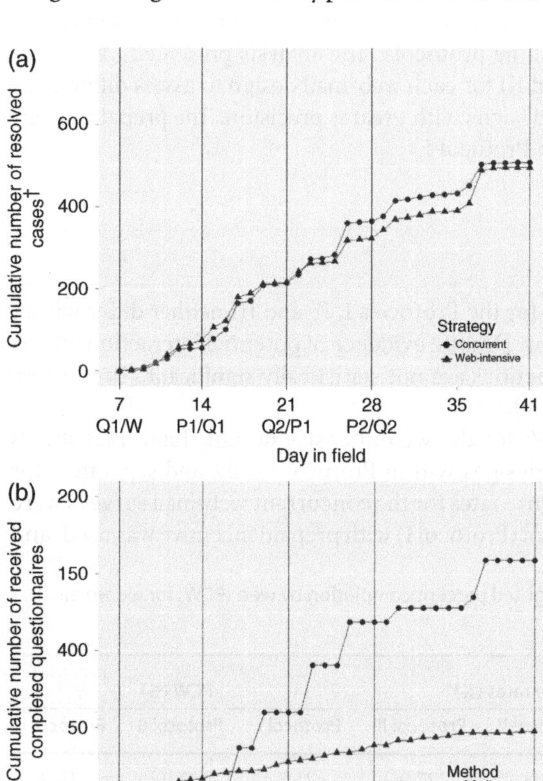

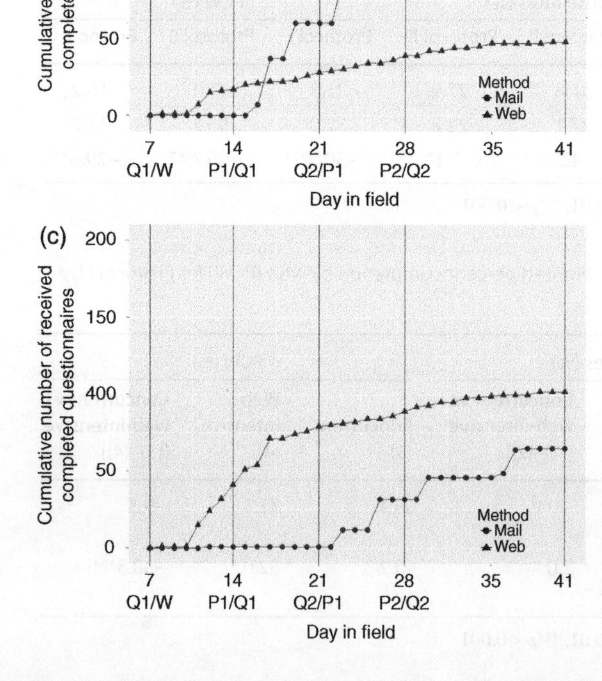

Figure 14.1 Cumulative number of resolved cases† and number of received completed questionnaires by web-mail strategy and day (Protocol I, with incentive condition). (a) Overall. (b) Concurrent. (c) Web-intensive. †Received completed questionnaires, refusals, ineligible households, and noncontacts.

Protocol I was coupled with prepaid incentive. Resolved cases include completed questionnaires, refusals, ineligible households, and noncontacts. The horizontal lines indicate the mailing dates for first questionnaire with web link or first mailing with web link invitation (Q1/W), the postcard reminder following the first questionnaire mailing or first questionnaire following the first web link invitation (P1/Q1), the second questionnaire mailing or the reminder postcard following the first questionnaire (Q2/P1), and the second reminder postcard mailing following the second questionnaire mailing or the second questionnaire mailing (P2/Q2). The 41st day in the field marks the end of data collection for the web-mail surveys.

Figure 14.1a shows that the concurrent and web-intensive designs yielded a similar pattern of completion times over the study period. The return rate increases fairly steadily throughout the field period, a reassuring result comparing favorably with the flow of completions in the landline telephone design.

Figures 14.1b and 14.1c show the number of received completed questionnaires across the survey field period by concurrent and web-intensive design, respectively, for mail and web returns separately. The patterns for the mail and web returns are the same in both web-mail designs. There is a time lag in the mail return rate due to the delay in providing the questionnaire in the web-intensive strategy. The number of web completes follows a steep incline after the first invitation and does not seem to be affected by the follow-up mailing. On the other hand, the number of mail completions continues to increase gradually after each follow-up mailing. As seen above, the number of web completions is markedly higher for the web-intensive strategy. Neither web-mail design showed a timeliness advantage over the other because of relatively low PCWs.

Table 14.4 compares web-mail design demographic characteristics (collapsing across mail protocols) to the March 2011 CPS. The "prepaid incentive" comparisons are based on collapsing across Protocols I, II, and III, while the "no incentive" comparisons are only possible for Protocol I.

The SOC's raking ratio weighting procedure forces weighted demographic distributions to be nearly identical[1] for age, sex, income, and region. Hence, it is more informative to examine unweighted web-mail design distributions to CPS to understand how the ABS web-mail design obtains response from various groups, especially those that cooperate poorly in the landline telephone survey.

In the incentive condition, significant differences are found in the same direction for both web-mail designs: higher percentages of males, college graduates, whites, married adults, and one-adult households compared to the CPS. That is, ABS web-mail is recruiting a sample that has higher proportions of these groups than would be expected in the population. At the same time, web-mail designs recruited lower percentage of Blacks, Hispanics, and no-child households. A contributing factor to the Hispanic differences was that only an English-language questionnaire was available in the ABS web-mail surveys. Both the concurrent and web-intensive designs underrepresent 18–34-year olds and over-represent persons 65 years or older. That is, these unweighted comparisons show an underrepresentation in ABS web-mail samples of younger, Black, never married, and no-child household respondents.

The results for the no incentive condition are similar in character, with an underrepresentation as well of younger, Black, Hispanic, and never married persons. In general, though, there appears to be an overall beneficial effect from the incentive on the sample composition. Using the incentive tended to reduce the extent to which more educated adults and non-Hispanic whites are overrepresented.

Table 14.5 presents weighted estimates of the percentage of economic attitudes for the prepaid incentive condition, collapsing across mailing protocols. The concurrent and

1 The raking ratio weighting benchmark is the 2011 CPS.

Table 14.4 Comparisons of demographic characteristics between SOC ABS web-mail survey respondents[a] and Current Population Survey (CPS)[b] estimates

Characteristic	CPS	Prepaid incentive condition				No incentive condition			
		Concurrent $n = 652$	Web-intensive $n = 564$	CPS vs. Concurrent	CPS vs. Web-intensive	Concurrent $n = 118$	Web-intensive $n = 119$	CPS vs. Con-current	CPS vs. Web-intensive
	(1)	(2)	(3)	(1) vs. (2)	(1) vs. (3)	(4)	(5)	(1) vs. (4)	(1) vs. (5)
Male	45.0	56.1	55.9	*	*	59.3	61.3	*	*
Age: 18–34 years	22.2	12.9	12.2	*	*	11.0	10.1	*	*
Age: 65+ years	21.3	26.5	30.1	*	*	22.0	37.0		*
Income: $0–$19 999	17.6	15.0	15.3			12.7	13.5		
Income: $100 000+	22.8	19.9	22.9			18.6	23.5		
College or more	39.6	45.6	51.4	*	*	55.9	55.5	*	*
White except Hispanic	71.9	79.0	80.5	*	*	82.2	84.0	*	*
Black except Hispanic	11.2	7.2	7.3	*	*	5.1	3.4	*	*
Hispanic	12.0	6.8	5.5	*	*	7.6	5.9		*
Homeowner	66.9	75.3	76.1	*	*	79.7	83.2	*	*
Married	51.5	64.4	65.3	*	*	67.0	70.6	*	*
Never married	20.8	12.0	12.6	*	*	11.9	8.4	*	*
No-child household	82.0	73.9	76.2	*	*	78.0	84.9		
One-adult household	27.5	34.7	34.9	*	*	36.8	24.6		*

*$p < 0.05$

a) Data across protocol variations are combined for the prepaid incentive condition.

b) Current Population Survey, March 2011, adult household heads and partners.

Table 14.5 Economic attitudes (%) and standard errors for the ABS web-mail designs, combined data (Protocols I, II, and III), prepaid incentive condition.

Economic attitude	Concurrent	Web-intensive
	$n = 652$	$n = 564$
Economy is better than a year ago	21.7 (1.7)	24.9 (2.0)
Economy will be better next year	26.3 (1.9)	25.4 (2.0)
Government is doing a good job	9.3 (1.2)	10.6 (1.4)
Unemployment will be more next year	24.4 (1.8)	20.5 (1.9)
Interest rates will go up in the next 12 months	38.2 (2.1)	34.5 (2.2)
Now is a good time to buy major household items	55.0 (2.2)	60.5 (2.4)
Now is a good time to buy a house	75.0 (1.9)	79.5 (1.9)
Now is a good time to sell a house	6.9 (1.2)	5.5 (1.1)
Now is a good time to buy a vehicle	58.2 (2.2)	59.7 (2.4)

Table 14.6 Economic attitudes (%) and standard errors for SOC ABS web-mail designs, no incentive condition under mailing Protocol I.

Economic attitudes	Concurrent	Web-intensive
	$n = 119$	$n = 122$
Economy is better than a year ago	11.0 (3.3)	16.4 (5.6)
Economy will be better next year	10.6 (3.0)	20.3 (5.9)
Government is doing a good job	2.5 (1.5)	1.1 (1.1)
Unemployment will be more next year	23.8 (4.6)	32.6 (5.3)
Interest rates will go up in the next 12 months	36.3 (5.4)	32.4 (6.2)
Now is a good time to buy major household items	51.8 (5.7)	56.6 (6.3)
Now is a good time to buy a house	72.6 (5.3)	70.2 (6.2)
Now is a good time to sell a house	5.7 (3.2)	5.5 (2.3)
Now is a good time to buy a vehicle	50.0 (5.7)	51.9 (6.5)

web-intensive strategies yielded similar results across all the economic attitudes: half of the estimates are larger for the concurrent strategy. Similarly, the weighted attitudinal estimates did not differ between concurrent and web-intensive designs in the no incentive condition either (Table 14.6).

14.4 Discussion

The findings in this chapter expand the web-mail survey literature with new results from randomized mode experiments for a voluntary general population national survey by mail and web. ABS concurrent and web-intensive designs show comparable response rates, demographic composition, and substantive findings.

With no differences in the sample compositions between the concurrent and web-intensive designs, the web-intensive design appears only to shift mail respondents to the web without attracting other groups of people to survey participation. Since the differences in the overall response rates are not significant, there is no evidence that supports an expectation that the concurrent design outperforms the web-intensive one in terms of response rates. Importantly, this study does not include comparisons to a mail-only design, where empirical evidence suggests that overall response rates would be higher than web-mail designs (Holmberg et al. 2010; Messer and Dillman 2011; Millar and Dillman 2011; Smyth et al. 2010; Werner and Forsman 2005).

Increasing the proportion of web completes could have reduced cost and increased the speed of data collection. But the PCWs (unweighted percent completions by web) in this study were not high enough for these size samples, even in the web-intensive design, to realize substantial cost and timeliness advantages through web data collection.

Previous studies have sought to encourage greater web completion by varying the sequence and the visibility of the web response option. These alternative design features have tended to produce the desired increases in the PCW (Holmberg et al. 2010; Messer and Dillman 2011; Tancreto 2012). The findings here indicate that providing the web option earlier in the process (e.g. in an advance letter) or delaying the availability of the mail questionnaire achieves a higher PCW. But the disadvantage is that the web-intensive design requires a longer survey field time to increase response rates due to more mailings.

Further increases in response rates for the web options may require understanding the mail-retrieval and mail-opening processes people use. Previous research among nonrespondents to web-mail and mail-only surveys showed that failure to respond is related to failure to open the mail envelope (Nichols 2012). As many as 40% of nonrespondents in the Nichols study reported that they did not receive a mailing, and as many as 20% reported that they had received the mailing but did not open it. Nichols (2012) also determined that a significant proportion of the mail respondents did not know about the web option, as did Holmberg, Lorenc, and Werner's follow-up (2010) where only half of the initial mail recipients reported noticing the web option. Thus, a large proportion of web-mail households may not be aware of the web option at all. Future research should thus also examine survey presentation design which might be manipulated to motivate respondents to see and switch to the web option.

While the expected cost advantages of a web survey are not fully realized here as a result of low PCWs, the web-mail strategy does rely on a paper questionnaire with inherent survey administration weaknesses relative to a telephone survey. Although cost considerations overall strongly favor web-mail over telephone data collection, there can be considerable difficulty converting items from telephone to self-administered data collection, as illustrated in Elkasabi et al. (2014).

We have also found several other limitations of web-mail surveys for the SOC. No satisfactory mechanism exists for selecting an eligible respondent at random in a household level self-administered survey. The literature on web-mail designs does not point to a clearly superior method (Battaglia et al. 2008; Link et al. 2008; Olson and Smyth 2014; Olson et al. 2014; Yan et al. 2015). Survey length may also be an issue. The study reported here only included core SOC items, amounting to less than one-half of the full set of items asked in the telephone survey.

Further investigation is needed to find strategies to increase web-mail response rates. The very low participation in the web-mail surveys among 18–34-year olds is troubling. Younger persons responded at higher rates to the web option, but this was only a switch to the web option without improving the overall response rate among this group. Greater use of technology such as mobile applications or using Quick Response (QR) codes in invitation letters in general population surveys may attract younger people to respond at higher rates. But there is at present little evidence to support such a conjecture. How technology could be used to this end is a clear area of potential research that could give web surveys distinct advantages over current survey technologies.

References

Abraham, K.G., Maitland, A., and Bianchi, S.M. (2006). Nonresponse in the American Time Use Survey who is missing from the data and how much does it matter? *Public Opinion Quarterly* 70 (5): 676–703.

Battaglia, M.P., Link, M.W., Frankel, M.R. et al. (2008). An evaluation of respondent selection methods for household mail surveys. *Public Opinion Quarterly* 72 (3): 459–469.

Beebe, T.J., McAlpine, D.D., Ziegenfuss, J.Y. et al. (2012). Deployment of a mixed-mode data collection strategy does not reduce nonresponse bias in a general population health survey. *Health Services Research* 47 (4): 1739–1754.

Church, A.H. (1993). Estimating the effect of incentives on mail survey response rates: a meta-analysis. *Public Opinion Quarterly* 57 (1): 62–79.

Couper, M.P. (2000). Review: Web surveys: a review of issues and approaches. *Public Opinion Quarterly* 64 (4): 464–494.

Couper, M.P. (2008). Web surveys in a mixed-mode environment. In: *Designing Effective Web Surveys*, 353–356. New York, NY: Cambridge University Press.

Couper, M.P. (2011). The future of modes of data collection. *Public Opinion Quarterly* 75 (5): 889–908.

De Leeuw, E.D. (2008). Choosing the method of data collection. In: *International Handbook of Survey Methodology* (ed. E.D. De Leeuw, J.J. Hox and D.A. Dillman), 113–135. Hove, East Sussex, UK: European Association of Methodology.

Dillman, D.A., Smyth, J.D., and Christian, L.M. (2009). *Internet, Mail, and Mixed-Mode Surveys: The Tailored Design Method*, 3e. Hoboken, NJ: Wiley Publishing.

Elkasabi, M., Suzer-Gurtekin, Z.T., Lepkowski, J.M. et al. (2014). A comparison of ABS mail and RDD surveys for measuring consumer attitudes. *The International Journal of Market Research* 56 (6): 737–756.

Fricker, R.D. and Schonlau, M. (2002). Advantages and disadvantages of internet research surveys: evidence from the literature. *Field Methods* 14: 347–367.

Griffin, D.H., Fischer, D.P., and Morgan, M.T. (2001). Testing an Internet Response Option for the American Community Survey. Presented at the Annual Conference of the American Association for Public Opinion Research. http://148.129.75.8/acs/www/Downloads/ACS/Paper29.pdf.

Groves, R.M. and Peytcheva, E. (2008). The impact of nonresponse rates on nonresponse bias: a meta-analysis. *Public Opinion Quarterly* 72 (2): 167–189.

Groves, R.M., Floyd, J.F., Couper, M.P. et al. (2009). *Survey Methodology*, 2e. Hoboken, NJ: Wiley Series in Survey Methodology.

Heberlein, T.A. and Baumgartner, R. (1978). Factors affecting response rates to mailed questionnaires: a quantitative analysis of the published literature. *American Sociological Review* 43 (4): 447–462.

Holmberg, A., Lorenc, B., and Werner, P. (2010). Contact strategies to improve participation via the web in a mixed-mode mail and web survey. *Journal of Official Statistics* 26 (3): 465–480.

Iannacchione, V. (2011). The changing role of address-based sampling in survey research. *Public Opinion Quarterly* 75 (3): 556–575.

Iannacchione, V., Staab, J., and Redden, D. (2003). Evaluating the use of residential mailing lists in a metropolitan household survey. *Public Opinion Quarterly* 62 (2): 202–210.

Link, M.W. and Lai, J.W. (2011). Cell-phone-only households and problems of differential nonresponse using an address-based sampling design. *Public Opinion Quarterly* 75 (4): 613–635.

Link, M.W., Battaglia, M.P., Frankel, M.R. et al. (2006). Address-based versus random-digit-dial surveys: comparison of key health and risk indicators. *American Journal of Epidemiology* 164 (10): 1019–1025.

Link, M.W., Battaglia, M.P., Frankel, M.R. et al. (2008). A comparison of address-based sampling (ABS) versus random-digit dialing (RDD) for general population surveys. *Public Opinion Quarterly* 72 (1): 6–27.

Messer, B.L. and Dillman, D.A. (2011). Surveying the general public over the Internet using address-based sampling and mail contact procedures. *Public Opinion Quarterly* 75 (3): 429–457.

Messer, B.L., Edwards, M., and Dillman, D. (2012). Determinants of item nonresponse to web and mail respondents in three address-based mixed-mode surveys of the general public. *Survey Practice* 5 (2).

Millar, M.M. and Dillman, D.A. (2011). Improving response to web and mixed-mode surveys. *Public Opinion Quarterly* 75 (2): 249–269.

Mohorko, A., De Leeuw, E.D., and Hox, J.J. (2013). Internet coverage and coverage bias in Europe: developments across countries and over time. *Journal of Official Statistics* 29 (4): 609–622.

Montaquila, J.M., Michael Brick, J., Williams, D. et al. (2013). A study of two-phase mail survey data collection methods. *Journal of Survey Statistics and Methodology* 1 (1): 66–87.

Morris, J.B. Jr, (2015). First Look: Internet Use in 2015. Web log post. National Telecommunications and Information Administration, United States Department of Commerce, 21 March 2016. Web 15 November 2016. https://www.ntia.doc.gov/blog/2016/first-look-internet-use-2015 (accessed 15 November 2016)

Nichols, E. (2012). The April 2011 American Community Survey Internet Test: Attitudes and Behavior Study Follow up, 2012 American Community Survey Research and Evaluation Report Memorandum Series ACS12-RER-16. http://www.census.gov/acs/www/Downloads/library/2012/2012_Nichols_01.pdf (accessed 7 October 2012)

NTIA (2016). Digital Nation Data Explorer. Digital Nation Data Explorer|NTIA. National Telecommunications and Information Administration, United States Department of Commerce, 27 October 2016. Web 15 November 2016. https://www.ntia.doc.gov/data/digital-nation-data-explorer (accessed 15 November 2016)

Olson, K. and Smyth, J.D. (2014). Accuracy of within-household selection in web and mail surveys of the general population. *Field Methods* 26 (1): 56–69.

Olson, K., Stange, M., and Smyth, J. (2014). Assessing within-household selection methods in household mail surveys. *Public Opinion Quarterly* 78 (3): 656–678.

Pew Research Center (2017a). *Internet/Broadband Fact Sheet*. Washington, D.C.: Pew Research Center http://www.pewinternet.org/fact-sheet/internet-broadband/ (accessed 5 January 2017).

Pew Research Center (2017b). *Tech Adoption Climbs Among Older Adults*. Washington, D.C.: Pew Research Center http://www.pewinternet.org/2017/05/17/tech-adoption-climbs-among-older-adults/ (accessed 17 May 2017).

Peytchev, A., Ridenhour, J., and Krotki, K. (2010). Differences between RDD telephone and ABS mail survey design: coverage, unit nonresponse, and measurement error. *Journal of Health Communication: International Perspectives* 15: 117–134.

Schneider, S.J., Cantor, D., Malakhoff, L. et al. (2005). Telephone, Internet and paper data collection modes for the Census 2000 short form. *Journal of Official Statistics* 21 (1): 89–101.

Schneider, K.L., Clark, M.A., Rakowski, W., and Lapane, K.L. (2012). Evaluating the impact of non-response bias in the behavioral risk factor surveillance system (BRFSS). *Journal of Epidemiology and Community Health* 66 (4): 290–295.

Schwarz, N., Strack, F., Hippler, H.-J., and Bishop, G. (1991). The impact of administration mode on response effects in survey measurement. *Applied Cognitive Psychology* 5 (3): 193–212.

Smyth, J.D. and Pearson, J. (2011). Internet survey methods; a review of strengths, weaknesses, and innovations. In: *Social and Behavioral Research and the Internet; Advances in Applied Methods and Research Strategies (European Association of Methodology Series)* (ed. M. Das, P. Ester and L. Kaczmirek), 11–44. New York, NY: Routledge Taylor & Francis Group.

Smyth, J.D., Dillman, D.A., Christian, L.M., and O'Neill, A.C. (2010). Using the Internet to survey small towns and communities: limitations and possibilities in the early 21st century. *American Behavioral Scientist* 53 (9): 1423–1448.

Smyth, J.D., Olson, K., and Millar, M.M. (2014). Identifying predictors of survey mode preference. *Social Science Research* 48: 135–144.

Tancreto, J. (2012). Revised-2011 American Community Survey Internet Tests: Results from First Test in April 2011. http://www.census.gov/acs/www/Downloads/library/2012/2012_Tancreto_01.pdf (accessed 21 June 2012).

Tourangeau, R., Rips, L.J., and Rasinski, K. (2000). Mode of data collection. In: *The Psychology of Survey Response*, 289–312. Cambridge, England: Cambridge University Press.

Wagner, J., Arrieta, J., Guyer, H., and Ofstedal, M.B. (2014). Does sequence matter in multi-mode surveys: results from an experiment. *Field Methods* 26 (2): 141–155.

Werner, P. (2005). On the cost-efficiency of probability sampling based mail surveys with a web response option. Dissertation Thesis. Linköping University.

Werner, P. and Forsman, G. (2005). Mixed mode data collection using paper and web questionnaires. In: *AAPOR-ASA Section on Survey Research Methods*, 4015–4017.

Yan, T., Tourangeau, R., and McAloon, R. (2015). A Meta-Analysis of Within-Household Respondent Selection Methods. 2015 Federal Committee on Statistical Methodology (FCSM) Statistical Policy Seminar.

Part VI

Introduction to Section on Special Surveys

Michael W. Traugott[1] *and Edith D. de Leeuw*[2]

[1] *Institute for Social Research, University of Michigan, Ann Arbor, MI, USA*
[2] *Department of Methodology & Statistics, Utrecht University, Utrecht, the Netherlands*

Experimental research in surveys has – from the first split ballot experiments in the early dates of polling (Singer and Presser 1989, p. 97) – mainly focused on cross-sectional surveys. Many of these early experiments were concerned with question wording and order effects. But as survey methodology developed, new topics of interest arose; and experiments were implemented to study interviewer effects, mode effects, and response inducing measures; for some excellent early examples see Singer and Presser (1989).

Two fields in survey research are underrepresented in the general methodological literature. The first concerns establishment or business surveys, which are usually different in nature and methodology from surveys of individuals or households. For an overview of the survey process in business surveys, see Snijkers et al. (2013). The second is longitudinal surveys, which offers special challenges and opportunities for implementation and analysis. For an overview of the complexities of longitudinal surveys and special methodology, see Lynn (2009). This section on special surveys contains two chapters that address issues in the design of experiments in these fields: experiments with the measurement of objective qualities in establishment surveys and experiments performed when individuals are the units of analysis in longitudinal surveys.

In the first chapter, Lynn and Jäckle emphasize the unique opportunities that longitudinal surveys offer for experimentation. They start with an insightful overview of experimental designs and the way these different designs can be implemented in longitudinal surveys. They further distinguish between experiments carried out over time (longitudinal experiments) and experiments conducted in longitudinal surveys. Their primary focus is on probability-based panels that serve as a platform for experimentation and offer scholars in various disciplines a way to implement experiments in the panel's ongoing field work. Furthermore, they provide an overview of international probability-based panels and the opportunities they offer to scholars worldwide. The Understanding Society Innovation Panel in the United Kingdom and experiments implemented in it provide clear illustrations of the advantages longitudinal surveys offer for experimentation. Randomized experiments in a longitudinal context offer many advantages, but also challenges in design and analysis. Lynn and Jäckle also discuss in depth the need for careful design and some of the special pitfalls to avoid in longitudinal panel experimentation. They not only advocate using more experimentation in longitudinal surveys and panels and

Experimental Methods in Survey Research: Techniques that Combine Random Sampling with Random Assignment, First Edition.
Edited by Paul J. Lavrakas, Michael W. Traugott, Courtney Kennedy, Allyson L. Holbrook, Edith D. de Leeuw, and Brady T. West.

Companion Website: www.wiley.com/go/Lavrakas/survey-research

valuing the strength of this approach, but also point out that this comes at the price of more complexity in design and implementation.

In the second chapter, Willimack and McCarthy discuss how establishment surveys differ from household surveys and the resulting challenges for implementing experiments in them. They give an excellent meta-analytic review of the limited experimental research in establishment surveys, commenting on how they often involve only single-factor designs focused most commonly on increasing response rates. They further discuss what makes experimental research in establishment surveys so special, describe key consideration in experimentation, and illustrate this with state-of-the-art examples of experiments. Willimack and McCarthy also argue for an expanded use of simulations deployed in secondary analysis of existing surveys in lieu of true experimental designs, but acknowledge that this is only a satisficing solution when the use of strong experimental designs would be clearly preferable.

Collectively, these two chapters provide important contributions to the design and implementation of more complex experiments. Although the emphasis in this section is on special surveys, the methodological guidelines and discussion will also profit those who plan to do experiments in cross-sectional household surveys.

References

Lynn, P. (ed.) (2009). *Methodology of Longitudinal Surveys*. New York: Wiley.

Singer, E. and Presser, S. (1989). *Survey Research Methods: A Reader*. Chicago: University of Chicago Press.

Snijkers, G., Haraldsen, G., Jones, J., and Willimack, D. (2013). *Designing and conducting Business Surveys*. New York: Wiley.

15

Mounting Multiple Experiments on Longitudinal Social Surveys: Design and Implementation Considerations

Peter Lynn and Annette Jäckle

University of Essex, Institute for Social and Economic Research, Colchester CO4 3SQ, UK

15.1 Introduction and Overview

There are now various longitudinal surveys that are used for experimentation. Several of these studies specifically invite proposals from external researchers. The longitudinal aspect adds a further dimension to the value of randomized experiments (designed to maximize internal validity) mounted in probability surveys (designed to maximize external validity): the repeated measurement of sample members over time can be used to test effects on inherently longitudinal outcomes, to take the histories of sample members into account in analyzing experimental outcomes, to analyze the long-term effects of experimental treatments on outcomes measured in later waves, or to exploit within as well as between respondent allocations to treatments by repeating experiments across waves. The unique value of experimentation in longitudinal surveys is matched by unique challenges in successfully designing and implementing experiments in a longitudinal context.

This chapter summarizes key methodological features and challenges based on experiences with the *Understanding Society* Innovation Panel, a probability-based household panel with annual interviews that exists solely for the purposes of experimentation and methodological development. The aim is to raise awareness of unique issues that arise when mounting multiple independent experiments on the same survey vehicle and, particularly, when the survey and the experiments are longitudinal.

Section 15.2 provides an overview of the types of experiments that can be carried in longitudinal surveys, Section 15.3 discusses the distinction between longitudinal experiments and experiments in longitudinal studies, and Section 15.4 provides an overview of international longitudinal studies that are used as platforms for experimentation. Section 15.5 provides further information on the design of the *Understanding Society* Innovation Panel and the types of experiments that have been implemented on it, while Sections 15.6 and 15.7 discuss how to avoid confounding and how to allocate units to treatments, respectively. Section 15.8 discusses the introduction of refreshment samples, which is a particular feature of longitudinal surveys that can strengthen the experimental setting. The final section provides a discussion of key lessons learned and possible future methodological developments.

Experimental Methods in Survey Research: Techniques that Combine Random Sampling with Random Assignment, First Edition.
Edited by Paul J. Lavrakas, Michael W. Traugott, Courtney Kennedy, Allyson L. Holbrook, Edith D. de Leeuw, and Brady T. West.

Companion Website: www.wiley.com/go/Lavrakas/survey-research

15.2 Types of Experiments that Can Be Mounted in a Longitudinal Survey

Different types of experiments can be embedded in a longitudinal survey. Table 15.1 presents an overview, where the horizontal axis indicates a hypothetical sequence of interviews at different points in time. Each design can be used to address different types of (longitudinal) research questions.

The classic experimental design, as described in standard handbooks such as Campbell and Stanley (1963) or De Vaus (2001), is a *pre-test post-test* control group design (1): the outcome of interest is measured before and after the exposure to treatment and a randomized control group is measured at the same points in time but not exposed to the treatment. To estimate the treatment effect the changes in the outcomes of the treatment and control groups are compared. Other information collected about participants in the pre-test interview can be used substantively to study whether treatment effects depend on pre-existing characteristics of study participants, or methodologically to select sub-samples with certain characteristics for the experiment, or to test for and estimate the impact of differential attrition in the experimental conditions (see Farrington et al. 2010). If the topic of the experiment does not require prior measures of either the outcome or covariates, the pre-test measurement may be omitted: in a *post-test* design (2) the treatment effect is estimated by comparing the outcomes of the treatment and the control groups.

In many survey experiments, the exposure to treatment and measurement of the outcome occur simultaneously: for example, in experiments testing the effects of reversing the order of response options in survey questions, participants are exposed to the treatment when they answer the question that measures the outcome. In this case, the classic experimental design becomes a *pre-test test* design (3) and the post-test design becomes a *cross-sectional test* (4). Cross-sectional tests can equally well be carried out in a single cross-sectional survey, as they do not make use of the repeated measures nature of longitudinal data.

Table 15.1 Typology of experiments in longitudinal surveys.

Type of experiment		R	Interview 1		Interview 2	Interview 3		Interview 4
1	Pre-test and post-test	T	O	X	O			
		C	O		O			
2	Post-test	T		X	O			
		C			O			
3	Pre-test test	T	O		XO			
		C	O		O			
4	Cross-sectional test	T			XO			
		C			O			
5	Multiple post-tests	T	(O)	X	O	O		O
		C	(O)		O	O		O
6	Multiple pre-tests	T	O		O	O	X	O
		C	O		O	O		O
7	Repeated tests	T	O		XO	XO		XO
		C	O		O	O		O

Notes: R = randomization, T = treatment group, C = control group, O = observation or measurement of outcome, X = exposure to treatment.

Further standard experimental designs include *multiple post-tests* (5) or *multiple pre-tests* (6) where participants are interviewed multiple times before or after the treatment exposure. *Multiple pre-test* studies can be used to identify existing change, or time trends, in the outcome before the treatment exposure. *Multiple post-test* studies (sometimes referred to as growth designs) can be used to estimate long-term effects of treatments, for example, to study how outcomes evolve with age, to identify delayed effects or to compare short and long-term effects (i.e. the growth trajectory of the treatment effect). Compared to studies with a single post-test interview, multiple post-test studies can identify and distinguish immediate lasting effects of a treatment, immediate but short-lived effects, delayed lasting effects, delayed undesirable effects on an existing trend, no effects of a treatment because of a pre-existing trend and haphazard oscillation (see figure 24.1 in Farrington et al. 2010). For example in criminology multiple post-tests are used to study the effect of interventions such as counseling or training of pre-school or school age children. Follow-up interviews may take place at irregular intervals over several decades, to study outcomes such as criminal activity, drug use, educational attainment, and labor market outcomes (Farrington et al. 2010). Similarly, in developmental epidemiology preventive trials with long-term follow-ups are for example used to test ways of reducing the risk of mental health problems in children of divorce, with behavioral or learning problems or whose parents are being treated for depression (Brown and Liao 1999).

Repeated tests (7) can be used to study how long-term treatments should best be applied to maximize effectiveness. For example clinical dynamic treatment trials are sequences of randomized trials, where at each stage the randomizations may depend on outcomes of previous stages, with the aim of developing optimum sequences of treatments by exploiting carry-over effects (Chakraborty and Murphy 2014). Repeated experiments can also be used to study how reliable and reproducible estimated treatment effects are if the conditions of the experiment are varied (see John and Quenouille 1977), or to study longitudinal outcomes.

In repeated designs the treatment conditions can either be held the same across implementations (between subject designs), or crossed (within subject designs). With cross-over designs each subject is observed in multiple treatment conditions and the same outcome is measured in each condition, such that each subject contributes multiple scores. Compared to between-subject designs this offers two advantages (Maxwell and Delaney 2004). Firstly, the repeated measures mean that a smaller number of subjects is needed to reach a certain level of statistical power. This is a clear advantage when the costs of recruiting subjects is high, in terms of money, time, or effort. Secondly, as each subject serves as his or her own control, variability in individual differences between subjects is removed from the error term, which increases statistical power.

15.3 Longitudinal Experiments and Experiments in Longitudinal Surveys

Not all longitudinal experiments, where data are collected about respondents at two or more points in time, are mounted on longitudinal surveys. There are many examples of free-standing field experiments that include follow-up surveys designed to test the long-term effects of treatments or interventions in economics (e.g. Aguila et al. 2015; Dupas and Robinson 2013), education (e.g. Hu et al. 2007), psychology (e.g. Acredolo 1978; Yeager et al. 2013), health (e.g. Marcus 1982; Olds et al. 2004), criminology (e.g. Belfield et al. 2006; Boisjoli et al. 2007; Ellickson and Bell 1990; McCord 2003), computer science (e.g. Lee et al. 2012; Wiedenbeck et al. 2005), market research (e.g. Aaker et al. 2004; Bolton and Drew 1991), and management research (Dvir et al. 2002; Workman and Bommer 2004). Such experimental studies correspond to designs (2) or (5) in Table 15.1.

Where experiments are embedded in a pre-existing longitudinal survey, two scenarios can be distinguished. In the first scenario, the researchers responsible for a longitudinal survey may carry out an experiment on their survey to inform design decisions. For example before switching from annual to biennial interviewing, the National Longitudinal Survey of Youth 1979 implemented an experiment to see how less frequent interviews would affect the quality of recall data (Pierret 2001). A sub-sample of respondents were asked to report on events in the past two years, rather than the year since the previous interview, simulating the two-year interviewing schedule and allowing comparison with data collected in an annual interview schedule. Similarly, in the 2013 wave of the UK National Child Development Study an experiment was conducted to test the effects of introducing web as a main mode of interviewing on attrition and measurement (Brown and Hancock 2015). A random control group was assigned to telephone only interviewing. The rest of the sample was invited to complete the survey online and nonrespondents were followed up by telephone interviewers.

In the second scenario, the longitudinal survey is a multipurpose vehicle forming part of the infrastructure for academic research. This chapter focuses on the latter: multipurpose longitudinal surveys that are used for multiple experiments.

15.4 Longitudinal Surveys that Serve as Platforms for Experimentation

Table 15.2 provides a summary of international longitudinal surveys of probability samples that are used as multipurpose platforms for social scientists to collect experimental data. There also exist a small number of commercial longitudinal panels based on probability samples that are occasionally used for social science experiments, but we focus here on the panels that exist primarily for experimental research purposes:

- *The Innovation Panel*: a household panel study in Great Britain that is part of the UK Household Longitudinal Study: *Understanding Society*
- *The SOEP Innovation Sample (SOEP-IS)*: a household panel study in Germany that is part of the German Socio-Economic Panel study
- *The LISS* (*Longitudinal Internet Studies for the Social sciences*) *panel*: a probability-based online panel in the Netherlands
- *The GESIS Panel*: a probability-based mixed mode (online-mail) panel in Germany
- *The ELIPSS* (*Étude Longitudinale par Internet Pour les Sciences Sociales*) *panel*: a probability-based online panel in metropolitan France
- *The American Life Panel*: an online panel in the United States that grew out of studies exploring the opportunities for Internet interviewing in the Health and Retirement Study
- *The Understanding America Study (UAS)*: a probability-based online panel in the United States

All of the panel surveys are scientific infrastructure projects enabling academic researchers in the social sciences to collect data. The funding for four of the surveys (the Innovation Panel, SOEP-IS, GESIS panel, and ELIPSS) is such that standard data collection is free for proposers; while the other panels' proposers have to pay for the data collection. The Innovation Panel and the SOEP-IS are modeled on their "parent" household panel surveys. The American Life Panel originally derived from the Survey of Consumer Attitudes, but later added new samples based on face-to-face interviewing and address based sampling. The other panels are free-standing surveys.

Table 15.2 Overview of longitudinal surveys that regularly field experiments.

Survey	UKHLS innovation panel (IP)	SOEP innovation sample (SOEP-IS)	LISS panel	GESIS panel	ELIPSS	American Life panel	Understanding America study (UAS)
Funder	Economic and Social Research Council; nonstandard elements funded by proposer	Leibniz-Association; nonstandard elements funded by proposer	Costs paid by proposers of experiments	Leibniz Association	Agence Nationale de la Recherche	Costs paid by proposers of experiments	Costs paid by proposers of experiments
Geographical coverage	Great Britain	Germany	Netherlands	German speaking population in Germany	France (metropolitan area)	US	US
Sample units	All members of sampled households	All members of sampled households and new members	All members of sampled households	Individuals	Individuals	Individuals, but other household members invited to participate	All members of sampled households
Sample design	Clustered, stratified sample based on postal addresses	Random route	Probability sample of households from population register	Random sample drawn from municipal population registers	Two stage random sample of individuals in households listed in 2011 census	Sample members recruited from multiple sources with different probability sample designs	Clustered, stratified sample of households based on postal addresses
Frequency	Annual since 2008	Annual since 2011	Monthly since 2007	Every two months since 2014	Monthly since 2012	Once or twice a month since 2006	Once or twice a month, depending on demand
Modes	CAPI; mixed mode experiments with CATI and Web	CAPI; experiments with smartphones and web	Web; households without Internet access are loaned computer and broadband connection	Web and mail; paper questionnaire sent to those unable/unwilling to participate online	Web; participants are loaned a tablet with mobile Internet connection	Web; households without Internet access are loaned computer and broadband connection	Web; households without Internet access are loaned computer and broadband connection
Website	https://www.understandingsociety.ac.uk/documentation/innovation-panel	www.diw.de/soep-is	https://www.lissdata.nl/Home	http://www.gesis.org/en/services/data-collection/gesis-panel/	http://quanti.dime-shs.sciences-po.fr/en/	https://alpdata.rand.org	https://uasdata.usc.edu/surveys

Sampling methods differ between the studies, but all are probability based. The Innovation Panel, SOEP-IS, the LISS panel, and UAS interview all members of households, while the other panels are samples of individuals. Sample sizes vary, but all studies regularly add refreshment samples. The frequency of interviewing varies: interviews are annual in the Innovation Panel and SOEP-IS, every two months in the GESIS panel, monthly in the LISS and ELIPSS panels and twice a month in the American Life Panel and UAS. Both the Innovation Panel and the SOEP-IS are primarily CAPI surveys, with some experimental testing of other modes. All other surveys are primarily online surveys, where sample members without Internet are either sent a paper questionnaire (GESIS panel), or loaned a tablet (ELIPSS) or computer (LISS, American Life Panel, UAS) with broadband access. Design and fieldwork procedures are summarized in Blom et al. (2015) for LISS and ELIPSS, Bosnjak et al. (2017) for the GESIS panel, Al Baghal and Jäckle (2016) for the Innovation Panel and Richter and Schupp (2015) for the SOEP-IS.

All panels have carried experiments testing aspects of questionnaire design and question wording. Several of the panels have also experimentally tested survey procedures such as different modes of data collection, respondent incentives, or audio-recording vs. writing in responses to open ended questions. Other types of experiments include information treatments, experiments to measure risk attitudes, financial decision making or decision making under uncertainty, or factorial surveys where the content of vignettes and allocation to respondents rely on randomization.

Researchers wishing to implement experiments on any of these panels submit proposals and draft questionnaires which are peer reviewed and assessed for scientific merit. Successful proposals are implemented by the survey teams, except for the GESIS panel where proposers have to program the questionnaire themselves using the guidelines provided by GESIS. All studies provide scientific open access to the data collected on their panel. Table 15.2 provides links to further documentation for each of the studies.

15.5 The *Understanding Society* Innovation Panel

The *Understanding Society* Innovation Panel is a platform for longitudinal methods research and social science experiments. It is an integral part of the design of the *Understanding Society* survey funded by the UK Economic and Social Research Council. Its purpose is to develop key innovations in survey methods and content that will ensure the future success of the *Understanding Society* survey, and more broadly to advance knowledge in the social sciences and in the methodology of designing longitudinal surveys. In addition to experimentation in the annual interviews, the focus of this chapter, the Innovation Panel is used as a base for Associated Studies collecting data using new and innovative mixed method approaches (see https://www.understandingsociety.ac.uk/research/get-involved/associated-studies).

The design of the Innovation Panel is based on the main *Understanding Society* survey. It consists of an original sample of around 2500 persons, clustered within households, first fielded in 2008, plus refreshment samples of around 700 persons added in 2011 and again in 2014. Attrition rates are documented in the User Guide at https://www.understandingsociety.ac.uk/documentation/innovation-panel. The sample is a stratified, clustered sample of all persons resident in Great Britain, excluding northernmost Scotland. An equal probability sample of addresses was drawn from the UK Postcode Address File and all residents at selected addresses at the time of Wave 1 became sample members (see Lynn 2009). Refreshment samples are added by selecting additional addresses from the existing primary sampling units (PSUs).

All sample members are eligible for annual interviews and followed if they move within Great Britain. Wave 1 was fielded in 2008 and data from each wave are deposited the following year with the UK Data Service (https://discover.ukdataservice.ac.uk/catalogue/?sn=6849), from where they are available to researchers. New household members are eligible for interviews as long as they live with a sample member, but not followed if they move out. To maintain contact with participants and update addresses, a between-wave-mailing is sent out. The mailing includes a report of research findings, an address confirmation slip that respondents are asked to return, and materials to encourage registration with the participant website (https://www.understandingsociety.ac.uk/participants).

The modes of data collection in Wave 1 were CAPI with a paper self-completion module for adults (aged 16+), and a paper self-completion questionnaire for youth aged 10–15. Wave 2 included an experiment where for a random two-thirds of households interviews were first attempted by telephone and nonrespondents were followed up by CAPI interviewers. The control group were interviewed by CAPI, as in Wave 1 (see Lynn 2013). Waves 5–9 included an experiment where two-thirds of households were first invited to complete the survey online and nonrespondents were followed up by CAPI interviewers. The control group were again interviewed in CAPI (Bianchi et al. 2017; Jäckle et al. 2015). The mode of the self-completion module was experimentally varied in Waves 4–6, with a random half of respondents interviewed in CAPI allocated to a CASI version and the control maintaining the paper self-completion.

The initial waves of the Innovation Panel were used for development and testing of the main *Understanding Society* survey. Since Wave 4 experiments are selected through an open competition: proposers submit a case for support including specification of the study design and draft questionnaires; the survey team assesses feasibility and costs; a review panel including external reviewers assesses the scientific merit and value for money and suggests a ranking of proposals; the *Understanding Society* Executive Team decide which proposals to accept. The criteria for the competition are described at https://www.understandingsociety.ac.uk/innovation-panel-competition.

Over the first nine waves, a total of 42 unique experiments have been fielded in the Innovation Panel, some of which are replicated over multiple waves. The experiments have included experiments with survey procedures (such as the mode experiments described above, experiments with the value of respondent incentives, or with the format and content of between wave mailings), experiments with generic questionnaire design issues (such as the use of show cards, the labeling or direction of scales, or the wording of dependent interviewing questions), and experiments with questionnaire design to measure specific concepts (such as testing ways of measuring consumption or wealth, life satisfaction, identity, or self-assessed disability). In addition, there have been vignette studies, studies with randomized information treatments, and some nonexperimental methodological studies, for example, measuring finger lengths as indicators of prenatal testosterone exposure, or testing time use diaries. Chapters 4 and 24 of this book are based on Innovation Panel experiments. All experiments implemented to date are described in the User Guide, including references to resulting publications (https://www.understandingsociety.ac.uk/documentation/innovation-panel). For each wave of the Innovation Panel an *Understanding Society* Working Paper is published which documents the rationale, design, and early findings from each experiment (Al Baghal 2015; Al Baghal 2016; Al Baghal 2014; Burton 2012 ; Burton 2013; Burton et al. 2011; Burton et al. 2008; Burton et al. 2010).

15.6 Avoiding Confounding of Experiments

Experimentation relies on randomized allocation of observational units (sample members) to treatments in order to ensure that the effect of treatment is not confounded with any other factor that could influence the observed outcomes. Pure random allocation (effectively simple random sub-sampling) ensures the absence of confounding *on average* (i.e. invoking the expected value under the sub-sampling distribution), but the sample may not be well balanced between treatments, due to random sampling (allocation) variance. This risk is particularly great when there are many potential confounding factors, as when many experiments are carried out on a longitudinal survey. The larger the number of potential confounding factors (other experiments) the greater the chance of observing severe imbalance with respect to at least one of those factors, under simple random allocation.

With respect to any particular potential confounding factor, balance can be ensured by using stratified random allocation rather than simple random allocation, where the potential confounding factor acts as the stratification variable.[1] For example if the outcome of interest is expected to be strongly influenced by the participant's age, participants could be listed in age order before allocating alternately to treatment and control groups, thus ensuring a similar age distribution in each of the two groups. In the longitudinal survey context, it is in principle therefore possible to achieve balance with respect to any other experiments administered at the current wave or at any previous wave, or with respect to any survey data collected previously. However, in practice there is a limit to the number of factors for which this can be done. Suppose, we wish to allocate sample members to one of two treatment groups, A1 and A2, with an equal sample size to be allocated to each group. Table 15.3 illustrates the extent of imbalance that can arise with simple random allocation (upper panel) and how this imbalance can be removed with stratified random allocation (lower panel). The imbalance between the two treatment groups for experiment A is shown with respect to the treatments for three other experiments B, C, and D, which have four, three, and two treatment groups respectively. These may be experiments that were carried at previous waves of the survey, or they may be planned for the same wave as experiment A. In either case, in order to use stratified random allocation the allocation to treatments for B, C, and D must have been made before the allocation to treatments for A. The distribution in the upper panel was obtained by allocating 504 sample units to each of A1 and A2 randomly, without regard to the distribution of the other three experimental indicators.

The potential for imbalance to affect observed outcomes can be seen in the upper panel of Table 15.3. For example treatment group A1 contains a higher proportion of sample units allocated to D2 than treatment group A2. A simple comparison of the outcome between groups A1 and A2 will confound the effect of A2–A1 with a small proportion of the effect of D1–D2. This

Table 15.3 Sample distributions generated by two alternative allocation methods.

	B1	B2	B3	B4	C1	C2	C3	D1	D2
Sample distribution: four experiments, simple random allocation									
A1	133	130	119	122	161	181	162	238	266
A2	119	122	133	130	175	155	174	266	238
Sample distribution: four experiments, stratified random allocation									
A1	126	126	126	126	168	168	168	252	252
A2	126	126	126	126	168	168	168	252	252

1 The strata are referred to as "blocks" in the classical experimental design literature (Addelman 1969), though in the survey context it is not necessary to assume that the stratification is explicit rather than implicit.

can be overcome either by controlling for the effect of D1–D2 in the analysis (for example, by carrying out a weighted analysis or by including the experiment D allocation as a covariate) or by controlling through design, as in the lower panel of the table. The distribution in the lower panel was obtained by sorting the sample units by the cross-classification of the other three experimental indicators before allocating alternately. The appendix provides `Stata` syntax for implementing each of these two allocation methods.

With modest sample sizes, allocation cannot be fully controlled by design for more than a few factors. In the *Understanding Society* Innovation Panel, with more than 40 experiments, any one experiment is likely to be unbalanced with respect to the majority of other experiments. In this situation, it becomes important to identify the experiments that are most likely to affect the outcome of interest and to at least stratify the allocation with respect to those experiments. For example at Wave 4, 11 new experimental manipulations were to be introduced. Fully crossing all 11 would have been impossible as this would lead to 45 056 experimental groups with a sample size of 2445 to allocate. Instead, full orthogonality was restricted to subsets of the 11 experiments that were likely to influence the same outcomes. Four of the experiments were explicitly designed to influence unit nonresponse rates (Burton 2012). A fully crossed experimental design was used to allocate sample units to these experiments, to ensure that the impact of each on unit nonresponse could be separately identified. As the experiments had 11, 4, 2, and 2 treatments respectively, this involved randomly assigning the 2445 units to 176 groups. While the other seven experiments could conceivably have had some impact on unit nonresponse, this was felt to be unlikely as they mainly involved manipulations to question wording or question placement, designed to influence measurement. Similarly, for each of the seven measurement experiments, the allocation was crossed with two or three other experiments that could conceivably have affected the measurement outcomes of interest. Complete confounding was avoided for every combination of experiments out of the 11 (and, indeed, with experiments carried at previous waves) by always assigning randomly within the groups defined by the crossing of experiments and making each assignment independently. However, for some of these combinations (the ones that were not expected to influence common outcomes) the sample distribution can be somewhat unbalanced due to the play of random chance.

In some cases, the experiments that are amongst the most likely to influence a particular outcome of interest may be ones that were carried at an earlier wave. An example is a set of five experiments concerned with measurement of change between waves. These particular experiments were introduced at Wave 3 and outcome measures were comparisons of responses given at Wave 3 with those given at Wave 2. However, Wave 2 had involved an experimental allocation to mode treatments and the mode could have affected the response given at Wave 2 to some of the relevant questions. The allocation to the Wave 3 measurement of change experiments was therefore fully crossed with the allocation to the Wave 2 modes experiment.

15.7 Allocation Procedures

In this section, we discuss two other important considerations regarding the allocation of sample units to treatments. Given that survey samples are often hierarchically structured in some way, the first consideration concerns the choice of the level at which to assign units to treatments. The second consideration comes into play when an experiment involves treatment at multiple waves. The researcher must decide how to allocate sample units to multiwave combinations of treatments (design (7) in Table 15.1).

15.7.1 Assignment Within or Between Households and Interviewers?

In the *Understanding Society* Innovation Panel, sample units are individual persons, but these are clustered within households, and households are clustered in turn within both PSUs and interviewer assignments (which are strongly correlated with each other, but not identical). It is possible to allocate experimental treatments at the level of PSU, interviewer, household, or individual. Each may have advantages and disadvantages, depending on the nature of the experiment. In principle, statistical power is greatest when allocation is made at the lowest level (individuals, in our case) for reasons that are analogous to those set out in the previous section regarding confounding and balance. Allocating individuals to treatments, with stratification by household and PSU, will ensure that all PSUs (and as many households as possible) are represented in each treatment group, so that the power to observe a treatment effect is not reduced by systematic differences between PSUs. However, there are a number of reasons why allocating at a higher level will sometimes be preferable.

The effects of some treatments may be contaminated if respondents are aware that some people received different treatments to them. In the case of the *Understanding Society* Innovation Panel (as with several of the other surveys outlined in Section 15.4 above) the clustering of individuals within households makes it quite likely that many respondents will be aware of the treatment received by other household members – at least for some types of experiments – so contamination effects are a serious concern. For that reason, most experiments have been allocated at the household level, so that all respondents in the same household receive the same treatment. For example in the online/CAPI mode experiment introduced at Wave 5 and described in Section 15.5 above, the reaction of an individual to the single-mode face-to-face protocol might be different if they knew that someone else in their household was offered an opportunity that they themselves were not offered, to complete the survey online. Furthermore, in this example the treatment of interest to the experiment proposers (the scientific leadership team of *Understanding Society*) was inherently a household-level treatment. They wished to compare the existing single-mode design with one designed to reduce survey costs by getting a proportion of households to participate entirely online. This specific objective of getting everyone in the household to participate online can only be achieved if everyone in the household is invited to participate online, so a design in which only some household members receive such an invite was simply not of interest. There are other examples of treatments that are inherently household-level. One of these is part of a series of experiments on respondent incentives: one treatment involves offering each individual in the household an additional incentive payment conditional on every individual in the household participating. The motivation for assessing this treatment is that it might increase the proportion of sample households for which data is successfully obtained from every individual.

There are at least two situations in which allocation to treatment may be best done at the level of interviewer assignment, rather than allocating households within each assignment. First, some experimental manipulations must be administered by the interviewer, such as when the interviewer has to present respondents with alternative versions of survey materials. In this situation, interviewers are less likely to make mistakes and the administration is more likely to be smooth and efficient if each interviewer only has one version to administer to all his or her respondents. The *Understanding Society* Innovation Panel learned this lesson the hard way at Wave 1 with an experiment in which show cards were to be shown to half the sample, but not the other half. Assignment to treatments was crossed with interviewers, to minimize any interviewer effect on the results, but it turned out that many interviewers, once equipped with a set of show cards, found it hard to remember that they should not always hand the cards to the respondent at the start of the interview (the usual procedure on the survey and on most other surveys).

The second situation in which allocation to treatment may be best done at the level of interviewer assignment is when one or more of the treatments is specifically designed to change interviewer behavior in some way. An example would be any treatment that should affect calling patterns in a face-to-face survey. Calls are not made independently for each unit in an assignment. Rather, an interviewer will often make additional call attempts while they are in the area visiting other sample units. Thus, for the treatment to be a realistic replication of how it would work if applied to a whole survey, all units in an interviewer assignment should receive the same treatment. At Wave 4 of the *Understanding Society* Innovation Panel an experiment was run in which a proportion of sample households were offered, via the advance letter, the opportunity to telephone their interviewer to make an appointment at a convenient time rather than waiting for the interviewer to visit them. Brown and Calderwood (2014) found only a very small reduction in the number of interviewer calls required to complete the interviews – a finding which could have been affected by the fact that each interviewer assignment included some treated cases and some control cases.

15.7.2 Switching Treatments Between Waves

The longitudinal survey context gives researchers the possibility of mounting experiments over two or more waves. There are a number of situations in which this can be desirable, and a number of possible multiwave designs. For example each sample unit could continue to receive the same treatment at each wave; each sample unit could switch from one treatment to another at the next wave; or treatments could be assigned randomly at each wave, without regard to the treatment assigned previously. The most appropriate choice should depend on the objectives of the experiment and the extent to which errors in the outcome variable(s) are likely to be correlated between waves.

For example, suppose there are two alternative treatments, A and B, to be compared. Using a within-subject design controls the between-subject component of variance and hence improves the accuracy of estimates of the effect of B rather than A. If the treatments are of a kind that cannot be both administered in the course of the same interview, a within-subject option in a longitudinal survey context is to administer one treatment at one wave and the other at the next wave. However, if all sample members are administered treatment A at wave t and treatment B at wave $t + 1$, this risks confounding the relative effects of treatment B with (a) a real change in the outcome between t and $t + 1$, and (b) a "priming" effect caused by having previously been administered treatment A. To avoid such confounding, a crossover design can be used. In a crossover design, one group of respondents would receive treatment A at wave t and treatment B at wave $t + 1$, while another group would receive treatment B at wave t and treatment A at wave $t + 1$. If the errors are uncorrelated between waves (or, more realistically, have very low correlation) this design should maximize the precision of estimates of the effect of B rather than A without confounding (priming).

For some research questions, the relevant treatments are themselves inherently longitudinal. Consider, for example, the choice of question wording or response options for a question that is to be repeated at each wave of a longitudinal survey for the purpose of measuring change. Suppose there are two candidate versions of a question, labeled A and B. The researcher wishes to know whether it is better to repeat version A at each wave or to repeat version B at each wave. If the purpose is to inform development of a new survey, with no *a priori* reason for preferring either question version, a simple repeated test design (type 7 in Table 15.1) could be implemented, in which one group is asked version A at each wave and another group is asked version B. If the experiment shows that repeated use of version B is superior, the survey will adopt that version. But what if the experiment is to inform an existing survey, in which version

A is currently used? The researcher might also need to assess the effect of transitioning from version A to version B. Thus, a third treatment group could be introduced in which version A is administered initially, with the treatment switching to version B after a number of waves.

Sometimes, the accuracy of both cross-sectional and longitudinal measures is of importance. If the survey questions in the example of the previous paragraph are about the level of savings held by a household, the answers could be used either to construct a (cross-sectional) measure of current savings or a (longitudinal) measure of change in savings since the previous wave. Even if the survey designs under consideration are only those that involve repeating the same question at each wave, a crossover design might provide a more accurate estimate of the quality of the cross-sectional measure.

Recognizing the competing design implications of different research objectives, an experiment carried at Waves 3 and 4 of the *Understanding Society* Innovation Panel involved four treatment groups: one group was administered version A at both waves, one was administered version B at both waves, one was administered version A at Wave 2 and version B at Wave 3, and the final group was administered version B at Wave 2 and version A at Wave 3. The experiment concerned several measures of change. For example one of the questions involved was designed to ascertain for how long the respondent had lived at their current address. Version A of this question asked (of people who had not lived at their current address their whole life) "In what month and year did you move to this address?" while version B asked "How long have you lived at this address?" The partial crossover in the experimental design enhances the precision of estimates of differences between the two question versions in cross-sectional measures, while the two simple repeated test treatments allow comparison of measures of change when either of the questions versions is repeated. A similar four-treatment design was used for an experiment at Waves 2 and 3 in which show cards were used with half of the respondents at each wave.

15.8 Refreshment Samples

The *Understanding Society* Innovation Panel introduces an additional sample, known as a refreshment sample[2], each three years. To date, refreshment samples have been added at Waves 4, 7, and 10. The main reason for doing this is to maintain the size of the panel, but an additional advantage is that the practice adds an extra dimension to experiments mounted on the panel.

Time in sample may affect respondents' familiarity with the survey questions, trust in the interviewer/survey, and knowledge of the topic(s) of the survey. These changes may affect the responses that are given to survey questions, producing "panel conditioning" (Struminskaya 2015; Warren and Halpern-Manners 2012). Furthermore, sample members at later waves of a panel may tend to be easier to contact and more co-operative than at earlier waves (Uhrig 2008; Watson and Wooden 2012). Thus, the results of any experiment designed to affect either survey responses or fieldwork outcomes may depend on the survey wave at which the experiment is conducted. The strength of a design with regular refreshment samples is that the extent and nature of dependency on survey wave can be estimated by comparing outcomes between samples. For example, an experiment mounted at Wave 8 will be administered to three samples consisting of respondents for whom it is their second, fifth and eighth wave of participation. This strength was exploited in the analysis of an experiment with targeted advance letters carried out at Wave 6 of the *Understanding Society* Innovation Panel (Lynn 2016). In the CAPI single-mode

2 On other surveys, similar additional samples are sometimes known as replenishment samples, refresher samples, or top-up samples.

part of the sample, the targeted letters improved response rates significantly for the refreshment sample (who had only participated in two previous waves) but not for the original sample (who had participated in five previous waves).

Another strength of a design with regular refreshment samples is that it may be possible to control for any possible exposure of respondents to relevant stimuli or experimental treatments at previous waves. For example, an experiment regarding ways to introduce a new, sensitive, topic to respondents clearly requires a context in which the respondents have not previously been asked questions on that topic. Having run such an experiment once, the findings may suggest a further line of enquiry that would require further experimentation. But further experimentation on the same sample would not provide a realistic setting. The existence of a new refreshment sample on which the first experiment had not been administered would provide an opportunity for the second experiment to be carried out in broadly the same context as the first (same survey).

15.9 Discussion

Randomized experiments mounted on probability-based longitudinal surveys have considerable strengths. The randomization provides internal validity, while probability sampling provides external validity. In this high-validity context, the longitudinal design provides opportunities to study dynamics in both the outcomes of experimental treatments and in the treatments themselves, as well as opening up the possibility of both treatments and analysis being cognizant of past experiences, prior characteristics or even past survey behavior. The range of design types that are possible with a longitudinal survey context are outlined in Section 15.2 of this chapter. A particular strength of the longitudinal survey context is the variety of repeated test designs that are possible, defined by whether and how treatments are varied within participants across waves (discussed in Section 15.7). Another advantage is the potential provided by regular refreshment samples (Section 15.8) to study the effect of time-in-sample and to control for previous exposure to similar treatments. Many examples of experimental studies that take advantage of these multiple strengths can be found amongst the experiments that have been mounted on the *Understanding Society Innovation Panel* (described in Section 15.5).

However, these strengths come at the price of challenge and complexity in design and implementation. The complexity increases when multiple independent experiments are to be carried on the same survey, and over many waves. This chapter has outlined some of these challenges and complexities and has demonstrated some of the ways in which the challenges can be met in order to ensure the success of the experiments. A key design objective is to avoid confounding between experiments and to maximize the statistical power of experiments. In the context of longitudinal surveys such as those described in Section 15.4, this is particularly challenging because of the large number of experiments carried on the same survey and because of the evolving nature of the experimentation: experiments in later waves are not yet conceptualized at the time of the design of earlier waves. In Section 15.6, we have described how confounding can be avoided through the use of stratified random allocation, where the treatment groups for other experiments constitute the strata. We have also discussed (Section 15.7) issues involved in choosing the level at which randomized allocation should take place (PSU, interviewer, household or individual). There is a balance to be struck between statistical and practical considerations and we have mentioned examples that demonstrate why neglecting the latter will not necessarily benefit the former.

Truly longitudinal experimentation on probability-based longitudinal surveys is still an evolving methodology. There are relatively few longitudinal surveys designed for this purpose and there is very little literature on design issues. Research would benefit from further study of the relative advantages of different longitudinal designs for different analytical purposes. For example, there is little guidance to be found on when a simple crossover design should be preferred to a crossover design with repeated-treatment groups, or how best to determine the optimum group sizes in the latter design. Analysis methods too are under-developed: for example, standard error estimation that takes into account stratified random allocation to treatments within a survey with a complex design. While the strengths of randomized experiments mounted on probability-based longitudinal surveys are truly considerable, work remains to ensure that study designs can take full advantage of these strengths.

15.A Appendix: Stata Syntax to Produce Table 15.3 Treatment Allocations

Simple random allocation to two groups:

```
ge rand=runiform()
sort rand
ge treatA=1 if trunc((_n-1)/2)==trunc(_n/2)
recode treatA .=2
```

Stratified random allocation, where stratification is by the treatment groups for three other experiments:

```
ge rand=runiform()
sort treatB treatC treatD rand
ge treatA=1 if trunc((_n-1)/2)==trunc(_n/2)
recode treatA .=2
```

References

Aaker, J., Fournier, S., and Brasel, S.A. (2004). When good brands do bad. *Journal of Consumer Research* 31 (1): 1–16.

Acredolo, L.P. (1978). Development of spatial orientation in infancy. *Developmental Psychology* 14 (3): 224.

Addelman, S. (1969). The generalized randomized block design. *The American Statistician* 23 (4): 35–36.

Aguila, E., Kapteyn, A., and Smith, J.P. (2015). Effects of income supplementation on health of the poor elderly: the case of Mexico. *Proceedings of the National Academy of Sciences* 112 (1): 70–75.

Al Baghal, T. (2015). Understanding Society innovation panel wave 7: results from methodological experiments. Understanding Society Working Paper 2015-03. Colchester: University of Essex.

Al Baghal, T. (2016). Understanding Society innovation panel wave 8: results from methodological experiments. Colchester: University of Essex.

Al Baghal, T. (ed.) (2014). Understanding Society innovation panel wave 6: results from methodological experiments. Understanding Society Working Paper 2014-4. University of Essex.

Al Baghal, T. and Jäckle, A. (2016). Understanding Society: the UK Household Longitudinal Study innovation panel, waves 1–8, user manual. Colchester: University of Essex.

Belfield, C.R., Nores, M., Barnett, S., and Schweinhart, L. (2006). The high/scope Perry preschool program cost–benefit analysis using data from the age-40 followup. *Journal of Human Resources* 41 (1): 162–190.

Bianchi, A., Biffignandi, S., and Lynn, P. (2017). Web-face-to-face mixed mode design in a longitudinal survey: effects on participation rates, sample composition and costs. *Journal of Official Statistics* 33 (2): 385–408.

Blom, A.G., Bosnjak, M., Cornilleau, A. et al. (2015). A comparison of four probability-based online and mixed-mode panels in Europe. *Social Science Computer Review* 34 (1): 8–25.

Boisjoli, R., Vitaro, F., Lacourse, E. et al. (2007). Impact and clinical significance of a preventive intervention for disruptive boys. *The British Journal of Psychiatry* 191 (5): 415–419.

Bolton, R.N. and Drew, J.H. (1991). A longitudinal analysis of the impact of service changes on customer attitudes. *The Journal of Marketing* 55 (1): 1–9.

Bosnjak, M., Dannwolf, T., Enderle, T. et al. (2017). Establishing an open probability-based mixed-mode panel of the general population in Germany: the GESIS panel. *Social Science Computer Review* 36 (1): 103–115.

Brown, C.H. and Liao, J. (1999). Principles for designing randomized preventive trials in mental health: an emerging developmental epidemiology paradigm. *American Journal of Community Psychology* 27 (5): 673–710.

Brown, M. and Calderwood, L. (2014). Can encouraging respondents to contact interviewers to make appointments reduce fieldwork effort? Evidence from a randomized experiment in the UK. *Journal of Survey Statistics and Methodology* 2 (4): 484–497.

Brown, M. and Hancock, M. (2015). *National Child Development Survey. 2013 Follow-up: A Guide to the Datasets*. London: Institute of Education.

Burton, J. (2012). Understanding Society innovation panel wave 4: results from methodological experiments. *Understanding Society Working Paper 2012-06*. Colchester: University of Essex.

Burton, J. (2013). Understanding Society innovation panel wave 5: results from methodological experiments. *Understanding Society Working Paper 2013-06*. Colchester: University of Essex.

Burton, J., Budd, S., Gilbert, E., Jäckle, A. et al. (2011). Understanding Society innovation panel wave 3: results from methodological experiments. *Understanding Society Working Paper 2011-05*. Colchester: University of Essex.

Burton, J., Laurie, H., and Uhrig, S.C.N. (2008). Understanding Society: some preliminary results from the wave 1 innovation panel. *Understanding Society Working Paper 2008-03*. Colchester: University of Essex.

Burton, J., Laurie, H., and Uhrig, S.C.N. (2010). Understanding Society innovation panel wave 2: results from methodological experiments. *Understanding Society Working Paper 2010-04*. Colchester: University of Essex.

Campbell, D.T. and Stanley, J.C. (1963). *Experimental and Quasi-Experimental Designs for Research*. Boston: Houghton Mifflin Company.

Chakraborty, B. and Murphy, S.A. (2014). Dynamic treatment regimes. *Annual Review of Statistics and Its Application* 1: 447–464.

De Vaus, D.A. (2001). *Research Design in Social Research*. London: Sage.

Dupas, P. and Robinson, J. (2013). Savings constraints and microenterprise development: evidence from a field experiment in Kenya. *American Economic Journal: Applied Economics* 5 (1): 163–192.

Dvir, T., Eden, D., Avolio, B.J., and Shamir, B. (2002). Impact of transformational leadership on follower development and performance: a field experiment. *Academy of Management Journal* 45 (4): 735–744.

Ellickson, P.L. and Bell, R.M. (1990). Drug prevention in junior high: a multi-site longitudinal test. *Science* 247 (4948): 1299–1305.

Farrington, D.P., Loeber, R., and Welsh, B.C. (2010). Longitudinal-experimental studies. In: *Handbook of Quantitative Criminology* (ed. A.R. Piquero and D. Weisburd), 503–518. New York: Springer.

Hu, P.J.-H., Hui, W., Clark, T.H.K., and Tam, K.Y. (2007). Technology-assisted learning and learning style: a longitudinal field experiment. *IEEE Transactions on Systems, Man and Cybernetics, Part A: Systems and Humans* 37 (6): 1099–1112.

Jäckle, A., Lynn, P., and Burton, J. (2015). Going online with a face-to-face household panel: effects of a mixed mode design on item and unit non-response. *Survey Research Methods* 9 (1): 57–70.

John, J.A. and Quenouille, M.H. (1977). *Experiments: Design and Analysis.* High Wycombe: Charles Griffin & Co. Ltd.

Lee, M.K., Forlizzi, J., Kiesler, S. et al. (2012). Personalization in HRI: a longitudinal field experiment. In: *Human-Robot Interaction (HRI), 2012 7th ACM/IEEE International Conference on: IEEE*, 319–326.

Lynn, P. (2009). Sample design for Understanding Society. *Understanding Society Working Paper 2009-01.* Colchester: University of Essex.

Lynn, P. (2013). Alternative sequential mixed-mode designs: effects on attrition rates, attrition bias, and costs. *Journal of Survey Statistics and Methodology* 1: 183–205.

Lynn, P. (2016). Targeted appeals for participation in letters to panel survey members. *Public Opinion Quarterly* 80 (3): 771–782.

Marcus, A.C. (1982). Memory aids in longitudinal health surveys: results from a field experiment. *American Journal of Public Health* 72 (6): 567–573.

Maxwell, S.E. and Delaney, H.D. (2004). *Designing Experiments and Analysing Data: A Model Comparison Perspective.* London: Lawrence Erlbaum Associates.

McCord, J. (2003). Cures that harm: unanticipated outcomes of crime prevention programs. *The Annals of the American Academy of Political and Social Science* 587 (1): 16–30.

Olds, D.L., Kitzman, H., Cole, R. et al. (2004). Effects of nurse home-visiting on maternal life course and child development: age 6 follow-up results of a randomized trial. *Pediatrics* 114 (6): 1550–1559.

Pierret, C.R. (2001). Event history data and survey recall: an analysis of the national longitudinal survey of youth 1979 recall experiment. *Journal of Human Resources* 36 (3): 439–466.

Richter, D. and Schupp, J. (2015). The SOEP Innovation Sample (SOEP-IS). *Schmollers Jahrbuch* 135: 389–400.

Struminskaya, B. (2015). Respondent conditioning in online panel surveys: results of two field experiments. *Social Science Computer Review* 34 (1): 95–115.

Uhrig, S.C.N. (2008). The nature and causes of attrition in the British Household Panel Survey. *ISER Working Paper 2008-05.* Colchester: University of Essex.

Warren, J.R. and Halpern-Manners, A. (2012). Panel conditioning in longitudinal social science surveys. *Sociological Methods & Research* 41 (4): 491–534.

Watson, N. and Wooden, M. (2012). The HILDA survey: a case study in the design and development of a successful household panel study. *Longitudinal and Life Course Studies* 3 (3): 369–381.

Wiedenbeck, S., Waters, J., Birget, J.-C. et al. (2005). PassPoints: design and longitudinal evaluation of a graphical password system. *International Journal of Human-Computer Studies* 63 (1): 102–127.

Workman, M. and Bommer, W. (2004). Redesigning computer call center work: a longitudinal field experiment. *Journal of Organizational Behavior* 25 (3): 317–337.

Yeager, D.S., Miu, A.S., Powers, J., and Dweck, C.S. (2013). Implicit theories of personality and attributions of hostile intent: a meta-analysis, an experiment, and a longitudinal intervention. *Child Development* 84 (5): 1651–1667.

16

Obstacles and Opportunities for Experiments in Establishment Surveys Supporting Official Statistics

Diane K. Willimack[1] *and Jaki S. McCarthy*[2]

[1] *United States Department of Commerce, U.S. Census Bureau, 4600 Silver Hill Road, Washington, DC 20233-0001, USA*
[2] *United States Department of Agriculture, National Agricultural Statistics Service (USDA/NASS), 1400 Independence Avenue, SW, Washington, DC 20250-2054, USA*

16.1 Introduction

Survey methods research supporting official statistics can have far-reaching impacts. The European Union Statistical Office, Eurostat (1958), states "Democratic societies do not function properly without a solid basis of reliable and objective statistics". In the United States, official statistics guide spending by Federal, state, and local governments to ensure adequate facilities for the well-being of their citizens – for example, schools and hospitals, transportation infrastructure, public services, food security, and public safety. Official economic indicators move markets, and publicly available official statistics provide a "gold standard" against which various types and sources of information are compared. Thus, research to improve or alter survey methods that support official statistics is often scrutinized, and their impact must be transparent. For example, before revised labor force questions, which support estimation of the US unemployment rate, were implemented in the Current Population Survey in 1994, research during 1987–1993 was documented extensively in the public domain via papers and presentations at numerous statistical conferences (for a summary, see Rothgeb et al. 1992).

Conducting randomized experiments is generally challenging for most surveys, but may be exacerbated by some unique considerations in establishment surveys. In this chapter, we discuss several challenges when undertaking experiments in establishment surveys that support official statistics. We begin by summarizing some characteristics of establishment surveys that differ from surveys of households and individuals. We provide a review of available literature describing randomized experiments associated with establishment surveys. We then describe key features of establishment surveys, along with their implications, that impact the design and implementation of experiments among this target population. We offer examples from our organizations' research to demonstrate mitigation strategies, and close with some thoughts about future experimentation in establishment surveys.

Experimental Methods in Survey Research: Techniques that Combine Random Sampling with Random Assignment, First Edition.
Edited by Paul J. Lavrakas, Michael W. Traugott, Courtney Kennedy, Allyson L. Holbrook, Edith D. de Leeuw, and Brady T. West.

Companion Website: www.wiley.com/go/Lavrakas/survey-research

16.2 Some Key Differences Between Household and Establishment Surveys

There are several key differences between establishment surveys and general population surveys of households and individuals that impact survey design, collection practices, and statistical procedures. Because of these differences, procedures used for and conclusions drawn from experiments in surveys of the general population may not hold for establishments. Here we focus on differences pertinent for conducting experiments in establishment surveys; for a more thorough discussion of differences between household and establishment surveys, the reader is referred to Snijkers et al. (2013).

For official statistics in the United States, the Federal Committee on Statistical Methodology (1988) defines an establishment as an economic unit, typically at a single physical location, where business is conducted or services or industrial activities are performed. While statistical organizations in other countries pose slight variations on this definition, mainly in terms of entities or enterprises made up of one or more establishments (Snijkers et al. 2013), "establishment survey" is generally defined as a survey collecting information from or about establishments or economic units comprised of establishments (Cox and Chinnappa 1995). We use this broader definition, which includes hospitals, schools, institutions, organizations, farms, and government agencies, along with "businesses" in the traditional sense.

16.2.1 Unit of Analysis

Establishment surveys, particularly, those supporting official statistics, typically collect objective information about the establishment or organization, and not information about personal attitudes and behaviors of their owners, managers, and employees. In this chapter, we limit our focus to surveys where the establishment, rather than its individual members, is the unit of analysis. Thus, surveys of teachers collecting their individual practices, or satisfaction of business employees, for example, are out of scope of this chapter. Establishment surveys do involve people answering questions, just like general population surveys. However, "individuals within these organizations are surveyed only as spokespersons for the organizations" (Cox and Chinnappa 1995, p. 3).

16.2.2 Statistical Products

Many establishment surveys supporting official statistics measure the economy and contribute to national income and product accounts. Key statistics typically produced from establishment surveys are totals, such as total revenue, total production, total employment, and/or total payroll, rather than means or proportions common in social surveys. Economic indicators support the goal of monitoring the status of the economy, such that time series and the ability to measure change are paramount, requiring consistent measurement, sometimes at the expense of "better" measurement.

16.2.3 Highly Skewed or Small Specialized Target Populations

Establishment survey target populations often have highly skewed distributions with a small number of very large establishments that contribute disproportionately to statistical estimates of totals. Obtaining data from these establishments may be critical to the quality and accuracy of statistical products. In addition, some establishment survey populations may be small and limited, for example, the population size for a survey of state governments in the United States is 50.

16.2.4 Sample Designs and Response Burden

As a consequence of these skewed or specialized populations, large or unique units are often selected with certainty into survey samples, and included in multiple independent, cross-sectional surveys. For example the largest hog producers in the United States are selected with certainty for every hog and pig inventory survey conducted by the United States Department of Agriculture (USDA). Although new survey samples are selected annually, some hog operations have been included in these surveys for many years.

Large enterprises operating in multiple industrial sectors are likely to be selected into multiple industry-specific surveys, as well as omnibus surveys such as structural business surveys, and separate independent cross-sectional surveys on different topics. For example, a large vertically integrated multinational firm operating in manufacturing, wholesale, and retail sectors could be included in surveys of factory production, inventories and supply chain logistics, consumer sales, employment, and foreign direct investment.

Moreover, surveys that support estimates of change require longitudinal designs, such that large establishments are not only selected with certainty but they are also surveyed repeatedly, exacerbating their survey burden. In the United States, a large manufacturer may be selected into samples for the several monthly or annual surveys conducted by the Census Bureau, the Bureau of Labor Statistics, or the Bureau of Economic Analysis, along with surveys focusing on specific topics, such as research and development, energy consumption, management practices, and so on. An evaluation of the burden placed on establishments by the National Agricultural Statistics Service (NASS) during a four-year-period showed that, while many were contacted infrequently, a significant minority of them were sampled for NASS surveys 10 times or more (McCarthy et al. 2006). Indeed, one establishment was contacted 103 times during the four-year period!

16.2.5 Question Complexity and Availability of Data in Records

The content of establishment survey questions is often technical, using jargon, accounting terminology, or other terms requiring precise definitions. As a result, the formulation of survey questions is often complex, with explanatory clauses, statements, or lists of specific attributes to include or exclude, and often includes detailed instructions that may be lengthy and wordy themselves. In addition, questions are often repetitive, requesting the same information about a number of related items. Survey questionnaires may resemble forms or tables – for example, employment counts, wages, and hours worked (columns) requested for different types of employees (rows) – rather than a series of questions and answers.

In addition, the data requested by these complex questions may not match establishment records, as technical concepts and definitions required for official statistics often deviate from those required by accounting standards, legal or regulatory requirements, or the needs of business managers for making decisions about business operations and strategic planning. The work needed to reconcile the mismatch between requested data and data found in business records is a major source of reporting burden for business survey respondents (Haraldsen 2010).

16.2.6 Labor-intensive Response Process

The response process in establishment surveys is more complex than in general population surveys, not only because of this data mismatch but also the need to gather data from multiple sources and people distributed throughout a company, where data are located to support business processes. Survey response represents a tangible cost to the business, with no associated production (Sudman et al. 2000; Willimack and Nichols 2010; Willimack and Snijkers 2013), because staff resources and work hours must be dedicated to identify data sources and gather information from other departments throughout the company (Bavdaz 2010). There may be an explicit cost in small businesses, where record-keeping may be contracted out to accounting professionals.

16.2.7 Mandatory Reporting

Many establishment surveys worldwide are mandatory, because of their importance to their countries' national accounts (Cox and Chinappa 1995). In the decentralized US federal statistical system, however, a number of very important establishment surveys are voluntary. For example except for the Census of Agriculture, most of the surveys conducted by NASS are voluntary, even those considered market sensitive; likewise for most of the Census Bureau's economic indicator surveys, which provide monthly or quarterly statistics that monitor the US economy. Since voluntary establishment surveys rely on the same skewed target population for the same reasons – response by large entities is needed to ensure high quality statistics – these surveys must compete with mandatory surveys for respondent cooperation.

16.2.8 Availability of Administrative and Auxiliary Data

Administrative data or other secondary data sources are often available for businesses and organizations. In many countries, business registry information is required by law, and data may be shared among different government agencies. Despite restrictions on data sharing among US agencies, with a few exceptions (see Confidential Information Protection and Statistical Efficiency Act 2002), the longitudinal nature of many establishment surveys enables statistical agencies to accumulate substantial amounts of historical data for their target populations. These statistical organizations may have the benefit of knowing a lot about both respondents and nonrespondents, facilitating statistical methods.

16.3 Existing Literature Featuring Establishment Survey Experiments

Relative to a larger body of establishment survey literature (see Snijkers et al. 2013), experiments in establishment surveys appear to be uncommon. While we note that our search was not exhaustive, our findings are based on 74 articles or reports found in peer-reviewed journals, conference papers or presentations, or unpublished internal reports. In a few cases, the same research experiments were described in multiple papers, and this duplication was not culled from our counts. Additionally, a few papers describe multiple separate experiments, and we did not count these separately. Many of the journal articles we found were in fields such as organizational behavior, management, and marketing, or associated with specific disciplines, such as public health and education, suggesting that these sources may be more fruitful for learning about peer-reviewed experimental research in business surveys, than are sources in statistical or survey research.

We now summarize this literature with respect to establishment population types, research purposes, and the types of factors manipulated. A list of the papers included in our literature review is available on the website associated with this text at www.wiley.com/go/Lavrakas/survey-research.

16.3.1 Target Populations

Table 16.1 identifies the types of target populations studied by the experiments described in these 74 papers. Survey experiments involving physicians were very common, and we narrowed

Table 16.1 Distribution of articles by type of target population.

Type of target population	
Physician's offices	7
Hospitals/health-care facilities (other than physician's offices)	4
Universities/schools	5
Farms	10
Other specific business subpopulation types or industries (e.g. manufacturers, financial institutions, government entities, organizations/associations, small businesses, etc.)	17
General businesses (not specifically targeted)	31
Total	74

our search, focusing our efforts on finding and reviewing articles involving other types of establishments. These included health-care facilities, universities, and farms. In addition, a number of articles described experiments with homogeneous business subpopulations, such as specific industries, small businesses, or organizational members of a trade association. The remaining 31 articles pertain to the business population more generally.

16.3.2 Research Purpose

Table 16.2 lists the purposes of the research and classifies the factors evaluated in the establishment survey experiments we reviewed. By and large, most of these establishment survey experiments evaluated strategies for improving response rates, the most common of which involved assessment of different communication strategies. Other research questions included assessing uptake of alternative collection modes, and evaluating effects of survey features on measurement error, data quality, and/or response burden.

16.3.3 Complexity and Context

The majority of these experiments relied on single-factor designs. A few multiple-factor designs manipulated two or three factors, while one very large experiment had 128 combinations of

Table 16.2 Purpose of experiments and experimental factors manipulated.

		Factor(s) manipulated in experimental treatment(s)[a]			
Research purpose (dependent variable)	**Number[a]**	**Incentives**	**Reporting mode**	**Contact/ communication strategy**	**Questionnaire or form design**
Response rates	66	28	8	51	8
Response timeliness/ respondent burden	11	1	1	9	2
Mode adoption	7	2	3	5	—
Data quality/ measurement error	15	2	2	7	6

a) The sum is greater than the total number of articles reviewed, as some articles/experiments assessed more than one research purpose and/or varied more than one experimental factor.

treatments. Nearly, all of these establishment survey experiments were embedded in production data collection. A few experiments were carried out in standalone collections, pretesting methodologies for later full production surveys.

16.4 Key Considerations for Experimentation in Establishment Surveys

Conducting experiments in establishment surveys is challenging for a number of reasons related to differences between general population surveys and establishment surveys discussed in Section 16.2. Readers interested in conducting randomized experiments in an establishment survey should consider these differences and interactions among them, and gauge potential consequences for their experimental designs, implementation, data analyses, and interpretation.

We first consider these issues in the context of the following typical strategies for conducting randomized experiments within sample surveys of establishments, while examples using these approaches are provided in Section 16.5:

1) Embed the experiment within production survey data collection;
2) Conduct a separate standalone experiment using sampled cases within a one-time survey collection (e.g. for piloting or research purposes);
3) Conduct a standalone experiment using nonsample cases alongside a production survey;
4) Remove large cases from eligibility for the experiment; and
5) Employ risk mitigation strategies and alternatives.

16.4.1 Embed the Experiment Within Production

For any target population, a clear benefit of this strategy is that effects of experimental treatments can be evaluated under production survey conditions. In addition, the costs associated with the research can be minimized, in comparison to conducting a standalone experiment. Moreover, realistic production costs can be monitored so that the cost effectiveness of experimental treatments can be measured.

However, experiments by their very nature are testing some treatment for which the outcome/impact on subjects is unknown. Thus, unknown and unforeseeable negative treatment effects may jeopardize the validity and accuracy of statistics generated by the survey. This represents a significant downside risk of embedding the experiment within production.

This risk is particularly consequential in establishment surveys, where estimates of totals are the primary statistical products, and data from the largest entities are critical for accurate, reliable official statistics. Experimental treatments that may threaten survey participation and the quality of data reported by these entities could jeopardize key economic statistics. For example, consider testing alternative question versions for a particular variable. Like in household surveys, it may be difficult to evaluate which question version better reflects the true values without some sort of validation study, so which one should be used to generate the official estimate? In addition, sample sizes for the control and each experimental treatment may be insufficient to support reliable population estimates, as sample weights must be expanded accordingly, increasing variances. This, in turn, may lead to confidence intervals too wide to permit detection of statistically significant differences between/among estimates from the experimental treatment(s) and the control.

Consider also large entities included in production experiments. These are likely to be included in future data collections, and negative experiences in experiments may jeopardize future survey participation. In addition, survey organizations may not want to implement experimental procedures that establishments find appealing if the experiment proves them to be ineffective or cost prohibitive. Thus, survey organizations are hesitant to field experimental procedures they cannot guarantee for future use among these critical establishments.

16.4.2 Conduct Standalone Experiments

While this strategy clearly protects official statistical products, undertaking a standalone experiment may be costly to the survey organization, as systems must be set up to conduct the survey, and resources are required to design, build, and carry out the survey. Conducting a standalone experiment often competes for the same staff resources already devoted to survey operations. However, if research procedures do not mimic production survey processes closely, the validity of conclusions may be jeopardized.

Besides extra workload for the survey organization, any standalone experiment adds burden to already heavily burdened establishment survey respondents, concern for which is evident in National Statistical Institutes' efforts to actively measure and reduce establishment survey burden (Bavdaz et al. 2015). The additional burden of standalone survey experiments may jeopardize cooperation and response quality in production surveys, particularly for multisurveyed large businesses. There may also be ethical considerations for a standalone survey with controlled experiments, particularly if data are not going to be used for official statistics.

16.4.3 Use Nonsample Cases

The additional burden associated with standalone survey experiments may be relieved by selecting cases not currently included in production surveys. However, experimenting with establishments not included in ongoing surveys may fail to mimic key aspects of production data collection, as these establishments may differ in substantive ways from those included in many samples. There may be differences in their characteristics; for example, they may be much smaller than production sample cases. Nonsample cases may exhibit different survey-taking behavior compared to establishments that are surveyed multiple times, who are familiar with survey data collection instruments and procedures and may have developed reporting routines. Nonsample cases may also differ in their attitudes and familiarity with survey organizations and statistical products.

In surveys with "take all" strata, there are no nonsample cases to include; likewise for small populations. For example, in USDA surveys of farm operators, for some commodities there are fewer than 100 establishments in the target population. Few, if any, of these farm operations will be excluded in surveys that estimate those commodities, leaving few or no nonsample cases for separate experiments.

16.4.4 Exclude Large Establishments

While this strategy does not directly jeopardize their essential contribution to official statistics or their cooperation on the many production surveys for which they are selected, establishment survey researchers clearly cannot make statistical inferences to large units based on results of experiments where they have been excluded. However, in many countries, large units receive individualized handling from statistical organizations in order to aid/maintain their cooperation and response quality, because of their exceptional characteristics (Brown 2016; Brady 2016; Geisen and Vaasen-Otten 2016; Vella 2016). Thus, their exclusion from controlled experiments is of little consequence, as most experimental results would likely not be applied to these units anyway.

16.4.5 Employ Risk Mitigation Strategies and Alternatives

One primary obstacle to embedding experiments in production surveys is the potential for jeopardizing the necessary contribution of the large entities. The impact may be alleviated through mitigation strategies set up to be implemented based on predetermined progress indicators. From the outset, process and quality indicators, such as response rates or edit failures, should be monitored for all experimental groups, with thresholds that would automatically invoke alternative procedures or interventions. Then, if the experimental treatments appear to be having an adverse effect on key quality indicators, these contingency plans would be readily available for implementation.

We have also noted that conducting experiments outside of production, in a standalone survey setting, is quite costly, may not replicate the production survey conditions, and adds burden to an already heavily burdened target population. Likewise, nonproduction cases may not exhibit the same survey behavior as establishments commonly sampled for establishment surveys that have formed reporting routines.

An alternative research strategy may take advantage of the fact that many establishment surveys contain a longitudinal design, as establishments remain in sample for multiple iterations, to support measurement of change along with level. Surveys with sample rotations may consider using sample units rotated out of production as experimental groups. While this strategy has been used in general population surveys, our literature review suggests that it is less common in establishment surveys.

As for excluding large units from experimental research, we have already pointed out that they often receive individualized handling due to their importance for official statistics, and likely would not be subjected to "successful" experimental treatments subsequently put into production. Thus, embedding experiments in production surveys may be a viable option, with the caveat that large entities be excluded.

Nevertheless, identifying and developing strategies to improve response from and reduce burden for large establishments and entities remains a high priority for establishment survey collection, warranting research using other methods to gauge effectiveness against costs. These large units may be handled as case studies rather than be included in randomized experiments. Case study outcomes can augment results obtained from experimental treatments tested with the remaining sample units.

Because any use of establishment survey sample units will have some impact on production procedures and the activities of survey personnel, a key risk mitigation strategy is for researchers to work closely with survey managers and production staff. Production staff may be tapped for ideas about problems that experimental research can help solve. In addition, a clear understanding of existing production procedures is important for designing experimental procedures that can be implemented in practice. Production staff identify processes and activities that may need to be altered or adapted for experimental procedures. They also help work through how experiments may impact the current surveys or others that include these same establishments in their samples. Finally, production staff will be responsible for implementing procedural changes into production surveys based on experimental results. Involving production staff throughout the experimental process will pay huge dividends for researchers by including operational considerations in both experiments and any potential application of their results.

Table 16.3 summarizes the strengths and weaknesses for the strategies we have discussed. Taking any of these approaches has advantages best realized by careful consideration of their weaknesses.

Table 16.3 Strengths and weaknesses of strategies for conducting experiments in establishment surveys.

Strategy	Strengths	Weaknesses
Embed experimental design in production survey	• Effects of experimental treatments can be evaluated under production survey conditions. • Costs may be reduced relative to conducting a standalone experiment. • Additional burden is minimized or eliminated.	• Unknown and unforeseeable negative treatment effects may adversely affect production statistics generated by the survey. • Costs may be increased relative to normal survey procedures. • Negative treatment effects may adversely impact key individual establishments that are important in future data collections. • Experimental procedures may not be able to be applied to cases ascribed special handling. • Not realizing the experimental context, establishments important for future data collections may naively expect future use of "experimental procedures." • Survey production processes may become more complicated.
Standalone experiment	• Official statistical products will not be impacted. • Regular survey production schedules and constraints are not pertinent and do not have to be met.	• Costs are typically higher than costs for embedding experimental conditions in a production setting. • Production survey conditions may not be fully replicated. • Organizations may not want to impose additional burden on businesses if the collected data will not be used for published statistics, because establishment survey response process is labor intensive and represents a tangible nonproductive cost to the business. • For small or rare (sub)populations, all cases may be needed for the production survey, leaving no nonsample establishments available for experimental research.
Use nonproduction cases alongside production survey	• Official statistical products will not be impacted. • Existing production systems and processes can be used. • Additional cost is minimal.	• Nonproduction cases may have very different reporting behavior, thus results may not pertain to production cases – for example establishments, particularly large ones, are multisurveyed, and thus may develop familiarity and routines for responding to surveys. • Organizations may not want to impose additional burden on businesses if the collected data will not be used for published statistics, because establishment survey response process is labor intensive and represents a tangible nonproductive cost to the business. • For small or rare (sub)populations, all cases may be needed for the production survey, leaving no nonsample establishments available for experimental research.
Exclude large units from experimental design	• Their contribution to official statistics or their cooperation on the many production surveys for which they are selected are not directly jeopardized. • These cases may not be subjected to alternative treatments in production, since they may receive special handling from statistical organizations to aid/maintain their cooperation and response quality.	• Statistical inferences based on experimental results cannot convey to large units. • Experimental results may not apply to large proportions of population estimates accounted for by large units.

16.5 Examples of Experimentation in Establishment Surveys

In this section, we present examples that illustrate the issues described in this chapter. For each example, we provide an overview of the experiment, and describe strategies used to address key experimental considerations.

16.5.1 Example 1: Adaptive Design in the Agricultural Resource Management Survey

NASS recently began developing and evaluating adaptive survey design strategies, whereby alternative data collection procedures may be assigned to different sample subgroups identified using available paradata or other auxiliary information, such as establishment characteristics like size and industrial sector, or response history from previous data collections. While adaptive or responsive design procedures may be used to improve data quality and survey estimates, and/or contain costs, measurement of their impact relies on experimental controls.

One of NASS's more challenging surveys is the Agricultural Resource Management Survey (ARMS), a multiphase survey collecting detailed information on production practices, finances, and farm economics, which provides a critical source of information for USDA and other agricultural policy makers. ARMS' response rates are below 80%, requiring the agency to demonstrate the potential for nonresponse bias, in accordance with the US Office of Management and Budget's Standards and Guidelines for Statistical Surveys (OMB 2006). Thus, the ARMS survey administration team was eager to identify methods for improving response, and they agreed to conduct experiments testing alternative nonresponse reduction strategies.

ARMS data collection typically consists of an initial mailing of the questionnaire with subsequent field follow-up by interviewers. Because the survey is quite lengthy, interviewers usually call nonrespondents to schedule appointments for in-person interviews. Using an adaptive design approach, researchers designed an experiment testing whether an alternative survey recruitment strategy would improve response rates among sample units considered less likely to respond (Wilson et al. 2016). Sample units meeting this criterion – that is, likely nonrespondents – were identified using predictive models that had been developed based on Census of Agriculture data (Earp et al. 2014).

Researchers proposed procedures that eliminated the mail contact for cases predicted to be nonrespondents, and have interviewers do initial in-person visits, providing token incentives, to recruit survey participation. Within the production survey, half of the predicted nonrespondents were assigned to the alternative data collection procedures and half received the existing data collection procedures.

16.5.1.1 Experimental Strategy: Embed the Experiment Within the Production Survey

Because ARMS imposes considerable burden on respondents, a standalone experiment was not a viable option for testing alternative data collection procedures. Instead, researchers worked closely with survey staff to embed procedures within the production survey.

To foster effective collaboration, researchers were encouraged to interact with production staff and processes as much as possible. They assisted with operational survey editing, presented previous research findings, background, and current research plans at workshops for production staff, and took every opportunity to spend time with their counterparts in survey operations. This helped develop relationships with production staff, both professionally and personally, and enabled researchers to understand intricacies and idiosyncrasies of ARMS usual production processes. Researchers learned how and by whom experimental groups would be marked and removed from the mailing, how the incentives and other materials for these cases needed to be distributed to NASS' twelve field offices and interviewers, how information relevant to the experiment would be captured and stored for analysis, and so on.

Researchers were responsible for several tasks to implement experimental procedures – providing specifications for random assignment of treatments to sample cases, developing field instructions and training materials, ordering necessary materials, providing them to field staff, and the like – minimizing additional work for production staff to implement the experiment.

16.5.1.2 Experimental Strategy: Exclude Large Units from Experimental Design

The procedures for these adaptive design experiments also included the removal of the largest and most difficult cases, which are typically handled on a case-by-case basis in data collection. These cases were not included in either the treatment group or the control group, but were marked for field office review. Field staff were also permitted to remove other cases being handled with unique procedures, such as assignment to a specific interviewer, combining data collection across surveys, etc. Cases of these types were handled equivalently in both the treatment and control groups.

The first experimental test of adaptive design for the ARMS survey provided lessons for future embedded experiments. While there was support from most of the production staff, the instructions were not clearly understood by all field staff, and procedures were not uniformly implemented. In addition, procedures to contain costs were found to be overly restrictive. Experimenting within the production survey was critical for identifying these problems.

The embedded field experiment was necessary to evaluate the impact of alternative data collection procedures on changes in current field procedures. Sample handling logistics for any new procedures needed to be feasible within production systems, if they were to be successfully implemented in future cycles. Some changes to field procedures were necessary to accommodate that.

Ultimately, the experiment showed that the alternative procedures, while they could be implemented in production, did little to improve response rates, were more expensive than current procedures, and had minimal impact on data quality. Therefore, the procedures were not implemented in the production survey. The experimental results clearly demonstrated whether alternative procedures were operationally feasible and sufficiently effective, and whether they should be adopted.

16.5.2 Example 2: Questionnaire Testing for the Census of Agriculture

NASS conducts the Census of Agriculture (COA) every five years, for reference years ending in 2 and 7. All known and potential agriculture operations producing and selling $1000 or more are included. Self-administered forms are mailed to potential respondents to collect production, inventory, economic, and demographic information (*For additional information about the COA, see* www.agcensus.usda.gov).

16.5.2.1 Experimental Strategy: Conduct a Standalone Experiment

NASS conducts experiments to test alternative questionnaires and procedures for the COA in a standalone environment, rather than in the production survey, for several reasons. First, since the COA is conducted only once every five years, results from any experiments embedded within a production COA would then have to wait another five years to be implemented in the subsequent COA. Second, as a census, all known and potential farming operations – nearly three million units – are included, so there are no units outside the COA available for parallel experiments.

For these reasons, NASS conducts several experiments with standalone samples prior to COA data collections. Prior to the 2017 COA, for example, NASS conducted the COA content test, a large field test (McCarthy 2017). The content test had several split sample comparisons of different proposed COA questionnaire versions. Alternative versions included different question ordering, format, and content. While the content test replicates much of the operational survey procedures, it also diverges in some significant ways. For example unlike the COA, reporting for the content test is not mandatory, and nonresponse follow-up (NRFU) was more limited than in the COA.

16.5.2.2 Experimental Strategy: Exclude Large Establishments

While the sample designs for the experiments support direct comparison of response rates and item edit and imputation rates, concern for respondent burden led to several concessions for selecting samples from the full COA population. Any establishments selected for nine NASS surveys conducted in the months immediately preceding and following the COA testing were removed from the population eligible for sampling. In addition, each year, NASS tracks the number of surveys for which each list frame record has been selected. Any records with a high number of survey contacts in the current year were also excluded. Finally, as is typical for any research related to NASS surveys, field offices reviewed the COA Content Test samples and removed individual establishments with prior special handling arrangements for data collection, and likely would have removed the high burden establishments in this process. Although these exclusions impact the generalizability of the results, this is a concession often considered in establishment surveys for the reasons discussed previously. It should be noted that, although the COA Content Test was a standalone survey with embedded experiments, production staff treated it as they would any other production survey, and it used existing production data collection and processing systems.

Because samples for the Content Test did not fully represent the COA population, the objective of the split sample experiments was not to produce population estimates. Instead, experiments were designed to compare alternative questions or procedures to determine which were better for inclusion in the COA. Determination of optimal data collection procedures or questionnaire versions was based on comparisons of response rates, and item edit or imputation rates, as appropriate, among experimental treatments.

16.5.3 Example 3: Testing Contact Strategies for the Economic Census

Like NASS, the US Census Bureau conducts a mandatory Economic Census (EC) of nonfarm businesses and organizations every five years, for reference years ending in 2 and 7. Since it collects detailed financial information for each individual physical location within a company, EC response burden can be heavy, particularly for large companies with hundreds, even thousands, of individual establishments (For more information about the Economic Census, see https://www.census.gov/programs-surveys/economic-census.html).

Like the COA, results from embedded experiments cannot be implemented until the next EC, five years later, and may be obsolete by that time. Unlike the COA, however, the EC does not conduct standalone field tests, because the additional respondent burden would be unacceptable. Even a medium-sized company with fewer than 100 establishments would be heavily burdened by the EC's detailed reporting requirements.

Unlike previous economic censuses, the 2017 Economic Census was administered via the web only, with no paper questionnaires available for reporting. Without a standalone EC field test, researchers and production staff were anxious to identify alternative opportunities for experimentally testing various contact strategies to help ensure adequate response rates with minimal cost.

16.5.3.1 Experimental Strategy: Embed Experiments in Production Surveys

A substantial amount of qualitative research worldwide suggests that large multiunit companies have remarkably similar approaches to making decisions about participating in and completing government-sponsored surveys, regardless of the organizational level or amount of detail requested (Snijkers et al. 2013). Based on this assumption, the Census Bureau tested a number of different contact strategies by embedding randomized experiments in several enterprise-level annual surveys in the years prior to the 2017 EC, thus enabling observation of response behavior under typical production survey conditions. Alternatives included differing types and timing of mail contacts, envelope design, and motivational messages, which were tested in the 2014 Annual Survey of Manufactures (ASM), 2014 Annual Retail Trade Survey, 2014 Annual Wholesale Trade Survey, and the 2015 Services Annual Survey (SAS), among others. Descriptions and results of several of these experiments can be found in Tuttle (2016) and Tuttle et al. (2018), and a summary is available on the website associated with this text at www.wiley.com/go/Lavrakas/survey-research.

16.5.4 Example 4: Testing Alternative Question Styles for the Economic Census

The Census Bureau also tested alternative designs for questions asking respondents to identify and specify revenue-generating products or services not explicitly listed in the questionnaire – that is "other/specify" text, also known as "write-ins." A key component of the economic census is an itemized collection of goods and services provided by each physical location, along with associated revenues. This is also a major source of response burden, inhibiting pilot testing of alternative question designs under normal survey conditions.

Instead, the Census Bureau's SAS questionnaire was adapted to mimic product line collection on a limited scale in a small number of industries. Like other annual economic surveys, the SAS sample design consists of a five-year longitudinal panel, with annual controls for new companies or those going out-of-business. The 2016 SAS featured a newly selected sample. Although large companies selected with certainty remain in the new sample, a number of smaller businesses were rotated out.

16.5.4.1 Experimental Strategy: Use Nonproduction Cases Alongside a Production Survey

These "dropped" nonproduction cases presented an opportunity for experimentation because their use would not impact survey estimates. Moreover, these cases were experienced respondents, familiar with survey-taking, with routines for gathering and reporting data. Thus, their response behavior was considered indicative of business survey respondents more generally.

More than 800 single unit businesses, rotating out of the sample from six services sector industries, were identified and assigned to one of two treatment conditions, consisting of alternative versions of questions for obtaining detailed product information. These industries, beauty/nail salons and auto repair shops, were selected because they not only provide services, but often also sell associated retail products. The alternative question versions did not explicitly list all possible retail products. Instead, they were purposely set up to encourage "write-ins," in order to obtain a complete accounting of all products and services provided by these businesses, along with their associated revenues.

Data collection and follow-up activities were conducted alongside the production survey to minimize costs and control problems during implementation. Nevertheless, complexities occurred because the experimental cases still required processing activities outside of production applications – for example, excluding chronic nonrespondents, identifying and excluding cases selected for other surveys, creating separate files, coding flags that directed experimental cases onto different paths in the web survey instrument and into separate output files, providing separate infrastructure for assisting respondents, and other processing activities. Like the earlier NASS example, involving production staff in the planning was critical to identifying procedures and resolving potential glitches prior to implementation. A more complete description of the experimental procedures, question versions, data analysis, and results can be found in Willimack et al. (2018).

16.5.5 Example 5: Testing Adaptive Design Alternatives in the Annual Survey of Manufactures

Researchers conducted a series of experiments investigating the efficacy of subsampling establishment survey nonrespondents in an adaptive design context, rather than conduct follow-up operations on all nonresponding units. While this is a fairly common practice in demographic surveys, it was not often utilized in establishment surveys at the Census Bureau.

This research began with simulation studies to determine optimal subsampling rates that would provide acceptable levels of variance for statistical estimates of totals (Kaputa et al. 2014). Next, a contact strategies experiment was embedded in the 2014 ASM, to determine which of two alternative follow-up strategies to use with a subsample of nonrespondents.

16.5.5.1 Experimental Strategy: Exclude Large Units

The study was restricted to businesses with only one establishment, to avoid complexities associated with subsampling among multiunit businesses. Both response rates and response timing were evaluated (Kaputa and Thompson 2016; Thompson and Kaputa 2015).

16.5.5.2 Experimental Strategy: Embed the Experiment Within the Production Survey

A second experiment was conducted with single unit establishments in the 2015 ASM eligible for the second NRFU, in a manner that enabled evaluation of data quality for two different cost-saving adaptive follow-up strategies. The goal was to identify the least-cost adaptive NRFU method that would maintain data quality relative to the standard – and expensive – full ASM NRFU.

Two treatments were systematically assigned to eligible NRFU cases within domains based on industry and measure of size, to maintain comparability within the skewed target population. The treatment groups were: (i) a control group using the current certified mail follow-up procedure and (ii) an experimental group slated for subsampling based on optimal allocation techniques tested earlier, and further divided to use different NRFU techniques, as follows:

T_{target}: Targeted cases were identified using an optimized allocation procedure that selected larger systematic samples in domains with lower initial response; these cases received reminder letters via certified mail.

T_{compl}: The complement of T_{target}; these cases received reminder letters via standard mail.

This design enabled the following comparisons:

A. Control vs. ($T_{target} + T_{compl}$), where all units received some form of NRFU, with the targeted cases receiving the more expensive follow-up procedure (certified mail) and the remainder receiving the less expensive follow-up procedure (standard mail).

B. Control vs. T_{target}, simulating a scenario where only the targeted cases received follow-up, and nontargeted cases receive no further contact, and thus remain nonrespondents.

In this design strategy, *all* nonresponding sample units received follow-up, mitigating concerns of ASM survey managers, while also permitting experimental evaluation of two adaptive design scenarios that would save cost over the current procedure. Scenario B mimics a more traditional version of subsampling common in general population surveys, while Scenario A ensures that all units receive some form of NRFU, a more acceptable approach among establishment survey practitioners. See Kaputa et al. (2017) for results comparing the effects of these two adaptive design scenarios on response, respondent sample balance, and quality of collected data.

16.6 Discussion and Concluding Remarks

Experimental research in establishment surveys appears infrequently in survey methodology literature, much less often than experiments in surveys of households or individuals, aside from papers and presentations at specialized establishment survey conferences. Instead, peer-reviewed literature describing experiments in business or organizational surveys tends to appear in journals associated with disciplines such as administrative science, organizational behavior, management, or marketing.

Moreover, by far, the most common research question addressed using randomized experiments concerned methods for reducing nonresponse. Many of these experiments consist of rather simplistic single-factor designs. Even recent experimental research conducted by our respective organizations continues to focus on response rates (Marquette et al. 2015; Tuttle 2016; Wilson et al. 2016; Tuttle et al. 2018), and variation of multiple factors is rare.

Randomized experiments appear to be underutilized as a technique for improving establishment survey methods. This is likely because considerations unique to establishment surveys outlined in this chapter make these experiments not only more difficult to field, but they also face different sources of risks to target populations and statistical outputs, particularly with respect to producing official statistics. Instead, as can be seen in the broader establishment survey research literature (see Snijkers et al. 2013), a variety of qualitative and quantitative methodologies, observational studies, comparisons over time, and simulation studies are used to evaluate new or alternative procedures.

However, some research questions may be nearly impossible to address without conducting experiments embedded in production or otherwise carried out under essential survey conditions. Differences between household and establishment surveys require that adjustments be made to strategies for conducting experiments. These differences also demonstrate the critical need to conduct randomized experiments with establishment survey populations.

Consideration of the issues discussed in this chapter can help realize the potential for conducting experiments in establishment surveys. To that end, we have offered suggestions and examples of strategies for mitigating seeming obstacles to experimentation, and demonstrated the need for researchers to work closely with survey production staff to understand production intricacies. In addition, we note that compromises may be necessary for practical implementation of experimental treatments. While consequences for inferences must be noted, these need not deter experimental research among establishment surveys. In fact, strategies to mitigate differences from household surveys may sometimes replicate common survey production procedures, such as excluding large units that receive special handling, and thus have less impact on research conclusions for the reasons we have discussed.

Being able to address perceived obstacles to experimentation in the establishment survey environment opens up many opportunities for developing and undertaking experimental research in this area. Indeed, improvements to the field of establishment survey methodology will only be stronger by including results from randomized experiments.

Acknowledgments

The authors gratefully acknowledge helpful review comments from our editors, Edith de Leeuw and Michael Trauggot, along with useful feedback from Carma Hogue (retired), US Census Bureau, and Jennifer L. Beck, National Science Foundation. We also recognize Brian Kriz and Timothy Lee, students/interns associated with the Joint Program in Survey Methodology, University of Maryland, for substantial assistance with the literature review. Any views expressed are those of the authors and not necessarily those of the US Census Bureau or USDA's National Agricultural Statistics Service.

References

Bavdaz, M. (2010). The multidimensional integral business survey response model. *Survey Methodology* 36 (1): 81–93.

Bavdaz, M., Giesen, D., Cerne, S. et al. (2015). Response burden in official business statistics: measurement and reduction practices of national statistical institutes. *Journal of Official Statistics* 31: 559–588.

Brady, C. (2016). Respondent outreach practices at the U.S. Census Bureau. Presentation at the Fifth International Conference on Establishment Surveys. Alexandria, VA: American Statistical Association.

Brown, P. (2016). Respondent advocacy at statistics New Zealand. *Proceedings of the Fifth International Conference on Establishment Surveys.* Alexandria, VA: American Statistical Association.

Confidential Information Protection and Statistical Efficiency Act (CIPSEA) (2002). Public Law 107-347, 44 United States Code 3501, U.S. Government Printing Office, Washington, D.C. https://www.gpo.gov/fdsys/pkg/PLAW-107publ347/pdf/PLAW-107publ347.pdf (accessed 4 August 2017).

Cox, B.G. and Chinnappa, B.N. (1995). Unique features of business surveys. Chapter 1 in *Business Survey Methods* (ed. B.G. Cox, D.A. Binder, B.N. Chinnappa, et al.), 1–17. New York, NY: Wiley.

Earp, M., Mitchell, M., McCarthy, J., and Kreuter, F. (2014). Modeling nonresponse in establishment surveys: using an ensemble tree model to create nonresponse propensity scores and detect potential bias in an agricultural survey. *Journal of Official Statistics* 30: 701–719.

Eurostat (1958). About Eurostat/overview. http://ec.europa.eu/eurostat/about/overview (accessed 4 August 2017).

Federal Committee on Statistical Methodology (1988). Quality in establishment surveys. Statistical Policy Working Paper 15, U.S. Office of Management and Budget, Washington, DC.

Giesen, D. and Vaasen-Otten, A. (2016). Response burden management for business surveys at statistics Netherlands. *Proceedings of the Fifth International Conference on Establishment Surveys.* Alexandria, VA: American Statistical Association.

Haraldsen, G. (2010). Reflections about the impact business questionnaires have on the perceived response burden and the survey quality. Paper presented at the European SIMPLY Conference on Administrative Simplification in Official Statistics, Ghent, Belgium (December 2–3).

Kaputa, S.J. and Thompson, K.J. (2016). Adaptive design strategies for nonresponse follow-up in economic surveys. *Proceedings of the Fifth International Conference on Establishment Surveys.* Alexandria, VA: American Statistical Association.

Kaputa, S.J., Bechtel, L., Thompson, K.J., and Whitehead, D. (2014). Strategies for subsampling nonrespondents for economic programs. *Proceedings of the Joint Statistical Meetings*, Survey Research Methods Section. Alexandria, VA: American Statistical Association.

Kaputa, S.J., Thompson, K.J., and Beck, J.L. (2017). An embedded experiment for targeted nonresponse follow-up in establishment surveys. *Proceedings of the Joint Statistical Meetings*, Survey Research Methods Section. Alexandria, VA: American Statistical Association.

Marquette, E., Kornbau, M.E., and Toribio, J. (2015). Testing contact strategies to improve response in the 2012 Economic Census. In: *Proceedings of the Joint Statistical Meetings*, Government Statistics Section, 2212–2225. Alexandria, VA: American Statistical Association.

McCarthy, J. (2017). Multi-use field testing: examples from the 2017 Census of Agriculture dry run. *Proceedings of the Joint Statistical Meetings*, Survey Methods Research Section. Alexandria, VA: American Statistical Association.

McCarthy, J., Beckler, D., and Qualey, S. (2006). An analysis of the relationship between survey burden and nonresponse: if we bother them more, are they less cooperative? *Journal of Official Statistics* 22 (1): 97–112.

Rothgeb, J.M., Polivka, A.E., Creighton, K.P., and Cohany, S.R. (1992). Development of the proposed revised Current Population Survey. *Proceedings of the Joint Statistical Meetings*, Survey Research Methods Section. Alexandria, VA: American Statistical Association.

Snijkers, G., Haraldsen, G., Jones, J., and Willimack, D.K. (2013). *Designing and Conducting Business Surveys*. Hoboken, NJ: Wiley.

Sudman, S., Willimack, D.K., Nichols, E., and Mesenbourg, T.L. Jr., (2000). Exploratory research at the U.S. Census Bureau on the survey response process in large companies. *Proceedings of the Second International Conference on Establishment Surveys*, 327–337. Alexandria, VA: American Statistical Association.

Thompson, K.J. and Kaputa, S.J. (2015). Investigating nonresponse subsampling in an establishment survey through embedded experiments. *Proceedings of the Federal Committee on Statistical Methodology Research Conference*, Federal Committee on Statistical Methodology, Washington D.C. (December 1–3). https://nces.ed.gov/FCSM/2015_research.asp (accessed 15 February 2017).

Tuttle, A.D. (2016). Experimenting with contact strategies to aid adaptive design in business surveys. *Proceedings of the Joint Statistical Meetings*, Survey Research Methods Section. Alexandria, VA: American Statistical Association.

Tuttle, A.D., Beck, J.L., Willimack, D.K. et al. (2018). Experimenting with contact strategies in business surveys. *Journal of Official Statistics* 34 (2): 365–395.

U.S. Office of Management and Budget (2006). Standards and guidelines for statistical surveys. https://obamawhitehouse.archives.gov/sites/default/files/omb/inforeg/statpolicy/standards_stat_surveys.pdf (accessed 18 January 2017).

Vella, M.A. (2016). Business respondent advocacy at statistics Canada: where we've been and where we're going? *Proceedings of the Fifth International Conference on Establishment Surveys*. Alexandria, VA: American Statistical Association.

Willimack, D.K. and Nichols, E.M. (2010). A hybrid response process model for business surveys. *Journal of Official Statistics* 26 (1): 3–24.

Willimack, D.K. and Snijkers, G. (2013). The business context and its implications for the survey response process. Chapter 2 in *Designing and Conducting Business Surveys* (ed. G. Snijkers, G. Haraldsen, J. Jones and D.K. Willimack), 39–82. Hoboken, NJ: Wiley.

Willimack, D.K., Linares, K.A., Kriz, B., and Beck, J.L. (2018). Experimenting with alternative question designs for "other, specify" product information in establishment surveys. *Proceedings of the Federal Committee on Statistical Methodology Research and Policy Conference*, Federal Committee on Statistical Methodology, Washington D.C. (March 7–9). https://nces.ed.gov/fcsm/2018_research.asp (accessed 22 March 2019).

Wilson, T., McCarthy, J., and Dau, A. (2016). Adaptive design in an establishment survey: targeting, applying and measuring 'optimal' data collection procedures in the Agricultural Resource Management Survey. *Proceedings of the Fifth International Conference on Establishment Surveys*. Alexandria, VA: American Statistical Association.

Part VII

Introduction to Section on Trend Data

Michael W. Traugott[1] *and Paul J. Lavrakas*[2]

[1] *Institute for Social Research, University of Michigan, Ann Arbor, MI, USA*
[2] *NORC, University of Chicago, 55 East Monroe Street, Chicago, IL 60603, USA*

Surveys with extended cross-sectional time series like the American National Election Study (ANES) and the General Social Survey (GSS) that repeat the same questions to produce trend data play an important role in helping us to understand how opinions, attitudes, and behavior change over time. However, the utility of such trend data is only as good as quality of the questions themselves, accounting for wording, response categories, and order. This creates a tension between using the best measures and maintaining a time series over an extended period of time. The two chapters in this section analyze and describe the results of experimentation with questions in these longitudinal survey data collections.

Smith and Son describe a series of experiments conducted with the spending items in the GSS over a 30-year period, starting with the 1984 survey experiments for all 11 items in the series. They discuss this work in the context of the need for replication while at the same time being sensitive to the changing meaning and understanding of specific language elements. The experimental results are analyzed in terms of terse vs. succinct versions of the descriptions of the spending categories, evaluations of temporal changes in response patterns over time, and an analysis of relevant subgroup differences.

Holbrook, et al. take an unusual approach to investigate this by combining the analysis of survey data with a summary of discussions they had with 24 survey methods experts. The end result is a set of principles and guidelines for experimenting with question wording in longitudinal surveys. In one of their case studies, they reanalyze experiments in the 2008 and 2012 ANES involving the political efficacy items to see whether new items reduce measurement error and produce improved measures of the concept in terms of construct validity. In their second case study, they investigate split-ballot experiments with the wording of a question on spending on "welfare" or "assistance to the poor" in the GSS, and they find inconsistent results for a reduction in associations with racial attitudes. Finally, they have four recommendations for those who are considering experimenting with questions in longitudinal surveys.

Experimental Methods in Survey Research: Techniques that Combine Random Sampling with Random Assignment, First Edition.
Edited by Paul J. Lavrakas, Michael W. Traugott, Courtney Kennedy, Allyson L. Holbrook, Edith D. de Leeuw, and Brady T. West.

Companion Website: www.wiley.com/go/Lavrakas/survey-research

17

Tracking Question-Wording Experiments Across Time in the General Social Survey, 1984–2014

Tom W. Smith and Jaesok Son

Center for the Study of Politics and Society, NORC at the University of Chicago, 1155 East 60th Street Chicago, IL 60637, USA

17.1 Introduction

Replication is a central component of the scientific method (Popper 1959). As Rand and Wilensky (2006) note, "One of the foundational components of the scientific method is the idea of reproducibility. In order for an experiment to be considered valid, it must be replicated." This applies to the social sciences as much as to the physical sciences. As Ansolabehere et al. (1999) (see also Barabas and Jerit 2010) state regarding political science, "external replication of experiments is essential…" In fact, in the case of the social sciences, the need for replication is even greater than in the physical world, since societies continually change and the meaning of the words and phrases that form questions also morph. Yet as Schuman and Presser (1981) observed, "survey research does not have a tradition of replicating results in order to establish their reliability and generality beyond a single survey." This is especially true when it comes to studies comparing substantive variants of question wordings or what Schuman and Presser (1981) call "the tone of wording."

Based on their large set of experiments, including several exact replications of question-wording experiments, Schuman and Presser (1981) concluded that "shifts in marginals that reflect aggregate attitudes change almost certainly both create and destroy question effects" and "time is a more potent variable in relation to question effects than we or others have recognized." They further note that "real changes over time" plausibly explain why several of their experiments failed to replicate.

But despite their warnings, both research by Schuman and Presser (1981) and colleagues (Schuman and Bobo 1988; Schuman and Scott 1989; Presser 1990; Schuman 2002) and others (Bishop et al. 1983; Smith 1987; Rasinski 1989) have found that question-wording effects generally do replicate over time. The occasional failures to replicate are important to note, but are more the exception rather than the rule.

To test the temporal stability of question-wording effects (including noneffect results as well as statistically significant effects), the spending-priority question-wording experiments on the General Social Survey (GSS) are examined. The GSS is a full-probability, in-person sample of adults living in households in the United States (Smith et al. 2015).

Experimental Methods in Survey Research: Techniques that Combine Random Sampling with Random Assignment, First Edition.
Edited by Paul J. Lavrakas, Michael W. Traugott, Courtney Kennedy, Allyson L. Holbrook, Edith D. de Leeuw, and Brady T. West.

Companion Website: www.wiley.com/go/Lavrakas/survey-research

17.2 GSS Question-Wording Experiment on Spending Priorities

Since 1984, the GSS has carried out wording experiments involving its national, spending-priorities items. The battery asks the following:

> We are faced with many problems in this country, none of which can be solved easily or inexpensively. I'm going to name some of these problems, and for each one I'd like you to tell me whether you think we're spending too much money on it, too little money, or about the right amount. First, (READ ITEM A) are we spending too much, too little, or about the right amount on (ITEM A).

This question has been on all GSSs since 1973. In 1984, there were three versions of the traditional 11-item scale, each administered to a random third of the sample. These were: (1) the standard wordings that had been used since 1973 (e.g. NATEDUC), (2) a variant with all terse wordings (the Y questions, e.g. NATEDUCY), and (3) a variant with all verbose wordings (the Z questions, e.g. NATEDUCZ). For the full-wordings of all items see Appendix 1 or Smith, Marsden, and Hout (2015). The standard wordings blended together terse versions that only mention a spending area (e.g. "Foreign aid," "Space exploration program") with verbose versions that indicated an area and promised some improvement or accomplishment (e.g. "Improving and protecting the environment," "Solving the problems of the big cities"). The terse-only versions (Y) replaced terse items used in the standard items with alternative terse formulations (e.g. "The military, armaments, and defense" changed to "National defense") and replaced verbose wordings with terse wordings (e.g. "Improving and protecting the environment" being changed to "The environment"). The verbose-only version generally retained the verbose wordings among the standard wordings and replaced terse wordings with verbose wordings (e.g. "Space exploration" with "Advancing space exploration"). One of the verbose-to-terse adaptations also dealt with the problem that the standard crime item ("Halting the rising crime rate") was based on the often counter-factual idea that the crime rate was increasing by eliminating that assertion in both the terse wording ("Law enforcement") and the verbose variant (Reducing crime). The all-verbose, Z-version was used only in 1984, but all subsequent GSSs (1985–2014) continued the split-ballot administration of the standard and Y-versions, each being fielded on random subsamples.

In addition, four new items were added to the scale making a total of 15 items starting in 1984 ("Highway and bridges," "Social security," "Mass transportation," and "Parks and recreation"). These items were asked in only terse versions, appeared after the standard, 11-items, and were not experimentally manipulated. Subsequently, other items have been added to the scale: "Assistance for children" starting in 2000, "Supporting scientific research" beginning in 2002, and "Developing alternative energy sources" since 2010. These items were successively added to the end of the list and were not asked with variant wordings. Thus, the current scale has 18 items, 11 of which appear in two experimental versions (standard wordings and terse-only wordings) and the seven that follow the initial 11 and appear in only one version. The seven added items are not involved in the experiments discussed herein and are not discussed further. For trends on all 18 spending items see Smith (2015b).

17.3 Experimental Analysis

The initial 1984 experiments revealed a number of moderate-to-large wording differences that have been analyzed in Smith (1984, 1987) and Rasinski (1989). Consistent with the general GSS practices of maintaining consistent measurement across time (Smith 2006) and of replicating experiments over time (Smith 1987; Schuman and Scott 1989), and specifically because several of the spending areas did show notable differences in support across the experimental wordings (Smith 1987; Rasinski 1989), the standard/X versions and the terse-only/Y versions have been asked on all GSSs from 1984 to the present.

Table 17.1 shows the results for the experiments on the 11 items for the 20 GSSs conducted in 1984–2014. Overall, these represent about 41 350 cases (exact Ns varying between 41 247 and 41 428 due to missing values). The percentages in the first columns are the difference in the percent saying "Too little is being spent" on the standard version minus the percent saying "Too little" on the Y variant. "Don't knows" are retained in the base. The second columns report the probability that the differences comparing the two versions are statistically significant. The probability levels are based on the four categories ("Too little," "About right," "Too much," and "Don't know"), not just the differences in the "Too little" percentages. The bottom two rows have the cumulative percentage difference and probability level, and the cumulative sample size. All probability estimates have been adjusted for sample design effects.

First, the main effects of the question-wording experiments are considered. The original three-way experiments in 1984 (Smith 1984) and subsequent replications found somewhat mixed results comparing the verbose versions promising some positive development or outcome to the terse versions only mentioning spending areas. But verbose versions did generally produce more support for spending than the terse versions did. As Table 17.1 indicates, the cumulative 1984–2014 figures show that "Improving the nation's education system" generated less support for more spending than "Education" (−3.5 percentage points) as did "Improving

Table 17.1 Differences on the question-wording experiments, 1984–2014.

	Space		Environment		Health		Cities		Crime		Drugs	
Year	% *D*	*p*	% *D*	*p*	% *D*	*p*	% *D*	*p*	% *D*	*p*	% *D*	*p*
1984	+1.2	0.119	+5.2	0.005	+2.9	0.778	+28.2	0.000	+12.8	0.003	+15.8	0.002
1985	+2.3	0.507	−5.7	0.052	+4.7	0.051	+20.2	0.000	+6.9	0.056	+8.2	0.015
1986	+2.6	0.161	+3.7	0.313	−0.5	0.210	+30.3	0.000	+12.9	0.000	+5.1	0.044
1987	+0.2	0.062	+4.7	0.367	+2.9	0.536	+18.2	0.000	+17.2	0.000	+7.4	0.030
1988	−3.2	0.320	−0.1	0.766	−1.5	0.088	+25.2	0.000	+14.7	0.000	+11.0	0.000
1989	+1.5	0.575	+3.8	0.105	−2.3	0.004	+27.1	0.000	+12.4	0.000	+10.1	0.006
1990	+0.6	0.833	+0.5	0.252	+4.8	0.105	+30.0	0.000	+12.0	0.000	−2.2	0.486
1991	−0.7	0.742	+0.7	0.216	+0.9	0.605	+26.7	0.000	+10.5	0.000	+0.2	0.422
1993	+1.6	0.069	−2.1	0.869	+2.5	0.002	+32.0	0.000	+13.2	0.000	+5.9	0.101
1994	+1.3	0.288	−2.9	0.080	+2.5	0.002	+32.5	0.000	+12.2	0.000	+7.6	0.001
1996	−1.4	0.440	−2.3	0.459	+3.8	0.009	+31.7	0.000	+10.0	0.000	+4.8	0.002
1998	−1.8	0.459	−1.6	0.225	−1.5	0.061	+26.9	0.000	+6.4	0.003	+5.9	0.008
2000	+1.3	0.436	−2.5	0.538	+2.0	0.033	+22.8	0.000	+8.6	0.000	+8.5	0.000
2002	+0.1	0.603	−4.4	0.006	+0.9	0.017	+25.5	0.000	+8.6	0.000	+4.4	0.014
2004	+0.9	0.541	−1.1	0.524	+0.5	0.008	+22.5	0.000	+3.2	0.006	+3.9	0.023
2006	−1.2	0.503	+1.1	0.255	+0.8	0.003	+24.8	0.000	+5.9	0.027	+8.7	0.000
2008	−1.8	0.008	+0.2	0.683	+0.2	0.000	+24.2	0.000	+6.3	0.001	+8.2	0.016
2010	+1.5	0.660	−5.5	.160	−0.7	0.000	+22.1	0.000	+10.4	0.000	+6.5	0.002
2012	−2.9	0.557	+0.8	0.134	+1.4	0.006	+24.3	0.000	+9.2	0.005	+11.8	0.000
2014	−0.6	0.065	−2.9	0.015	+1.0	0.002	+24.5	0.000	+13.9	0.000	+8.3	0.003
All Years	−0.1	0.029	−1.1	0.000	+1.1	0.000	+26.1	0.000	+9.8	0.000	+6.7	0.000
N	41 428		41 403		41 393		41 354		41 374		41 371	

Table 17.1 (Continued)

	Education		Race		Defense		Foreign aid		Welfare	
Year	% *D*	*p*	% *D*	*P*	% *D*	*p*	% *D*	*p*	% *D*	*p*
1984	−1.0	0.074	+11.8	0.000	−0.8	0.775	+1.5	0.025	−38.1	0.000
1985	−5.0	0.022	+3.3	0.253	−1.7	0.270	−1.0	0.437	−46.3	0.000
1986	−7.2	0.032	+12.5	0.000	−1.5	0.031	+1.8	0.042	−38.9	0.000
1987	−5.7	0.122	+8.6	0.010	−5.4	0.083	+1.7	0.408	−45.5	0.000
1988	−5.2	0.033	+9.2	0.000	−2.4	0.272	0.0	0.259	−45.0	0.000
1989	−5.1	0.005	+5.5	0.019	− 0.5	0.980	−1.9	0.157	−43.5	0.000
1990	−0.4	0.977	+9.4	0.002	+1.7	0.592	−0.2	0.250	−44.1	0.000
1991	+0.3	0.901	+4.1	0.110	+1.0	0.411	+1.3	0.382	−41.4	0.000
1993	−3.6	0.439	+12.3	0.000	−1.5	0.261	+2.1	0.026	−46.3	0.000
1994	−1.7	0.283	+7.4	0.000	−0.1	0.885	+0.2	0.585	−45.0	0.000
1996	−5.7	0.003	+9.9	0.000	−0.1	0.997	+0.2	0.015	−39.0	0.000
1998	−3.2	0.297	+7.6	0.000	+0.1	0.928	+0.5	0.001	−45.0	0.000
2000	−3.8	0.207	+4.1	0.000	−2.3	0.441	+1.1	0.001	−42.7	0.000
2002	−4.3	0.077	+7.3	0.000	−3.7	0.014	−2.2	0.002	−45.7	0.000
2004	−5.2	0.002	+7.4	0.000	+1.7	0.719	+1.7	0.009	−46.1	0.000
2006	−0.7	0.237	+9.8	0.000	−2.2	0.051	+1.0	0.029	−43.9	0.000
2008	−5.7	0.175	+6.2	0.002	−2.4	0.146	+1.7	0.000	−44.5	0.000
2010	−4.1	0.074	+4.9	0.005	+2.9	0.155	+1.8	0.036	−43.2	0.000
2012	−1.7	0.685	+11.2	0.000	−0.8	0.323	+2.8	.005	−42.9	0.000
2014	−5.3	0.010	+6.2	0.000	+2.0	0.518	+2.1	0.000	−42.8	0.000
All Years	−3.5	0.000	+7.7	0.000	−0.7	0.001	+0.8	0.000	−43.8	0.000
N	41 408		41 247		41 385		41 362		41 389	

Notes: % *D* = difference in percentage points between % "Too little" on the standard version minus % "Too little" on new/variant version.
Chi-square tests were conducted to check independence between question wordings and responses, and *p*-values are reported here. The test was performed after having adjusted for design effects. It is based on comparisons across four categories: "Too little," "About right," "Too much," and "Don't know." 0.000 stands for $p = 0.000$ or smaller.

and protecting the environment" compared to "The environment" (−1.1 points). But the verbose versions attracted greater support for more spending in five comparisons: for Health (+1.1 points), Drugs (+6.7 points), Race/blacks (+7.7 points), Crime (+9.8 points), and Cities (+26.1 points). The original experimental comparisons involving the verbose z-versions further supported the result that the promising wordings generally garner more support than terse wordings (Smith 1984). This common, but not universal, pro-spending impact of verbose wordings that promised some benefit or improvement is noteworthy since it influences the ranking of spending priorities and thus affects the validity of results and the reliability of inter-topical comparisons (Smith 2015b).

The terse-to-terse experiments were mostly designed to produce greater consistency in the presentation of spending across areas rather than testing substantively different descriptions. This was true for the "Space exploration program" vs. "Space exploration"; "The military, armaments, and defense" vs. "National defense"; and "Foreign aid" vs. "Assistance to other countries." These produced only small differences. As Table 17.1 shows, the cumulative, standard vs. variant differences were −0.1 points for Space, −0.7 points for Defense, and +0.8 points for Foreign aid/assistance. The last terse-to-terse experiment changed "Welfare" to "Assistance to the poor." As Table 17.1 indicates, "Welfare" garnered much less support for spending than "Assistance to the poor" (−43.8 points). Analysis of this result finds that these are fundamentally different stimuli to respondents (Smith 1987; Rasinski 1989; Bishop 2005; Huber and Paris 2013). Clearly, these two very different wordings both must be considered to more fully understand public attitudes toward governmental anti-poverty programs.

Second, the question-wording interactions with time are examined. Overall, an initial examination finds a great deal of consistency in the direction and magnitude of the wording effects. The cumulative, pooled experimental differences were all statistically significant in a chi-squared test (Table 17.1). But an ANCOVA test found no statistically significant main effects for Space ($F(1,34\,295) = 1.059$, $p = 0.303$) and Environment ($F(1,34\,295) = 0.933$, $p = 0.334$) and no association with time for Race/blacks ($F(1,34\,182) = 0.656$, $p = 0.418$). But given 20 samples with an n of over 41 000 cases, achieving statistical significance is not a very high threshold to cross. Many of the cumulative differences are quite modest: Space (−0.1 points), Defense (−0.7 points), Foreign aid/assistance (+0.8 points), Environment (−1.1 points), and Health (1.1 points). But the others range from moderate to very large: Education −3.5 points, Drugs +6.7, Race/blacks +7.7, Crime +9.8, Cities + 26.1 points, and Welfare/assistance −43.8 points.

Table 17.2 summarizes some of the annual patterns for the spending items. Three items showed few statistically significant annual differences, differences in both positive and negative directions, and almost all small differences. Space only once out of 20 comparisons achieved statistical significance in the annual comparisons and showed only small differences in both the positive and negative directions (+12/−8). Defense had two statistically significant differences and mostly small differences in both directions (+6/−14). Environment had three statistically significant differences with an almost equal balance of mostly small positive and negative differences (+9/−11).

Five items showed almost all statistically significant differences, all or all but one in the same direction, and almost all moderate-to-large. Drugs had 17 statistically significant differences

Table 17.2 Summary of annual experiments.

	Years significant[a)]	+/−/0 signs[b)]	Absolute range in differences[c)]	Pooled difference[d)]
Space	1	12/8/0	5.8	−0.1
Environment	3	9/11/0	10.9	−1.1
Health	12	15/5/0	7.1	+1.1
Cities	20	20/0/0	14.3	+26.1
Crime	19	20/0/0	20.4	+9.8
Drugs	17	19/1/0	18.0	+6.7
Education	8	1/19/0	7.5	−3.5
Race/blacks	19	20/0/0	15.8	+7.7
Defense	2	6/14/0	8.3	−0.7
Foreign aid/assist.	13	15/4/1	5.0	+0.8
Welfare/poor	20	0/20/0	8.2	−43.8

a) The number of years the wording difference made a statistically significant difference. Please refer to the p column in Table 17.1.
b) The number of positive-difference years, the number of negative-difference years, and the number of no-difference years are shown in order. Refer to the "%D" column in Table 17.1.
c) The range of percentage differences in absolute values. Refer to the "%D" column in Table 17.1.
d) The difference in percentages when all years are combined. Refer to the "All years" row in Table 17.1.

that were mostly moderate to large and positive in 19 years. Race/blacks had statistically significant differences in 19 years and all were moderate to large and in the positive direction. Crime had 19 statistically significant differences, almost all were large, and all 20 were in the positive direction. Cities had 20 statistically significant differences, all positive, and all large. Welfare/Poor had 20 statistically significant differences, all negative, and all large.

Another three items showed mixed, intermediate patterns. Foreign aid/Assistance had 13 statistically significant differences, mostly in the positive direction (+15/−4/zero 1). Health had 12 statistically significant differences, almost all small and mostly positive (+15/−5). Education had eight statistically significant differences in the small to moderate range, and in the negative direction in 19 years.

Four spending areas – Space, Environment, Race/blacks, and Defense – can be considered to show no meaningful differences and/or no notable variations in differences over time. The first two had been found to have had no statistically significant main difference in the ANCOVA model. Race/blacks had no statistically significant variation across years in the ANCOVA model. Defense had a very small cumulative difference, only two statistically significant annual differences, and a mix of positive and negative differences.

Three spending areas – Cities, Crime, and Foreign aid/Assistance – had statistically significant annual variation and evidence of temporal patterns. Crime had a cumulative difference of +9.8 points, but a dip in the 1998–2008 period to +6.5 points compared to +12.3 points before 1998 and +11.2 points in 2010–2014. Cities had a cumulative gap of +26.1 points and a bulge to +29.6 points in 1989–1998 compared to +24.2 points before and after. Foreign aid/Assistance showed increasing differences over time. Using the "Too much" category rather than the seldom-selected "Too little" category shows an increase from −3.6 points in 1984–1994, to −6.3 points in 1996–2006, and then to −9.6 points in 2008–2014. For the Cities and Crime, the changes have been moderate in magnitude and temporary with the increases and decreases not being permanent. For Foreign aid/Assistance there has been a general increase in the size of the wording effect over time (although there is still also a notable amount of year-to-year fluctuation). The increases in the wording effect come from fact that opposition to "Foreign aid" has diminished more than has opposing "Assistance to other countries." It is possible that "Assistance to other countries" has been understood differently over time, perhaps increasingly encompassing elements such as US involvement in military action in other countries. In each of these cases, there was no meaningful change in the direction of the wording effects.

Four spending areas – Health, Drugs, Education, and Welfare/Poor – showed statistically significant annual variation, but no temporal pattern that suggests a trend or a substantive explanation for the variation. For Health the cumulative difference is small (+1.1 points), and direction occasionally changes (+15/−5). The differences also fluctuated more before 1998 that since 1998. The largest negative difference (−2.3 points) and the largest positive difference (+4.8 points) were in adjoining years (respectively 1988 and 1989) for an absolute range of 7.1 points. Since 1998, the only negative difference was −0.7 point and the highest positive was +2.0 for a range of 2.7 points. For Drugs, Education, and Welfare/Poor, the direction of the differences were consistent (with 2 exceptions out of 60 comparisons). The ranges in differences were 7.5 points for Education, 8.2 points for Welfare/Poor, and 18.0 points for Drugs.

There are two likely explanations for the statistically significant, but temporally unpatterned fluctuations in differences. First, the total survey error perspective indicates that there is more random variation than is explained by sampling variance even when design effects are taken into consideration (Smith 2005, 2011). Since the wording of the items and the placement of the items in the questionnaire have been unchanged, probably the most likely source of additional, random variance comes from interviewers. They can misread items, emphasize some items or some response options more than others, mishear responses, miscode responses, and trigger interviewer–respondent interaction effects. These in turn increase both random and systematic error. Second, media coverage and political discussions may frame and present spending areas differently from year-to-year. Specifically, terms used by the paired wordings for each spending area may be used more or less frequently and more or less favorably, thereby leading to increases or decreases in the wording differences. Thus, it is likely that some of the variation in question-wording differences not explained by sampling variance is artifactual, resulting from additional sources of random and/or systematic measurement error and some is substantial resulting from real, but transitory, fluctuations in the language and nature of media coverage and political discourse. Overall, the impacts of these factors on these spending areas are small and have not produced any apparent temporal pattern or basic change in the direction or magnitude of the question-wording effects.

Third, conditional, question-wording effects or interactions across subgroups are explored. Question-wording effects are usually assumed to be a function of the wordings themselves and apply uniformly across respondents. But just as there are conditional context effects that vary in magnitude and even direction across subgroups (Smith 1982, 1991), so there can be conditional, question-wording effects (Schuldt & Konrath 2011).

There are thousands of other variables on the GSS that could be used to test for conditional wording effects. Two groups of candidates were selected as potentially interactive. First, 10 variables that had a direct substantive tie to specific spending areas were examined. For example race was used with Race/blacks, community type with Cities, and confidence in education with Education. In general, the expected main effects were found. Blacks were more supportive of spending on Race/blacks than whites were. Residents of large central cities were more in favor of spending for "Big cities" than residents of small communities and rural areas were. Those with less confidence in the educational system supported more spending for education. Since these groups differed in their support of spending in the linked areas, it was plausible that they could also differ in their responses to the various question wordings.

Table 17.3 indicates there were statistically significant differences for 6 of the 10 substantively related variables. The Race/blacks question-wording effect was larger for whites than for blacks and others. The Welfare/Poor effect was greater for those with middle household incomes than those with high or low incomes. The Education effect was greatest for those with a great deal of confidence in education, notably diminished among those with only some confidence, and was very small and in reversed direction for those with hardly any confidence. Health had the largest effect among those with hardly any confidence, a smaller positive effect for those with only some confidence, and a negative effects for those with a great deal of confidence. Drugs showed a larger effect among those for the legalization of marijuana than among those opposed to legalization. Crime had its largest effect among those thinking courts were too harsh, an intermediate effect for those saying courts were about right in their sentencing, and those wanting courts to be harsher showed a somewhat smaller effect. These six comparisons showed highly, statistically significant differences, and with the exception of Welfare/Poor with household income had effects that monotonically changed across subgroups (e.g. decreasing in size across the courts responses). There were no statistically significant interaction for Space and confidence in the scientific community, Defense and confidence in the military, Crime and being afraid of walking along at night, and Cities and community type.

Second, Table 17.4 shows that there were statistically significant differences for 4 of the 11 comparisons by political-party identification. While party identification was not explicitly tied to spending areas (e.g. as race was to spending on Race/blacks), the political parties and their adherents have well-established differences toward government spending in general and support for many programs in particular. The expected main effects were generally found. For example, Democrats were more supportive than Republicans were on spending for

Table 17.3 Conditional question-wording effects: question-wording differences by selected variables, pooled – 1984–2014.

			Race			
	Whites	Blacks	Others	Difference (white–black)	*p*	
Race/blacks	+8.2	+5.9	+4.8	3.4	0.000	
			Household income			
	Low	Low-mid	Mid-high	High	Difference (low-high)	*p*
Welfare/poor	−41.7	−46.1	−46.5	−42.4	0.7	0.000
			Confidence in institution[a]			
	Great deal confidence	Only some confidence	Hardly any confidence	Difference (great deal-hardly any)	*p*	
Education/education	−6.5	−1.9	+0.9	7.4	0.001	
Health/medicine	−3.4	+2.0	+7.3	10.7	0.000	
Space/Sci. Com.	−0.5	−0.6	−1.6	1.1	0.620	
Defense/military	+0.8	−0.9	−1.4	2.2	0.260	
			Legalization of marijuana[b]			
	Legalize marijuana	Don't legalize	Difference (legal–illegal)	*p*		
Drugs	+3.2	+7.9	4.7	0.000		
			Courts dealing with criminals[c]			
	Courts too harsh	Courts about right	Courts not harsh enough	Difference (too harsh–not harsh enough)	*p*	
Crime	+15.8	+11.0	+9.6	6.2	0.000	
			Afraid to walk at night[d]			
	Afraid to walk	Not afraid to walk	Difference (afraid–not afraid)	*p*		
Crime	+10.4	+9.9	0.5	0.638		
			Residential place[e]			
	Large central city	Suburb of LCC	Ex-urbia of LCC		Difference (LCC-open)	*p*
Cities	+26.2	+27.2	+24.7			
	Medium central city	Suburb of MCC	Ex-urbia of MCC			
Cities	+24.0	+27.7	+25.7			
	Small city	Town	Small town	Open country		
Cities	+30.0	+26.3	+23.5	+23.0	+3.2	.363

Notes: Differences are reported in absolute values. The interaction effects between wording differences and selected variables were tested using the two-way ANOVA (adjusted for the design effect) and the results are reported here (*p*). The question wordings of the selected control variables are as follows:

a) I am going to name some institutions in this country. As far as the people running these institutions are concerned, would you say you have a great deal of confidence, only some confidence, or hardly any confidence at all in them? Education/Scientific community/Military/Medicine.
b) Do you think the use of marijuana should be made legal or not?
c) In general, do you think the courts in this area deal too harshly or not harshly enough with criminals?
d) Is there any area right around here – that is, within a mile – where you would be afraid to walk alone at night?
e) Within an SMSA and a large central city (over 250 000); a suburb of a large central city; an unincorporated area of a large central city; Within an SMSA and a medium central city (50 000–250 000); a suburb of a medium central city; an unincorporated area of a medium central city; a small city (10 000–50 000); a town (25 000–9999; incorporated place less than 2500 or designated unincorporated area 1000+; open country, unincorporated area.

Table 17.4 Conditional question-wording effects: Question-wording differences by political party, pooled – 1984–2014.

Spending areas	Democrats	Independents	Republican	Difference (dem–rep)	*p*
Space	−0.3	−0.8	+0.4	0.7	0.713
Environment	−1.3	+1.6	−2.0	0.7	0.088
Health	+0.6	+2.5	+0.9	0.3	0.265
Cities	+27.0	+24.4	+26.2	0.8	0.025
Crime	+12.5	+15.5	+3.9	8.6	0.000
Drugs	+5.3	+7.7	+8.1	2.8	0.033
Education	−3.7	−2.5	−3.7	0.0	0.315
Race/blacks	+8.6	+7.9	+6.6	2.0	0.478
Defense	−1.1	+0.3	−0.4	0.7	0.942
Foreign aid/assist.	+0.8	+0.8	+0.7	0.1	0.863
Welfare/poor	−48.0	−47.2	−37.4	10.6	0.000[a)]

Differences are reported in absolute values. The interaction effects between wording differences and selected variables were tested using the two-way ANOVA (adjusted for the design effect) and the results are reported here (*p*).

a) Some ANOVA and loglinear models show this as statistically significant and others do not, so this relationship is more tentative than others.

Welfare/Poor and Republicans more in favor of spending for Defense than Democrats were. Table 17.4 shows statistically significant, conditional wording effects for four effects, but only the partisan differences on Drugs, Crime, and Welfare/Poor are notable. On Drugs, Democrats show a somewhat smaller question-wording effect than either Independents or Republicans (respectively +5.3 points vs. +7.7 and +8.1 points). While the difference in partisan differences is small (8.1 vs. 5.3 or 2.8 points), it is quite consistent over the years. In the 20 annual comparisons, Democrats showed the smallest difference 13 times and the largest only once, while Independents and Republicans had the largest difference respectively in 9 and 10 years. Likewise, in multivariate models, Timberlake, Rasinski, and Lock (2001) have shown that the Drugs-wording effect is not constant across social groups, being larger among non-Democrats and nonliberals. On Crime, all partisan groups are more supportive of spending for "Halting the rising crime rate" than for "Law enforcement," but the boost is higher for Democrats (+12.5 points) and Independents (+15.5 points) than for Republicans (+3.9 points). The difference in partisan differences was 8.6 points (or 11.6 points if Independents and Republicans were compared). Similarly, while all parties are much less supportive of spending on "Welfare" than for "Assistance to the Poor," the difference is larger for Democrats (−48.0 points) and Independents (−47.2 points) than for Republicans (−37.4 points). The difference in partisan differences was 10.6 points.

Finally, question-wording effects may be both conditional and interact with time. One example will illustrate this situation. As Table 17.4 indicates, Democrats and Independents have across all years pooled had larger question-wording effects on "Halting the rising crime rate" vs. "Law enforcement" than Republicans have had. Table 17.5 reveals that this conditional, question-wording effect has varied over time. In 1984–1994, the average partisan differences were fairly similar in magnitude (+13.6 points for Democrats, +14.6 points for Independents, and +10.4 for Republicans). In 1996–2012, the difference in partisan differences became large with +10.6 points for Democrats, +16.3 points for Independents, and −0.7 for Republicans for

Table 17.5 Conditional question-wording effects by time: the effect of party identification on crime spending.

Year	Democrats	Independents	Republicans
1984	+9.3	+14.0	+16.0
1985	+3.4	+15.1	+9.1
1986	+12.3	+16.2	+10.8
1987	+17.3	+21.0	+13.6
1988	+18.1	+0.7	+15.8
1989	+15.8	+6.9	+11.1
1990	+13.2	+16.8	+9.5
1991	+10.5	+15.5	+8.0
1993	+18.2	+25.4	+4.9
1994	+17.6	+14.7	+4.7
1996	+13.2	+21.3	+0.7
1998	+12.0	+12.1	−1.3
2000	+8.3	+19.4	+2.4
2002	+15.4	+12.5	−3.2
2004	+4.6	+17.2	−4.2
2006	+10.0	+11.1	−2.3
2008	+11.6	+15.1	−2.8
2010	+10.1	+17.9	+4.0
2012	+10.2	+19.8	+0.1
2014	+17.9	+15.2	+7.4
All	+12.5	+15.5	+3.9

an absolute difference in differences of 11.3–17.0 points. This switch was foreshadowed by a monotonic decrease in the question-wording effect for Republicans from +15.8 points in 1988 to −1.3 points in 1998. Thus, Republicans no longer differed in their support for more spending to combat crime in response to "Halting the rising crime rate" and "Law enforcement," while for Democrats and Independents these different formulations continued to make a difference.

17.4 Summary and Conclusion

Overall, wording effects appear to be mostly stable across 20 years of the 11 replicated spending experiments. This is consistent with results from the handful of measurement experiments cited above that have been replicated across time and a larger body of research on the forbid/not allow wording effect (Schuman and Presser 1981; Hippler and Schwarz 1989; Holleman 2000; Schuman 2002; Reuband 2003; Bishop 2005; Frankovic 2007; Smith 2015a). But, while temporal stability generally prevails, it is not a given. While the direction of question-wording effects did not change sign nor reliably switched from showing an effect to showing no effect, the magnitude of several effects did vary across time. In addition, many question-wording effects were conditional, varying in size, and sometimes in direction across subgroups. Moreover, conditional effects can interact with time as the crime and political party temporal interaction demonstrates. This cautions researchers to consider that even well-established measurement effects may not be constant across time.

The repeating of experiments concerning question wordings has several benefits. First, it follows from the basic principle of the scientific method of replicating experiments to confirm scientific results. Second, replication increases the precision of results by reducing sampling variance. This is especially important when the experiment is designed to calibrate two measures such as when one indicator is to replace another measure (Smith 2006). The question-wording differences on spending figures show that annual estimates can range substantially (e.g. from −5.7 points to +5.2 points for Environment). If either of those extremes had been solely relied upon rather than the cumulative differences of −1.1 points, the question-wording effect for Environment would have been seriously distorted and calibration would have been errant. This is true for all of the 11 question-wording experiments. Third, replication enhances time-series analysis in several ways. The meaning of words and phrases are not constants and do change over time. For example, many standard racial terms have changed over time such as from colored to negro to black to African American (Smith 1992) and a Gallup question asked in the mid-1950s ("From what you have heard, read, or think, which large city in the United States has the gayest night life?") would have a very different meaning today. Replication can detect and quantify such changes. In addition, the results presented above show that differences can be so pronounced (e.g. Cities and Welfare/Poor) that the two experimental wordings essentially represent different measures tracking separate time series. Repeating experiments in effect established two distinct time series, both of which are valuable to monitor over time. Finally, replication allows the pooling of experiments. This greatly facilitates the study of question-wording effects among subgroups in general and especially the detection of conditional, question-wording effects. Replication should be a standard tool in survey research's methodological toolbox.

17.A National Spending Priority Items

> We are faced with many problems in this country, none of which can be solved easily or inexpensively. I'm going to name some of these problems, and for each one I'd like you to tell me whether you think we're spending too much money on it, too little money, or about the right amount. First, (READ ITEM A) are we spending too much, too little, or about the right amount on (ITEM A).

Items covering 11 areas have been asked in every GSS since 1973 (The Space Exploration Program, Improving and Protecting the Environment, Improving and Protecting the Nation's Health, Solving the Problems of the Big Cities, Halting the Rising Crime Rate, Dealing with Drug Addiction, Improving the Nation's Education System, Improving the Condition of Blacks, The Military, Armaments, and Defense, Foreign Aid, Welfare).

Since 1984, experiments have been conducted and 11 alternative wordings for the original spending items have been asked on random subsamples (Space Exploration, The Environment, Health, Assistance to Big Cities, Law Enforcement, Drug Rehabilitation, Education, Assistance to Blacks, National Defense, Assistance to Other Countries, Assistance to the Poor).

In 1984 only, the Z version included 11 alternative, verbose wordings (Advancing Space Exploration, Improving and Protecting the Environment, Improving and Protecting the Nation's Health, Solving the Problems of the Big Cities, Reducing Crime, Reducing Drug Addiction, Improving the Nation's Education System, Improving the Conditions of Blacks, Strengthening National Defense, Helping Other Countries, Caring for the Poor).

References

Ansolabehere, S.D., Iyengar, S., and Simon, A. (1999). Replicating experiments using aggregate and survey data: the case of negative advertising and turnout. *American Political Science Review* 93: 901–909.

Barabas, J. and Jerit, J. (2010). Are survey experiments externally valid? *American Political Science Review* 104: 226–242.

Bishop, G.F. (2005). *The Illusion of Public Opinion: Fact and Artifact in American Public Opinion Polls*. Lanham, MD: Rowman & Littlefield.

Bishop, G.F., Oldendick, R.W., and Tuchfarber, A.J. (1983). Effects of filter questions in public opinion surveys. *Public Opinion Quarterly* 47: 528–546.

Frankovic, K. (2007). Forbid or allow? *CBS News* (3 October 3). www.cbsnews/news/forbid-or-allow

Hippler, H.-J. and Schwarz, N. (1989). Not forbidding isn't allowing: the cognitive basis of the forbid-allow asymmetry. *Public Opinion Quarterly* 50: 87–96.

Holleman, B. (2000). *The Forbid/Allow Asymmetry*. Amsterdam: Rodopi.

Huber, G.A. and Paris, C. (2013). Assessing the programmatic equivalence assumption in question wording experiments: understanding why americans like assistance to the poor more than welfare. *Public Opinion Quarterly* 77: 385–397.

Popper, K.R. (1959). *The Logic of Scientific Discovery*. New York: Harper and Row.

Presser, S. (1990). Measurement issues in the study of social change. *Social Forces* 68: 856–868.

Rand, W. and Wilensky, U. (2006). Verification and validation through replication: a case study using Axelrod and Hammond's ethnocentricism model. Paper presented to the Conference of the North American Association for Computational Social and Organizational Sciences, South Bend.

Rasinski, K.A. (1989). The effect of question wording on public support for government spending. *Public Opinion Quarterly* 53: 388–396.

Reuband, K.-H. (2003). The allow-forbid asymmetry in question wording – a new look at an old problem. *Bulletin of Sociological Methodology* 80: 1–10.

Schuldt, J.P., Konrath, S.H., and Schwarz, N. (2011). 'Global warming' or 'climate change'? Whether the planet is warming depends on the question wording. *Public Opinion Quarterly* 75: 115–124.

Schuman, H. (2002). Sense and nonsense about surveys. *Contexts* 1: 40–47.

Schuman, H. and Bobo, L. (1988). Survey-based experiments on white racial attitudes towards residential integration. *American Journal of Sociology* 94: 273–299.

Schuman, H. and Presser, S. (1981). *Questions and Answers in Attitude Surveys*. New York: Academic Press.

Schuman, H. and Scott, J. (1989). Response effects over time: two experiments. *Sociological Methods and Research* 17: 398–408.

Smith, T.W. (1982). Conditional order effects. GSS Methodological Report No. 20. Chicago: NORC, May.

Smith, T.W. (1984). A preliminary analysis of methodological experiments on the 1984 GSS. GSS Methodological Report No. 30. Chicago: NORC.

Smith, T.W. (1987). That which we call welfare by any other name would smell sweeter: an analysis of the impact of question wording on response patterns. *Public Opinion Quarterly* 51: 75–83.

Smith, T.W. (1991). Thoughts on the nature of context effects. In: *Context Effects in Social and Psychological Research* (ed. N. Schwarz and S. Sudman), 163–184. New York: Springer-Verlag.

Smith, T.W. (1992). Changing racial labels: from colored to negro to black to African American. *Public Opinion Quarterly* 56: 496–514.

Smith, T.W. (2005). Total survey error. In: *Encyclopedia of Social Measurement* (ed. K. Kempf-Leonard), 857–862. New York: Academic Press.
Smith, T.W. (2006). *Formulating the Laws of Studying Societal Change, Version 2.2.* Chicago: NORC.
Smith, T.W. (2011). Refining the total-survey error perspective. *International Journal of Public Opinion Research* 23: 464–484.
Smith, T.W. (2015a). Three forbid-allow experiments on the 1989 GSS. NORC report.
Smith, T.W. (2015b). Trends in national spending priorities, 1973–2014. NORC report.
Smith, T.W., Marsden, P.V., and Hout, M. (2015). *General Social Survey Cumulative Codebook: 1972–2014.* Chicago: NORC. http://gss.norc.org/documents/codebook/GSS_Codebook.pdf.
Timberlake, J.M., Rasinski, K.A., and Lock, E.D. (2001). Effects of conservative sociopolitical attitudes on public support for drug treatment spending. *Social Science Quarterly* 82 (1): 184–196.

18

Survey Experiments and Changes in Question Wording in Repeated Cross-Sectional Surveys

Allyson L. Holbrook[1], David Sterrett[2], Andrew W. Crosby[3], Marina Stavrakantonaki[4], Xiaoheng Wang[4], Tianshu Zhao[4], and Timothy P. Johnson[5]

[1] *Departments of Public Administration and Psychology and Survey Research Laboratory, University of Illinois at Chicago, Chicago, IL 60607, USA*
[2] *NORC at the University of Chicago, Chicago, IL 60603, USA*
[3] *Department of Public Administration, Pace University, New York, NY 10038, USA*
[4] *Department of Public Administration, University of Illinois at Chicago, Chicago, IL 60607, USA*
[5] *Department of Public Administration and the Survey Research Laboratory, University of Illinois at Chicago, Chicago, IL 60607, USA*

18.1 Introduction

Repeated cross-sectional surveys serve as a valuable resource for researchers, practitioners, and both private and public organizations (cf., Caplow et al. 2001). Surveys such as the American National Election Studies (ANES) and the General Social Survey (GSS) provide longitudinal data for researchers in political science, sociology, and other social science disciplines. Other repeated cross-sectional surveys are used to collect important data about health (e.g. the Behavioral Risk Factor Surveillance System, National Health and Nutrition Examination Survey, National Health Interview Survey, the National Survey on Drug Use and Health, the National Survey of Family Growth, the Collaborative Psychiatric Epidemiology Studies, the European Social Survey), and the behaviors, and characteristics of a nation or more than one nation (e.g. the Current Population Survey, the American Housing Survey, the Consumer Expenditure Survey, the Survey of Income and Program Participation, the Residential Energy Consumption Survey, the Survey of Consumer Finance, the Survey of Consumer Attitudes and Behavior). Although some of these surveys may include elements of panel surveys (i.e. where the same individuals are interviewed more than once), each of these surveys involves periodically sampling and interviewing fresh national samples.

These longitudinal surveys are extremely impactful and often well-funded. As such, the organizations conducting the surveys are motivated to obtain the highest quality data. Because many of these data collections have been ongoing for decades, there have been substantial methodological and substantive developments during the time they have been administered that have implications for reducing a number of sources of survey error. Specifically, there has been much research on reducing measurement error, particularly in the area of improving questionnaire design. Ongoing research about survey satisficing (e.g. Chang and Krosnick 2009; Holbrook et al. 2003; Krosnick 1991; Krosnick et al. 1996; Narayan and Krosnick 1996), response formats and scales (e.g. Saris et al. 2010; Schwarz et al. 1985; see Tourangeau et al. 2000 for

Experimental Methods in Survey Research: Techniques that Combine Random Sampling with Random Assignment, First Edition.
Edited by Paul J. Lavrakas, Michael W. Traugott, Courtney Kennedy, Allyson L. Holbrook, Edith D. de Leeuw, and Brady T. West.

Companion Website: www.wiley.com/go/Lavrakas/survey-research

a review) and other aspects of question wording (e.g. Holbrook et al. 2007; Krosnick et al. 2002; Schwarz 1996) have substantially changed conventional wisdom about question wording in surveys in the last 30 years (see Schaeffer and Dykema 2011; Sudman et al. 1996 for reviews). The other primary motivation for changing question wording is in response to societal or cultural changes in definitions, terminology, or meaning. Word choices that are appropriate for data collected at one time may become inappropriate at a later time either because the meaning of the words change or are replaced with other terminology. As such, one might imagine that repeated cross-sectional surveys would quickly and frequently change question wording to adjust their methods to conform with current methodological best practices and language/terminology in order to minimize measurement error.

However, strong arguments also exist for not changing question wording in these types of surveys. In a number of cases, questions have been fielded for long periods of time, and one of their major advantages is that they allow longitudinal analysis including tracking of trends (e.g. examining public support for legalized abortion or attitudes toward immigration) and analysis of changes in the relationship among variables (e.g. the predictors of Presidential candidate choice). Furthermore, evidence regarding question wording in particular suggests that small changes in question wording can have a significant impact on survey responses (e.g. Schuman and Presser 1981). In fact, changes in question wording that affect survey responses have led researchers to conclude incorrectly that the opinions, beliefs, or behaviors of the American public have changed (Menard 2008). A well-known example is the book *The Changing American Voter* published in 1976, which showed substantial changes in the American electorate that later research suggested were likely almost entirely due to changes in the ANES surveys starting in 1964 (e.g. Bishop et al. 1978; Sullivan et al. 1978). Similarly, changes in how variables are coded may also be misinterpreted as a substantive change (Smith 2005). As such, even if newer measures successfully reduce measurement error, they may not be desirable if changes in measures introduce attributional ambiguity about changes in trends over time.

As a result of these competing methodological motivations to use the best available measures and to maintain consistency over time, changing the methodology of repeated cross-sectional surveys (and panel surveys which have similar competing impetuses) has been widely discussed as a challenge to repeated cross-sectional surveys that are used to track trends over time (e.g. Tourangeau 2004). Despite near universal agreement that this is a major challenge facing survey researchers, there is a notable lack of research laying out guidelines for either making or empirically testing these kinds of changes. The goal of this chapter is to address these issues. First, we describe what repeated cross-sectional surveys are and their uses. We then briefly review how changes in both questionnaire design best practices and language norms impact question wording over time. Next, we summarize the results of discussions about this topic with more than 20 experts in the field of survey methods, many of whom have been involved in designing and conducting repeated cross-sectional surveys. Then, we present two case studies in which we analyze the results of experiments in two large nationally representative repeated cross-sectional surveys. Finally, we suggest some best practices and steps for implementing and testing question wording experiments in repeated cross-sectional surveys.

18.2 Background

18.2.1 Repeated Cross-Sectional Surveys

Repeated cross-sectional surveys are a type of longitudinal data collection where a fresh cross-sectional sample is periodically drawn and surveyed using similar methods and instrumentation. This is in contrast to a panel study where the same respondents are interviewed at each wave of the survey. Many of the surveys listed earlier in this chapter combine elements of both cross-sectional and panel surveys. For example, the Presidential election cross-sectional surveys conducted by the ANES involve surveying respondents both before and after the Presidential election. The challenges associated with changing the methodology, and specifically the question wording used in repeated cross-sectional surveys, are similar to those faced with panel surveys, although the latter face additional challenges as well. Both types of longitudinal surveys are often used to track changes over time. In particular, repeated cross-sectional surveys are used to track trends over time in attitudes, behaviors, and beliefs (e.g. Guest and Wierzbicki 1999; Schuman et al. 2007; Shapiro and Mahajan 1986).

18.2.2 Reasons to Change Question Wording in Repeated Cross-Sectional Surveys

18.2.2.1 Methodological Advances

A number of advances in survey methods have been designed to reduce different sources of error identified in the Total Survey Error perspective (see Groves et al. 2009). One major focus has been on reducing measurement error, particularly through the wording and design of survey questions (e.g. Schwarz 1999). There have been enormous advances in question wording and questionnaire design in the last 50 years, and it would be impossible to summarize them here. Most of these advances, however, have their roots in a relatively small set of theories about the processes by which respondents answer survey questions. The CASM or Cognitive Aspects of Survey Methodology Movement is perhaps the most influential theory that has contributed to the questionnaire design literature (e.g. Sirken et al. 1999a,b). This movement is based on the theory that respondents go through a set of standard cognitive steps in answering survey questions (see Tourangeau et al. 2000 for a review). One theory based in the CASM movement is the theory of survey satisficing (Krosnick 1991) that argues that when respondents are motivated and able, they go through four cognitive steps to answer survey questions completely and thoroughly. In contrast, when respondents are not able or willing, they engage in survey satisficing in which they do not complete the retrieval and integration steps and this results in reduced survey data quality (e.g. Saris et al. 2010; Holbrook et al. 2003).

Surveys are also conversations that take place in a social context. As such, survey data may be affected by conversational norms and conventions (e.g. Holbrook et al. 2000; Schwarz 1996; Schwarz et al. 1985; Schwarz et al. 1991), social processes such as self-presentation concerns (e.g. Holbrook and Krosnick 2010) and the characteristics and behaviors of others who are present including the interviewer (e.g. Foucault et al. 2009; West and Peytcheva 2014). Finally, research suggests that survey responses may be influenced by aspects of the question's context, including more narrow factors like question order (Schwarz et al. 1991) as well as broader environmental factors such as the weight of a clipboard on which one is completing the survey (Jostman et al. 2009) or the weather during the interview (e.g. Williams and Bargh 2008; Ijzerman and Semin 2010). Evidence about the impact of the broader context comes predominantly from laboratory research and has not been fully tested in surveys. The implications of these latter perspectives for questionnaire design have also not been fully developed, but one could imagine that some types of questions are less affected by context than others (e.g. Chandler et al. 2012) and that survey researchers would be interested in designing surveys that are unaffected by context.[1] These various advances in questionnaire design can lead researchers to recognize that the items that have been employed to monitor or track social trends over time may suffer

1 A large body of research has also examined the visual appearance of self-administered questionnaires, both paper-and-pencil instruments and web surveys (Dillman et al. 2008), but many of these advances are less applicable to interviewer-administered surveys.

from some of the design flaws that have been identified by ongoing research. To understand how this is handled in practice, we (i) explored the research literature on this topic, which was surprisingly sparse and (ii) reached out to notable survey experts regarding their experiences and advice on this matter.

18.2.2.2 Changes in Language and Meaning

Other motivations for changing survey question wording involve changes in language or meaning that occur over time, or changes that occur in official language or definitions. Changes in meaning or terminology can occur spontaneously and informally as language evolves over time. For example, many questions about African-Americans in surveys today asked in the past about Blacks or Negros (Smith 1992). Changes in meaning may also be linked to specific events. For example, in the second case study, we'll examine researchers who wanted to move away from using the term "welfare" because it had become associated with race in the wake of polarized political debate about the policy. Other changes may be a function of evolution in official or legal terminology. For example, the term "mental retardation" was recently changed to "intellectual disability" in diagnostic guidelines (American Psychiatric Association 2013) and federal regulations (Social Security Administration 2013). These changes are important because if the meaning of a term changes over time, a question using that term may measure something different at two different points in time even if question wording is the same. Similarly, the question may measure something it does not intend to measure, or the question may not be consistent with important official or legal definitions.

18.2.3 Current Practices and Expert Insight

We reached out via email to 30 personal contacts who are experts in survey methodology to learn about their experiences confronting the need to consider revising survey items that are used in tracking polls.[2] Of these, 24 responded (see Acknowledgments) and provided useful information, either by telephone or email. Based on their generosity and insights, we identified four general alternatives for addressing this problem in practice. Several nonexperimental approaches were considered. One of these is to avoid changing the wording of survey questions, if at all possible, in order to maintain the comparability of tracking data. Several experts indicated that this is typically the preferred option of survey clients: "nobody we ever worked for ever wants to change," said one expert. Indeed, "It is difficult to overcome the desire to maintain the trend," commented another. An alternative option is to change the wording entirely and/or abruptly once a new question version deemed to be superior is identified and accept the consequences of an interruption in the time trend being monitored. Few experts endorsed this approach, but some acknowledged that it may sometimes be necessary. One remarked that "changing wording without evidence is a bad idea." Yet, another expert felt that "sometimes you just have to make a change and accept the discontinuity of a trend." A third approach is to begin collecting parallel trend data by continuing to employ the original question wording and additionally introducing the new survey item, either concurrently or perhaps alternating the use of each, depending on the frequency with which tracking data are being collected.

2 Initial emails to experts included a variation on the following text: "I am working on a project that involves examination of approaches that researchers have been taking when dealing with longitudinal data collection. Specifically, the dilemma many of us have faced where a fixed survey question is being tracked over time and we learn that revisions to the question's wording may be necessary to address measurement problems that have come to light. The dilemma being whether to change the wording to improve measurement, or maintain the original wording so that the longitudinal time series is not lost. Have you ever had this experience? If so, how have you dealt with it in practice?"

In addition to these nonexperimental strategies, a fourth approach, specifically recommended by a majority of the experts, involves the use of one or more split-ballot experiments to compare empirically an existing question wording to a new and presumably improved version of the survey item. Split-ballots can be implemented as part of repeated cross-sectional data collections or as part of stand-alone experiments designed specifically to assess the impact of the question wording change. The data produced can be used to (i) estimate the magnitude of the effects on trend data that could be expected if a switch to a revised question was subsequently made, (ii) ask follow-up questions to understand better how respondents interpret various forms of the survey item and how these may influence responses, (iii) examine associations between original and alternative question versions with measures known to be predictive of the construct of interest in order to evaluate their comparability, and (iv) calibrate or adjust the time series for the introduction of new question wording, should a decision be made to transition to the new survey item. Current conventional wisdom is that the optimal solution to this general problem would be conditioned on the specifics of each situation.

We also asked these experts about literature regarding the use of split-ballot experiments to change question wording in repeated cross-sectional surveys. A number of published studies exist that examine the impact of revising various survey design features on self-reports in repeated cross-sectional surveys (cf. Cantor and Lynch 2005; Kindermann et al. 1997; Schuman and Presser 1981; Zablotsky et al. 2015), and a number of studies have examined split-ballot question wording experiments conducted in repeated cross-sectional surveys (e.g. Holbrook and Krosnick 2013; Hougland et al. 1992). However, little research exists that directly addresses the use of split-ballot experiments to revise the wording of survey questions for surveys used to track attitudes, behaviors, or beliefs over time.

18.3 Two Case Studies

In order to provide some examples of how experiments have been used to investigate the effects of revisions to question wording in repeated cross-sectional surveys, we report two case studies examining question wording experiments conducted in the ANES in 2008 and 2012 surveys and the GSS between 1984 and 2014.

18.3.1 ANES

18.3.1.1 Description of Question Wording Experiment

The experiment we analyzed in the ANES was conducted in 2008 and 2012 and involved four items used to measure political efficacy. The standard wording of these questions uses agree–disagree questions. Researchers have identified a number of problems with this format, including acquiescence response bias whereby respondents agree with statements regardless of content (e.g. Schuman and Presser 1981). Because of the problems identified with this question format, researchers recommend the use of what are called item-specific response options as an alternative (e.g. Saris et al. 2010). These involve response options that represent the underlying construct.

In 2008, respondents in the preelection survey were randomly assigned to receive either the standard agree–disagree items or the revised construct specific items (see Table 18.1 for both sets of questions). Respondents who were successfully reinterviewed after the election were asked the same questions (i.e. those asked the standard questions in the preelection survey were also asked the standard questions in the postelection survey and those asked the revised

Table 18.1 Question wording of political efficacy measures.

Question	Standard condition (STD)	Revised condition (REV)
1. Government too complicated	"Sometimes, politics and government seem so complicated that a person like me can't really understand what's going on." Do you agree strongly, agree somewhat, neither agree nor disagree, disagree somewhat, or disagree strongly with this statement?	How often do politics and government seem so complicated that you can't really understand what's going on? All the time, most of the time, about half the time, some of the time, or never?
2. Respondent understands politics	"I feel that I have a pretty good understanding of the important political issues facing our country." Do you agree strongly, agree somewhat, neither agree nor disagree, disagree somewhat, or disagree strongly with this statement?	How well do you understand the important political issues facing our country? Extremely well, very well, moderately well, slightly well, or not well at all?
3. Public officials don't care/how much do officials care?	"Public officials don't care much what people like me think." Do you agree strongly, agree somewhat, neither agree nor disagree, disagree somewhat, or disagree strongly with this statement?	How much do public officials care what people like you think? A great deal, a lot, a moderate amount, a little, or not at all?
4. No say in government/how much can affect government	"People like me don't have any say about what the government does." Do you agree strongly, agree somewhat, neither agree nor disagree, disagree somewhat, or disagree strongly with this statement?	How much can people like you affect what the government does? A great deal, a lot, a moderate amount, a little, or not at all?

questions in the preelection survey were also asked the revised questions in the postelection survey).

The 2012 ANES involved both a traditional face-to-face household survey as well as an Internet survey. For the purposes of consistency, we limited our analysis to the face-to-face cases, which represented 35% of the total sample. Respondents in the preelection survey were randomly assigned to receive either the standard agree–disagree items or the revised construct specific items as in 2008. However, respondents interviewed face-to-face were asked about efficacy using the other set of questions in the postelection. This arrangement means respondents asked the standard items in the preelection survey were asked the revised items in the postelection survey, and those asked the revised items in the preelection survey were asked the standard items in the postelection survey.

18.3.1.2 Data Collection

We use ANES Time Series data from 1952 to 2012. The ANES Time Series surveys have been conducted since 1948 and cover voting behavior in elections. The survey uses a national sample of eligible voters in the United States. Although the survey is typically administered every two years, we focus only on Presidential election years for consistency. Almost all of these surveys were conducted face-to-face. In cases of parallel surveys completed in a different mode (e.g. a parallel telephone survey was conducted in 2000, and a parallel Internet survey was conducted in 2012), our analyses were limited to those surveyed face-to-face. In several years, some postelection surveys were conducted via telephone. In these cases, we included these interviews as long as the respondents were part of the original fresh cross-section interviewed face-to-face in the preelection survey. Finally, in cases where the survey involved reinterviews of participants from previous surveys (e.g. respondents from a previous survey were reinterviewed in the preelection survey in addition to a fresh cross section), we only included those respondents included in the fresh cross-sectional sample in our analyses. Methodological details vary somewhat from year to year, and details for each study can be found in http://www.electionstudies.org.

18.3.1.3 Measures

Political Efficacy Political efficacy is the extent to which individuals believe they do or can influence the political process (Niemi et al. 1991), and it has been consistently measured in the ANES. Between 1952 and 1984, respondents were asked whether they agreed or disagreed with the statements presented in questions 1, 3, and 4 shown in column 2 of Table 18.2 (question 2 was not asked until 2008). From 1988 until 2004, respondents were offered the same statements, but were instead asked whether they disagreed strongly, disagreed somewhat, neither agreed

Table 18.2 Descriptive statistics for political efficacy standard (STD) and revised (REV) items.

	2008				2012			
	Pre		Post		Pre		Post	
	STD	REV	STD	REV	STD	REV	STD	REV
Government too complicated								
Agree strongly/all the time (%)	25	10	22	8	25	9	29	8
Agree somewhat/most of the time	44	23	46	24	41	25	41	27
Neither agree nor disagree/about half the time	10	26	11	30	8	27	8	24
Disagree somewhat/some of the time	12	33	13	35	16	32	14	32
Disagree strongly/never	8	7	7	3	10	6	7	9
Mean	0.33	0.51	0.34	0.51	0.36	0.51	0.33	0.52
N	1151	1169	1039	1053	1046	998	942	977
R's understands politics								
Disagree strongly/not well at all (%)	3	6	3	6	4	6	3	9
Disagree somewhat/slightly well	10	17	11	19	8	15	12	21
Neither agree nor disagree/moderately well	11	48	11	53	9	52	9	47
Agree somewhat/very well	50	22	59	18	48	21	55	19
Agree strongly/extremely well (1)	26	7	16	4	30	7	21	5
Mean	0.72	0.52	0.69	0.49	0.73	0.52	0.70	0.47
N	1147	1166	1038	1052	1044	1001	939	980
Officials don't care/how much officials care								
Agree strongly (1)/not at all(5) (%)	24	15	17	13	24	18	22	13
Agree somewhat (2)/a little (4)	33	32	43	35	32	36	43	37
Neither agree nor disagree (3)/a moderate amount (3)	16	38	17	41	19	36	14	39
Disagree somewhat (4)/a lot (2)	22	9	20	8	19	7	18	7
Disagree strongly (5)/a great deal (1)	4	6	3	3	6	4	2	4
Mean	0.37	0.40	0.37	0.38	0.38	0.35	0.34	0.38
N	1144	1162	1040	1048	1039	989	942	975

(Continued)

Table 18.2 (Continued)

	2008				2012			
	Pre		Post		Pre		Post	
	STD	REV	STD	REV	STD	REV	STD	REV
No say in govt./how much can affect govt.?								
Agree strongly (1)/not at all (5) (%)	21	13	18	15	17	16	17	12
Agree somewhat (2)/a little (4)	24	29	32	35	25	37	31	39
Neither agree nor disagree (3)/a moderate amount (3)	11	29	10	29	12	27	11	28
Disagree somewhat (4)/a lot (2)	30	16	32	12	30	13	32	12
Disagree strongly (5)/a great deal (1)	14	13	9	9	15	8	9	7
Mean	0.49	0.46	0.45	0.42	0.50	0.40	0.47	0.41
N	1145	1163	1040	1051	1043	997	941	971
Index mean	0.48	0.47	0.46	0.45	0.49	0.45	0.46	0.45
Alpha reliability coefficient for 4-item efficacy index	0.44	0.54	0.50	0.48	0.56	0.55	0.58	0.55
Index *N*	1033	1046	1035	1041	1034	981	959	937
Correlation with postreports								
Government too complicated	0.48	0.44			0.42	0.35		
R understands politics	0.43	0.49			0.43	0.45		
Officials don't care/how much do officials care	0.39	0.38			0.31	0.38		
N say in government/how much can affect government	0.47	0.48			0.30	0.39		
Overall index	0.57	0.58			0.51	0.53		

nor disagreed, somewhat agreed, or strongly agreed. In 2008 and 2012, half of respondents were randomly assigned to receive all four of the questions shown in column 2 of Table 18.1, and half were asked a set of revised items that used construct-specific response formats for the following questions (shown in column 3 of Table 18.1). We recoded each of these variables to range from 0 to 1, where 0 indicated minimum political efficacy and 1 indicated maximum political efficacy.[3] For example, a respondent that strongly disagreed with a negative statement about efficacy such as "public officials don't care what people like me think," would be coded as a 1.

For the 2008 and 2012 analyses of the experiment, we constructed an index from all four items which ranged from 0 to 1 where higher values indicated greater efficacy.[4] For the trend

3 Throughout the chapter, we code variables to range from 0 to 1. We do so for consistency, index construction, and ease in interpreting regression coefficients. Coding variables in this way allows us combine items with different numbers of scale points into indices so that each item is weighted equally and to interpret regression coefficients as the total change in the dependent variable that corresponds to change across the whole range of the independent variable.

4 Researchers who have examined a larger set of efficacy measures have found that questions 1 and 2 measure internal political efficacy (a respondent's belief about his or her political competence) while questions 3 and 4 measure external political efficacy (beliefs about the system allowing citizens' influence; Niemi et al. 1991). Because we had only a small set of items we combined all four (three in the trend analyses) questions into a single index. We also conducted analyses with the 2008 and 2012 data with the four items separated into two indices for internal and

analyses, we averaged the recoded variables for questions 1, 3, and 4 together into a political efficacy index as well as examining the three items separately.

Political Mobilization We also constructed an index of political mobilization. The specific wording of these questions changed somewhat over time, but in each year, respondents were asked questions about their political activities. These were used to construct an index of how many of the following behaviors they had done during the campaign: (1) influence the vote of others, (2) attend political meetings or rallies, (3) work for a party of candidate, (4) display a candidate button or sticker during the campaign, and (5) donate money to a party or candidate. These self-reports were used to construct a variable coded to range from 0 if a respondent had engaged in none of the behaviors, 0.20 if s/he had engaged in 1 behavior, 0.40 if s/he had engaged in 2 behaviors, 0.60 if s/he had engaged in 3 behaviors, 0.80 if s/he had engaged in 4 behaviors, and 1 if s/he had engaged in all 5 behaviors.

Voter Turnout Turnout was coded using self-reported turnout from the postelection survey (exact question wordings varied). We coded the vote variable into a binary measure, where 0 indicates the respondent did not vote in the November general election and 1 indicates the respondent did vote.

Sex We coded sex where 0 represents female and 1 represents male respondents.

Race/Ethnicity We coded race as white/nonwhite, where 0 represents white respondents and 1 represents all nonwhite respondents including Black or African-American, American Indian or Alaska Native, Asian, Native Hawaiian or other Pacific Islander, and all other specified nonwhite respondents. In terms of ethnicity, we created another binary variable called "Latino," where 1 represents respondents who identified as Spanish, Hispanic, or Latino and 0 represents all others.

Age and Age Squared Age was coded as a continuous variable. Respondents ranged in age from 17 to 90 in both 2008 and 2012, and the variable was recoded to range from 0 (17 years old) to 1 (90 years old or older) in each survey. In order to detect possible nonlinear relationships, we included an age squared variable that is simply the age variable multiplied by itself. This continuous variable also ranges from 0 to 1.

Income The income variable was based on annual household income. In the 2008 survey, it was originally an ordinal variable with 25 categories that were coded from 1 (none or less than $2999) to 25 ($150 000 and over); in the 2012 survey, it was initially an ordinal variable with 28 categories that were coded from 1 (under $5000) to 28 ($250 000 and over). We recoded this family income into a continuous variable in both years, which was ranged from 0 (none or less than $2999) to 1 ($150 000 and over) in the 2008 survey and was ranged from 0 (under $5000) to 1 ($250 000 or more) in the 2012 survey.

Education Educational attainment was coded into a categorical variable, where 1 represents respondents that have completed 0–8 grades of school, 2 represents respondents with 9–12 grades, 3 represents respondents that have completed a high school diploma, 4 represents respondents with more than a high school diploma but without a college degree, 5 represents respondents with junior or community college level degree, 6 represents respondents with bachelor degrees, and 7 represents respondents with advanced/graduate degrees.

---

external efficacy and the results were not appreciably different (this is not particularly surprising since although internal and external efficacy have different antecedents, both are associated with increased participation).

Year A variable coded 1 for respondents from the 2012 survey and 0 for respondents from the 2008 survey was constructed to control for year of survey in analyses where the 2008 and 2012 data were combined.

18.3.1.4 Analysis

We first examined the distribution and means of each of the four items, the overall four-item political efficacy index, and the 2-item internal and external political efficacy item indices for the preelection and postelection surveys for 2008 and 2012. We also examined the alpha reliability coefficient for each index, and we examined the association of the pre- and postelection measures for each single item and all the indices we constructed. Note that in 2008, this represents the test–retest reliability, but in 2012 this is the association between the standard and revised measure.

In order to test whether the revised items reduced the association of political efficacy with education (as an indicator of satisficing), we combined the data across years and regressed each efficacy index on a dummy variable for year, race, ethnicity, age, age squared family income, educational attainment, a dummy variable for question version (standard = 0, revised = 1), and the interaction between educational attainment and question version. In order to compare the predictive validity of the standard and revised items, we regressed the campaign activism index on political efficacy (separate analyses were conducted for the overall index and the internal and external efficacy indices), question version, year of study, demographic variables, and the interaction between political efficacy and question version. A similar logistic regression analysis was run to predict self-reports of voter turnout with political efficacy, question version, year of study, demographic variables, and the interaction between political efficacy and question version.

18.3.1.5 Results

Effect of Experiment The distribution of responses and the mean for each item and the overall standard and revised efficacy indices for the preelection and postelection surveys for both 2008 and 2012 are shown in Table 18.2. An examination of the distribution of the four items shows clear evidence of acquiescence. There were consistently differences between the distributions of the standard and revised items such that the standard items tended to show greater proportions of respondents who selected the "agree strongly" and "agree somewhat" options (relative to the parallel items in the revised questions). Furthermore, the distribution of the revised items appeared to be closer to a normal distribution, whereas the distribution of the standard items tended to be skewed. Instances where the means for the standard and revised items are significantly different are highlighted in gray in Table 18.2. The means for the standard and revised items were consistently significantly different for the first and second items, but less consistently different for the third and fourth items. The index mean was only significantly different for one of the four tests (the preelection survey in 2012).

The alpha reliability coefficients for the standard and revised indices were not consistently different for the standard and revised sets of items. This is perhaps not surprising because three of the four standard items are coded in the same direction (i.e. agreement with the statement indicates less political efficacy) so acquiescence would increase the correlation between these items. Interestingly, the test–retest reliability correlation (correlation with post items in 2008 where respondents were given the same items in both the pre- and postelection surveys) were generally higher than the correlations between the standard and revised versions of the questions (correlations with postitems in the 2012 data where respondents were given different items in the pre- and postelection surveys).

One purpose of the revised items was to reduce measurement error at least due to acquiescence response bias. Examining means and reliability coefficients does not provide direct evidence that this was successful. In order to test this more directly, we next examined the demographic correlates of political efficacy. If acquiescence bias affects the standard measure of efficacy (particularly since three of the four items are coded in the same direction where agreement means less efficacy), we would predict that education, a known predictor of acquiescence and satisficing behavior more broadly (e.g. Holbrook et al. 2003), would be more strongly associated with efficacy when measured using the standard question wording than when measured using the revised question wording.

Table 18.3 shows the results of analyses predicting political efficacy (combined for the 2008 and 2012 surveys) using demographic variables, experimental condition, and the interactions between experimental conditions and demographics. The interaction between condition and

Table 18.3 Demographic predictors of political efficacy (standard errors shown in parentheses).

	Survey wave (2008 and 2012 combined)					
	Preelection			Postelection		
	ALL	STD	REV	ALL	STD	REV
Year 2012	−0.01 (0.01)	0.02 (0.01)	−0.03* (0.01)	−0.01 (0.01)	−0.01 (0.01)	−0.003 (0.01)
Male	0.02* (0.01)	0.02+ (0.01)	0.02* (0.01)	0.02 (0.01)	0.02 (0.01)	0.02* (0.01)
Non-white	0.02+ (0.01)	0.02+ (0.01)	0.03** (0.01)	0.01 (0.01)	0.01 (0.01)	0.04* (0.01)
Hispanic	−0.01 (0.01)	−0.01 (0.01)	0.03* (0.01)	−0.02 (0.01)	−0.02 (0.01)	0.01 (0.01)
Age (0–1)	−0.13+ (0.08)	−0.14+ (0.08)	−0.08 (0.08)	−0.01 (0.07)	−0.01 (0.07)	0.05 (0.07)
Age squared	0.12 (0.08)	0.13 (0.08)	0.15+ (0.08)	−0.003 (0.09)	−0.004 (0.09)	−0.04 (0.07)
Family income	0.07* (0.02)	0.08* (0.02)	0.02 (0.02)	0.03 (0.02)	0.03 (0.02)	0.02 (0.02)
Educational attainment	0.21** (0.02)	0.20** (0.02)	0.12** (0.02)	0.17** (0.02)	0.17** (0.02)	0.16** (0.02)
Revised questions	0.01 (0.03)			−0.04 (0.03)		
Revised * male	−0.0002 (0.02)			0.01 (0.01)		
Revised * non-white	0.01 (0.02)			0.03+ (0.02)		
Revised * hispanic	0.03+ (0.02)			0.03* (0.01)		
Revised * age	0.04 (0.11)			0.06 (0.10)		
Revised * age squared	0.04 (0.11)			−0.03 (0.11)		
Revised * family income	−0.03 (0.03)			−0.01 (0.03)		
Revised * education	−0.09* (0.03)			−0.002 (0.03)		
N	3689	1867	1822	3814	1769	1825
R^2	0.09	0.11	0.06	0.08	0.07	0.09

$+p < 0.10$, $^*p \leq 0.05$, $^{**}p \leq 0.01$, $^{***}p < 0.001$

education is significant for the preelection political efficacy measure ($b = -0.09$, SE = 0.03, $p < 0.01$), but not for the postelection measure ($b = -0.002$, SE = 0.03, ns). For the preelection measure of efficacy, educational attainment was a stronger predictor of efficacy when the standard measure was used ($b = 0.20$, SE = 0.02, $p < 0.01$) than when the revised version was used ($b = 0.12$, SE = 0.02, $p < 0.01$). This is consistent with the hypothesis that the revised item reduced satisficing behavior.

If the revised measure contains less measurement error, it should show greater predictive validity. One of the most consistent findings regarding political efficacy is that it predicts political activism – behaviors that people engage in to communicate their opinions and preferences to government officials (e.g. Campbell et al. 1954; Rosenstone and Hansen 1993; Verba and Nie 1972). We next tested whether the revised items had higher predictive validity than the standard items by assessing the extent to which each measure of political efficacy predicted an index of campaign mobilization. The results of these analyses are shown in the left-hand panel of Table 18.4. Another form of activism is turning out to vote. We conducted parallel analyses predicting whether or not respondents had voted using the same set of predictors. These analyses are shown in the right-hand panel in Table 18.4. The interaction between the

Table 18.4 Unstandardized regression coefficients predicting index of campaign participation and voter turnout measured postelection (standard errors shown in parentheses).

	Campaign activism (2008 and 2012)						Voter turnout (2008 and 2012)					
	Preelection			Postelection			Preelection			Postelection		
	ALL	STD	REV	ALL	STD	REV	ALL	STD	REV	ALL	STD	REV
Year 2012	−0.11* (0.04)	−0.16* (0.06)	−0.06 (0.05)	−0.11** (0.04)	−0.10 (0.06)	−0.13* (0.05)	−0.06 (0.13)	−0.17 (0.17)	0.06 (0.18)	−0.03 (0.12)	−0.01 (0.17)	−0.04 (0.17)
Political efficacy	0.72** (0.13)	0.72** (0.13)	0.93** (0.14)	1.11** (0.17)	1.12** (0.17)	0.86** (0.16)	2.79** (0.40)	2.88** (0.42)	3.65** (0.42)	2.18** (0.48)	2.19** (0.50)	3.32** (0.51)
Male	−0.02 (0.03)	−0.03 (0.05)	−0.002 (0.05)	−0.02 (0.03)	−0.01 (0.05)	−0.02 (0.05)	−0.38* (0.11)	−0.48* (0.15)	−0.28 (0.17)	−0.33* (0.11)	−0.31* (0.15)	−0.37* (0.16)
Non-white	0.14** (0.04)	0.14* (0.05)	0.14* (0.05)	0.14** (0.04)	0.18* (0.05)	0.11* (0.05)	0.28* (0.14)	0.38* (0.18)	0.18 (0.19)	0.29* (0.13)	0.59* (0.18)	−0.001 (0.19)
Hispanic	−0.09* (0.04)	−0.06 (0.04)	−0.13* (0.05)	−0.08* (0.04)	−0.12* (0.05)	−0.04 (0.04)	−0.35* (0.13)	−0.51* (0.19)	−0.19 (0.15)	−0.30* (0.12)	−0.11 (0.17)	−0.48* (0.17)
Age (0–1)	−0.01 (0.29)	0.08 (0.35)	−0.11 (0.44)	−0.10 (0.28)	−0.41 (0.40)	0.21 (0.36)	2.33* (0.87)	1.73 (1.19)	3.07 (1.27)	1.99* (0.85)	1.94 (1.14)	2.08 (1.24)
Age squared	0.46 (0.32)	0.38 (0.41)	0.56 (0.49)	0.57+ (0.32)	0.92* (0.45)	0.22 (0.41)	0.20 (0.97)	1.06 (1.42)	−0.79 (1.43)	0.44 (0.95)	0.77 (1.31)	0.12 (1.39)
Family income	−0.04 (0.08)	0.003 (0.08)	−0.08 (0.13)	−0.03 (0.08)	−0.001 (0.11)	−0.05 (0.11)	1.15** (0.22)	1.24** (0.31)	1.09* (0.31)	1.23** (0.23)	1.20** (0.31)	1.23** (0.31)
Educational attainment	0.43** (0.08)	0.42** (0.10)	0.45** (0.12)	0.40** (0.08)	0.36* (0.11)	0.44** (0.12)	2.06** (0.26)	1.88** (0.39)	2.24** (0.38)	2.11** (0.26)	2.30** (0.40)	1.95** (0.34)
Revised questions	−0.07 (0.07)			0.10 (0.09)			−0.28 (0.23)			−0.44 (0.28)		
Revised * political efficacy	0.20 (0.28)			−0.23 (0.22)			0.88+ (0.54)			10.04 (0.67)		
N	3596	1816	1780	3813	1769	1824	3591	1815	1776	3805	1764	1821
R^2	0.09	0.09	0.09	0.10	0.11	0.09						

+$p < 0.10$, *$p \leq 0.05$, **$p \leq 0.01$, ***$p < 0.001$.

revised question wording and political efficacy was marginally significant in only one case (for campaign activism using the postelection measure of efficacy; $b = 0.88$, $SE = 0.54$, $p < 0.10$). However, in three of the four cases, the association between political efficacy and activism was stronger for the revised items than for the standard items (see row 2 of columns 2 and 3, 5 and 6, 8 and 9, and 11 and 12), providing very weak evidence that the revised items resulted in greater validity.

Overall, our analyses consistently show that the standard items resulted in different estimates of efficacy compared to the standard items, but we found little evidence to show the items were successful at reducing measurement error due to acquiescence response bias (consistent with Lelkes and Weiss 2015).

18.3.1.6 Implications for Trend Analysis

In order to examine the potential effect on trend analyses, we next looked at trends in the three efficacy items and the efficacy index (excluding item #2 in Table 18.1, which was not consistently asked before 2008) in Presidential election years from 1952 until 2012. The solid lines in the graphs below indicate the means for the standard items (and the three-item index of the standard items), and the dotted lines indicate the means for the revised items (and the three-item index of the revised items).

What is evident from these trend lines is the value of repeating experiments multiple times. The effect of the experiment on trends is difficult to assess given that the experiment has only been conducted twice and one of the efficacy items was not included in the 2004 survey (government and politics too complicated). Given the small number of data points in which

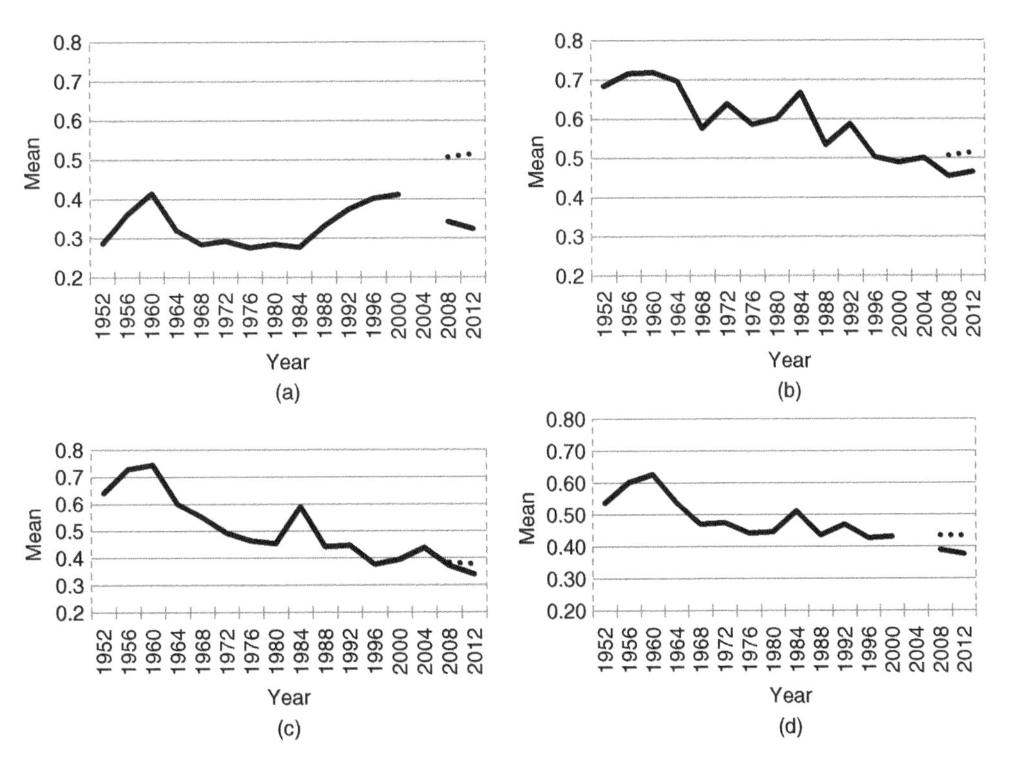

Figure 18.1 (a) Government and politics too complicated (higher values = less complicated). (b) People like Respondent have a say what government does (higher values = more say). (c) Government cares about people like me (higher values – Govt. cares more). (d) Efficacy index (higher values = higher efficacy).

the revised items were asked; however, it appears that changing the question wording had a varying effect on the items. The difference between the revised and standard items was much larger for the "too complicated" question (Figure 18.1a), and the standard and revised items also may be showing differing trends for these items (with efficacy measured using the revised item going up from 2008 to 2012 but going down slightly and efficacy measured using the standard item going down). The differences are much smaller for the other two items and the index (see Figures 18.1b–d), and trends appear to be similar between 2008 and 2012.

18.3.2 General Social Survey (GSS)

18.3.2.1 Purpose of Experiment

The second experiment we analyze has been conducted on every GSS since 1984. This experiment is part of a larger experiment involving the wording of a dozen questions about spending priorities that have been included on the GSS since 1973. For the 1984 survey, researchers decided to include a split-sample, alternative wording experiment for all these questions (Smith 1984). Specifically, we focus on one item in which researchers had particular concerns and that is the question about spending on "welfare." GSS investigators were concerned that "'welfare' had become too negative a term connotating waste and loafing and therefore was not serving as a measure of support for the 'safety net' that it had originally been designed to tap" (Smith 1984, p. 12). Moreover, research in the 1970s and early 1980s showed that the term welfare had become a racially charged word, raising concerns "that attitudes of the mass public toward welfare spending are closely tied to racial attitudes" (Wright 1977, p. 728). The GSS methodological report following the 1984 survey explains that the investigators "believe that the term 'welfare' is essentially pejorative, bringing forth to many people's minds images of abuse of public assistance rather than the notion of public assistance itself" (Smith 1984, p. 15). The experimental question to be compared to the welfare item asked about "assistance to the poor" in an effort to minimize the extent to which reports were influenced by racial attitudes. We focus specifically on the impact of this experiment among Whites and whether it was successful at reducing the association with racial attitudes relative to the welfare item.

18.3.2.2 Data Collection

We use GSS data from 1973 to 2014 to examine trends, but most of our analyses focus on data collected between 1984 and 2014. The GSS has been conducted at least every two years since 1972 and covers a variety of social, political, and economic topics. NORC at the University of Chicago, with support from the National Science Foundation, conducts the GSS with a nationally representative sample of adults in the United States. This in-person survey monitors trends in American attitudes, behaviors, and attributes across time, and methodological details of the survey can be found at http://gss.norc.org/.

18.3.2.3 Measures

Spending Experiment Before 1984, respondents were asked: "We are faced with many problems in this country, none of which can be solved easily or inexpensively. I'm going to name some of these problems, and for each one I'd like you to tell me whether you think we're spending too much money on it, too little money, or about the right amount. Are we spending too much, too little, or about the right amount on welfare?" Between 1984 and 2014, half of respondents were randomly assigned to be asked this same question and the other half were randomly assigned to be asked about spending on "assistance to the poor." For our analyses, both items were coded 0 for respondents who said "too much," 0.5 for respondents who said "about the right amount," and 1 for respondents who said "too little." A dummy variable was created that was coded 0 if a respondent was asked the traditional welfare question and 1 if s/he was asked the experimental assistance to the poor question.

Racial Prejudice Nine different racial attitudes indices were calculated. These indices were drawn from the literature and reflect changes in the way racial prejudice has been conceptualized and operationalized over time (see Schuman et al. 2007). The GSS and other organizations have done extensive work to develop and validate the various measures of racial prejudice used over time (e.g. Smith 2002). The nine indices and the items included in each are shown in Table 18.5. Each individual variable was recoded to range from 0 to 1 with the most *tolerant or positive* attitude coded as 0 and the most *intolerant or negative attitude* toward Blacks coded as 1. For example, on the question of whether the respondent would favor a law against racial intermarriage, answering "yes" was coded as 1 and "no" was coded as 0. Variables were then averaged into indices of racial attitudes. A respondent needed to have valid responses to at least three of the items in each scale to be given a valid index score. For each of the indices shown below, the alpha reliability coefficient was calculated for each year in which all items in the index were asked. This alpha reliability coefficient (or the range of alphas if all scale items were asked in multiple years) is shown in parentheses after the measure name. Below each measure name, we have provided a reference article in which this measure of racial prejudice was used and the alpha reliability coefficient of the measure (or range of alphas) from that reference (if one was reported).

Gender Gender was observed by the interviewer and was coded 0 for females and 1 for males.

Race/Ethnicity Our analyses focused only on White respondents, but ethnicity was coded via a dummy variable coded 1 if the respondent identified as Latino or Hispanic and 0 if s/he did not.

Age and Age Squared Age was coded as a continuous variable. Respondents ranged in age from 18 to 89, and the variable was recoded to range from 0 (18 years old) to 1 (89 years old). In order to detect possible nonlinear relationships, we included an age squared variable that is simply the age variable multiplied by itself. This continuous variable also ranges from 0 to 1.

Income The income variable was based on annual household income, and it is an ordinal measure with 12 categories that are coded from 0 (less than \$1000) to 12 (more than \$25 000).

Education The education variable was based on the highest year of schooling respondents completed. The variable is coded from 0 (0 years) to 1 (20 or more years).

Year In order to control for years, a dummy variable was coded for each year of the survey experiment: 1984, 1985, 1986, 1988, 1989, 1998, 1991, 1993, 1994, 1996, 1998, 2000, 2002, 2004, 2006, 2008, 2010, 2012, and 2014.

18.3.2.4 Analysis

Because one intent of the revised assistance to the poor item was to reduce the extent to which the item was associated with racial attitudes or racial prejudice, we focused only on White respondents in our analyses (which is common in research studying racial attitudes). We first examined descriptive statistics for the experimental and control questions for White respondents for the years in which the experiment was conducted. In order to test whether the experimental assistance to the poor item reduced the association with racial attitudes, we regressed responses to the welfare/assistance to the poor item on demographic variables (gender, age,

Table 18.5 Racial attitudes measures.

Social distance feelings index (Alpha: 0.239)

Bobo et al. (2012)

1. Would you yourself have any objection to sending your children to a school where a few of the children are Negroes/Blacks/AAs? IF NO OR DON'T KNOW: Where half of the children are Negroes/Blacks/AAs? IF NO OR DON'T KNOW: Where more than half of the children are Negroes/Blacks/AAs?

2. If a Negro with the same income and education as you have, moved into your block, would it make any difference to you?

3. How strongly would you object if a member of your family wanted to bring a (Negro/Black) friend home to dinner? Would you object strongly, mildly, or not at all?

Racial principles index (Alpha: 0.717)

Bobo et al. (2012)

1. Do you think Negroes should have as good a chance as white people to get any kind of job, or do you think white people should have the first chance at any kind of job?

2. Was the high school you attended all black, integrated – but mostly black, or integrated – but mostly white?

3. If your party nominated a Negro/Black/AA for President, would you vote for him if he were qualified for the job?

4. Do you think there should be laws against marriages between Negroes/Blacks/AAs and whites?

5. Here are some opinions other people have expressed in connection with Negro/Black–white relations. Which statement on the card comes closest to how you, yourself, feel? White people have a right to keep Negroes/Blacks/AA out of their neighborhoods if they want to, and Negroes/Blacks/AA should respect that right.

6. Suppose there is a community-wide vote on the general housing issue. There are two possible laws to vote on. Which law would you vote for? (owner decides, can't discriminate)

Traditional prejudice: social distance index (Alpha: 0.557–0.749)

Taylor and Mateyka (2011), Reported Alpha: 0.694

1. Now I'm going to ask you about different types of contact with various groups of people. In each situation, would you please tell me whether you would be very much in favor of it happening, somewhat in favor, neither in favor nor opposed to it happening, somewhat opposed, or very much opposed to it happening? Living in a neighborhood where half of your neighbors were blacks?

2. How about having a close relative or family member marry a black person?

Racial resentment (Alpha: 0.413–0.573)

Taylor and Mateyka (2011), Reported Alpha: 0.517

1. Here are some opinions other people have expressed in connection with Negro/Black–white relations. Which statement on the card comes closest to how you, yourself, feel? The first one is Negroes/Blacks/AAs shouldn't push themselves where they're not wanted.

2. Do you agree strongly, agree somewhat, neither agree nor disagree, disagree somewhat, or disagree strongly with the following statement: Irish, Italians, Jewish and many other minorities overcame prejudice and worked their way up. Blacks should do the same without special favors. *Traditional Prejudice: Stereotyping Index* (Alpha: 0.418–0.449)

Taylor and Mateyka (2011), Reported Alpha: 0.724

1. Now I have some questions about different groups in our society. I'm going to show you a seven-point scale which the characteristics of people in a group can be rated. In the first statement, a score of 1 means that you think almost all of the people in that group are "rich." A score of 7 means that you think almost everyone in the group are "poor." A score of 4 means that you think that the group is not toward one end or another, and of course you may choose any number in between that comes closest to where you think people in the group stand. [WEALTH RATINGS] The second set of characteristics asks if people in the group tend to be hardworking or if they tend to be lazy.

2. [ASKED IMMEDIATELY AFTER #1 ABOVE] Where would you rate Whites on this scale? Where would you rate Blacks in general on this scale?

Table 18.5 (Continued)

3. Do people in these groups tend to be unintelligent or tend to be intelligent? Where would you rate Whites in general on this scale?

4. [ASKED IMMEDIATELY AFTER #3 ABOVE] Where would you rate Blacks in general on this scale?

5. The next set asks if people in each group tend to be violence prone or if they tend not to be prone to violence? Where would you rate Whites in general on this scale?

6. [ASKED IMMEDIATELY AFTER #5 ABOVE] Where would you rate Blacks in general on this scale?

Attributions for racial inequality index (Alpha: 0.396–0.597)

Taylor and Mateyka (2011), Reported Alpha: 0.515

1. On the average, Negroes/Blacks/AAs have worse jobs, income, and housing than white people. Do you think these differences are because most Negroes/Blacks/AAs don't have the chance for education that it takes to rise out of poverty?

2. Do you think these differences are because most Negroes/Blacks/AAs have less inborn ability to learn?

3. Do you think these differences are mainly due to discrimination?

4. Do you think these differences are because most Negroes/Blacks/AAs just don't have the motivation or will power to pull themselves up out of poverty?

Traditional prejudice: emotion index (Alpha: 0.608)

Taylor and Mateyka (2011), Reported Alpha: 0.662

1. In general, how warm or cool do you feel toward African Americans?

2. In general, how warm or cool do you feel toward white or Caucasian Americans?

3. In general, how close do you feel to Blacks?

4. And in general, how close do you feel to Whites?

Symbolic racism (Alpha: 0.666)

Tarman and Sears (2005), Reported Alphas: 0.59–0.86

1. Do you agree strongly, agree somewhat, neither agree nor disagree, disagree somewhat, or disagree strongly with the following statement: Irish, Italians, Jewish, and many other minorities overcame prejudice and worked their way up. Blacks should do the same without special favors.

2. On the average, Negroes/Blacks/AAs have worse jobs, income, and housing than white people. Do you think these differences are mainly due to discrimination?

3. On the average, Negroes/Blacks/AAs have worse jobs, income, and housing than white people. Do you think these differences are because most Negroes/Blacks/AAs just don't have the motivation or will power to pull themselves up out of poverty?

4. Do you think that blacks get more attention from government than they deserve? Would you answer much more attention from government than they deserve, more attention than they deserve, about the right amount of attention, less attention than they deserve, or much less attention from government than they deserve? *Old-fashioned racism* (Alpha: 0.657)

Tarman and Sears (2005), Reported Alpha: 0.54

1. Do you think there should be laws against marriages between Negroes/Blacks/AAs and whites?

2. Do you think these differences are because most Negroes/Blacks/AAs have less inborn ability to learn?

3. How about having a close relative or family member marry a black person?

4. Here are some opinions other people have expressed in connection with Negro/Black-white relations. Which statement on the card comes closest to how you, yourself, feel? White people have a right to keep Negroes/Blacks/AA out of their neighborhoods if they want to, and Negroes/Blacks/AA should respect that right.

5. If your party nominated a Negro/Black/AA for President, would you vote for him if he were qualified for the job?

6. Do people in these groups tend to be unintelligent or tend to be intelligent? Where would you rate Blacks in general on this scale?

Note: The alphas reported for the GSS data above (after each scale name) only include data from years in which all the questions in the scale were asked.

age squared, income, race/ethnicity, and education), experimental condition (1 = assistance to the poor, 0 = welfare), racial prejudice, and the interaction between racial prejudice and experimental condition.

18.3.2.5 Results

Effect of Experiment We began by looking at descriptive statistics for the welfare and assistance to the poor item among White respondents. In every year that the experiment was conducted, support for spending for assistance to the poor among Whites was greater than support for spending for welfare (Table 18.6).

Although it is clear that the revised wording impacted responses to the question, this evidence does not provide support for the assumption that the revised assistance to the poor question was less affected by racial attitudes (because people see welfare as a more racialized issue than assistance to the poor). In order to test this more directly, we conducted analyses predicting responses to the spending for welfare/assistance to the poor with demographic variable, the dummy variable for experimental condition, racial attitudes, and the interaction between racial attitudes and experimental condition. We ran separate analyses for each of the nine racial attitudes measures. The results of these analyses are shown in Table 18.7 (each analysis included

Table 18.6 Support for welfare spending and assistance to the poor among whites.

	Spending on welfare					Spending on assistance to the poor				
Year	Too little (1) (%)	Right amount (0.5) (%)	Too much (0) (%)	Mean	*N*	Too little (1) (%)	Right amount (0.5) (%)	Too much (0) (%)	Mean	*N*
1984	24.38	35.86	39.76	0.42	471	63.17	25.56	11.27	0.76	472
1985	18.88	34.69	46.43	0.36	719	65.79	24.26	9.95	0.78	762
1986	23.36	34.51	42.13	0.41	700	63.17	28.09	8.74	0.77	717
1987	27.57	31.40	41.02	0.43	568	71.14	20.64	8.21	0.81	1179
1988	24.68	32.54	42.78	0.41	685	70.24	22.85	6.91	0.82	734
1989	24.17	32.70	43.13	0.41	720	68.06	23.40	8.54	0.80	741
1990	24.45	37.12	38.43	0.43	635	68.74	24.53	6.72	0.81	666
1991	23.95	36.73	39.32	0.42	707	66.35	24.45	9.20	0.79	720
1993	17.13	25.56	57.31	0.30	756	64.45	22.99	12.56	0.76	769
1994	13.08	24.55	62.37	0.25	1448	59.27	25.20	15.53	0.72	1403
1996	15.29	26.65	58.07	0.29	1394	55.06	26.32	18.62	0.68	1366
1998	16.01	37.73	46.26	0.35	1317	62.66	25.89	11.45	0.76	1390
2000	20.81	39.85	39.35	0.41	1331	64.26	24.21	11.53	0.76	1363
2002	20.86	37.84	41.30	0.40	1314	67.33	24.51	8.16	0.80	1375
2004	23.59	35.00	41.41	0.41	1360	69.86	23.72	6.42	0.82	1389
2006	25.12	36.54	38.34	0.43	1434	69.76	22.42	7.81	0.81	1472
2008	25.38	36.38	38.24	0.44	955	70.30	21.78	7.92	0.81	998
2010	23.15	34.49	42.36	0.40	980	67.51	22.54	9.94	0.79	996
2012	19.13	33.31	47.57	0.36	957	63.91	25.98	10.11	0.77	935
2014	19.82	29.62	50.57	0.35	1224	63.62	24.27	12.11	0.76	1233
1984–2014	20.86	31.57	47.57	0.37	32928	65.73	24.18	10.09	0.78	20680

Table 18.7 Regression coefficients for models predicting support for welfare/assistance for poor people with racial attitudes.

	Racial attitudes index								
	Social distance	Racial principles	Traditional prejudice: social distance feelings	Racial resentment	Traditional prejudice: stereotyping	Attributions for racial inequality	Emotion	Symbolic racism	Old fashioned racism
Male	−0.11* (0.04)	−0.10* (0.05)	−0.16** (0.05)	0.16 (0.09)	−0.13** (0.04)	−0.08* (0.03)	−0.15*** (0.43)	−0.11* (0.04)	−0.11** (0.04)
Hispanic	0.33** (0.10)	NA	NA	0.33 (0.29)	0.33** (0.10)	0.34** (0.11)	0.35*** (0.91)	0.34** (0.11)	0.32** (0.10)
Age squared	−0.28 (0.54)	0.59 (0.65)	0.66 (0.74)	−1.43 (1.17)	−0.47 (0.54)	0.01 (0.43)	−1.16* (0.53)	−0.58 (0.57)	0.50 (0.47)
Age	0.22 (0.52)	−1.14+ (0.60)	−1.31+ (0.69)	1.23 (1.10)	0.20 (0.52)	−0.35 (0.41)	1.03* (0.51)	0.38 (0.55)	−0.67 (0.45)
Income	−0.95*** (0.13)	−1.16*** (0.12)	−1.03*** (0.14)	−0.87** (0.25)	−1.00*** (0.13)	−0.99*** (0.09)	−1.09*** (0.13)	−1.05*** (0.13)	−1.11*** (0.11)
Education	−0.74*** (0.16)	−0.45* (0.18)	−0.32 (0.20)	−0.95** (0.31)	−0.54** (0.16)	−1.18*** (0.13)	−0.19 (0.15)	−1.32*** (0.17)	−0.56*** (0.14)
Racial attitudes	−1.09*** (0.14)	−0.84*** (0.13)	−0.53* (0.22)	−1.79*** (0.24)	−0.79* (0.31)	−1.66*** (0.08)	−0.59*** (0.16)	−1.91*** (0.10)	−0.93*** (0.13)
Condition (0 = welfare, 1 = assist the poor)	0.97*** (0.11)	2.07*** (0.06)	2.18*** (0.08)	1.41*** (0.20)	2.11*** (0.21)	1.93*** (0.06)	2.05*** (0.09)	1.75*** (0.10)	1.99*** (0.07)
Racial attitudes × condition	0.15 (0.19)	0.16 (0.17)	−0.15 (0.32)	1.01** (0.32)	−0.19 (0.45)	0.45*** (0.11)	−0.08 (0.23)	0.60*** (0.14)	0.25 (0.18)
N	10 434	8 315	6 359	2 483	10 290	17 194	10 417	10 134	13 878

$+p < 0.10$, $^{*}p < 0.05$, $^{**}p < 0.01$, $^{***}p < 0.001$.
Note: All analyses included dummy variables to control for year of survey.

dummy variables to control for year and only included data from years between 1984 and 2014 in which the questions included in the racial attitudes measure were included).

Not surprisingly, racial attitudes (coded so that higher values mean greater racial prejudice) were negatively and consistently associated with support for spending. The coefficient for the experimental condition variable was positive denoting greater support for spending when the assistance to the poor wording was used than when the welfare wording was used. Finally, for three of the nine measures of racial attitudes (racial resentment: $b = 1.01$, SE $= 0.32$, $p < 0.01$; attributions for racial inequality: $b = 0.45$, SE $= 0.11$, $p < 0.01$; and symbolic racism: $b = 0.61$, SE $= 0.14$, $p < 0.01$), the impact of racial attitudes was significantly weaker (less negative as represented by the positive interaction coefficients) for the assistance to the poor item than for the welfare item. These results provide mixed evidence as to whether asking about spending on "assistance to the poor" is less associated with racial attitudes than asking about spending on "welfare." The alternative question has a weaker association with certain measures of racial attitudes (e.g. symbolic racism), but it has a similar relationship with other racial measures (e.g. old fashioned racism) as the original question.

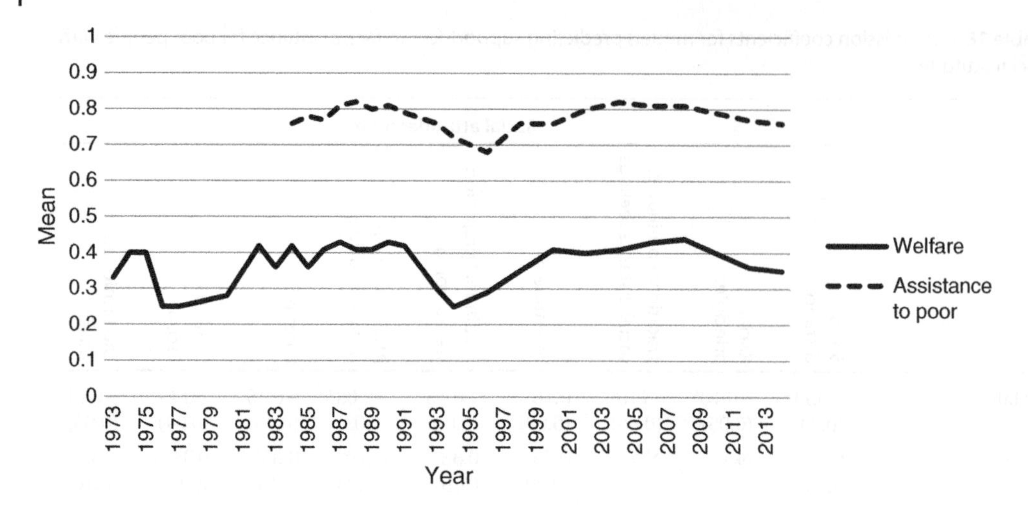

Figure 18.2 Support for spending.

Implications for Trend Analysis We next examined the implications of the experiment for trends in support for spending (again limiting our analyses to Whites). The results are shown in Figure 18.2. Although there is a consistently large mean difference between the welfare and assistance to the poor items (with support for spending being greater for assistance to the poor than for welfare), the two items show remarkably similar patterns in terms of trends. At the year level, the correlation between the two conditions was very large ($r = 0.80$, $N = 20$), further suggesting that changes in the two measures are remarkably consistent. There appears to be somewhat more variability in responses to the welfare question over time. One potential explanation for this is that the term welfare is often used in media coverage (rather than "assistance to the poor") and that opinions toward spending on "welfare" may be more volatile over time than those about spending on "assistance to the poor" because the former are more affected by such media coverage than the latter. Supporting this argument, the change in support in the early 1990s, while apparent across both measures, is larger for the welfare item. Indeed, the biggest differences between the two question wordings appear in the mid-1990s (1993 and 1994, specifically) when welfare reform was on the national agenda prior to the welfare reform bill signed in 1996 by then-President Bill Clinton. The larger difference across conditions observed during this time could reflect greater politicization of the term welfare during the media debate. The lower support for welfare spending observed during the late 1970s to early 1980s is also consistent with this explanation as this was during a time when Reagan was actively campaigning for the Republican nomination and vocal in speaking out against welfare.

18.4 Implications and Conclusions

Based on our case studies and interviews with experts, we would make several recommendations to researchers who are considering using an experiment to assess the impact of question wording changes in repeated cross-sectional surveys.

(1) The first is that the experiment has a clear purpose and goal that it is designed to achieve, and that this purpose be made clear in the documentation for the survey. This would seem obvious, but we note that sometimes the logic for such experimentation (i.e. the reason behind it) is not as well documented as the actual experimental manipulation itself.
(2) We would also recommend that, as in the two examples we examined here, the experiment be included as part of the main data collection when possible. This is because testing the question wording in a separate survey often has limitations. It is often expensive and because of the cost, sometimes such separate studies are done on a small scale or using a different methodology that does not fully test the question wording change in circumstances that mirror those in the repeated cross-sectional survey (e.g. a different mode). A second reason to include the experiment in the main data collection is that it helps to ensure that the experiments methodology will be documented and its data preserved because main study data tend to be better documented and archived than do pilot studies.
(3) A third recommendation is that the experiment be conducted multiple times if possible. It is unclear what the optimal number of times is, but the ANES data analyzed above suggest that repeating the experiment more than twice would be beneficial for assessing its effects and also for trend analyses. Conducting the experiment multiple times also allows researchers to analyze the experiment across waves and maximize power to detect differences.
(4) Fourth, we recommend designing surveys to maximize researchers' ability to assess the success of the experimental question wording. This may require including additional measures to assess convergent or divergent validity or other approaches (e.g. validation).
(5) In evaluating the impact of the experiment, we encourage researchers to examine more than just descriptive statistics and to assess associations with other key variables (and statistics such as scale reliability if appropriate). This may be useful not only for the surveys in which the experiment was conducted but also potentially over time (e.g. one could imagine circumstances in which it makes sense to examine changes in associations over time).
(6) Finally, we encourage researchers to document the experiment and to archive the data from such experiments to maximize their impact.

18.4.1 Limitations and Suggestions for Future Research

This chapter provided examples of several analytic strategies that can be employed as part of split-ballot experiments designed to investigate the effects that revisions to question wording might have on social trends being monitored as part of repeated cross-sectional social surveys. We focused on the use of split-ballot experiments to examine the effect of changing question wording in order to reduce measurement error. Other potential uses of split-ballot experiments, though, such as employing follow-up questions or probes to understand how respondents interpret various forms of a survey question, or to calibrate or adjust a time series for the introduction of new question wording, were not examined. Furthermore, we focused on examining the effect of question wording experiments on measurement error, but question wording could also influence other sources of error such as item nonresponse. Although it is clear from our consultations with multiple experts that each of these strategies has been employed in practice, there are unfortunately very few examples to be found in the published literature. We strongly recommend that researchers work to disseminate more broadly their empirical findings relevant to the application of split-ballot experiments for this purpose.

Acknowledgments

We thank the following colleagues for communicating with us, either by phone or email, regarding the topic of this chapter, and sharing their knowledge and experiences:

George Bishop
Norman Bradburn
David Cantor
Rich Clark
Jennifer Dykema
Ed Freeland
Cathy Haggerty
Liz Hamel
Jim Hougland
Scott Keeter
John Kennedy
Ron Langley
Lynn Langton
Mary Losch
Peter Miller
David Moore
Bob Oldendick
Michael Planty
Rob Santos
Andy Smith
Tom W. Smith
John Stevenson
Mike Traugott
Santa Traugott
Tim Triplet

References

American Psychiatric Association (2013). *Diagnostic and Statistical Manual of Mental Disorders: DSM-5*. Washington, DC: American Psychiatric Association.

Bishop, G.F., Tuchfarber, A.J., and Oldendick, R.W. (1978). Change in the structure of American political attitudes: the nagging question of question wording. *American Journal of Political Science* 22: 250–269.

Bobo, L.D., Charles, C.Z., Krysan, M., and Simmons, A.D. (2012). The real record on racial attitudes. In: *Social Trends in American Life: Findings From the General Social Survey Since 1973* (ed. P. Marsden), 38–83. Princeton, NJ: Princeton University Press.

Campbell, A., Gurin, G., and Miller, W.E. (1954). *The Voter Decides*. Evanston, IL: Row Peterson & Co.

Cantor, D. and Lynch, J.P. (2005). Exploring the effects of changes in design on the analytical uses of the NCVS data. *Journal of Quantitative Criminology* 21 (3): 293–319.

Caplow, T., Hicks, L., and Wattenberg, B.J. (2001). *The First Measured Century: An Illustrated Guide to Trends in America, 1900-2000.* Washington, DC: AEI Press.

Chandler, J., Reinhard, D., and Schwarz, N. (2012). To judge a book by its weight you need to know its content: knowledge moderates the use of embodied cues. *Journal of Experimental Social Psychology* 48: 948–952.

Chang, L. and Krosnick, J.A. (2009). National surveys via RDD telephone interviewing vs. the Internet: comparing sample representativeness and response quality. *Public Opinion Quarterly* 73: 641–678.

Dillman, D.A., Smyth, J.D., and Christian, L.M. (2008). *Internet, Mail, and Mixed-Mode Surveys: The Tailored Design Method*, 3e. New York: Wiley.

Foucault, B., Aguilar, J., Miller, P., and Cassell, J. (2009). Behavioral correlates of rapport in survey interviews. *Paper presented at the annual conference of the American Association for Public Opinion Research*, Hollywood, FL (14 May 2009)

Groves, R., Fowler, F., Couper, M. et al. (2009). *Survey Methodology*, 2e. New York: Wiley.

Guest, A.M. and Wierzbicki, S.K. (1999). Social ties at the neighborhood level: Two decades of GSS evidence. *Urban Affairs Review* 35: 92–111.

Holbrook, A.L., Green, M.C., and Krosnick, J.A. (2003). Telephone vs. face-to-face interviewing of national probability samples with long questionnaires: comparisons of respondent satisficing and social desirability response bias. *Public Opinion Quarterly* 67: 79–125.

Holbrook, A.L. and Krosnick, J.A. (2010). Social desirability bias in voter turnout reports: tests using the item count technique. *Public Opinion Quarterly* 74: 37–67.

Holbrook, A.L. and Krosnick, J.A. (2013). A new question sequence to measure voter turnout in telephone surveys: results of an experiment in the 2006 ANES pilot study. *Public Opinion Quarterly* 77: 106–123.

Holbrook, A.L., Krosnick, J.A., Carson, R.T., and Mitchell, R.C. (2000). Violating conversational conventions disrupts cognitive processing of attitude questions. *Journal of Experimental Social Psychology* 36: 465–494.

Holbrook, A.L., Krosnick, J.A., Moore, D., and Tourangeau, R. (2007). Response order effects in dichotomous categorical questions presented orally: the impact of question and respondent attributes. *Public Opinion Quarterly* 71: 325–348.

Hougland, J.G., Johnson, T.P., and Wolf, J.G. (1992). A fairly common ambiguity: comparing rating and approval measures in public opinion polling. *Sociological Focus* 25 (3): 257–271.

Ijzerman, H. and Semin, R.R. (2010). Temperature perceptions as a ground for social proximity. *Journal of Experimental Social Psychology* 46 (6): 867–873.

Jostman, N., Lakens, D., and Schubert, T. (2009). Weight as an embodiment of importance. *Psychological Science* 20: 1169–1174.

Kindermann, C., Lynch, J., and Cantor, D. (1997). *Effects of the redesign on victimization estimates. Bureau of Justice Statistics. National Crime Victimization Survey. NCJ-164381.* http://www.bjs.gov/content/pub/pdf/ERVE.PDF.

Krosnick, J.A. (1991). Response strategies for coping with the cognitive demands of attitude measures in surveys. *Applied Cognitive Psychology* 5: 213–236.

Krosnick, J.A., Holbrook, A.L., Berent, M.K. et al. (2002). The impact of "no opinion" response options on data quality: non-attitude reduction or an invitation to satisfice? *Public Opinion Quarterly* 66: 371–403.

Krosnick, J.A., Narayan, S.S., and Smith, W.R. (1996). Satisficing in surveys: initial evidence. In: *Advances in Survey Research* (ed. M.T. Braverman and J.K. Slater), 29–44. San Francisco: Jossey-Bass.

Lelkes, Y. and Weiss, R. (2015). Much ado about acquiescence: the relative validity and reliability of construct-specific and agree-disagree questions. *Research and Politics* 2 (3): 1–8.

Menard, S. (2008). *Handbook of Longitudinal Research: Design, Measurement, and Analysis*. Burlington, MA: Elsevier.

Narayan, S. and Krosnick, J.A. (1996). Education moderates some response effects in attitude measurement. *Public Opinion Quarterly* 60: 58–88.

Niemi, R.G., Craig, S.C., and Mattei, F. (1991). Measuring internal political efficacy in the 1988 national election study. *American Political Science Review* 85 (4): 1407–1413.

Rosenstone, S.J. and Hansen, J.M. (1993). *Mobilization, Participation, and Democracy in America*. New York: Macmillan Publishing Company.

Saris, W., Revilla, M., Krosnick, J.A., and Shaeffer, E. (2010). Comparing questions with agree/disagree response options to questions with item-specific response options. *Survey Research Methods* 4: 61–79.

Schaeffer, N.C. and Dykema, J. (2011). Questions for surveys: current trends and future directions. *Public Opinion Quarterly* 75 (5): 909–961.

Schuman, H. and Presser, S. (1981). *Questions and Answers in Attitude Surveys: Experiments on Question Form, Wording, and Context*. Thousand Oaks, CA: Sage.

Schuman, H., Steeh, C., Bobo, L., and Krysan, M. (2007). *Racial Attitudes in America: Trends and Interpretations*. Cambridge, MA: Harvard University Press.

Schwarz, N. (1996). *Cognition and Communication: Judgmental Biases, Research Methods and the Logic of Conversation*. Hillsdale, NJ: Erlbaum.

Schwarz, N. (1999). Self-reports: how the questions shape the answers. *American Psychologist* 54: 93–105.

Schwarz, N., Hippler, H.J., Deutsch, B., and Strack, F. (1985). Response scales: effects of category range on reported behavior and subsequent judgments. *Public Opinion Quarterly* 49: 388–395.

Schwarz, N., Knauper, B., Hipler, H.J. et al. (1991). Numeric values may change the meaning of scale labels. *Public Opinion Quarterly* 55: 570–582.

Schwarz, N., Strack, F., and Mai, H.P. (1991). Assimilation and contrast effects in part-whole question sequences: a conversational logic analysis. *Public Opinion Quarterly* 55: 3–23.

Shapiro, R.Y. and Mahajan, H. (1986). Gender differences in policy preferences: a summary of trends form the 1960s to the 1980s. *Public Opinion Quarterly* 50: 42–61.

Sirken, M.G., Herrmann, D.J., Schechter, S. et al. (1999a). *Cognition and Survey Research*. New York: Wiley.

Sirken, M., Jabine, T., Willis, G. et al. (1999b). *A New Agenda for Interdisciplinary Survey Research Methods*. National Center for Health Statistics http://www.cdc.gov/nchs/data/misc/casm2pro.pdf.

Smith, T. W. (1984). *A preliminary analysis of methodological experiments on the 1984 GSS*. GSS Technical Report No. 49. Chicago: National Opinion Research Center.

Smith, T.W. (1992). Changing racial labels: from "colored" to "negro" to "black" to "African American". *Public Opinion Quarterly* 56: 496–514.

Smith, T.W. (2002). Measuring racial and ethnic discrimination. GSS Methodological Report No. 96. Chicago: National Opinion Research Center.

Smith, T.W. (2005). The laws of studying societal change. *Survey Research* 36 (2): 1–5. http://www.srl.uic.edu/publist/newsletter/2005/05v36n2.pdf.

Social Security Administration (2013). Change in terminology: "mental retardation" to "intellectual disability". *Federal Register* 78: 148.

Sudman, S., Bradburn, N., and Schwarz, N. (1996). *Thinking About Answers: The Application of Cognitive Processes to Survey Methodology*. San Francisco, CA: Jossey-Bass.

Sullivan, J.L., Piereson, J.E., and Marcus, G.E. (1978). Ideological constraint in the mass public: a methodological critique and some new findings. *American Journal of Political Science* 22: 233–249.

Tarman, C. and Sears, D.O. (2005). The conceptualization and measurement of symbolic racism. *The Journal of Politics* 67: 731–761.
Taylor, M.C. and Mateyka, P.J. (2011). Community influences on white racial attitudes: what matters and why? *The Sociological Quarterly* 52 (2): 220–243.
Tourangeau, R. (2004). Recurring Surveys: Issues an Opportunities: A Report to the National Science Foundation Based on a Workshop Held on March 28–29, 2003. www.nsf.gov/sbe/ses/mms/nsf04_211a.pdf.
Tourangeau, R., Rips, L.J., and Rasinski, K. (2000). *The Psychology of Survey Response*. Cambridge: Cambridge University Press.
Verba, S. and Nie, N.H. (1972). *Participation in America: Political Democracy and Social Equality*. New York: Harper and Row.
West, B.T. and Peytcheva, E. (2014). Can interviewer behaviors during ACASI affect data quality? *Survey Practice* 7 (5): http://surveypractice.org/index.php/SurveyPractice/article/view/280.
Williams, L.E. and Bargh, J.A. (2008). Experiencing physical warmth promotes interpersonal warmth. *Science* 322: 606–607.
Wright, G. (1977). Racism and welfare policy in America. *Social Science Quarterly* 57: 718–730.
Zablotsky, B., Black, L.I., Maenner, M.J. et al. (2015). *Estimated prevalence of autism and other developmental disabilities following questionnaire changes in the 2014 National Health Interview Survey*. National Health Statistics Reports, 87. http://www.cdc.gov/nchs/data/nhsr/nhsr087.pdf.

Part VIII

Vignette Experiments in Surveys

Allyson Holbrook[1] and Paul J. Lavrakas[2]

[1] *Departments of Public Administration and Psychology and the Survey Research Laboratory, University of Illinois at Chicago, Chicago, IL, USA*
[2] *NORC, University of Chicago, 55 East Monroe Street, 30th Floor, Chicago, IL 60603, USA*

Vignette experiments in surveys involve presenting respondents with one or more vignettes, which are brief descriptions or depictions of a target (e.g. a person or scenario). The content of these vignettes is the independent variable(s) that the researcher manipulates. Respondents are asked to provide one or more ratings of the target and these data serve as the dependent variable(s). One of the hallmarks of vignette research is that aspects of the description are varied independently so that the researcher can statistically assess the effect of various aspects of the description on respondents' ratings (Rossi and Nock 1982). A common (although not necessary) feature of vignette research is that respondents are exposed to and asked to rate more than one vignette so that the characteristics of the descriptions can be manipulated as a within-subjects variable. However, when the number of vignettes generated by the factorial design is large, respondents are often assigned to a subset of the total universe of possible vignettes (e.g. Atzmuller and Steiner 2010). The goal of this approach is to have all of the possible combinations of factors being manipulated in the vignettes evaluated by a subset of respondents while limiting the number of vignettes each respondent is being asked to rate.

The section contains two chapters. The first chapter by Auspurg, Hinz, and Walzenbach examines the role of survey mode in vignette studies. Comparing data from a self-administered and a face-to-face sample, it discusses to which extent the presence of an interviewer affects responses to vignette modules. From a theoretical point of view, interviewers can motivate respondents and assist with more complex questions. Their presence might affect the number of vignettes as well as the complexity of the information that can be presented. However, interviewers also raise survey costs, cause correlated error terms across respondents, and might even cause bias stemming from the social situation in which the survey is completed. The authors of this chapter review the literatures on vignettes and mode effects and examine how substantive answers, item nonresponse, response consistency, and also the use of response sets differ across modes. For their empirical analyses, the authors used data from a vignette study for which data was obtained in two different random samples: in one sample, respondents answered the vignettes in a completely self-administered survey, while the other was a self-administered module in a face-to-face setting where an interviewer was present.

Experimental Methods in Survey Research: Techniques that Combine Random Sampling with Random Assignment, First Edition.
Edited by Paul J. Lavrakas, Michael W. Traugott, Courtney Kennedy, Allyson L. Holbrook, Edith D. de Leeuw, and Brady T. West.

Companion Website: www.wiley.com/go/Lavrakas/survey-research

The second chapter by Eifler and Petzold compares responses to survey vignettes with data collected in a field observation study. One potential strength of vignette research is that they may be less subject to social desirability bias than direct self-reports because respondents are reacting to a hypothetical target instead of directly reporting their attitudes or behavior. A weakness of vignettes is that they tend to lack mundane realism (cf. Aronson and Carlsmith 1968), and it is unclear how well they predict actual behavior. This chapter addresses these issues by reviewing the relevant literature and presenting the results of studies that assess a socially undesirable behavior either through behavioral observation or responses to vignettes.

There are many directions for research using vignettes in surveys. Vignettes are an efficient way to conduct an experiment as part of a survey and have been used extensively to assess substantive research questions (e.g. the effect of candidate attributes on attitudes toward the candidate). Vignettes have also been used in methodological research, for example, by asking respondents their likelihood of participating in response to a series of vignettes describing survey invitations as an approach to study strategies for reducing survey nonresponse, but there are many additional ways that vignettes could be used to address methodological questions (cf. Lavrakas et al. 2000). Work by King and colleagues also demonstrates that vignettes can be used in surveys to control for individual-level or group-level differences in the interpretation of survey response scales (e.g. King et al. 2004; King and Wand 2004).

References

Aronson, E. and Carlsmith, J.M. (1968). Experimentation in social psychology. In: *The Handbook of Social Psychology*, vol. 2 (ed. G. Lindzey and E. Aronson), 1–79. Reading, MA: Addison-Wesley.

Atzmuller, C. and Steiner, P.M. (2010). Experimental vignette studies in survey research. *Methodology in European Journal of Research Methods for the Behavioral and Social Sciences* 6 (3): 128–138.

King, G. and Wand, J. (2004). Comparing incomparable survey responses: evaluating and selecting anchoring vignettes. *Political Analysis* 15 (1): 46–66.

King, G., Murray, C.J.L., Salomon, J.A., and Tandon, A. (2004). Enhancing the validity and cross-cultural comparability of measurement in survey research. *American Political Science Review* 98 (1): 191–207.

Lavrakas, P.J., Diaz-Castillo, L., and Monson, Q. (2000). Experimental Investigations of the Cognitive Processes Which Underlie Judgments of Poll Accuracy. Paper presented at the 55th annual conference of the *American Association for Public Opinion Research*, Portland OR.

Rossi, P.H. and Nock, S.L. (1982). *Measuring Social Judgments: The Factorial Survey Approach*. Beverly Hills, CA: Sage.

19

Are Factorial Survey Experiments Prone to Survey Mode Effects?

Katrin Auspurg[1], Thomas Hinz[2], and Sandra Walzenbach[2,3]

[1] *Ludwig Maximilian University of Munich, Germany*
[2] *University of Konstanz, Konstanz, Germany*
[3] *ISER/University of Essex, Colchester, UK*

19.1 Introduction

Over the past decades, factorial survey experiments that ask respondents to evaluate experimentally varied scenarios (i.e. *vignettes*) have become an increasingly popular tool in many subfields of the social sciences (for overviews, see Mutz 2011; Wallander 2009). While there is some literature on experimental design issues, such as the ideal number of experimental factors and levels, the effects of survey modes and respondent samples have not received much attention. In this chapter, we will use key concepts of the total survey error framework (Biemer 2010; Groves and Lyberg 2010) to study possible mode effects. We will compare two survey modes, namely, a mode with interviewer presence vs. a completely self-administered mode. Both survey modes have been applied to study the same substantive issue (fairness of earnings) and have been used in random respondent samples of the German residential population in 2009. We are mainly interested in the effects of interviewer presence on item nonresponse, inconsistency of responses and measurement errors such as response sets, and how these issues and possible further mode effects affect the substantive results gained by the factorial survey experiment in our case study.

We begin with a brief illustration of the factorial survey method (Section 19.2) and continue with a discussion of typical modes, design issues, and their connection to mode effects (Section 19.3). In the case study described in Section 19.4, two different survey modes are analyzed: a *face-to-face* interview where respondents filled in the factorial survey module themselves, but an interviewer was present for optional support, and a completely *self-administered* mode (where respondents could choose between completing a mail and a web survey). Do these modes differ in regard to data quality issues such as the proportion of nonresponse and response consistency? Does this lead to different substantive results? We conclude with a short summary and discussion of the practical implications (Section 19.5).

Experimental Methods in Survey Research: Techniques that Combine Random Sampling with Random Assignment, First Edition.
Edited by Paul J. Lavrakas, Michael W. Traugott, Courtney Kennedy, Allyson L. Holbrook, Edith D. de Leeuw, and Brady T. West.

Companion Website: www.wiley.com/go/Lavrakas/survey-research

19.2 Idea and Scope of Factorial Survey Experiments

Factorial survey experiments go back to Peter Rossi who originally used the method to study normative judgments (Rossi and Andersen 1982). Since then, they have been applied to a wide range of research topics, including punishment preferences for deviant behavior, measurements of social status, normative perceptions of fair earnings, and attitudes toward immigration. In factorial survey experiments, respondents' opinions and attitudes are retrieved by asking participants to evaluate scenarios (*vignettes*) in which the values (*levels*) of several characteristics (*dimensions*) are experimentally varied.[1] Respondents are frequently asked to evaluate the vignettes on an ordinal rating scale. For illustration, we refer to the factorial survey experiment used in our case study: respondents assessed the fairness of earnings of hypothetical employees with varied characteristics on an eleven-point rating scale ranging from *unfairly low* to *unfairly high* (see Figure 19.1).

By including various dimensions at once, vignettes can account for the fact that (normative) judgments typically require a simultaneous consideration of several factors. An independent but joint experimental variation of the factors allows the researcher to quantify their impact on the requested evaluations and to estimate trade-offs and interactions between the different factors. For instance, one could find out if respondents believe that men or women should get higher earnings, and if this is true even if they show the same levels of education, or if this holds particularly true in some occupational fields, such as typical male or female occupations. Because of the multifactorial design, it is also possible to estimate (monetary) trade-offs, such as just gender pay gaps (for more details, see Auspurg and Hinz 2015a).

Typically, respondents are asked to evaluate 10–30 vignettes, because these multiple ratings allow economizing research resources (many vignette judgments can be gathered with few respondents). However, there are cognitive restrictions: methodological studies have found that fatigue effects occurred when respondents had to evaluate more than 10 vignettes (Sauer et al. 2011). To get higher numbers of vignettes evaluated in a survey, researchers commonly work with different subsets of vignettes that are included into different versions of a questionnaire (these different vignette subsamples are called *decks* or *sets*; for more information on such design issues, see Section 19.3).

A crucial advantage of factorial survey experiments is that they combine randomized treatments with the benefits of surveys. Based on the randomized assignment of respondents to experimental stimuli (vignettes), the method offers a high internal validity for testing causal hypotheses. But while laboratory experiments usually suffer from small sample sizes and

A 60-year-old woman with vocational training works as a social worker.
Her monthly gross earnings total €2500 (before tax and extra charges).

Are the gross monthly earnings of this person fair or are they (from your point of view) unfairly high or low?

Unfairly low					Fair					Unfairly high
−5	−4	−3	−2	−1	0	1	2	3	4	5
□	□	□	□	□	□	□	□	□	□	□

Figure 19.1 Vignette implementation: example (varied dimensions underlined).

1 Note that there are also "vignette studies" where researchers do not vary any characteristics. We use the term *factorial surveys* for all designs where the vignette scenarios are varied in the form of a multifactorial experiment.

homogeneous groups of participants (Henrich et al. 2010), factorial survey modules can be implemented in any (large scale) survey.[2]

19.3 Mode Effects

19.3.1 Typical Respondent Samples and Modes in Factorial Survey Experiments

Factorial survey experiments have frequently used small convenience samples, such as university students (Wallander 2009). This might be adequate if one is solely interested in testing causal mechanisms based on the experimental vignette factors that do not involve any moderator (or mediator) variables on the part of respondents. More insights on potential moderator variables, and consequently more general conclusions, can, however, be gained with more heterogeneous respondent samples (Auspurg and Hinz 2015a). Random respondent samples provide more firm ground for inference statistics, and they are needed if one wants to describe distributions such as the amount of social consensus across a population. For instance, do older and younger respondents share the same (justice) principles? Are there indicators of a social cleavage, e.g. between respondents with leftwing or rightwing political positions (for such an application, see Hermkens and Boerman 1989)?

Empirical social research often builds on telephone surveys, as these provide a less cost-intensive alternative to face-to-face interviews. However, although there are a few factorial survey studies using this mode (e.g. Pager and Quillian 2005), most applications seem to be too complex to be completely administered on the phone. Typical applications involve various experimental dimensions that are combined into paragraph descriptions spanning several text lines. A full processing of so much information probably requires the respondents to read the vignettes themselves. One has at least to expect *recency-effects* (for those effects, see Krosnick and Alwin 1987) when vignettes are only presented orally – that is, respondents will likely only recall the last pieces of information given to them, which would give vignette dimensions presented later in the vignettes disproportionate weight on evaluations.

For all these reasons, it is clearly advisable to allow respondents to assess the vignette information themselves. This can be achieved using different modes, such as computer-assisted personal or self-administered interviews (CAPI and CASI), or paper and pencil interviews (PAPI). In personal interviews, the interviewer should be instructed to hand the questionnaire (computer) over to the respondents so that they can read and evaluate the vignettes themselves, while the interviewer remains present in the room to provide optional support. Following this advice, factorial survey modules are nearly always given in a self-administered manner. Researchers have nevertheless to make a decision which mode to use for their survey (completely self-administered or not).

Often this decision is linked to the techniques used to recruit respondents. We will focus on two different options that are frequently applied in social science research, and which we have used in our case study. (i) *Random route* procedures are often employed in countries where countrywide registers of the residential population do not exist and are implemented as a multi-stage sampling technique. Starting with randomly selected addresses within randomly selected

---

2 Although not the primary focus of this chapter, there are other survey methods apart from factorial surveys that rely on multifactorial experimental plans, namely, choice experiments and conjoint analyses (Auspurg and Hinz 2015b). In contrast to factorial surveys, they usually do not use full textual descriptions of scenarios (vignettes) but sum up characteristics in short keywords that are arranged in a table format. The crucial feature of choice experiments is that respondents make choices between options. Their answer is not a gradual one but a definite "yes" statement for the preferred option. In conjoint analyses, the typical task consists of simultaneously ordering options according to the respondent's preferences.

geographical areas, interviewers walk from house to house following a prescription that should ideally lead to a "random route" through the area (Bauer 2014). During this walk, dwelling units are listed. The final sampling step consists of a random selection of a person within the household, which is commonly realized by a Kish-selection grid. This sampling technique is typically combined with face-to-face interviewing, as the interviewer is already present with "a foot in the door." (ii) Respondents can alternatively be recruited via a short screening interview on the phone, where a *random digit dialing* technique is combined with a technique (such as a Kish-selection grid) to randomly select one target person in multiperson households. After that, for factorial surveys, all recruited respondents get invited to a self-administered mail and/or online survey (as indicated, it is not recommended to do factorial surveys completely on the phone). We will refer to this technique hereafter as the completely *self-administered* mode.

This second option, the completely self-administered mode, is in general cheaper than the random walk procedure in combination with a personal interview, but is often accompanied with lower response rates and more severe coverage problems (such as underrepresentation of less educated people or low-income households; see de Leeuw 2008: 127 et seq.; Groves et al. 2009: 153 et seq.; Sue and Ritter 2007: 7–8). However, factorial surveys are an experimental method, and the key factor for good experimental designs is not the random selection of respondents, but rather a well-established experimental design, combining an adequate set-up of vignettes which contain the experimental stimuli and are randomly allocated to respondents. If this is done well, coverage and nonresponse problems should not strongly impair valid conclusions on the causal impact of vignette dimensions (Mutz 2011). Sampling errors might still affect descriptive statistics, which one might try to correct through weighting procedures. So far, there seems to be nothing specific about factorial survey methods. Thus, the discussion boils down to the question of whether the presence of an interviewer has an impact on measurement issues, and if yes, what that impact is.

19.3.2 Design Features and Mode Effects

When setting up factorial surveys, researchers have to plan different steps. They have to select a number of experimental dimensions and levels for the vignette characteristics, and they have to choose a strategy for sampling vignettes and splitting them into different subsamples (i.e., *decks*). Some further design issues, such as use of response scales, also have to be decided. In the following, we will review the most important design features and discuss how they might be related to survey modes, with a focus on the presence of interviewers.[3]

19.3.2.1 Selection of Dimensions and Levels

Factorial surveys are primarily designed to test theories on the causal impact of experimental factors. Therefore, selecting the vignette dimensions should be done carefully. The recommendation is to use theories to derive the dimensions and specify the regression equations one wants to identify (including possible interactions effects). This helps not only to ensure that all relevant dimensions are included, but also to optimize the experimental design (for more details on this, see Section 19.3.2.2). Regarding the number of dimensions, the general advice is to use a mid-level of complexity of about 6–8 variable dimensions that all have about 2 or 3 different levels.[4] Theoretical assumptions that respondents are sensitive to complexity have been put forward under the term *satisficing*: respondents do not necessarily use maximum

3 For more detailed instructions see, for instance, Auspurg and Hinz (2015a) and Mutz (2011).

4 Lower numbers of levels help minimize the number of vignettes that are needed to achieve a certain amount of statistical power (for details see Auspurg and Hinz 2015a). Yet, one needs at least 3 different levels to estimate nonlinear relationships, and there might be further substantial reasons to specify more different vignette levels.

effort to answer survey questions, but might shortcut the process to provide satisfactory answers requiring less effort (Krosnick 1991; originally Simon 1957). According to Krosnick, satisficing is more likely when the respondents' cognitive ability and motivation are low, but also when the task complexity is high.[5] For the data used in our case study, Sauer et al. (2011) only found a lowered response consistency for very complex vignette modules (consisting of 12 instead of 5 or 8 variable dimensions). According to their results, participants first learn and get better at dealing with the question format, until a fatigue effect sets in after about the 12th vignette (the authors tested at most 30 vignettes per respondent).[6]

However, it was not investigated if these effects interact with the survey mode. In completely self-administered modules, respondents can simply drop out if they are overburdened. In contrast, in face-to-face interviews, social cooperation norms make it very unlikely that, once the interview has started, respondents will refuse to finish the questionnaire or one of its core modules (Meulemann 1993: 113 et seq.; Schnell 1997: 122 et seq., 144). This is also because respondents can easily rely on interviewer assistance in case they have problems fulfilling the task. Particularly if there are many vignettes and dimensions, however, respondents in a face-to-face mode also might suffer from fatigue effects or boredom. In that case, respondents might provide satisficing responses (i.e. select any level on the response scale or fade out some dimensions). Although there might be lower nonresponse in cases of interviewer assistance, at least for complex modules the data quality might be lower. This is also because respondents might experience pressure to evaluate the vignettes within a short time, so as not to keep the interviewer waiting. However, one might also argue that the absence of anonymity in the face-to-face situation makes respondents invest more cognitive effort into answering the vignette questions. It is therefore still an open question which mode shows a higher data quality.

19.3.2.2 Generation of Questionnaire Versions and Allocation to Respondents

In the overwhelming majority of factorial survey modules, the full *universe* of all possible combinations of dimensions and levels is too large to be completely administered. This means that the researcher needs techniques for selecting vignettes to be used in the survey and to allocate these vignettes to the different questionnaire versions. What is important is that, first, the chosen vignettes show minimum correlations of vignette dimensions (including their interaction effects), because otherwise one would no longer be able to disentangle their effects. Second, it is desirable that one is able to estimate the effects of the vignette dimensions independent of deck effects (which might stem from the specific composition of vignettes within the different decks) or characteristics of respondents. If characteristics of respondents get confounded (i.e., get strongly correlated) with vignette dimensions, the effects of these characteristics or respondents' cognitive abilities might be mistaken for substantial effects of the vignette dimensions (Sauer et al. 2011).

So far, most researchers have used random selections of vignettes that leave confounding structures simply to chance. One should better use *quota designs* that allow the researcher to plan the confounding structures beforehand (Atzmüller and Steiner 2010; Dülmer 2007, 2016).

Similar numbers of levels across dimensions help to avoid the so-called *number of levels effects* that have been found in choice experiments (De Wilde et al. 2008; Verlegh et al. 2002; Wittink et al. 1990): dimensions varying on a higher number of levels attracted disproportional attention, probably because they have been more eye-catching than dimensions showing less variation.

5 However, too few dimensions might also cause problems, because respondents might lack information to evaluate the vignettes and undercomplexity might risk causing boredom and fatigue effects (Auspurg et al. 2009).

6 These results are somewhat confirmed by Teti et al. 2016. The authors report that even elderly respondents aged up to 90 had no difficulty in responding consistently to a vignette module on relocation preferences that consisted of 10 vignettes with 6 (dichotomous) dimensions.

In particular, D-*efficient designs* are recommended. They combine the vignette dimensions in a way that makes the vignette sample as orthogonal (meaning that vignette dimensions are maximally uncorrelated) and balanced (in that all dimension levels occur with about the same frequency) as possible, while at the same time ensuring that all important parameters (effects of vignette dimensions) can be identified. Optimizing these features allows maximizing the precision (or power) with which effects of the vignette dimensions can be identified. The higher this statistical efficiency, the lower the number of respondents needed to achieve a given amount of precision (for details, see Auspurg and Hinz 2015a). Regarding mode effects, such sampling techniques might also help to decrease correlations of decks with interviewer effects in personal interviews. To be able to separate substantial from possible interviewer effects, one should at least avoid strong correlations of decks with interviewers. This can be achieved by randomly blocking decks to interviewers or by using *D*-efficient sampling techniques again (where the interviewers have to be treated as another blocking factor; the later technique would again increase the statistical efficiency of the experimental design in terms of a maximum orthogonality of substantive *and* methodological factors).

Whatever way is chosen to select vignettes, it is absolutely crucial that they are *randomly* assigned to respondents. In addition, when working with fixed decks, it is advisable to randomize the vignette order across respondents to avoid possible order effects.[7] All these designs likely pay out particularly in the case of heterogeneous respondent samples, as these are more prone to a confounding of respondents' effects with order or deck effects.

There is one additional issue to be considered. Often, combining all vignette dimensions and levels leads to implausible or even unrealistic vignette scenarios (for instance, in our case study, there could have been medical doctors without a university degree). It has been shown that too unrealistic combinations of levels lead respondents to ignore the respective dimensions for future evaluations (Auspurg et al. 2009: 86). Very unrealistic combinations should therefore be dropped. But there may nevertheless be some implausible combinations: eliminating all such combinations often triggers strong correlations of vignette dimensions, and it might also be interesting to get some evaluations on scenarios that expand reality (Rossi and Anderson 1982). In case respondents get irritated, the interviewer might inform them that those vignette cases were deliberately designed. In that regard, we expect interviewer presence to lead to lower nonresponse rates and/or to evaluations based on more different dimensions.

19.3.2.3 Number of Vignettes per Respondent and Response Scales

As already mentioned, it is recommended to restrict the number of vignettes per respondent to no more than 10 vignettes if one wants to prevent fatigue effects. Another decision parameter is survey time, particularly in face-to-face interviews, meaning that longer vignette modules increase the costs per interview ("time is money"). We can only provide some rules of thumb. According to our experience, for vignette modules with mid-complexity (8 dimensions and about 10 vignettes), respondents need an average of about five minutes to complete the vignette module (for more detailed statistics, see Auspurg and Hinz 2015a). In personal interviewers, respondents might show a higher willingness to evaluate long vignette modules, while at the same time counterweighting this lower nonresponse by satisficing strategies. That is, we expect effects that are similar to those assumed for the number of dimensions.

---

7 Such random orders can be achieved by means of software in computer-assisted surveys, but it is also possible to implement them in paper questionnaires (for some instructions, see Auspurg and Hinz 2015a). Methodological research showed that the effects of the order of vignette dimensions can also occur in complex designs (containing 12 variable dimensions), or if respondents feel unsure about the topic (Auspurg and Jäckle 2017). Randomizing the order of vignette dimensions across respondents – which can especially be implemented in computer-assisted modes – can help to neutralize such order effects.

Regarding response scales, sometimes two-step answering scales were employed (where, for example, respondents first rate whether the vignette earnings are fair or not, and then rate the amount of unfairness only if earnings are perceived as being unfair). Such scales, however, offer easy opt-out alternatives to respondents and thus enhance satisficing (Sauer et al. 2014). For this and other reasons, it is advisable to use standard response scales like the ordinal rating scale employed in our case study (see Figure 19.1 again), although alternatives have been implemented by some researchers (Jasso 2006; Wallander 2009: 511). Complex response scales (such as magnitude scales) probably work with interviewer assistance; they can definitely not be recommended for self-administered modes (Auspurg and Hinz 2015a: 64 et seq.)

19.3.2.4 Interviewer Effects and Social Desirability Bias

Turning more specifically to mode effects, the role of the interviewers has been widely debated (Groves et al. 2009: 141 et seq.; Loosveldt 2008). On the one hand, interviewers can clarify the respondents' role and provide explanations and assistance, which might at least be important in case of more demanding response tasks (as reported by Holbrook et al. 2003 for a standard survey). On the other hand, data collection in a social situation offers manifold ways for interviewers to compromise data quality and introduce systematic interviewer bias. Apart from conscious falsifications, misinterpretations of vague responses and inappropriate feedback, more subtle cues such as the interviewer's gender or dialect can also play a role. What is also important is that in face-to-face interviews the clustering of respondents within interviewers and sample points causes autocorrelations of error terms (Snijders 2005). These autocorrelations commonly decrease the efficiency (precision) of estimates (Groves et al. 2009: chapter 4.4; Auspurg and Hinz 2015a: chapter 5). However, an important concern of self-administered modes is the inability to control the circumstances under which the survey is completed (Zwarun and Hall 2014).

Research with factorial surveys is often targeted at sensitive issues, such as respondents' discriminatory attitudes or the reported likelihood of engaging in criminal behavior (for some applications, see Auspurg et al. 2017; Pager and Quillian 2005; Graeff et al. 2014). Factorial surveys are assumed to be less prone to social desirability bias than single-item questions, or even randomized response techniques that are particularly designed to reduce this bias. Auspurg et al. (2015) were the first to report promising empirical evidence on this topic. Compared to a direct question format, the vignette module yielded less socially desirable answers concerning a just gender wage gap.[8] However, it was not possible to assess if the results generalized to modes other than the employed face-to-face mode, and if there was still some social desirability bias left. Such bias might be induced by the interviewers' presence, even when respondents were allowed to answer the vignette questions themselves, because of a latent fear of negative consequences within a personal situation of low anonymity, and a higher awareness of social norms due to the presence of another person. There is so far only one study focusing on the second mechanism: in a split-ballot experiment, in which a student sample answered a factorial survey module on the fairness of earnings, the presence of an experimenter led to stronger preferences for equality (Liebig et al. 2015). Therefore, particularly in face-to-face interviews, one might

8 Similar results in a related scenario question format have previously been found in articles by Armacost et al. (1991) and Burstin et al. (1980). However, the authors varied none or only one dimension of their vignettes.

use techniques to reduce social desirability bias. For instance, one could vary the sensitive dimension only between (and not within) the vignettes presented to single respondents.[9]

19.3.3 Summing Up: Mode Effects

All in all, we expect higher nonresponse rates in the completely self-administered mode. This is because it is more difficult for respondents to ask for assistance if they are, for instance, irritated by some implausible vignette combinations, or have any other problems with the response task. In addition, the risks of being sanctioned for quitting the questionnaire are lower. For other data quality issues, it is more difficult to make clear predictions on mode effects. Respondents might be more likely to take shortcuts in more anonymous situations, but there are also good reasons to assume that respondents speed up and satisfice if an interviewer is sitting next to them. Social desirability bias should be more present in the face-to-face mode; therefore, in our application, we expect lower effects for the most sensitive dimension, that is, the vignette person's sex, in the face-to-face mode. In addition, the stronger clustering of responses in this mode (evaluations are nested within interviewers) should lead to a lower statistical efficiency (i.e. higher standard errors).

19.4 Case Study

While mode effects are a well-investigated field in general survey methodology (Berrens et al. 2003; Carini et al. 2003; Fisher and Herrick 2013; Malhotra and Krosnick 2007), empirical research on mode effects for factorial survey experiments is, to the best of our knowledge, completely lacking (with the only exception being the cited study of Liebig et al. 2015 on interviewer presence). All in all, it is an open question whether the higher survey costs related to personal interviews pay out in higher data quality. To find out, we run a case study with different survey modes.

19.4.1 Data and Methods

19.4.1.1 Survey Details

Our case study consists of two factorial survey experiments on the fairness of earnings in Germany that were run simultaneously in 2009, using the same questionnaire and factorial survey module. However, the two surveys differed in recrutiment procedure and data collection mode. Respondents either completed the questionnaire in a personal interview (*face-to-face mode* with an interviewer present) or as a totally self-administered survey, which could take place online or as a paper and pencil questionnaire (*self-administered mode*). Recruitment into those survey modes took place by drawing two separate random samples from the adult residential population in Germany. For the face-to-face mode, 129 regional sample points in Germany were selected. Sampled subjects were chosen by applying a random route strategy in combination with a Kish-selection grid. For this face-to-face sample, there were 82 interviewers, each conducting between 2 and 28 interviews (median value: 7). For the evaluation of the factorial survey module, the interviewers handed the computer over to the respondents so that

9 In a study by Auspurg et al. (2015), using an application on fairness of earnings, the results of a split with a pure between variation of the sensitive dimension (the vignette person's gender) did not differ from the split employing a within variation. This finding might, however, not generalize to more sensitive topics: in a factorial survey on discriminatory attitudes towards Muslims, answers from a between condition suffered less from social desirability bias than the ones from a within condition (Walzenbach and Hinz 2019).

they could read and evaluate the vignettes themselves, but the interviewers remained present in the room. After the factorial survey module, interviewers took over the computer again and went on with asking questions. For the self-administered data collection mode, respondents were recruited by telephone using random digit dialing. Within this group, respondents selected via the Kish-selection grid were offered two options to complete self-administered questionnaires.[10]

To test if mode effects interact with the complexity of factorial survey modules, the length of the sequence (10, 20, or 30 vignettes) and the number of dimensions (5, 8, or 12) were experimentally varied using split ballot experiments in both survey modes. Figure 19.1 shows a vignette example with 12 dimensions. Table 19.1 contains an overview of all experimentally varied dimensions and levels.

For implementation, a *D*-efficient sample of 240 vignettes was drawn from the vignette universe (which contained more than 1 million possible vignettes).[11] To standardize the experimental design across all experimental splits, the same vignette sample was used for all complexity conditions and irrelevant surplus dimensions were deleted for the splits with 5 and 8

Table 19.1 Vignette implementation: overview of dimensions and levels.

				Dimensions	Levels
Experimental split with 5 dimensions	Experimental split with 8 dimensions	Experimental split with 12 dimensions	1	Age	30/40/50/60 years
			2	Sex	Male/female
			3	Vocational training	Without degree/vocational degree/university degree
			4	Occupation (ordered according to their magnitude prestige scores)	Unskilled worker/door(wo)man/engine driver/clerk/hairdresser/social worker/software engineer/electrical engineer/manager/medical doctor
			5	Monthly gross earnings	500/950/1200/1500/2500/3800/5400/6800/10 000/15 000
			6	Experience	Little/much
			7	Job tenure	Entered recently/entered a long time ago
			8	Children	None/1/2/3/4
			9	Health status	No health problems/long-term health problems
			10	Performance	Below average/above average
			11	Economic situation of the firm	High profits/threatened by bankruptcy/solid
			12	Firm size	Small/medium/large

10 Respondents in the self-administered mode could decide themselves whether they preferred the paper or the online questionnaire. In the analyses, we will combine both variants, because we are mainly interested in the effects of sampling compositions resulting from the two different recruiting techniques and in interviewer effects.

11 Some implausible combinations such as medical doctors without a university degree were deleted (for details, Sauer et al. 2011).

dimensions. The number of vignettes, as well as the order in which the scenarios were displayed, was randomly assigned to respondents, and in the mode with interviewers, decks were randomly blocked to the different interviewers.[12]

19.4.1.2 Analysis Techniques

In the first step (Section 19.4.2), we will assess if the different modes are accompanied by different data quality issues: Do both modes differ in regard to nonresponse rates and indicators for measurement errors, such as respondents engaging in satisficing strategies (signaled by response sets or inconsistent judgments)? Even if there are some data quality issues, results from the factorial survey experiment, with its high potential to increase the internal and construct validity (Auspurg and Hinz 2015a), might turn out to be robust. Therefore, in the second step (Section 19.4.3), we explore differences in substantive results. These sections also include some analyses on social desirability bias, respondents' use of heuristics, and the possible loss of statistical efficiency in the face-to-face mode that is caused by interviews not being independent but clustered within interviewers and sample points.

When setting up our case study, we decided to apply the most typical survey settings, where data collection modes go hand-in-hand with different recruitment strategies (random walks in the personal interviews and phone recruitments in the completely self-administered mode). This makes it necessary to disentangle the possible mode effects we are interested in from sampling errors: our two survey splits might not only differ because of mode effects but also because of the composition of respondents. To separate both effects, we did some analyses on the composition of respondents (for details see analysis A1 in the online appendix).[13] Both modes/recruitment strategies showed some indications for sampling errors: in both splits, some groups of respondents were under- or overrepresented in comparison to the sociodemographic distributions expected in the general German population (these "true" distributions at the time when the survey took place were captured by census data; see analysis A1 in the online appendix for more details). We used different techniques to adjust for these sample selection errors and, more importantly, to adjust the composition of respondents in both survey splits. Respondent characteristics were used as controls in multivariate regressions, and in the analyses on the substantive effects of the vignette dimensions we tested if results changed when using post-stratification weights to adjust both samples to the general population. More details on the employed (regression) techniques are provided in the respective subsections.

19.4.2 Mode Effects Regarding Nonresponse and Measurement Errors

19.4.2.1 Nonresponse

Table 19.2 shows descriptive statistics about item nonresponse by survey mode. If we only compare the portions of respondents with at least one missing value in the vignette module, numbers first seem pretty similar in face-to-face interviews and the self-administered mode (10.3% vs. 9.4%). However, a closer look shows that there is no complete refusal to answer the factorial questionnaire if an interviewer was present, whereas 1.4% of the respondents in self-completion modes left out the whole module. Partial nonresponse is a bit higher for face-to-face respondents (10.3%) than for respondents who answered in the self-administered mode (8.0%).

Given that partial item nonresponse takes place, the quantity of missing values also differs by sampling mode. Respondents who answered the questionnaire in self-administered modes

12 However, this was done in a way that makes the data mostly balanced, meaning that, across interviewers, all different decks occurred with about the same frequency.
13 www.wiley.com/go/Lavrakas/survey-research.

Table 19.2 Item-nonresponse by survey mode.

	Face-to-face	Self-administered
	777 respondents	844 respondents
Respondents refusing to answer any vignette question	0.0% (0)	1.42% (12)
Respondents partly refusing to answer vignette questions	10.3% (80)	8.0% (67)
Missing vignettes in case respondent partly refused answers	10.6% (1.9)*	17.3% (2.9)*

Absolute numbers in parentheses.
*Difference significant on 5% level.
Data: respondents who at least saw the first vignette

show higher proportions of item-nonresponse than the face-to-face participants. This is true both for absolute frequencies (face-to-face: on average 1.9 vignettes per respondent; self-administered: 2.9) as well as for proportions relative to the number of displayed vignette questions (face-to-face: 10.6%, self-administered: 17.3%, difference significant according to undirected t-test: $p = 0.011$).

Table 19.3 shows the results of binary regressions on the missing values. We display average marginal effects (AMEs) under control of the respondents' sex, age, and educational background. Apart from the survey mode, the number of vignettes (10, 20, or 30) and the number of vignette dimensions (5, 8, or 12) were included as explanatory variables, both measuring the complexity of the factorial survey module. Model 1 contains only the cases with complete nonresponse. Since completely missing vignette modules only occurred in self-completion questionnaires, the corresponding mode dummy predicts the dependent variable perfectly, and the model is restricted to the self-administered sample. For the models on partial nonresponse, the dependent variable was coded 1 as soon as at least one of the vignette questions remained unanswered, irrespective of the exact amount of item nonresponse. Cases in which vignettes are partly missing are compared with those in which respondents showed no nonresponse whatsoever, meaning that all the vignette modules with complete nonresponse were excluded from the analysis.

Contradictory to what we found for complete nonresponse, the data collection mode has no noteworthy effect on partial nonresponse (Model 2, summarizing both modes). Concerning the complexity measures, the number of presented vignettes shows significant effects. Respondents seem to get tired of the relatively complex vignette questions if they become too numerous. In sets with 20 vignettes, nonresponse is already more likely than in sets with only 10 vignettes, and the effect is even larger and more significant for the 30 vignette questionnaire versions. If we consider partial nonresponse separately for the two modes (Models 2a and 2b), we can see that this result is mainly driven by the respondents who completed the questionnaire in a face-to-face situation. This might be a hint that respondents feel rushed when an interviewer is waiting for them to finish the vignette task. If respondents do not want to continue answering, producing missing values might simply be a more socially acceptable way to get through the vignette task than completely refusing it in front of the interviewer.

All in all, the presence of an interviewer seems to be a powerful strategy to prevent complete dropouts, although the exact same personal interview situation is likely to foster item nonresponse if the vignette module is too long. Irrespective of data collection mode, we did not find any effect of the number of displayed vignette dimensions on nonresponse.[14]

14 Additional analyses showed that there were no significant interaction effects between the number of vignettes and the number of dimensions.

Table 19.3 Item-nonresponse in the vignette module – regression tables.

	Logistic regressions (AMEs) (controlled for respondents' sex, age, and education)			
	(1) Complete nonresponse (self-admin)	**(2) Partial nonresponse (both samples)**	**(2a) Partial nonresponse (face-to-face)**	**(2b) Partial nonresponse (self-admin)**
Face-to-face (ref: self-administered)	*Omitted*	0.018		
		(0.015)		
20 vignettes (ref: 10)	−0.013	0.038 *	0.057 *	0.021
	(0.008)	(0.017)	(0.026)	(0.023)
30 vignettes (ref: 10)	−0.006	0.055 **	0.089 **	0.025
	(0.011)	(0.021)	(0.032)	(0.026)
8 dimensions (ref: 5)	0.001	−0.011	−0.024	0.001
	(0.010)	(0.017)	(0.027)	(0.022)
12 dimensions (ref: 5)	0.004	0.009	−0.009	0.024
	(0.010)	(0.018)	(0.028)	(0.023)
Intercept	−2.908 *	−2.911 ***	−2.932 ***	−2.823 ***
	(1.262)	(0.539)	(0.724)	(0.797)
Pseudo R^2	0.113	0.031	0.037	0.041
N respondents	767	1593	771	822

Standard errors in parentheses.
$+p<0.10$,* $p<0.05$,** $p<0.01$,*** $p<0.001$
Data: respondents who saw at least the first vignette; models on partial nonresponse do not consider cases with complete nonresponse and models on complete nonresponse ignore cases with partial nonresponse; additional dropouts occur due to missings on the control variables.

19.4.2.2 Response Quality

Table 19.4 sums up some indicators for response quality by survey mode. It is based on the cases that will be considered as valid from now on. This means that all unanswered vignette questions were dropped. Additionally, two more respondents (one from every sample mode) were excluded from further analyses since they always ticked the middle category on the answer scale, so that it can be doubted that they took the task seriously. We are therefore left with 12 699 vignette judgments from 776 face-to-face interviews and 13 537 evaluations from 831 completely self-administered questionnaires.

There are no differences in the proportion of vignettes rated with the middle category *fair* but the standard deviation of vignette evaluations is a bit higher in the face-to-face mode and this difference is highly significant in a two-tailed *t*-test ($p = 0.000$). These results are further illustrated in Figure 19.2, which shows the distribution of the vignette evaluations by survey mode. It is interesting to note that the extreme categories of the ordinal scale were more likely to be ticked in the face-to-face interviews than in the self-administered modes – which is probably the main mechanism responsible for the reported differences in standard deviations.

Traditionally, higher standard deviations are considered an indicator for good data quality in item batteries, suggesting that respondents do not satisfice by ticking the same response level many times. Frequently selecting only the two most extreme categories, however, might

Table 19.4 Response quality by survey mode.

		Face-to-face	Self-administered
		12 699 (by 776 resp.)	13 537 (by 831 resp.)
Proportion of vignettes rated as fair		16.5%	16.1%
Standard deviation of vignette evaluations[a)]	Mean	3.30*	3.05*
	Median	3.35	3.07
Proportion of unexplained variance if vignette judgments are regressed on first five vignette dimensions	Mean	7.5%*	8.8%*
	Median	4.8%	5.7%

*Difference significant at 5% level.

a) There was one respondent in each survey mode with only one valid judgment; thus, the standard deviations of their vignette evaluations could not be calculated.

Data: nonmissing vignette evaluations

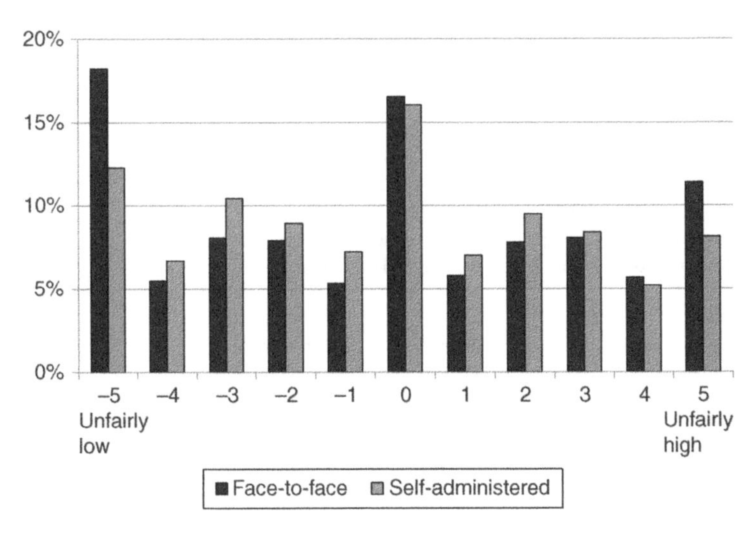

Figure 19.2 Vignette evaluations by survey mode.

represent another heuristic behavior. Apart from that, differences between survey modes are unsystematic, and even the proportion of vignettes rated as *fair* is almost equal in both samples.

To obtain another indicator of response quality, the vignette judgments were regressed on the five vignette dimensions that were evaluated by each respondent.[15] This procedure was undertaken separately for each respondent, so that the residual sum of squares could be calculated as an indicator of individual response consistency. Consistent behavior is expected to result in a good model fit, that is, a little proportion of variance left unexplained by the vignette dimensions. As can be seen in Table 19.4, the average proportion of unexplained variance was significantly higher in the self-administered mode ($p = 0.000$), suggesting that respondents answered more consistently if an interviewer was present.

15 The vignette dimensions 6–12 that were only answered by some of the respondents were intentionally omitted from the regression models for two reasons: to ensure comparability across the experimental splits and to avoid the systematic loss of cases with a high number of dimensions but with vignette sets that were too small to provide enough degrees of freedom to estimate the full model.

The proportion of unexplained variance on the respondent level can be used as the dependent variable in a regression model to see how the vignette module's complexity and survey mode influence the consistency of responses. Since the proportion of unexplained variance did not at all follow a normal distribution, we used a fractional regression model. In such models, no assumption about the distribution is made and effects are estimated on the conditional mean (for details, see Papke and Wooldridge 1996). The results of these regressions are shown in Table 19.5. To detect potential interaction effects of complexity and mode, the analyses were also run separately for the face-to-face interviews and the self-administered survey (Models 1 and 2).

An interesting first result is that the length of the factorial survey module has a negative effect on response consistency in both survey modes. However, the number of displayed vignette dimensions only seems to matter in the self-administered surveys. Compared to the experimental split with 5 dimensions, those respondents who had to evaluate vignettes with 12 dimensions show significantly higher shares of unexplained variance, and thus probably more inconsistent responses.[16]

The pooled model containing both samples (Model 3) confirms the mode effect already reported in the descriptive statistics in Table 19.4. Also under the control of respondent characteristics and task complexity, the result holds that the face-to-face group leaves lower proportions of variance unexplained and hence answers more consistently throughout the

Table 19.5 Response consistency in the vignette module – regression tables.

	Fractional response regression (AMEs) dependent variable: proportion of unexplained variance (controlled for respondent sex, age and education)		
	(1) Face-to-face	(2) Self-admin	(3) Both samples
20 vignettes (ref: 10)	0.096 ***	0.106 ***	0.100 ***
	(0.007)	(0.006)	(0.005)
30 vignettes (ref: 10)	0.114 ***	0.159 ***	0.138 ***
	(0.007)	(0.010)	(0.006)
8 dimensions (ref: 5)	−0.002	0.009	0.004
	(0.006)	(0.006)	(0.004)
12 dimensions (ref: 5)	0.000	0.022 **	0.012 **
	(0.006)	(0.007)	(0.005)
Face-to-face (ref: self-administered)			−0.014 ***
			(0.004)
Intercept	−3.412 ***	−3.259 ***	−3.218 ***
	(0.172)	(0.249)	(0.162)
Pseudo R^2	0.082	0.102	0.092
N Respondents	743	788	1531

Standard errors in parentheses.
$+p<0.1$, $^{*}p<0.05$, $^{**}p<0.01$, $^{***}p<0.001$.
Data: respondents who answered at least 10 vignettes.

16 Additional analyses showed that there were no significant interaction effects between the number of vignettes and the number of dimensions.

factorial survey module.[17] A higher consistency might, however, also be bought by using some heuristics, such as fading out dimensions. We will check in Section 19.4.3 if the effect sizes of vignette dimensions depend on data collection modes.

19.4.3 Do Data Collection Mode Effects Impact Substantive Results?

19.4.3.1 Point Estimates and Significance Levels

Leaving potential differences arising from varying complexity levels aside, the following analysis refers only to the group of respondents who answered 10 vignettes with 8 dimensions. As the main target of factorial surveys is to estimate the causal effects of vignette dimensions (Auspurg and Hinz 2015a; Rossi and Anderson 1982), measurement errors affecting these estimates would obviously be particularly problematic.

Figure 19.3 shows how the varied dimensions in the vignette texts influence the respondents' judgments on the justice of earnings (for the underlying regression models, see Table A2 in the online appendix). The plotted coefficients were obtained by regressing the respondents' vignette evaluations on the vignette dimensions of sex, age, educational background, occupation (higher numbers are related to a higher prestige score), earnings, occupational experience,

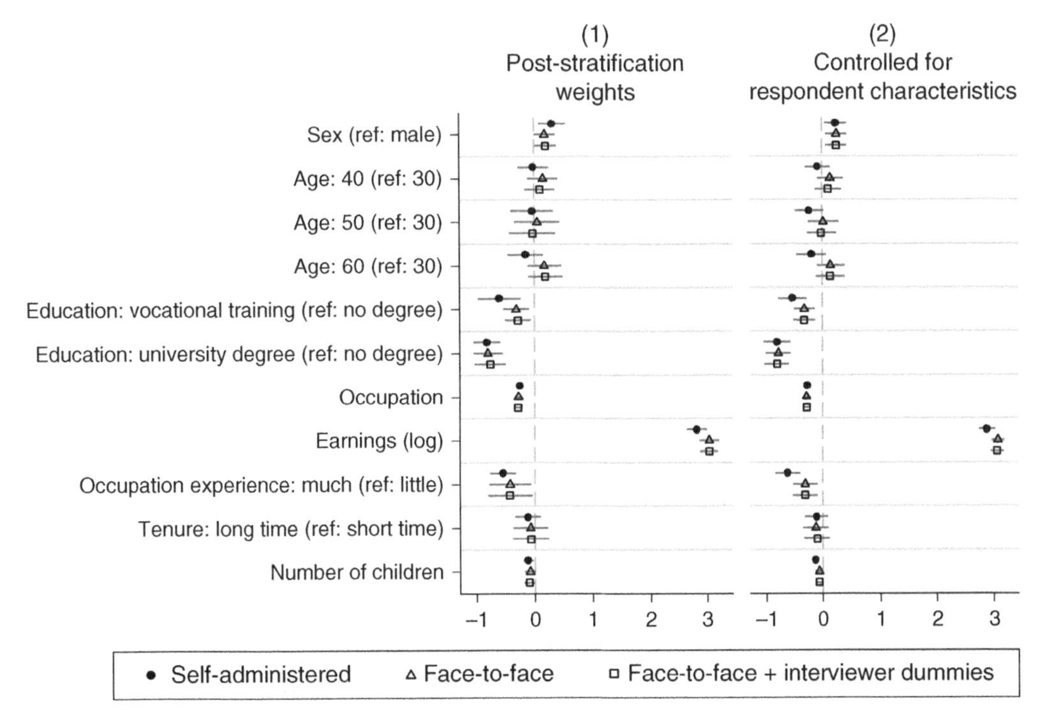

Figure 19.3 Regressions of fairness evaluations on vignette dimensions by survey mode.
Data: only respondents evaluating 10 vignettes with 8 dimensions; regression coefficients with 95% confidence intervals. For the 10 different occupations, prestige scores as measured by a common prestige scale in Germany were used (magnitude prestige scores; for these, see Christoph 2005). Positive (negative) values mean that earnings are evaluated as being unfairly too high (low) compared to the reference group.

17 Inconsistent responses certainly lead to larger residuals. However, there might also be other reasons for large residuals, such as misspecification errors (e.g. when respondents react in a nonlinear way to the age of the vignette persons). With the low number of degrees of freedom in respondent-specific regressions, it is hardly possible to test for such misspecification errors. We therefore use the proportion of unexplained variance only as a first hint for a lower response consistency.

length of tenure, and number of children. This was done separately for the self-administered survey (circles) and the face-to-face condition (triangles and squares). Since various judgments are nested within respondents (Hox et al. 1991), cluster-robust standard errors (on the level of respondents) were applied.

To differentiate between genuine data collection mode effects and mere sampling problems, it is necessary to take the sociodemographic discrepancies across the two samples into account. A common but controversial method to deal with this issue is the application of poststratification weights to adjust the drawn sample to the distribution of certain sociodemographic features known for the general population. An alternative approach would be to include only respondent characteristics as control variables in the regression model (for a discussion see Snijders and Bosker 2011: chapter 14). Both strategies were applied to our data. Model 1 (the left column in Figure 19.3) uses poststratification weights, while Model 2 (the right column) controls for the respective respondent characteristics (sex, age, and educational background). R^2-values were similar across models for both modes, with slightly higher shares of explained variance in the face-to-face condition (adjusted $R^2 = 0.712$ and 0.734) compared to the self-administered survey (adjusted $R^2 = 0.704$ and 0.710; see regression Table A2 in the online appendix).

First of all, one can see that results are plausible. For instance, vignette persons who have a higher educational degree, whose occupations show higher prestige, and who have more work experience should get higher gross earnings: the earnings are more likely to be evaluated as being too low (see the negative effects of these vignette variables). All these effects are at least significant at a 5% level, meaning that the 95% confidence intervals do not intersect with the zero-line. A comparison between the self-completion (circles) and the face-to-face mode (triangles) shows that the effects of most regressors differ either not at all or only moderately between the two modes. This is particularly true for the weighted data, where we only find one noteworthy difference: higher amounts of income are penalized more clearly in the face-to-face mode. However, the difference between modes marginally fails to reach the conventional 5% significance level ($p = 0.077$).[18] Once we control for respondent characteristics (see the second column of Figure 19.3), we observe some more severe modes effects. The interaction between income and mode reaches statistical significance ($p = 0.037$), with the income having more impact in the face-to-face mode, while job experience was more important in the self-completion mode ($p = 0.026$).[19]

Interestingly, we do not find any significant mode effects concerning the most sensitive of our vignette dimensions: the sex of the hypothetical employee. Respondents approve lower earnings for women, which is evidence for some kind of discrimination going on (for more details, see Auspurg et al. 2017), but they do so consistently and irrespective of interviewer presence. Only after poststratification weighting do we find a subtle cue that, in some cases, a reactive face-to-face mode might increase social desirability bias or activate norms of equality. If we translate the respective coefficients to the gender wage gaps perceived as just, respondents in the self-administered surveys would penalize women with a 10.5% deduction in mean wages but only 5.9% in the face-to-face sample.[20] However, we could not observe those differential estimations for the models that include respondent characteristics as controls. In

18 The significance level was obtained by specifying a full model containing the data from both survey modes, meaning that apart from the main effect for mode, interactions between mode and all the vignette dimensions could be estimated.

19 This is evidence against self-completers fading out more dimensions. Dimensions were not consistently rated more important in the face-to-face mode. This only was true for the earnings while the contrary occurred for job experience.

20 The just gender wage gaps were computed with the formula mean[income]*(exp(-_b[sex]/_b[income])-1) (for further details, see Auspurg and Jäckle 2017 or Auspurg and Hinz 2015a).

contrast, some rather robust differences were found in the regression intercepts: respondents seem to agree to higher baseline wages if an interviewer is present ($p = 0.077$ for the weighted data and $p = 0.003$ if respondent characteristics are controlled).

According to an overall Chow test, the null hypothesis that the effect sizes of vignette dimensions in both samples are equal was not rejected for Model 1 ($p = 0.238$), but Model 2 only barely failed to reach statistical significance ($p = 0.055$).[21] When the same analysis was run with the most and least complex vignette modules (the conditions that assessed 30 vignettes with 12 dimensions and 10 vignettes with 5 dimensions) as a robustness check, overall Chow tests indicated mode effects at least on the 10% significance level in all cases. They were particularly driven by differences in the baseline wages perceived as just: in the case of interviewer presence, respondents considered higher base wages as just (see the more negative intercepts in Table A2 in the online appendix). Moreover, above-average effort was rewarded significantly more in the self-administered condition than when an interviewer was present. (This finding could only be observed in the high complexity specification with 30 vignettes and 12 dimensions, because this dimension was not included in the less complex vignette versions with only 8 dimensions.)

19.4.3.2 Measurement Efficiency and Sources of Unexplained Variance

So far, we have ignored the fact that vignette judgments in the face-to-face mode are not only clustered within respondents but also within sample points and interviewers on a still higher level. This means that one can expect cluster (*design*) effects that decrease measurement efficiency. To address this issue, we specified an additional model that captures the hierarchical data structure more adequately by adding dummy variables for the different interviewers (see squares in Figure 19.3). In our case, however, the nested data structure does not seem to be accompanied by a high loss of efficiency: standard errors change only marginally when taking all three levels into account, and, as a consequence, estimates with and without fixed effects (dummies) for interviewers are very similar.

Beside a possibly biased estimation of significant effects, another potential consequence of ignoring the hierarchical data structure could be an erroneous attribution of the variance in vignette evaluations to respondents, although the observed results would partly describe differences between interviewers. To check how much variance can be ascribed to interviewer effects, we estimated random effects models for the experimental condition in which respondents answered 10 vignettes with 8 dimensions (random effect models allow for the estimation of the amount of unexplained variance that can be attributed to higher levels, i.e. respondents or interviewers). In a first step, random effects were specified only on the respondent level (similar to what was done in Figure 19.3). The second step enhanced the model for the face-to-face group to a three-level specification that also takes interviewer random effects into account. As before, the vignette evaluations were regressed on the vignette dimensions separately for both modes. Apart from the respondents' sociodemographic characteristics (sex, age, and educational background), however, two regional variables were added to the models: a dummy variable specifying if a respondent lived in the western or the eastern part of Germany and an ordinal indicator if their place of residence was located in a rural or an urban area. This was done to disentangle possible interviewer effects from confounded variables indicating regional variance.

Table 19.6 shows the calculated intra-class-correlations (ICCs) for these models. The ICCs indicate the fraction of unexplained variance that can be attributed to the respondent

21 Chow tests are Wald tests for the null hypothesis that several interactions between survey modes and vignette dimensions are jointly zero (Wooldridge 2003). In our case, the Chow test comprised interactions of the mode with all vignette variables and the main effect of the mode. When testing without the main effect, the p-value was 0.467 for the weighted data and 0.036 for Model 2.

Table 19.6 Random effects (RE) models: variance due to interviewers.

	Self-admin RE respondents (2 levels)	Face-to-face RE respondents (2 levels)	Face-to-face RE respondents + interviewers (3 levels)
ICC (respondents)	0.062	0.154	0.040
ICC (interviewers)			0.127

Data: only respondents evaluating 10 vignettes with 8 dimensions.
ICC: intraclass correlation.

and interviewer levels. At a first glance, the ICC for respondents is considerably lower in the self-administered surveys (6%) than in the face-to-face group (15.4%; see the first row in Table 19.6, which shows results of two-level models that only include random effects for respondents). However, as the three-level specification suggests, the lion's share of this variance is caused by differences in interviewers (or by other regional variables we have not controlled for). These results suggest that interviewers (or the small regional districts they cover with their random walks) cause much additional noise to the data that has to be accounted for to be able to achieve the main goals of factorial survey research: exploring the impact of vignette dimensions and the heterogeneity of respondents in regard to their judgment rules.[22]

19.5 Conclusion

This chapter has contributed to fill the void in methodological research on factorial surveys. Based on a total survey error approach, we discussed different design characteristics of factorial survey experiments as well as decisions about data collection mode and sampling with respect to possible measurement and representation errors. Empirically, the data analyzed came from a factorial survey module on the justice of earnings obtained through two different data collection modes. Respondents either completed the factorial survey module while an interviewer was present or as a self-administered survey. Recruitment into those modes took place after drawing two separate random samples.

Regarding mode effects, there were some differences in the prevalence of nonresponse. Nonresponse was particularly low in the face-to-face sample, but also the self-administered mode produced low nonresponse rates: in both modes, less than 3% of vignette evaluations were missing. It is noteworthy that interviewers in the face-to-face mode were very successful in convincing *all* respondents to answer at least parts of the vignette module and avoiding complete dropouts.

Regarding possible measurement errors, evidence was mixed. Respondents did not differ across mode in their use of evasive response sets (that is, ticking the middle or always the same response category). Small but significant differences were found with regard to the standard deviation of responses, such that face-to-face participants tended to make greater use of the

22 Obviously, respondents in self-administered data collection modes are also clustered in regional units, such as the (city) districts they live in. However, sampled with a random dialing procedure, the respondents are more evenly distributed across the whole regional area where the target population lives. Adding random effects for regional units (such as federal states) hardly changes the ICC measured for respondents. In contrast, the random walk procedure leads to a narrow clustering of respondents within smaller geographical units. In addition, interviewer effects cause autocorrelations of observations collected by single interviewers.

extreme categories on the answering scale. These small effects do not suggest substantial differences in heuristic response behavior across survey modes. However, response consistency was significantly higher if an interviewer was present during data collection. Sequences containing more than 10 vignettes were, irrespective of survey mode, related to higher proportions of unexplained variance and thus inconsistent responses, suggesting that the number of evaluation tasks should not exceed this threshold in any mode of data collection. At the same time, higher numbers of vignette dimensions were only problematic in terms of lower response consistency if there was no interviewer present.[23]

We also tested whether substantive results differ across modes: the experimental design might make factorial survey methods immune against the found measurement errors. First, concerning the causal impact of vignette dimensions, it has to be said that, overall, there were very few differences across data collection mode. There was some evidence for mode-specific perceptions of just baseline wages (in the case of interviewer presence, respondents supported somewhat higher base wages), whereas most of the vignette dimensions did not show any differential evaluations across mode. Depending on the model specification, either 1 or 2 out of the 8 vignette dimensions were found to differ by survey mode on a 10% significance level. Respondents in the self-administered mode more strongly acknowledged the job experience of the vignette person, while respondents in the face-to-face mode reacted more strongly to the earnings of the vignette person. An explanation for this latter finding could be that the presence of others motivates respondents to support higher earning equality (Liebig et al. 2015). However, it is difficult to decide if this higher preference for equality is triggered by the frame of a more cooperative situation when others are present or by social desirability bias. Obviously more research on the sources of this mode effect is needed.

Finally, adding interviewer dummies or random effects to account for the nested data structure hardly changed the importance of the vignette dimensions on the final judgment. Multilevel analyses, however, showed that the face-to-face survey mode induced some random noise due to observations not being independent from each other but interviews being clustered into interviewers and/or the small geographical areas where the interviewers recruit their respondents by means of the random walk. Following our random effects estimates, more than 10% of unexplained variance can be attributed to the level of interviewers. When not taking this source of heterogeneity into account, one might erroneously conclude that there are large differences between respondents' judgment styles.

To sum up, effect sizes in factorial surveys are somewhat sensitive to survey mode, whereas the distinction between significant and nonsignificant effects seems to be more robust across modes and hence more reliable. Self-administered and face-to-face interviews showed little difference in terms of item nonresponse and response consistency, and (apart from the possible social desirability bias due to the interviewer presence) both survey modes hardly differed in meaningful results. That is, the causal impact of vignette dimensions on the vignette evaluations was mostly the same.

Nonetheless, we refrain from generally recommending the use of less expensive online surveys without interviewer assistance. It is true that in the case of regional two-stage sampling, interviewers go hand in hand with an artificial between-respondents variation due to the clustering of respondents, and survey costs for this mode are also higher because of the higher unit costs for completed interviews. However, face-to-face surveys – if implemented properly – should still produce respondent samples with lower representation errors (Biemer

---

23 Further research should be undertaken to analyze whether part of the higher response consistency in the face-to-face mode is caused by respondents using heuristics, such as fading out some dimensions. Our first analysis on the aggregate level provided no evidence for this being the case; effect sizes of vignette variables were found to be very similar across both modes.

2001, see also our analysis A1 in the online appendix), and they yielded somewhat more consistent responses in our study. In addition, replications with further factorial survey modules would be desirable, dealing with more sensitive topics or being better designed to disentangle sample selection and mode effects. Although we used different techniques, such as poststratification weights, to adjust both survey splits with regard to the sociodemographic composition of respondents, some differences might have remained.

We conclude with a very general statement on factorial surveys. Due to their experimental setup, factorial survey modules produce high internal validity regarding the causal influence of vignette variables whenever the randomization of vignettes across respondents worked properly. This is true for both random and non-random samples of respondents, even though large random population samples increase the possibilities for testing possible moderator variables on the level of respondents (Aronson et al. 1998). This feature makes factorial survey designs attractive even if researchers have a somewhat biased random sample at hand or cannot build on a random sample of respondents at all.

References

Armacost, R.L., Hosseini, J.C., Morris, S.A., and Rehbein, K.A. (1991). An empirical comparison of direct questioning, scenario, and randomized response methods for obtaining sensitive business information. *Decision Sciences* 22 (5): 1073–1090.

Aronson, E., Wilson, T.D., and Brewer, M.D. (1998). Experimentation in social psychology. In: *The Handbook of Social Psychology* (ed. D.T. Gilbert, S.T. Fiske and G. Lindzey), 99–142. New York: McGraw-Hill.

Atzmüller, C. and Steiner, P.M. (2010). Experimental vignette studies in survey research. *Methodology: the European Journal of Research Methods for the Behavioral and Social Sciences* 6 (3): 128–138.

Auspurg, K. and Hinz, T. (2015a). *Factorial Survey Experiments*. Thousand Oaks: Sage.

Auspurg, K. and Hinz, T. (2015b). Multifactorial experiments in surveys: conjoint analysis, choice experiments, and factorial surveys. In: *Experimente in den Sozialwissenschaften. Sonderband 22 der Sozialen Welt* (ed. T. Wolbring and M. Keuschnigg), 291–315. Baden-Baden: Nomos.

Auspurg, K. and Jäckle, A. (2017). First equals most important? Order effects in vignette-based measurement. *Sociological Methods & Research* 46 (3): 490–539.

Auspurg, K., Hinz, T., and Liebig, S. (2009). Komplexität von Vignetten, Lerneffekte und Plausibilität im Faktoriellen Survey. *Methoden, Daten, Analysen* 3: 59–96.

Auspurg, K., Hinz, T., Liebig, S., and Sauer, C. (2015). The factorial survey as a method for measuring sensitive issues. In: *Improving Survey Methods. Lessons from Recent Research* (ed. U. Engel, B. Jann, P. Lynn, et al.), 137–149. New York: Routledge.

Auspurg, K., Hinz, T., and Sauer, C. (2017). Why should women get less? Evidence on the gender pay gap from multifactorial survey experiments. *American Sociological Review* 82 (1): 179–210.

Bauer, J.J. (2014). Selection errors of random route samples. *Sociological Methods & Research* 43 (3): 519–544.

Berrens, R.P., Bohara, A.K., Jenkins-Smith, H. et al. (2003). The advent of internet surveys for political research: a comparison of telephone and internet samples. *Political Analysis* 11 (1): 1–22.

Biemer, P.P. (2001). Nonresponse bias and measurement bias in a comparison of face to face and telephone interviewing. *Journal of Official Statistics* 17 (2): 295–320.

Biemer, P.P. (2010). Total survey error: design, implementation, and evaluation. *Public Opinion Quarterly* 74 (5): 817–848.

Burstin, K., Doughtie, E.B., and Raphaeli, A. (1980). Contrastive vignette technique: an indirect methodology designed to address reactive social attitude measurement. *Journal of Applied Social Psychology* 10 (2): 147–165.

Carini, R.M., Hayek, J.C., Kuh, G.D. et al. (2003). College student responses to web and paper surveys: does mode matter? *Research in Higher Education* 44 (1): 1–19.

Christoph, B. (2005). Zur Messung des Berufsprestiges. Aktualisierung der Magnitude-Prestigeskala auf die Berufsklassifikation ISCO88. *ZUMA-Nachrichten* 57: 79–127.

De Leeuw, E.D. (2008). Choosing the method of data collection. In: *International Handbook of Survey Methodology* (ed. E.D. de Leeuw, J.J. Hox and D.A. Dillman), 113–135. New York: Lawrence Erlbaum.

De Wilde, E., Cooke, A.D.J., and Janiszewski, C. (2008). Attentional contrast during sequential judgments: a source of the number-of-levels effect. *Journal of Marketing Research (JMR)* 45 (4): 437–449.

Dülmer, H. (2007). Experimental plans in factorial surveys: random or quota design? *Sociological Methods & Research* 35 (3): 382–409.

Dülmer, H. (2016). The factorial survey: design selection and its impact on reliability and internal validity. *Sociological Methods & Research* 45 (2): 304–347.

Fisher, S.H. III and Herrick, R. (2013). Old versus new: the comparative efficiency of mail and Internet surveys of state legislators. *State Politics & Policy Quarterly* 13 (2): 147–163.

Graeff, P., Sattler, S., Mehlkop, G., and Sauer, C. (2014). Incentives and inhibitors of abusing academic positions: analysing university students' decisions about bribing academic staff. *European Sociological Review* 30 (2): 230–241.

Groves, R.M. and Lyberg, L. (2010). Total survey error: past, present, and future. *Public Opinion Quarterly* 74 (5): 849–879.

Groves, R.M., Fowler, F.J., Couper, M. et al. (2009). *Survey Methodology*, 2e. Hoboken: Wiley.

Henrich, J., Heine, S.J., and Norenzayan, A. (2010). The weirdest people in the world? *Behavioral and Brain Sciences* 33 (2–3): 61–135.

Hermkens, P.L.J. and Boerman, F.A. (1989). Consensus with respect to the fairness of incomes: differences between social groups. *Social Justice Research* 3 (3): 201–215.

Holbrook, A.L., Green, M.C., and Krosnick, J.A. (2003). Telephone versus face-to-face interviewing of national probability samples with long questionnaires: comparisons of respondent satisficing and social desirability response bias. *Public Opinion Quarterly* 67 (1): 79–125.

Hox, J.J., Kreft, I.G.G., and Hermkens, P.L.J. (1991). The analysis of factorial surveys. *Sociological Methods & Research* 19 (4): 493–510.

Jasso, G. (2006). Factorial survey methods for studying beliefs and judgments. *Sociological Methods & Research* 34 (3): 334–423.

Krosnick, J.A. (1991). Response strategies for coping with the cognitive demands of attitude measures in surveys. *Applied Cognitive Psychology* 5 (3): 213–236.

Krosnick, J.A. and Alwin, D.F. (1987). An evaluation of a cognitive theory of response-order effects in survey measurement. *Public Opinion Quarterly* 51 (2): 201–219.

Liebig, S., May, M., Sauer, C. et al. (2015). How much inequality of earnings do people perceive as just? The effect of interviewer presence and monetary incentives on inequality preferences. *Methoden, Daten, Analysen (mda)* 9 (1): 57–86.

Loosveldt, G. (2008). Face-to-face interviews. In: *International Handbook of Survey Methodology* (ed. E.D. de Leeuw, J.J. Hox and D.A. Dillman), 201–220. New York: Lawrence Erlbaum.

Malhotra, N. and Krosnick, J.A. (2007). The effect of survey mode and sampling on inferences about political attitudes and behavior: comparing the 2000 and 2004 ANES to Internet surveys with nonprobability samples. *Political Analysis* 15 (3): 286–323.

Meulemann, H. (1993). Befragung Und Interview: Über Soziale und Soziologische Situationen der Informationssuche. *Soziale Welt* 44 (1): 98–119.

Mutz, D.C. (2011). *Population-Based Survey Experiments*. Princeton: Princeton University Press.

Pager, D. and Quillian, L. (2005). Walking the talk? What employers say versus What they do. *American Sociological Review* 70 (3): 355–380.

Papke, L.E. and Wooldridge, J.M. (1996). Econometric methods for fractional response variables with an application to 401(k) plan participation rates. *Journal of Applied Econometrics* 11 (6): 619–632.

Rossi, P.H. and Anderson, A.B. (1982). The factorial survey approach: an introduction. In: *Measuring Social Judgments: The Factorial Survey Approach* (ed. R.H. Rossi and S.L. Nock), 15–67. Beverly Hills: Sage.

Sauer, C., Auspurg, K., Hinz, T., and Liebig, S. (2011). The application of factorial surveys in general populations samples: the effects of respondent age and education on response times and response consistency. *Survey Research Methods* 5 (3): 89–102.

Sauer, C., Auspurg, K., Hinz, T. et al. (2014). Methods Effects in Factorial Surveys: An Analysis of Respondents' Comments, Interviewers' Assessments, and Response Behavior. *SOEPpaper* 629.

Schnell, R. (1997). *Nonresponse in Bevölkerungsumfragen: Ausmaß, Entwicklung und Ursachen*. Opladen: Leske & Budrich.

Simon, H. (1957). *Models of Man*. New York: Wiley.

Snijders, T.A.B. (2005). Power and sample size in multilevel linear models. In: *Encyclopedia of Statistics in Behavioral Science*, vol. 3 (ed. B.S. Everitt and D.C. Howell), 1570–1573. Chichester: Wiley.

Snijders, T.A.B. and Bosker, R.J. (2011). *Multilevel Analysis: An Introduction to Basic and Advanced Multilevel Modeling*, 2e. London: Sage.

Sue, V.M. and Ritter, L.A. (2007). *Conducting Online Surveys*. Los Angeles: Sage.

Teti, A., Gross, C., Knoll, N., and Blüher, S. (2016). Feasibility of the factorial survey method in aging research consistency effects among older respondents. *Research on Aging* 38 (7): 715–741.

Verlegh, P.W.J., Schifferstein, H.N.J., and Wittink, D.R. (2002). Range and number-of-levels effects in derived and stated measures of attribute importance. *Marketing Letters* 13 (1): 41–52.

Wallander, L. (2009). 25 Years of factorial surveys in sociology: a review. *Social Science Research* 38 (3): 505–520.

Walzenbach, S. and Hinz, T. (2019). Hiding sensitive topics by design? An experiment on the reduction of social desirability bias in factorial surveys. *Survey Research Methods* **13** (1): 103-121.

Wittink, D.R., Krishnamurthi, L., and Reibstein, D.J. (1990). The effect of differences in the number of attribute levels on conjoint results. *Marketing Letters* 1 (2): 113–123.

Wooldridge, J.M. (2003). *Introductory Econometrics. A Modern Approach*, 2e. Mason: Thomson.

Zwarun, L. and Hall, A. (2014). What's going on? age, distraction, and multitasking during online survey taking. *Computers in Human Behavior* 41: 236–244.

20

Validity Aspects of Vignette Experiments: Expected "What-If" Differences Between Reports of Behavioral Intentions and Actual Behavior

Stefanie Eifler[1,*] *and Knut Petzold*[2]

[1] *Department of Sociology, Catholic University of Eichstätt-Ingolstadt, 85072 Eichstätt, Germany*
[2] *Sociology Section, Ruhr-Universität Bochum, 44801 Bochum, Germany*

20.1 Outline of the Problem

20.1.1 Problem

Our contribution is focused on analyzing different aspects of the validity of vignette experiments. Vignettes are short descriptions of real-life situations that can be employed as stimuli in a survey to assess attitudes and intentions. On the one hand, vignettes have been used within the framework of the factorial survey approach following Rossi (1979; Rossi and Anderson 1982) in order to examine the effects of situational characteristics on attitudes or intentions when employing within-subjects designs (Auspurg and Hinz 2015; Jasso 2006; Wallander 2009). On the other hand, vignettes have been integrated into scenario techniques in order to analyze the effects of situational characteristics on attitudes or intentions when employing between-subjects designs. With regard to the latter, intentions are considered as approximations to actual behavior in real-life situations in many studies in which vignettes typically refer to norm-related situations (Caro et al. 2012; Mutz 2011 for an overview).

What the different applications have in common is the goal of designing vignettes that mirror situations in everyday experience as much as possible. Employing vignettes follows the tradition of the indirect measurement movement, which is anchored in the research on attitudes in social psychology initially begun by Campbell (1950). His approach departs from general measurement theory (Suppes and Zinnes 1963) by assuming that the difference between direct and indirect measurement relates to distinguishing nondisguised from disguised forms of capturing social phenomenon in a standardized way. Thereby, direct measurement procedures like Likert scales that consist of singular statements are termed nondisguised, and indirect measurement procedures based on scenarios are termed disguised. According to Campbell (1950), direct measurement procedures are problematic, because they force respondents to adapt their responses to predefined categories that oftentimes do not adequately represent the complexity of their "true" attitudes. Indirect measurement procedures avoid this problem by representing the social phenomena measured with descriptions of situations in the most possible realistic manner (cf. Liebig et al. 2015).

* Both authors contributed equally to this work.

Experimental Methods in Survey Research: Techniques that Combine Random Sampling with Random Assignment, First Edition. Edited by Paul J. Lavrakas, Michael W. Traugott, Courtney Kennedy, Allyson L. Holbrook, Edith D. de Leeuw, and Brady T. West.

Companion Website: www.wiley.com/go/Lavrakas/survey-research

The other major advantage of using vignettes in experiments (either within- or between subjects) is that they provide high internal validity. The researcher-controlled manipulation of the independent variables within different vignettes together with the randomization of respondents allows for the assessment of whether changes in elements of the vignettes (i.e. the independent variables) affect the attitudes or intentions measured in response to the vignettes.

However, there is an ongoing controversy concerning the external validity of vignettes. In their debate on the experimental process in social psychology, Aronson and Carlsmith (1968) refer to this undertaking as the *mundane realism* of experiments. The idea of mundane realism is closely related to validity aspects of measurement procedures, in particular to construct and external validity. Construct validity refers to "inferences about the higher-order constructs that represent sampling particulars," i.e. treatments, observations, persons, and settings, while external validity addresses "inferences about whether the cause-effect relationship holds over variation in persons, settings, treatment variables, and measurement variables" (Shadish et al. 2002, 38). Thus, both validity aspects refer to the generalizability of measurement procedures. According to these definitions, judging the external validity of vignette measurements necessarily requires the assumption of a sufficient level of construct validity. In other words, inferences about the stability of cause–effect relationships across real-life situations and respective hypothetical descriptions of situations require the assumption that all research elements represent the same higher-order constructs across these situations.

20.1.2 Vignettes, the Total Survey Error Framework, and Cognitive Response Process Perspectives

The analysis of vignettes from a methodological point of view is addressed within the *Total Survey Error* (*TSE*) framework (Groves et al. 2009). In this framework, the answer to a question is regarded as the sum of the "true value" and "errors" resulting from several sources. Currently, these sources are grouped into the categories of representation and measurement (Biemer et al. 2004; Groves et al. 2009). Representation errors refer to errors that are related to the relationship between the target population, the sampling frame, and the initial versus the final sample of respondents, while measurement errors are related to validity and reliability aspects and refer to errors due to interviewer behaviors/characteristics, respondents' behaviors/characteristics, survey instrument characteristics, survey mode, and processing errors (Groves et al. 2009).

At the center of measurement error are errors that come into play during the response process. According to the most widely accepted cognitive model of the response process (Tourangeau 1984; Tourangeau et al. 2000), a respondent has to interpret the question's content (interpretation), retrieve information from her or his memory (retrieval), form an opinion (judgment), and then bring the answer into line with the predefined response format (response selection). Correspondingly, a respondent, who has to judge a presented vignette, must interpret the question and retrieve associated information from memory before forming an opinion and selecting a response.

One critical question about vignettes is the extent to which measures of reported behavioral intentions in response to a vignette are likely to accurately represent and measure what respondents would actually do in a situation described in a vignette. This addresses a seeming mix of the issues of construct validity (i.e. how well something measures what it purports to measure) and of external validity (i.e. generalizability). In terms of construct validity, the issue is how accurately the reported behavioral intention data that are produced in the vignette conform to how people actually behave in the same real-life circumstances. In terms of external validity, one could argue that the high mundane realism of vignettes makes it likely that the cognitive process of answering questions about one's behavior in the situation (as described in the vignette) mirrors the cognitive process of making a decision about acting in the real-life situation described in the vignette, thereby providing high external validity between the vignette and the real-life circumstance the vignette is meant to represent.

Another important source of measurement error in the cognitive response process that may affect the correspondence between behavioral intentions in response to vignettes and actual behavior is the tendency of many respondents to report data that present themselves in a favorable manner. Giving a response to a self-report survey question (including a response to a vignette) that deviates from a "true value" and corresponds to the normative standard is called socially desirable response behavior (Edwards 1957). Until now, little research has been conducted to examine the extent to which vignettes are affected by social desirability response bias. On the one hand, one could argue that because they rely on self-reports, vignettes are as prone to socially desirable response behavior as are more direct self-reports. According to Tourangeau and Yan (2007), this might happen notably with vignettes requiring the respondents to give embarrassing responses by addressing personally sensitive situations, i.e. situations "(...) in which there are potential social consequences or implications, either directly for the participants in the research or for the class of individuals represented by the research" (Sieber and Stanley 1988, 49). On the other hand, we suspect that the concrete descriptions of everyday experiences provided in vignettes diminish the level of threat felt when addressing sensitive topics, so that respondents may tend to give more accurate (i.e. less socially desirable) responses. This, in turn, would increase the construct validity of self-reported measures of attitudes or intentions based on vignettes compared to more direct self-reports of these constructs (Alexander and Becker 1978; Finch 1987).

This study described in this chapter was designed to assess the extent to which vignettes are prone to socially desirable response behavior. In order to do so, we used vignettes that were designed to analyze the effects of situational characteristics on reported behavioral intentions for norm-related behaviors. Our purpose in doing so is to compare responses to vignettes with observed behavior in real-life situations as a reference point for external validity. It is assumed that differences between reported behavioral intentions from vignettes and behavior actually observed in real-life situations constitute an empirical baseline for judging the external validity of vignette responses. However, this strategy is based on the assumption that measurement procedures in both experiments receive a sufficient level of construct validity. Otherwise, a judgment of the external validity of vignette responses would not be possible. Starting from these ideas, the next part of the chapter refers to studies that compare the results of vignette-based studies with those of covert observations or field experiments.

20.1.3 State of the Art

Up to now, methodological research on vignette-based studies has covered a variety of topics, i.e. issues of the underlying experimental designs, vignette construction, judgment consistency, and applicability to specific research areas. As for issues of the underlying experimental designs, some studies compared design features and arrived at recommendations for the experimental designs of survey experiments (Dülmer 2007, 2016; Atzmüller and Steiner 2010; Ganong and Coleman 2006). As for issues of vignette construction, several studies combined vignette-based approaches with qualitative methods. For example, Jones (2014) used an Internet-mediated qualitative, narrative approach to provide in-depth analyses of vignettes. With regard to judgment consistency, influences of the complexity and the amount of detail of vignettes on responses were analyzed. One study revealed that response consistency depends upon the presentation order of treatments within a vignette, particularly in the case of complex vignettes (Auspurg and Jäckle 2015). In contrast to this, most studies came to the conclusion that judgment consistency is independent of the number of treatments, thus the complexity varied in vignettes (Auspurg et al. 2009a,b) and the elaboration, thus the amount of detail, of their description (Eifler and Petzold 2014), except when respondents have to judge a very large number of vignettes (Sauer et al. 2011). In addition, response consistency proved to be sensitive to respondents' characteristics that are associated with cognitive abilities, such as age and education (Sauer et al. 2011, 2014). However, the use of video clips can help to arouse interest in respondents, even in older participants (Caro et al. 2012). With regard to issues of the applicability of survey experiments to specific research areas, several authors extend the use of vignettes to the study of multiple and interrelated beliefs (Li et al. 2007) or moral reasoning (Mah et al. 2014).

The methodological studies mentioned so far mostly refer to aspects of reliability and internal validity. In contrast to this, the present study is devoted to the analysis of the *mundane realism* of survey experiments or, rather, aspects of the external validity (i.e. the generalizability) of vignettes. Studies comparing vignettes with covert observations, i.e., observations in which the observer does not participate in the observed activities and records observations from a completely unobtrusive location, indicate that vignettes do not lead to more externally valid measurements. For example, considerable deviations are revealed when comparing reported behavioral intentions with exhibited deviant behavior (theft by finding, small violations of norms in traffic) and when comparing socially desirable behavior (returning a lost letter) with actual behavior (Eifler 2007).

Other studies comparing vignette experiments with field experiments show that the level of actual behavior in the context of sensitive topics can hardly be determined with the help of vignettes, even though the relative effects of situational characteristics point in the same direction (Eifler 2010; Groß and Börensen 2009; Pager and Quillian 2005; Petzold and Eifler 2019; Petzold and Wolbring 2019). Other studies have demonstrated that vignette experiments enable valid predictions of behavior (Hainmueller et al. 2015; Nisic and Auspurg 2009).

Furthermore, some studies do not explicitly aim to test the validity of vignettes but nevertheless provide additional insights. For example, Raub and Buskens (2008) use various empirical approaches to investigate cooperation in problematic situations and tested different mechanisms in vignettes, in a survey, and in a laboratory experiment. Further studies focus on the perception of vignettes compared to the perception of more realistic situations, for instance, in the decision-making capacity in elderly patients with (and without) cognitive impairment (Vellinga et al. 2005). The authors conclude that they did not find any major differences between the hypothetical and the realistic situations. Schwalbe et al. (2004) use video vignettes with offenders to classify youths according to their risk of recidivism in juvenile justice and find that the vignette-based risk scores are significantly correlated with rearrest. In contrast, a study by Neff (1979) reveals that persons are differently evaluated to a significant extent when untrustworthiness is observed rather than when it is presented in a vignette.

The results of these studies can, in summary, be interpreted as examples illustrating the potentially problematic aspects of using vignettes in empirical social research: Especially when referring to attitudes and reported behavioral intentions in norm-relevant situations, the presentation of vignettes can trigger normative processes among the respondents that can promote the tendency of providing socially desirable answers. Correspondingly, Collett and Childs (2011) conclude that "(...) vignettes are good at measuring how people *should* react in a given situation, but not necessarily how they will respond (...)" (Collett and Childs 2011, 520).

Based on these results, the presumed advantage of the vignette methodology could turn out to be a disadvantage: it is reasonable to assume that concrete information can activate even more normative concepts and may therefore increase socially desirable responding behavior.

Against the backdrop of these considerations, we devised a research strategy to allow us to evaluate a vignette experiment by systematically comparing it with a field experiment in which actual behavior was measured in a way that is unaffected by social desirability.

20.1.4 Research Strategy for Evaluating the External Validity of Vignette Experiments

Our research design is based on the idea that the external validity of a vignette can be judged by systematically comparing reported behavioral intentions in response to a situation described by a vignette with observed behavior in the respective real situation. However, this comparison requires that measures of reported behavioral intentions and actual behavior achieve a high level of construct validity, i.e. refer to the same underlying theoretical construct. Against this background, we develop a research strategy in which a field experiment serves as a reference point for judging the external validity of a vignette experiment, assuming that a sufficient level of construct validity can be established in the course of research. Our research strategy makes use of general models of causal inference.

Analyzing the effects of the mode of data generation on measures of reported behavioral intentions and observed behavior corresponds to analyzing the *effects of causes* (Holland 1986). Within the framework of the counterfactual model of causal inference (Morgan and Winship 2007), sometimes referred to as the model of potential outcomes (Rubin 2005), two causal states called treatment and control are distinguished, which are each connected to one potential outcome. Thereby, the fundamental problem of causal inference arises. This describes the problem that a unit of observation can only always be subjected to one of the causal states at any point in time, thus leaving the counterfactual unknown (Holland 1986). The so-called counterfactual problem leads to the impossibility to observe causal effects on the level of individual units of measurement. Instead, so-called *average treatment effects* (ATEs) are estimated as differences between expected values/expectations (Holland 1986; Morgan and Winship 2007; Rubin 2005). In this context, the ATE is referred to as the "*what-if difference*" (Morgan and Winship 2007, 43) with regard to a certain outcome, if the unit of observation could have been examined in the treatment and control condition simultaneously. Two conditional treatment effects follow a different but related notion: The average treatment effect of the treated (ATT) is the ATE for the units of observation subjected to the treatment and the average treatment effect of the untreated (control state) (ATC) is the ATE for the units of observation not subjected to the treatment (Morgan and Winship 2007).

In general, modeling the differences of the outcomes in the treatment and control conditions as ATEs or conditional ATEs is tied to three crucial assumptions, which allow for the control of unobserved and observed heterogeneity in both conditions and which – as a consequence – allow the assumption of a sufficient level of construct validity:

- The units of observation in the field experiment and the vignette experiment are drawn from an *infinite superpopulation* or a very large finite population (Morgan and Winship 2007, 51).
- The units of observation are randomly assigned to either the treatment or control states through a mechanism of randomization.
- The independent variables and covariates are independent of one another – also referred to as the conditional independence assumption.

If the assumption of randomization is not met, it is possible to analyze differences between outcomes in the treatment and control conditions in accordance with the methodology of observational studies (Rosenbaum 2002). In an observational study, the researcher is unable to control the assignment of observational units to the treatment and control conditions (Cochran 1965). In such a situation, the control of unobserved heterogeneity through randomization is absent, and comparability between observational units in the treatment and control conditions first must be established. In addition, strategies of matching and stratification are used in order to adjust for overt biases by conditioning on observables (Rosenbaum 2002). Common procedures for harmonizing observational units in the treatment and control conditions are propensity score matching, optimal stratification, and the construction of optimal matched sets (Guo and Fraser 2010; Rosenbaum 2002; Rosenbaum and Rubin 1983). If comparability of observational units in the treatment and control conditions can be established using these procedures, the estimation of so-called sample average treatment effects (SATE), sample average treatment effects for the treated (SATT), and sample average treatment effects for the untreated (SATC) is possible (Guo and Fraser 2010; Morgan and Winship 2007). However, irrespective of whether compared outcomes are conditioned by unobserved heterogeneity through randomization or by observed heterogeneity through matching operations, two additional identifying, yet untestable, assumptions are necessary:

- The stable unit treatment value assumption[1] holds.
- Measurement invariance[2] is given for independent and dependent variables as well as covariates.

In the present study, a vignette experiment is compared to a field experiment. Within the framework of the counterfactual model of causal inference, differences between the outcomes of a vignette experiment and a field experiment can in principle be interpreted as ATEs. If the field experiment works as a benchmark, the vignette experiment constitutes the treatment condition and the field experiment the control condition.

Accordingly, a series of basic identifying assumptions are required in order to ensure that inferences can be made about the effect of the mode of data generation for differences between behavior in real-life situations and reported behavioral intentions when confronted with vignettes. First, we start from the assumption that both parts of the study are based on an identical target population but using different frames. Second, the population in the survey inferred upon deviates from the frame population, and distinguishing the frame population from the population inferred upon in a field experiment is neither reasonable nor possible. Third, different measures of covariates are used in the vignette experiment and in the field experiment, which poses a challenge to the assumption of measurement invariance.

Referring to concepts of validity, these assumptions can be captured by the concept of construct validity since they address inferences about higher-order constructs represented by treatments, persons, settings, and outcomes. In order to make inferences about whether reported behavioral intentions in vignettes can be generalized to actual behavior in real-life situations the assumptions are identifying, yet untestable. For testing the external validity of vignettes, it is a necessary presupposition to assume that the study particulars do represent the respective constructs accurately and stable with regard to real-life situations. In other words, construct validity is a necessary but untestable assumption if external validity is to be identified.

1 The stable unit treatment value assumption (SUTVA) was introduced by Rubin (1986): "SUTVA is simply the a priori assumption that the value of Y for unit u when exposed to treatment t will be the same no matter what mechanism is used to assign treatment t to unit u and no matter what treatments the other units receive" (Rubin, 1986, 961; see also Morgan and Winship, 2007, 37 for further references).

2 In measurement theory, the idea of construct validity is elaborated and called measurement invariance, measurement equivalence, lack of measurement bias, or absence of differential item functioning (Mellenbergh 1989; Holland and Wainer 1993; Millsap 2011). These statistical concepts refer to the situation where measurement instruments that are applied to different groups allow to assess a theoretical construct in a corresponding manner. According to several authors in the field of measurement theory, measurement invariance means that different measures are equally representing a theoretical construct of interest (see Millsap, 2011, in particular).

Though assumptions regarding construct validity of treatments, units, settings, and outcomes cannot be proved, certain methodological precautions can help to increase their conceptual appropriateness in a specific study. In case of a validation study, it is crucial for the aims of our study that the study particulars in both parts of the study are being made comparable as much as possible. The more comparable the study elements are, the more unquestionable are assumptions about an equivalent level of construct validity.

While both parts of our study – the vignette experiment and the field experiment – are "true" experiments in themselves and control for unobserved heterogeneity by means of randomization, the whole study follows the methodology of an observational study (Rosenbaum 2002). The observational units are not randomly assigned to one or the other part of the study. Therefore, the observational units are first aligned across the vignette and field experiments. Then pairs of observational units from the vignette and the field experiment are formed using propensity score matching (Rosenbaum and Rubin 1983).

Differences between the vignette and field experiments are then analyzed as "expected what-if differences" (Morgan and Winship 2007, 43) analogous to ATEs. Thereby, our interest lies especially on the expected what-if differences for the treated, meaning the differences in outcomes of observational units in the treatment condition, under the assumption that these units could be analyzed in both the vignette and the field experiment. If there is no difference between the outcomes for those analyzed in the vignette experiment, it is unlikely that a difference is present in the field experiment. However, the methodological approach as introduced by Morgan and Winship has not been applied to the validation of vignette experiments yet.

To facilitate the understanding of the differences between the vignette and field experiments and the comparative analysis corresponding to the counterfactual model of causal inference, in the following part of the chapter, both studies are described separately, before proceeding with the description of the comparative analysis by means of *expected what-if differences*.

20.2 Research Findings from Our Experimental Work

For our comparative study, we chose an everyday situation, which has previously been examined in field experiments and can easily be recreated using a vignette experiment. Our study is centered around the so-called horn-honking experiment at a traffic light, which was originally designed for studying the effects of social status on aggressive behavior (Doob and Gross 1968). In our study, the horn-honking experiment was conducted both as a field experiment and as a vignette experiment. Both experiments were based on the same 2 × 2-between-subjects design.

20.2.1 Experimental Design for Both Parts of the Study

In the horn-honking experiment, a person (the subject) in a car is blocked by an experimental car (i.e. so-called "frustrator car") at a traffic light when the signal turns green. The observed outcome variable is whether and when the subject (aggressor) shows an aggressive reaction (horn-honking). To operationalize the effects of social status, the type of frustrator car is varied. The experiment has been replicated many times using various additional treatments, such as age, sex, national stereotypes, or a sticker on the frustrator car identifying the driver as a learner (e.g. Deaux 1971; Diekmann et al. 1996; Ellison et al. 1995; Forgas 1976; Kenrick and MacFarlane 1986; Baxter et al. 1990; Yazawa 2004).

In our case study, the horn-honking experiment was conducted both as a field experiment and as a vignette experiment. As can be seen from Table 20.1, a 2 × 2-factorial between-subjects design was used, varying the social status of the frustrator car by car type (low status: small car

Table 20.1 2 × 2-factorial design of field experiment and vignette experiment.

Factor 1: Social status of frustrator car	Factor 2: Group membership of frustrator car	
	In-group member	Out-group member
Low status	11	12
High status	21	22

vs. high status: luxury car) and the group membership of the frustrator car (in-group member: local license plate, out-group member: nonlocal license plate). The assumption was that a low status and out-group membership for the experimental car will be the condition most likely to stimulate aggressive behavior in the driver of the blocked car, i.e. more frequent horn-honking and a shorter latency from the traffic light turning green to the beginning of horn-honking.

20.2.2 Data Collection for Both Parts of the Study

Both the field experiment and the vignette experiment were conducted simultaneously in late winter/early spring of 2014 in Ingolstadt, a city in South Germany. The city of Ingolstadt has about 130 000 inhabitants and is influenced by the large German car manufacturer Audi.

20.2.2.1 Field Experiment

Procedure and Independent Variables The field experiment was conducted on five randomly sampled Monday–Friday days between 24 February and 2 April at a crossroad with a light signaling system. The experimental conditions were realized randomly across the five working days. Therefore, no systematic associations between the experimental cars and the days of data collection exist. The field experiment took place between 10 a.m. and 5 p.m. in order to capture variations in traffic and thus to provide a comparable stress level of drivers. Sunny days were chosen in accordance with the weather forecast to eliminate overly large weather effects.

The crossroad was chosen in a way that allowed an efficient and discreet realization of the field experiment. Criteria for this selection were good visibility from the opposite side of the road, a suitable volume of traffic in order to have enough time for positioning, the opportunity to return to the crossroad without attracting attention, and appropriate time intervals in order to achieve a maximum number of runs. The selected crossroad was well observable from two opposite supermarket parking sites, allowed the experimental car to turn and be replaced inconspicuously through three different ways, and permitted a complete trial run in approximately one minute, so that almost every single signal phase could be used.

To ensure the validity of the observations, five experimenters were involved. The first was driving the experimental car, paying attention solely to the traffic light and the subjects' reaction. The second, seated on the front seat, was responsible for measuring the occurrence of an aggressive reaction and the latency with a stopwatch. The third, on the back seat, observed drivers' characteristics and features of possible passengers. Finally, two observers in a third car, parking on an opposite supermarket parking site, judged the cars' characteristics (period of construction, social status) and the cars behind using binoculars. The experimenters did not rotate positions during data collection, in fact neither between nor within the experimental conditions. Thus, all five experimenters were identical persons fulfilling identical tasks in all experimental conditions. Hence, although the experimenters were aware of the respective experimental conditions, any biases linked to individual observers are ruled out through stabilization. In addition, all experimental cars were equipped with a toned rear window to prevent eye contact between the subjects and the experimenters.

The independent variables were operationalized as follows: The social status of the frustrator car (factor 1) was measured by different types of vehicles representing different levels of social status. The largest car, a sports utility vehicle (SUV) manufactured by Audi (Audi Q7), was used to measure high social status. Since subjects might have perceived very small sized cars as the second car of a household with a high social status, a medium sized car manufactured by Volkswagen (VW Golf) served as a measure of low social status. Both cars differ remarkably in their rears' width and height. Group membership of the frustrator car (factor 2) was measured through varying license plates. A local license plate indicated in-group membership, a nonlocal license plate out-group membership.

The internal validity of the field experiment is ensured by randomly assigning the subjects to one of the four experimental conditions and by controlling for a number of covariates. All covariates were observed by one of the experimenters involved in the data collection (see above). In particular, the experimenter recorded the subjects' estimated age (categories: <25; 25–35; 36–50; 51–65; >65) and gender, the type (categories: small car; mid-size car; luxury car) and period of manufacturing (categories: <1990; 1990–1999; 2000–2010; >2010) of the subjects' cars, the number of further passengers, and the number of further cars behind the subjects' cars.[3]

Sample The drivers of cars and vans passing the selected crossing during the period of the experiment served as the experimental units. Drivers of trucks, buses, motorcycles, police cars, and ambulances were not included in the study. The subject's cars were assigned by appearance at the traffic light to the experimental conditions triggered during a driver's everyday activities. As all observations on different days were made during a whole time span (between 10:00 am and 5:00 pm), we were able to capture systematic variations in the volume of traffic (e.g. rush hour). Table 20.2 shows the absolute and relative frequencies of the observations across experimental conditions.

It follows from Table 20.2 that the proportions are well balanced, even though the experimental condition "high social status paired with a nonlocal license plate" contains a slightly larger amount of subjects. As mentioned above, five Monday–Friday days were randomly sampled from all days of the period of data collection. In order to check the quality of the randomization,

Table 20.2 Number of observations in experimental groups (percents shown in parentheses).

Factor 1: social status of frustrator car	**Factor 2: group membership of frustrator car**		**Total**
	In-group member	Out-group member	
Low status	88 (24.3%)	85 (23.5%)	173 (47.8%)
High status	85 (23.5%)	98 (27.0%)	189 (52.2%)
Total	179 (50.5%)	183 (49.5%)	362 (100%)

3 Additional information concerning the univariate distributions of the covariates are provided at the website related to the book (Table A1).

the bivariate distributions of the observed covariates across the four experimental conditions were examined. No differences across conditions in subjects' age or gender or the number of persons in the subjects' car, the number of cars behind the subjects' car or the origin of the subjects' car (local or nonlocal) were found. Only the vehicle class ($\chi^2 = 18.200, p = 0.006$) and the year of construction of the subjects' car ($\chi^2 = 22.055, p = 0.009$) differ slightly across the experimental conditions.[4] Although these differences ought not to distort the treatment effects, we will control for them statistically through a multivariate analysis.

Dependent Variables The dependent variables were measured as follows: Using the horn or the headlight flashers was considered as aggressive responses to being blocked by the experimental car. One experimenter in the experimental car recorded the occurrence of such an aggressive response and measured its latency during each green phase with a stop watch. In total, 91.4% of all subjects have shown any reaction to the blocking situation, what lasted 10.8 seconds on average within a range between 2 and 29 seconds.[5]

20.2.2.2 Vignette Experiment

Procedure and Independent Variables The vignette experiment was conducted within the framework of a mail survey, exclusively designed for the validation study. Data collection was based on a self-administered questionnaire (SAQ), which was sent at one point in time. No announcement letter and no reminder could be sent due to financial restrictions.

The SAQ was entitled "Car and Traffic – Your Opinion" and consisted of four sections: In the first section, subjects were asked to report their general experience with traffic, i.e. the availability of an own driving license, ownership and usage of a car, its model, class and year of manufacturing, as well as frequency, distances, and motivation of driving. The last question addressed experienced traffic jam at the time of the survey. In sections three and four, attitudes toward traffic in general and personality traits were measured, and information on age, gender, and socioeconomic background were collected. The completion of the SAQ took, by pretesting, around 12 minutes.

The important part of the SAQ was a vignette capturing the horn-honking experiment. It was presented in the second section due to the highest attention and lowest fatigue effects at this stage of the SAQ (cf. Auspurg and Hinz 2015). The vignette was introduced as follows: "Below we present a hypothetical situation at a traffic light and ask for your assessment. Please try to get as much into the situation as possible. How would you react spontaneously to this situation in your everyday life?"

In the vignette, the independent variables were realized using different verbalizations. Figure 20.1 shows the vignette wording including the experimental manipulations. According

> Imagine that on a working day you are driving your car in Ingolstadt and approach a traffic light. The traffic light is on red and waiting in front of you is only a single vehicle. It is a **small/luxury car** with a **non-local/local license plate**. When the traffic light turns green, instead of driving away, the car does not move.

Figure 20.1 Vignette wording.

---

4 Additional information concerning the univariate distributions of the covariates across experimental conditions are provided at the website related to the book (Table A2).

5 Additional information concerning the univariate distributions of the dependent variables are provided at the website related to the book (Table A3).

to the underlying 2×2-factorial between-subjects-design, each subject had to judge one vignette.

The internal validity of the vignette experiment is ensured by randomly assigning the subjects to one of the four vignette versions and by statistically controlling for covariates. For the random assignment of subjects to one of the four experimental conditions, random numbers were added to the postal addresses, which were put in a random order. The random numbers were then used to assign the questionnaires one after the other to the four experimental conditions. All covariates were measured within the SAQ.[6]

Sample All adult citizens aged at least 18 with a main or secondary residence registered in Ingolstadt at the cutoff date 27 January 2014, formed the frame population ($N = 107809$) of the vignette experiment. A stratified random sample ($n = 2000$) was drawn, which was stratified proportionally to the subpopulations of 12 city districts. Table 20.3 provides an overview of reasons for unit nonresponse and the resulting response rate according to the standard definitions of the American Association for Public Opinion Research (AAPOR 2015, 27–33).

Table 20.3 shows that noneligibility reduces the register sample to $n = 1984$ cases. Noneligible persons were not available at the registered addresses, as they have most likely deceased or moved away. The majority of the systematic nonresponse amounting to 71.82% is based on refusal; only one person was removed from the sample due to a self-reported age of 17 years. Few of the selected persons personally cancelled their participation in the survey. Instead, the majority are denied by returning an empty questionnaire. A total of 2.42% of the questionnaires had to be excluded from further analyses because they were incomplete. Finally, $n = 465$ cases returned a completed questionnaire, corresponding to a response rate of 23.43%. In contrast to the field experiment, where none of the subjects could avoid participation once arriving at the crossroad, the sample of the survey experiment could be biased through unit nonresponse. However, due to the random assignment of the subjects to one of the four vignette conditions in the questionnaire, the internal validity should not be affected by unit nonresponse. Our finding that there were minimal differences in subjects' characteristics across experimental conditions suggests that differential unit nonresponse was not likely a problem (although we did not

Table 20.3 Unit nonresponse and response rate.

	Frequencies	Percent
Register sample	2000	100.00%
Not eligible, returned		
Not eligible respondents	16	0.80%
Screened out of sample	1	0.05%
Cleaned register sample	1984	100.00%
Eligible, unit nonresponse		
Refusal	1425	71.82%
Explicit refusal	5	0.25%
Implicit refusal	3	0.15%
Too incomplete	85	2.42%
Completed questionnaire, response rate	465	23.43%

6 Additional information concerning the univariate distributions of the covariates are provided at the website related to the book (Table A4).

Table 20.4 Number of completed questionnaires in experimental groups (percentages in parentheses).

Factor 1: social status of frustrator car	Factor 2: group membership of frustrator car		Total
	In-group member	Out-group member	
Low status	109 (23.4%)	115 (24.7%)	224 (48.2%)
High status	125 (26.9%)	116 (24.9%)	241 (51.8%)
Total	234 (49.7%)	231 (50.3%)	465 (100%)

directly assess unit nonresponse across conditions). Nevertheless, unit nonresponse results in a specific sample composition and might therefore be related to the external validity. This aspect is discussed thoroughly in Sections 2.4 and 2.5.

The completed questionnaires are to a large extent evenly distributed across the four experimental conditions (see Table 20.4). In order to evaluate the quality of the randomization, the distributions of all covariates across the four experimental conditions were examined. No noteworthy differences across conditions in subjects' age or gender or the number of persons in the subjects' car, the number of cars behind the subjects' car, the origin of the subjects' car (local or nonlocal), the vehicle class, and the year of construction of the subjects' car were found.[7] Therefore, covariates' distributions across the experimental conditions indicate a successful randomization also in the vignette experiment. We will stress the quality of randomization further by statistical control in multivariate analysis.

Dependent Variables The dependent variables were measured as follows: Using the horn or the headlight flashers was considered as aggressive responses to being blocked by the experimental car. The occurrence and latency of aggressive responses were measured. Subjects were asked to report what they would do in the situation described in the vignette. The wording of the questions is as follows: "What would you do?" with the response categories "Nothing," "I use the headlight flasher," "I use the horn." The number of subjects reporting that they would show one of the aggressive responses was counted as a measure of the occurrence of aggressive responses. Because the category of using the headlight flasher was very rarely used and any actual reaction was measured in only one category in the field experiment, the category was combined with horn usage to one single category.[8] Only a minority (22.4%) reported that they would not react at all to the blocking situation, while most (77.6%) reported that they would react. Respondents who reported they would react were asked to report the latency of their aggressive response. They were asked to estimate the relative time until reaction with the question "How fast do you react?" on a response scale ranging from 1 to 10 with the endpoints labeled "Immediately" and "After a long hesitation."[9] The mean reported latency was close to the midpoint of the scale (5.65).

7 Additional information concerning the univariate distributions of the covariates across experimental are provided at the website related to the book (Table A5).

8 The category was chosen by only 16.3% of all participants. The results of all subsequent data analyses remain, by testing, substantially stable, even when the data of these participants are coded as missing values.

9 The categories "Immediately" to "After a long hesitation" were coded with "1" and "10" subsequent to data collection. Answer scales were presented without numeric labels in order to avoid incongruities. No middle category was used in order to prevent satisficing. Instead, we used verbal anchors at the extreme points of the scale (cf. Krosnick and Fabrigar 1997). Additional information concerning the univariate distributions of the dependent variables are provided at the website related to the book (Table A6).

20.2.3 Results for Each Part of the Study

In the following paragraph, we report our analyses of the treatment effects, i.e. the effects of the independent variables "social status" and "group membership" on the dependent variables "occurrence and latency of aggressive behavior" separately for both parts of the study. The two main threats to treatment effects, unobserved and observed heterogeneity, are accounted for by randomization and stepwise inclusion of covariates. After modeling the occurrence of aggressive behavior using logistic regression, the average marginal effects were estimated (cf. Mood 2010). For those who exhibited or reported the occurrence of aggressive behavior, the latency of an aggressive response was measured with a stopwatch in the field experiment and with a rating scale in the vignette experiment. From a strict measurement theoretical point of view, the latency measured in the field experiment requires the estimation of negative binomial regression models that account for the skewness of the distribution, while the vignette experiment requires the estimation of ordinary least square (OLS) regressions. Therefore, both OLS regressions and negative binomial regressions were estimated for the field experimental data. As the results are comparable with regard to the direction, the strength, and the significance of the estimated effects, the results of OLS regressions could be reported due to their easier interpretability and comparability.

20.2.3.1 Field Experiment

The analysis of the dependent variables leads to the conclusion that the occurrence of aggressive behavior is neither predicted by the social status nor by the group membership of the frustrator car, while the latency of aggressive behavior is predicted by the social status of the frustrator car. Higher status frustrator cars are significantly later sanctioned in comparison to lower status cars (Table 20.5). It also follows from Table 20.9 that all effects remain largely stable across the stepwise regression models controlling for different sets of observed covariates. This can be taken as a clear indication for successful randomization.

20.2.3.2 Vignette Experiment

The analysis of the effects on the dependent variables leads to the conclusion that both the occurrence and latency of aggressive behavior are predicted by the group membership of the frustrator car. An in-group car, i.e. a car with a local license plate, increases the likelihood of an aggressive response and diminishes its latency (Table 20.6). Table 20.6 also shows that all

Table 20.5 Effects of status and group membership on occurrence and latency of aggressive behavior.[a)]

	Occurrence			Latency		
	Model 1a	Model 2a	Model 3a	Model 1b	Model 2b	Model 3b
Status frustrator car (high, ref. low)	−0.704 (−1.75)	−0.744 (−1.73)	−0.753 (−1.74)	1.511* (2.24)	1.501* (2.16)	1.596* (2.29)
Group Membership frustrator car (local, ref. nonlocal)	0.316 (0.82)	0.335 (0.81)	0.384 (0.91)	0.267 (0.39)	0.122 (0.17)	0.006 (0.01)
Observations	362	362	362	329	329	329
(Pseudo) R^2	0.019	0.057	0.078	0.015	0.047	0.070

a) Model 1a, Model 2a, Model 3a: Logit estimation, average marginal effects, z-values in parentheses; Model 1b, Model 2b, Model 3b: OLS estimation, b coefficients, t-values in parentheses; Model 1a,b without control variables; Model 2a,b controlled for aggressor (age, sex, car class, year of construction); Model 3a,b controlled for aggressor (age, sex, car class, year of construction, origin of car, persons in car); $^{*}p < 0.05$, $^{**}p < 0.01$, $^{***}p < 0.001$.

Table 20.6 Effects of status and group membership on occurrence and latency of aggressive behavior.[a)]

	Occurrence			Latency		
	Model 1a	**Model 2a**	**Model 3a**	**Model 1b**	**Model 2b**	**Model 3b**
Status frustrator car (high, ref. low)	−0.128 (−0.54)	−0.154 (−0.63)	−0.164 (−0.67)	−0.097 (−0.37)	−0.104 (−0.39)	−0.112 (−0.41)
Group membership frustrator car (local, ref. nonlocal)	0.554* (2.30)	0.565* (2.22)	0.565* (2.22)	−0.554* (−2.11)	−0.580* (−2.16)	−0.577* (−2.15)
Observations	408	408	408	317	317	317
(Pseudo) R^2	0.013	0.066	0.067	0.015	0.037	0.051

a) Model 1a, Model 2a, Model 3a: Logit estimation, average marginal effects, z-values in parentheses; Model 1b, Model 2b, Model 3b: OLS estimation, b coefficients, t-values in parentheses; Model 1a,b without control variables; Model 2a,b controlled for aggressor (age, sex, car class, year of construction); Model 3a,b controlled for aggressor (age, sex, car class, year of construction, jam today, relationship, own children); $^{*}p < 0.05$, $^{**}p < 0.01$, $^{***}p < 0.001$.

effects remain largely stable across the stepwise regression models controlling for different sets of observed covariates. This indicates that the random selection of subjects into the treatment groups was successful.

20.2.4 Systematic Comparison of the Two Experiments

In our design, the data obtained in field experiment should hold as a behavioral benchmark in comparison to reported behavioral intentions in the vignette experiment. For this purpose, we consider participation in the vignette experiment as mode-treatment, while participants' observed behavior in the field experiment serves as approximation to the counterfactual for reported behavioral intentions. We will do so by propensity score matching (Rosenbaum and Rubin 1983), i.e. by combining so-called twins of both experiments.[10] However, referring to the field outcomes as approximated counterfactuals of the measured outcomes in survey requires a number of assumptions regarding construct validity of settings, treatments, outcomes, and units, which will briefly be introduced. Only if these conceptual assumptions are accepted, can the outcomes of the field experiment serve as potential outcomes of the survey experiment (Section 20.1) so that its external validity can be identified. In order to increase the appropriateness of the assumption of stable and equal construct validity across both experiments, we applied a number of methodological precautions, which increase the comparability of the study particulars.

20.2.4.1 Assumptions and Precautions About the Comparability of Settings, Treatments, and Outcomes

A comparison requires assumptions about construct validity according to settings, treatments, observed outcomes, and also covariates. With regard to the setting, we choose an everyday

10 The propensity score approach is a statistical approach and was suggested by Rosenbaum and Rubin (1983) to facilitate the analysis of data from observational studies or quasi-experimental designs with nonequivalent groups, i.e. studies in which randomization is not possible. On the basis of covariates that are likely to have influenced the selection to either treatment or control group, a propensity score is calculated by means of a logistic regression for every participant in the study. According to Shadish et al., the propensity score corresponds to "the probability of being selected in the treatment (versus control) group" (Shadish et al. 2002, 162). The propensity scores are used to guide a pair-matching procedure: Subjects with equal propensity scores in the treatment and control group are identified as so-called *twins* and included in subsequent data analyses (see Shadish et al. 2002 for further references).

situation at a blocked green traffic light and assume that this situation facilitates a high level of construct validity due to its ubiquitous character. As this situation can be relatively easily described and presented in a vignette, we assume that we reach a high level of construct validity of the setting.

We must also assume that there is no systematic difference in the construct validity of the observed outcomes between participants. This means that persons generally react similarly when asked to answer on a scale in a presented vignette and when confronted in a real situation. For the comparison of the latency, the survey answer scale was adapted to the time scale of the field experiment, so that the length between two scale points formally corresponds to a three second time interval on the time scale.

Regarding the construct validity of treatments, we must suppose an equal amount of treatment stimulation induced by real frustrator cars and reported behavioral intentions induced by the descriptions of frustrator cars in the vignettes. This is a far-reaching assumption, since real cars are linked to specific brands and models with actual license plates, while more general categories were used to describe the cars in the survey.

Finally, for the comparison also the entire set of covariates need to represent the same constructs across both experiments. While the covariates gender, the age of the participants, cars and the class of the car were self-reported in the vignette experiment, experimenters judged them in the field experiment using merely rough categories. Accordingly, we categorized the continuous ages of participants in the vignette experiment into the same categories used in the field experiment.

All measurement processes in the field experiment and the vignette experiment may also contain processing errors resulting from coding, which are presumed to be negligible.

20.2.4.2 Assumptions and Precautions About the Comparability of Units

Following Groves et al. (2009), inference errors in surveys can also occur in the area of representation resulting from sampling errors that lead to differences in target, frame, and covered populations and from overcoverage and undercoverage through self-selection. These errors address issues of the construct validity of units as different units of observation may represent different higher-order constructs. While each of the separate studies is "true" experiments that allow us to estimate isolated treatment effects, comparing the effect between the experiments can only take place in accordance with the methodology of observational studies (Rosenbaum 2002). Since fundamental differences in data collection procedures lead to different sample compositions, heterogeneities between the experimental effects can be related to sample variances (Winship and Morgan 1999). The results may, thus, be biased by limited comparability of the construct validity of persons rather than by limited construct validity of setting, treatments, and outcomes.

Both experiments were conducted simultaneously in the same city and are therefore generally related to the same target population: Inhabitants of Ingolstadt that principally should have a driving license (at least 18 years old). However, the covered populations do indeed differ. The frame population and the covered population ought to be almost identical for the vignette experiment (apart from complete nonresponse), while this cannot be assumed for the field experiment. Not all inhabitants over 18 years and even not only inhabitants passed the crossroad, so that overcoverage and undercoverage errors emerged. Moreover, we can estimate the sampling error in the vignette experiment, due to a systematic stratified probability selection strategy, while the sampling error cannot be estimated in the field experiment, due to a convenience sampling strategy. The different sampling strategies might thus result in varying but not revealed sampling errors. It can hardly be claimed that all persons of the frame population had the same probability to pass the intersection during the experimental time. Finally, while the possibility of nonresponse error is nonignorable in the vignette experiment, as potential respondents could decide to refuse, it is zero in the field experiment, because, once blocked at the traffic light, participants did not have the option to refuse participation.

As a consequence, additional assumptions and precautions regarding the comparability of the concepts represented by the units in both experiments are necessary to test for external validity. We applied a three-step strategy in order to achieve the best possible unit comparability using the observed heterogeneity. First, we excluded all participants in the field experiment from the comparison who owned a car with a nonlocal license plate because they are ineligible units regarding the target population of both experiments. Second, we apply nonparametric propensity score matching in order to reduce sampling bias (Rosenbaum and Rubin 1983). Third, we use nonparametric test methods allowing us to give up assumptions about parameters in the target population.

The differences between the vignette experiment and the field experiment can only be claimed to be causal, if appropriate counterfactual values are approximated. For this reason, we estimate *expected what-if-differences for the treated* given the abovementioned assumptions (Section 20.1) and using conditioning variables. So-called twins are constructed using a propensity score that reduces the multidimensionality of conditioning variables to one dimension (Rosenbaum and Rubin 1983). We estimated the propensity score by a logit model using the conditioning variables such as age, gender, year of construction, and the class of the participants' car (aggressor). Only participants who have twins with similar propensity scores are considered. Twin outcomes were assigned by one-to-one matching with replacement due to a large common support (cf. Morgan and Harding 2006), which leads to reduced standardized biases. Regarding the occurrence of any reaction, the outcome values of 192 field twins were matched to 450 participants of the vignette experiment. Accordingly, regarding the latency until any reaction, 177 outcomes of field twins who have shown any reaction were used to approximate the counterfactual outcomes of 357 survey participants who reported any intention that they would react in the situation. Accepting the conditional independence assumption, once conditioned on observed variables, all sample differences between participants in the vignette experiment and in the field experiment are balanced and the remaining outcome difference is due to the mode of data collection, i.e. the kind of experiment.

20.2.4.3 Results

First, we consider the distributions of the reported occurrence and the latency and their approximated counterfactuals (Table 20.7). Given the prerequisites above, it is noticeable that about three-quarters of the respondents exhibit a propensity of occurrence in the survey, while actually more than 90% of them would react in a real situation. Nevertheless, the majority of participants show a reaction in both factual and counterfactual. Furthermore, for the majority of subjects, any reactive behavior would occur during the first 10 seconds in the real experiment, while their reported latency is considerably longer with a larger variance. However, the latency measures differ between the vignette and field experiment. In addition, the reported latency in the vignette experiment is distorted by the latency it takes the subjects to respond to an unknown extent. The latter could not be measured within the framework of a survey.

We analyze the amount of overreporting and underreporting by summarizing all participants with corresponding and diverging outcomes and approximated counterfactuals. Krumpal (2013) points to the common "the-more-the-better assumption" in studies aiming on uncovering social desirability in which the true value is normally unknown. In contrast, our study design allows for the analysis of both overreporting and underreporting.

For the analysis of occurrence, we grouped the participants by corresponding or diverging reaction or nonreaction. Regarding the latency, we used the mean and median to create

Table 20.7 Distributions of outcomes in vignette experiment and matched approximated counterfactuals.

	Vignette experiment	Counterfactual (approximated)
Occurrence of aggressive response		
Occurrence (=1)	350 (77.78%)	431 (95.78%)
No occurrence (=0)	100 (22.22%)	19 (4.22%)
Total	450	450
Latency of aggressive response		
Range	2–29	2–29
Mean	15.630	10.541
Standard deviation	6.964	5.478
25% quartile	11	6
50% quartile	17	10
75% quartile	20	13
Skewness	−0.068	0.855
Kurtosis	2.367	3.321

analogous groups, whereby we used two strategies. First, we considered the scale values to be directly comparable. That is, we use the mean/median of the approximated counterfactuals and compared the reported latency by grouping. Second, we only used the categories below and above average calculated separately from the reported and approximated counterfactual latency. This approach discards the direct comparability of both scales but allows for a relative interpretation below or above average regarding scale-specific mean/median. The second strategy is thus the more rigorous comparison and used for interpretation.

As one can see in Table 20.8, almost three-quarters of the survey participants would react in the experimental situation, if they would have participated in the field experiment. Also six subjects would correctly not have shown any reaction. On the contrary, more than one-fifth reported no intended reaction, while it is very likely that they would react in a real situation, and only a small minority reported that they intended to react, while they would not have done so in a field experiment. Thus, we observe a clear indication of underreporting in the survey. Detailed analysis across the four experimental conditions reveals that this pattern is stable, while the largest occurrence of underreporting can be found, if the frustrator car has a nonlocal

Table 20.8 Overreporting and underreporting groups in occurrence (complete).

Occurrence	Vignette Experiment			
	Yes		No	
Field experiment	*N*	%	*N*	%
Yes	337	74.89	94	20.89 (underreporting)
No	13	2.89 (overreporting)	6	1.33

Table 20.9 Overreporting and underreporting groups in latency (complete).

	Vignette experiment			
	Below average		(Equal) above average	
Field experiment	***N***	**%**	***N***	**%**
Below average	Mean field: 71	19.89	Mean field: 277	77.59
	Median field: 71	19.89	Median field: 277	77.59
	Means separate: 169	47.34	Means separate: 179	50.14
	Medians separate:169	47.34	Medians separate:179 (underreporting)	50.14
(Equal) above average	Mean field: 1	0.28	Mean field: 8	2.24
	Median field: 1	0.28	Median field: 8	2.24
	Means separate: 4	1.12	Means separate: 5	1.40
	Medians separate:4 (overreporting)	1.12	Medians separate: 5	1.40

Note: Scale field: mean = 10.54; median = 10; scale survey: mean = 15.63; median = 17.

license plate. Under this condition, the proportion of underreporting is 26.4%. If the frustrator car has a local license plate, in turn, underreporting is with 15.1% at lowest. In contrast, there is no substantial deviation in underreporting between the low status (20.6%) and the high status (21.1%) conditions.[11]

The same applies for the analysis of latency distributions (Table 20.9). Nearly one-half of the subjects report a below average latency, while the other half reported an (equal) above average latency, although they would react faster in field situation. At the same time, the amount of overreporting, i.e. faster reported behavioral intention compared to the approximated counterfactual, is negligibly low. Additional analyses have shown that this pattern holds also markedly true across all experimental conditions. Again, the largest amount of underreporting is found, if the car has a nonlocal license plate, where it ranges between 57.1% and 79.66% depending on comparison strategy (mean/median; joint/separate). In turn, given the local license plate condition, underreporting ranges between 43.3% and 75.6%. However, underreporting varies remarkably slightly across the status conditions (48.1–77.8% vs. 52.3–77.3%).[12]

In sum, participants reported less and later behavioral intentions in the survey compared to the field-based approximated counterfactual values. This stable underreporting in the survey can serve as clear indication of the importance of social desirability in surveys. However, this interpretation requires that our nontested presumptions on an equivalent level of construct validity across both experiments hold.

In a last step, we calculated the difference between the reported occurrence and its latency and the corresponding approximated counterfactuals. We tested the significance of differences using a test on the equality of matched pairs of observations for occurrence and the Wilcoxon matched-pairs signed-ranks test for latency. The results are displayed in Table 20.10.

All sample average effects become significant. Thus, taking all necessary prerequisites into account, the differences between reported values in the vignette-based survey and the field-based approximated counterfactuals are statistically reliable and thus relevant. The

11 Additional information concerning over- and underreporting in occurrence across experimental are provided at the website related to the book (Tables A7–A10).

12 Additional information concerning over- and underreporting in latency across experimental are provided at the website related to the book (Tables A11–A14).

Table 20.10 Expected what-if differences for the treated (vignette experiment).

Mode as treatment (vignette experiment)	Occurrence (no)		Latency	
	What-if difference[3]	Equality of matched-pairs[1] (p)	What-if difference	Wilcoxon matched-pairs[2] $z(p)$
Complete	0.180 ($n = 450$)	(0.000)	5.089 ($n = 357$)	9.338 (0.000)
Status high	0.163 ($n = 232$)	(0.000)	4.859 ($n = 185$)	6.561 (0.000)
Status low	0.197 ($n = 218$)	(0.000)	5.337 ($n = 172$)	6.630 (0.000)
Local	0.110 ($n = 219$)	(0.000)	3.761 ($n = 180$)	5.380 (0.000)
Nonlocal	0.246 ($n = 231$)	(0.000)	6.440 ($n = 167$)	7.644 (0.000)

Using Propensity Score Matching (pscore by logit model, one-to-one matching with replacement);
[1]Test on the equality of matched pairs of observations (H0: median of the differences is zero);
[2]Wilcoxon matched-pairs signed-ranks test (H0: both distributions are the same);
n differs from original sample due to being on support after matching;
[3] What-if difference: expected what-if difference for the treated analogous to average treatment effect for the treated (mean difference between observed and matched outcomes).

probability to underreport a behavioral intention is generally higher in the survey experiment. Consequently, the reported latency until occurrence is longer in the vignette experiment.

20.3 Discussion

The study presented was designed to assess the external validity of vignette experiments in terms of the extent to which causal effects of treatments on reported behavioral intentions in hypothetical situations are generalizable to effects of treatments on actual behavior in respective situations in the real life. For this purpose, we compare responses to vignettes to observed behavior in real-life situations as a reference point for external validity. In particular, we used vignettes that were designed to analyze the effects of situational characteristics on reported behavioral intentions for norm-related behaviors.

Both studies were designed as "true" experiments in themselves, that is, in each experiment, two stimuli were varied systematically at two levels, which results in a 2×2 experimental design, and respondents were assigned randomly to one of these experimental groups. In this way, we captured, by testing, unobserved heterogeneity and avoided any confounding factors, so that we can trust the estimated isolated treatment effects in separate experiments.

However, a comparison of the separate results challenges measurement and representation issues so that identifying assumptions regarding the construct validity of settings, treatments, outcomes, and units are required. Testing the external validity requires to assume that these study particulars represent in principle the same higher-order constructs across both experiments. Though these assumptions cannot be finally proved, a number of methodological precautions were undertaken in order to consider them more appropriate.

Considering the comparability of variables and settings construct validity, we used the same decision situation in the same city, i.e. a sanctioning situation at a blocked traffic light, similar treatments, and nonrepeated measurement of similar dependent variables. A comparison requires the assumption of measurement invariance in all variables: treatments, outcomes, and covariates. This is a far reaching assumption, since it particularly does mean that participants in survey do not systematically differ in their perception of treatments and principle reaction to the answer scales as they would in the field experiment. By adapting the categories in occurrence and range of latency, we ensured that the variables are technically comparable.

With regard to the comparability of units' construct validity, both experiments generally addressed the same target population. However, different sample compositions emerged due to different data collection strategies and since the participants could not be assigned randomly to one of the experimental designs. While the sample of respondents in the vignette experiment is primarily biased by nonresponse, in the field experiment, we covered ineligible units in the frame population. Therefore, we relied on the methodology of observational studies (Rosenbaum 2002) and conditioned the samples in both experiments by observed heterogeneities using propensity score matching, which requires the assumption of conditional independence. First, we excluded all car drivers with a nonlocal license plate from the sample. Then, we matched most similar twins' outcomes to the outcomes of the survey group while conditioning along with age, gender, year of construction, and class of the car. Finally, we calculated what-if-differences for the treated and used nonparametric testing methods to avoid assumptions about the target population.

Our comparison strategy revealed remarkable differences in the effects of treatments on reported behavioral intentions and effects of treatments on actual behavior. In particular, reported behavioral intentions in response to hypothetical vignettes seem to be systematically underreported. Given our identifying but nontestable assumptions of comparable construct validity of all study elements across both experiments, one must conclude that reported behavioral intentions obtained with vignettes are not externally valid in terms of their generalizability to actual behavior. However, without doubt, our prerequisites are the most critical points of the comparison and one may have serious concerns whether all study elements do in fact represent the same constructs in both experiments. In case one takes the possibility into account that one or more of the particulars, i.e. settings, treatments, outcomes, or units, represent different constructs across both experiments, this would provide an alternative explanation for effect differences. Yet, if the assumption of comparable construct validity is rejected, it is also impossible to clarify the external validity of vignettes.

As a consequence, the greatest efforts of upcoming studies testing the external validity of vignette experiments should be undertaken in making all elements of the vignette study and the control study as much comparable as possible. By doing so, the assumption of comparable construct validity of these elements is made as convincing as possible. Limitations of our study regarding the comparability of experiments' elements provide excellent venues for further studies.

For a high construct validity of the general setting, upcoming studies should ensure that participants can assess the presented situation realistically. For instance, in our case, it would not have been helpful to interview little children or persons without driver's license about their behavior as car drivers at a traffic light. Although all participants in our sample reported to have a driver's license, we asked for it as a precaution. Nonetheless, the situational understanding can be improved by a more detailed vignette description, containing information such as the daytime, the concrete location, or the actual traffic volume.

In our study, the assumption that treatments of both experiments represent the same constructs is questionable. In particular, while the treatments in the field experiment were concrete car characteristics, the respective treatments in the vignettes were just categories, because we could not expect that every participant knows the status and sizes of specific car models. In further studies, it may be an advantage to use more similar treatment constructions like the car manufacturer and regional information about the license plate.

One interesting possibility to increase the realism, and thus construct validity, of vignettes in both general situations and specific treatments could lie in the application of visual vignettes (cf. Eifler 2007; Havekes et al. 2013) or video vignettes (Caro et al. 2012). Compared to text vignettes, pictures, photos, or videos can encapsulate much more detailed information concerning a decision situation. This is especially true for everyday situations, where decisions are made primarily based on visual information, such as traffic lights or built environment.

We measured two outcome variables where the answer scales in survey differed from the field observations, what makes the assumption of one equivalently represented construct debatable. Regarding occurrence, the answer presented in the survey was simply whether or not one may use the horn or flashlights for sanctioning. In fact, participants can show many more variations of reaction, such as using the horn frequently or even leaving the car. Although this did not appear in our field study, it may be included as possible answers in the outcomes. Additionally, while we measured the latency with a subjective-oriented scale ranging from "immediately" to "after a long time," it might provide more insights to use a scale of time units, i.e. seconds. Further, it could be appropriate to use more detailed measurements of age and year of manufacturing in a field experiment. As it was a challenge for the experimenters to estimate the ages of persons and cars, one may use in upcoming studies more precise measurement methods, such as age determination of taken photos with computer support.

With regard to the comparability of construct validity of the units of observation across both studies, we controlled for differences in sample compositions by estimating a propensity score and matching similar twins. This meets the idea of conditioning on observed heterogeneity and still requires the conditional independence assumption. This assumption might be more plausible, if more covariates would have been used, which is however a challenge. One possible alternative solution would be conditioning on unobserved heterogeneity. Here, the participants have to be randomly assigned to the field experiment and the vignette experiment in order to capture differences in all their characteristics. In particular, the researcher needs a list of possible participants to assign randomly. Moreover, he needs control on access to the conditions. As one will quickly notice, this cannot be generally applied. In our case, a limited area of traffic, such as a work premises with a limited number of persons, would be required. If the persons are well-known persons, one could split them randomly to a vignette experiment or a field experiment while avoiding overcoverage and undercoverage by ineligible participants. Such a design could be addressed as an "experiment about experiments" (see Petzold and Wolbring for an example).

References

AAPOR (2015). *Standard Definitions. Final Dispositions of Case Codes and Outcome Rates for Surveys*, 8e. American Association for Public Opinion Research. https://www.aapor.org/Publications-Media/AAPOR-Journals/Standard-Definitions.aspx.

Alexander, C.S. and Becker, H.J. (1978). The use of vignettes in survey research. *Public Opinion Quarterly* 42 (1): 94–104.

Aronson, E. and Carlsmith, J.M. (1968). Experimentation in social psychology. In: *The Handbook of Social Psychology. Volume Two: Research Methods. Reading* (ed. G. Lindzey and E. Aronson), 1–79. Addison-Wesley.

Atzmüller, C. and Steiner, P.M. (2010). Experimental vignette studies in survey research. *Methodology: European Journal of Research Methods for the Behavioral and Social Sciences* 6 (3): 128–138.

Auspurg, K. and Hinz, T. (2015). *Factorial Survey Experiments*. London/Thousand Oaks: Sage Publications.

Auspurg, K., Hinz, M., and Liebig, S. (2009a). Komplexität von Vignetten, Lerneffekte und Plausibilität im Faktoriellen Survey. *Methoden – Daten – Analysen* 3 (1): 59–96.

Auspurg, K., Hinz, M., Liebig, S. et al. (2009b). Auf das Design kommt es an. Experimentelle Befunde zu komplexen Settings in Faktoriellen Surveys. *SoFid Methoden und Instrumente der Sozialwissenschaften* 2009 (2): 23–39.

Auspurg, K. and Jäckle, A. (2015). First equals most important? Order effects in vignette-based measurement. *Sociological Methods & Research* https://doi.org/10.1177/0049124115591016.

Baxter, J.S., Manstead, A.S.R., Stradling, S.G. et al. (1990). Social facilitation and driver behavior. *British Journal of Psychology* 81 (3): 351–360.
Biemer, P.P., Groves, R.M., Lyberg, L.E. et al. (2004). *Measurement Errors in Surveys*. Hoboken, NJ: Wiley.
Campbell, D. (1950). The indirect assessment of social attitudes. *Psychological Bulletin* 47 (1): 15–38.
Caro, F.G., Yee, C., Levien, S. et al. (2012). Choosing among residential options: results of a vignette experiment. *Research on Aging* 34 (1): 3–33.
Cochran, W.G. (1965). The planning of observational studies of human populations (with discussion). *Journal of the Royal Statistical Society, Series A* 128: 134–155.
Collett, J.L. and Childs, E. (2011). Minding the gap: meaning, affect, and the potential shortcomings of vignettes. *Social Science Research* 40 (2): 513–522.
Deaux, K.K. (1971). Honking at the intersection: a replication and extension. *Journal of Social Psychology* 84 (1): 159–160.
Diekmann, A., Lorenz, S., Jungbauer-Gans, M. et al. (1996). Social status and aggression. A field experiment about horn-honking responses analyzed by methods of survival analysis. *The Journal of Social Psychology* 136 (6): 761–768.
Doob, A.N. and Gross, A.E. (1968). Status of frustrator as an inhibitor of horn-honking responses. *Journal of Social Psychology* 76 (2): 213–218.
Dülmer, H. (2007). Experimental plans in factorial surveys: random or quota design? *Sociological Methods & Research* 35 (3): 382–409.
Dülmer, H. (2016). The factorial survey. Design selection and its impact on reliability and internal validity. *Sociological Methods & Research* 45 (2): 304–347.
Edwards, A.L. (1957). *The Social Desirability Variable in Personality Assessment and Research*. New York: Hold, Rinehart and Winston.
Eifler, S. (2007). Evaluating the validity of self-reported deviant behavior using vignette analyses. *Quality and Quantity* 41 (2): 303–318.
Eifler, S. (2010). Validity of a factorial survey approach to the analysis of criminal behavior. *Methodology: European Journal of Research Methods for the Behavioral and Social Sciences* 6 (3): 139–146.
Eifler, S. and Petzold, K. (2014). Der Einfluss der Ausführlichkeit von Vignetten auf die Erfassung prosozialer Einstellungen. Ergebnisse zweier Split-Ballot Experimente. *Soziale Welt* 65 (2): 247–270.
Ellison, P.A., Govern, J.M., Petri, H.L. et al. (1995). Anonymity and aggressive driving behavior: a field study. *Journal of Social Behavior and Personality* 10 (1): 265–272.
Finch, J. (1987). The vignette technique in survey research. *Sociology* 21 (1): 105–114.
Forgas, J.P. (1976). An unobtrusive study of reactions to national stereotypes in four European countries. *Journal of Social Psychology* 99 (1): 37–42.
Ganong, L.H. and Coleman, M. (2006). Multiple segment factorial vignette designs. *Journal Of Marriage and Family* 68 (2): 455–468.
Groß, J. and Börensen, C. (2009). Wie valide sind Verhaltensmessungen mittels Vignetten? Ein methodischer Vergleich von faktoriellem Survey und Verhaltensbeobachtung. In: *Klein aber fein! Quantitative Sozialforschung mit kleinen Fallzahlen* (ed. P. Kriwy and C. Gross), 149–178. Wiesbaden: VS Verlag für Sozialwissenschaften.
Groves, R.M., Fowler, F.J.J., Couper, M.P. et al. (2009). *Survey Methodology*, 2e. Hoboken: Wiley.
Guo, S. and Fraser, M.W. (2010). *Propensity Score Analysis: Statistical Methods and Applications*. Los Angeles, CA: Sage.
Hainmueller, J., Hangartner, D., and Yamamoto, T. (2015). Validating vignette and conjoint survey experiments against real-world behavior. *Proceedings of the National Academy of Sciences* 112 (8): 2395–2400.

Havekes, E., Coenders, M., and van der Lippe, T. (2013). Positive or negative ethnic encounters in urban neighbourhoods? A photo experiment on the net impact of ethnicity and neighbourhood context on attitudes towards minority and majority residents. *Social Science Research* 42: 1077–1091.

Holland, P.W. (1986). Statistics and causal inference. *Journal of the American Statistical Association* 81 (4): 945–960.

Holland, P.W. and Wainer, H. (1993). *Differential Item Functioning*. Hillsdale: Lawrence Erlbaum.

Jasso, G. (2006). Factorial survey methods for studying beliefs and judgements. *Sociological Methods & Research* 34 (3): 334–423.

Jones, P. (2014). Narrative vignettes and online enquiry in researching therapist accounts of practice with children in schools: an analysis of the methodology. *Counselling and Psychotherapy Research* 14 (3): 227–234.

Kenrick, D.T. and MacFarlane, S.W. (1986). Ambient temperature and horn honking: "a field study of the heat/aggression relationship". *Environment and Behavior* 18 (2): 179–191.

Krosnick, J.A. and Fabrigar, L.R. (1997). Designing rating scales for effective measurement in surveys. In: *Survey Measurement and Process Quality* (ed. L.E. Lyberg, P.P. Biemer, M. Collins, et al.), 141–164. New York, NY: Wiley.

Krumpal, I. (2013). Determinants of social desirability bias in sensitive surveys: a literature review. *Quality and Quantity* 47 (3): 2025–2047.

Li, J.A., Chihhung, C.E., and Jasso, G. (2007). Computerized multivariate factorial survey. *Conference Papers – American Sociological Association, 1*, New York (11–14 August 2007).

Liebig, S., Sauer, C., and Friedhoff, S. (2015). Using factorial surveys to study justice perceptions: five methodological problems of attitudinal justice research. SFB 882 Working Paper Series 47.

Mah, C.L., Taylor, E., Hoang, S. et al. (2014). Using vignettes to tap into moral reasoning in public health policy: practical advice and design principles from a study on food advertising to children. *American Journal of Public Health* 104 (10): 1826–1832.

Mellenbergh, G.J. (1989). Item bias and item response theory. *International Journal of Educational Research* 13 (2): 127–143.

Millsap, R.E. (2011). *Statistical Approaches to Measurement Invariance*. New York: Psychology Press.

Mood, C. (2010). Logistic regression: why we cannot do what we think we can do, and what we can do about it. *European Sociological Review* 26 (1): 67–82.

Morgan, S.L. and Harding, D.J. (2006). Matching estimators of causal effects. Prospects and pitfalls in theory and practice. *Sociological Methods & Research* 35 (1): 3–60.

Morgan, S.L. and Winship, C. (2007). *Counterfactuals and Causal Inference: Methods and Principles for Social Research*. Cambridge: Cambridge University Press.

Mutz, D.C. (2011). *Population-Based Survey Experiments*. Princeton: Princeton University Press.

Neff, J.A. (1979). Interactional versus hypothetical others: the use of vignettes in attitudes research. *Sociology and Social Research* 64 (1): 105–125.

Nisic, N. and Auspurg, K. (2009). Faktorieller Survey und klassische Bevölkerungsumfrage im Vergleich - Validität, Grenzen und Möglichkeiten beider Ansätze. In: *Klein aber fein! Quantitative Sozialforschung mit kleinen Fallzahlen* (ed. P. Kriwy and C. Gross), 211–245. Wiesbaden: VS Verlag für Sozialwissenschaften.

Pager, D. and Quillian, L. (2005). Walking the talk? What employers say versus what they do. *American Sociological Review* 70 (3): 355–380.

Petzold, K. and Eifler, S. (2019). Die Messung der Durchsetzung informeller Normen im Vignetten- und Feldexperiment. In: *Devianz und Subkulturen: Theorien, Methoden und empirische Befunde* (ed. I. Krumpal and R. Berger). Wiesbaden: Springer VS.

Petzold, K. and Wolbring, D. (2019). What can we learn from factorial surveys about human behavior? A validation study comparing field and survey experiments on discrimination. *Methodology* 15 (1): 19–30.

Raub, W. and Buskens, V. (2008). Theory and empirical research in analytical sociology: the case of cooperation in problematic social situations. *Analyse und Kritik* 30 (2): 689–722.

Rosenbaum, P.R. (2002). *Observational Studies*. New York: Springer.

Rosenbaum, P.R. and Rubin, D.B. (1983). The central role of the propensity score in observational studies for causal effects. *Biometrika* 70 (1): 41–55.

Rossi, P.H. (1979). Vignette analysis: uncovering the normative structure of complex judgments. In: *Qualitative and Quantitative Social Research: Papers in Honor of Paul F. Lazarsfeld* (ed. R.K. Merton, J.S. Coleman and P.H. Rossi), 176–186. New York: Free Press.

Rossi, P.H. and Anderson, A.B. (1982). The factorial survey approach: an introduction. In: *Measuring Social Judgments. The factorial Approach* (ed. P.H. Rossi and S.L. Nock), 15–67. Beverly Hills: Sage Publications.

Rubin, D.B. (1986). Which ifs have causal answers. *Journal of the American Statistical Association* 81 (396): 961–962.

Rubin, D.B. (2005). Causal inference using potential outcomes: design, modeling, decisions. *Journal of the American Statistical Association* 100 (469): 322–331.

Sauer, C., Auspurg, K., Hinz, T. et al. (2011). The application of factorial survey in general population samples: the effect of respondent age and education on response times and response consistency. *Survey Research Methods* 5 (3): 89–102.

Sauer, C., Auspurg, K., Hinz, T. et al. (2014). Methods effects in factorial surveys: An analysis of respondents' comments, Interviewers' assessments, and response behavior. *SOEPpapers* 629.

Schwalbe, C.S., Fraser, M.W., Day, S.H. et al. (2004). North Carolina Assessment of Risk (NCAR): reliability and predictive validity with juvenile offenders. *Journal of Offender Rehabilitation* 40 (1/2): 1–22.

Shadish, W.R., Cook, T.D., and Campbell, D.T. (2002). *Experimental and Quasi-Experimental Designs for Generalized Causal Inference*. Boston, NY: Houghton Mifflin Company.

Sieber, J.E. and Stanley, B. (1988). Ethical and professional dimensions of socially sensitive research. *American Psychologist* 43 (1): 49–55.

Suppes, P. and Zinnes, J.L. (1963). Basic measurement theory. In: *Handbook of Mathematical Psychology* (ed. D.R. Luce, R.R. Bush and E. Galanter), 1–76. New York/London: Wiley.

Tourangeau, R. (1984). Cognitive science and survey methods. In: *Cognitive Aspects of Survey Methodology: Building a Bridge Between Disciplines* (ed. T.B. Jabine, M.L. Straf, J.M. Tanur, et al.), 73–100. Washington, DC: National Academy Press.

Tourangeau, R., Rips, L.J., and Rasinski, K. (2000). *The Psychology of Survey Response*. Cambridge: Cambridge University Press.

Tourangeau, R. and Yan, T. (2007). Sensitive questions in surveys. *Psychological Bulletin* 133 (5): 859–883.

Vellinga, A., Smit, J.H., Van Leeuwen, E. et al. (2005). Decision-making capacity of elderly patients assessed through the vignette method: imagination or reality? *Aging and Mental Health* 9 (1): 40–48.

Wallander, L. (2009). 25 years of factorial surveys in sociology: a review. *Social Science Research* 38 (3): 505–520.

Winship, C. and Morgan, S.L. (1999). The estimation of causal effects from observational data. *Annual Review of Sociology* 25 (1): 659–706.

Yazawa, H. (2004). Effects of inferred social status and a beginning driver's sticker upon aggression of drivers in Japan. *Psychological Reports* 94 (3): 1215–1220.

Part IX

Introduction to Section on Analysis

Brady T. West[1] *and Courtney Kennedy*[2]

[1] *Survey Research Center, Institute for Social Research, University of Michigan, Ann Arbor, MI, USA*
[2] *Pew Research Center, Washington, DC, USA*

This section includes four chapters that speak to general issues related to the design and analysis of randomized experiments embedded within surveys. From a design perspective, one needs to evaluate whether a nationally representative probability sample will actually include enough sample units from important subgroups (and especially hard-to-reach subgroups) to enable efficient implementation of randomized experimental treatments for those subgroups. Chapter 21 by Klar and Leeper considers the merits of purposive sampling, in contrast to probability sampling, for this type of design problem and presents compelling empirical evidence of the benefits of this type of sampling in the context of studies of intersectional identity.

Chapter 22 by Tipton and colleagues also speaks to the importance of careful sample design when evaluating interactions between experimental factors and key population subgroups. This chapter provides excellent guidance on the design of formal probability samples that will ensure efficient estimation of *treatment effect heterogeneity* across population subgroups, which is all too often ignored when analyzing the results of experiments embedded within surveys. Average treatment effects may not generalize to all subgroups of a population, and this chapter provides an important contribution on design considerations when treatment effect heterogeneity across subgroups is of primary interest.

Chapter 23 by van den Brakel is among the most technical in this edited volume, but also one of the most important, as it presents a general statistical framework for analyzing data from a randomized experiment embedded within a larger probability sample that has complex design features (such as stratification and/or cluster sampling). This chapter speaks to both design strategies and inferential approaches that one should follow depending on the type of experimental design used, including point and variance estimation. The chapter concludes with a detailed example of applying these techniques and discusses software from the Netherlands (i.e. X-tool) that is capable of implementing the techniques.

Chapter 24 by Cernat and Oberski also speaks to both design and analysis considerations, discussing the benefits of within-person experimental designs for quantifying stochastic measurement errors in surveys. This chapter introduces the general "multitrait multierror" (MTME) framework for designing experiments that enable one to model these types of measurement errors associated with specific survey questions. The general framework presented

Experimental Methods in Survey Research: Techniques that Combine Random Sampling with Random Assignment, First Edition. Edited by Paul J. Lavrakas, Michael W. Traugott, Courtney Kennedy, Allyson L. Holbrook, Edith D. de Leeuw, and Brady T. West.

Companion Website: www.wiley.com/go/Lavrakas/survey-research

by the authors will allow researchers to design experiments yielding estimates of measurement error variance, and the authors provide Mplus syntax (with the online supplemental materials) that can be used to fit appropriate statistical models reflecting the MTME designs that can quantify these important sources of error.

Collectively, these chapters provide important contributions on intelligent design strategies and state-of-the-art analytic approaches for extracting as much meaningful, representative information as possible from experiments when they are embedded within surveys.

21

Identities and Intersectionality: A Case for Purposive Sampling in Survey-Experimental Research

Samara Klar[1] *and Thomas J. Leeper*[2]

[1] *School of Government & Public Policy, University of Arizona, Tucson, AZ 85721, USA*
[2] *Department of Methodology, London School of Economics and Political Science, London WC2A 2AE, UK*

21.1 Introduction

The 2016 US Presidential election – like every election that has come before it – was framed as a battle to win the support of groups (Tamman and Stephenson 2015). As the Iowa caucuses approached, the news media reported daily updates on how candidates were faring with various voting blocs. To Hillary Clinton's chagrin, Bernie Sanders was leading among millennials. Troublingly, for Donald Trump, Ted Cruz held the advantage among evangelicals. Demographic groups such as these do not simply provide arbitrary means for dissecting voter behavior. Identity groups drive preference formation among group members, causing particular identity-related interests to become salient and to subsequently lead to political choices.

A great deal of survey research is thus understandably devoted to investigating the differences in public opinion among various demographic groups. Famous examples include the widely studied gender gap in public opinion (for example, heterosexual women appear to express greater support for employment protection and adoption rights for gay Americans, as compared with heterosexual men [Herek 2002]), distinctions in policy preferences by race (e.g. black respondents are slightly less likely than whites to support a reduction in legal immigration [Citrin et al. 1997]), and, of course, differences in opinion across partisan groups (Leeper and Slothuus 2014), with Democratic and Republican respondents expressing sharply divergent opinions on everything from foreign policy (Gaines et al. 2007; Baum and Groeling 2009), to the economy (Lewis-Beck et al. 2008), to even a candidate's skin color (Caruso et al. 2009). Opinion often varies by groups in dramatic and unexpected ways. Research of this nature can be carried out using any sample that includes members of each demographic subgroup. Most nationally representative survey data accommodate this requirement except when groups are extremely scarce (e.g. adherents to particular schools of Buddhist teaching) or atypically difficult to contact (e.g. isolated individuals such as the few-hundred member Old Order Amish).

Meanwhile, social scientists have long known that individuals do not merely identify with one solitary identity group. Census data plainly show that a growing number of Americans identify as multiracial and speak multiple languages at home. And multiple cross-cutting identities at the individual level are becoming increasingly common in America. As traditional over-arching social structures decline, individuals are freer to cross boundaries that divide demographic

Experimental Methods in Survey Research: Techniques that Combine Random Sampling with Random Assignment, First Edition. Edited by Paul J. Lavrakas, Michael W. Traugott, Courtney Kennedy, Allyson L. Holbrook, Edith D. de Leeuw, and Brady T. West.

Companion Website: www.wiley.com/go/Lavrakas/survey-research

groups (Iyengar and Bennett 2008; Klar 2013). Indeed, most individuals find themselves at the intersection of multiple identity groups. It is the aptly named field of *intersectionality* that focuses on these cases. The term intersectionality refers to "both a normative theoretical argument and an approach to conducting empirical research that emphasizes the interaction of categories of difference (including but not limited to race, gender, class, and sexual orientation)" (Hancock 2007, 64). The academic search engine J-STOR archived its first article referencing intersectionality 26 years ago and, since then, nearly 1466 have appeared – with 35% (507) of them having been published in just the last five years.

Yet survey-*experimental* research has just barely flirted with intersectionality for one very practical reason: individuals at the crossroads of multiple identity groups are difficult to find. As we will discuss in this chapter, traditionally heralded methods of designing survey samples make it difficult for scholars to focus on the increasingly important topic of intersectionality. For example, the 2008 American National Election Studies included a sample of 2232 respondents – only 33 of whom identified at the intersection of both non-White and nonheterosexual. As we argue in this chapter, nationally representative samples rarely include sufficient samples of subgroups because these subgroups are, inherently, minorities. The oversampling of identifiable minority populations is possible, and indeed common, in large-scale social surveys (the ANES, for example, has oversampled African Americans and Latinos since 2008), yet once intersectionality is taken into account even oversampled subgroups will be small (as the data just cited make clear). This limitation places boundaries on the types of research questions that researchers tend to pursue. Considering the notion of "fitness for use" of any given sample (Biemer 2010), we argue that *purposive samples* can be useful for answering many research questions in the domain of intersectional identity, particularly when using experimental methods. Purposive samples are a type of nonprobability sample "that can be logically assumed to be representative of the population" by "applying expert knowledge of the population to select in a nonrandom manner a sample of elements that represents a cross section of the population" (Battaglia 2008a, p. 645). Compared to general population samples, purposive samples are often a practical choice, particularly when one considers external validity of research a feature of a collective research literature rather than a feature of individual studies.

We begin by reviewing the strengths and weaknesses of various approaches for obtaining samples of intersectional identity groups, and then provide some empirical evidence for the viability of general population samples for providing large numbers of respondents from intersectional identity groups. We conclude with a review of relevant studies and a checklist of considerations for researchers to weigh when deciding how to recruit participants for identity research.

21.2 Common Techniques for Survey Experiments on Identity

A survey experiment "involves an intervention in the course of an opinion survey" (Druckman and Kam 2011, p. 17) designed to measure the impact of that intervention on an outcome of interest. The flexibility of the logistics involved in administering survey experiments – which can be administered in nearly any setting – make them a popular method of experimental inquiry. For example, in a simple manipulation of identity salience, Transue (2007) randomly assigned a sample of Caucasian American and African American respondents from the Minneapolis–Saint Paul metropolitan area to one of the two conditions meant to prime subordinate group identities or superordinate national identities. In the treatment condition, respondents were first asked "How close do you feel to your ethnic or racial group?" while those in the control were asked "How close do you feel to other Americans?" After priming identity through these group identification questions, respondents were then asked for their support of a tax increase intended to improve educational opportunities for minority groups. The treatment effect was the difference in support for the policy in each experimental group.

The gold standard of survey research is often thought to be probability-based sampling of individuals from a sampling frame with high coverage of a target population (Mullinix et al. 2015). Using this population-based sample as the basis for experimental treatment is sometimes seen as an even more ideal research design (Mutz 2011). Deviations from this ideal are rightly seen as relying on (possibly untestable) assumptions about the representativeness of the resulting sample with respect to the inference population. Probability-based samples, however, face increasingly insurmountable practical challenges; among these are the familiar challenges of plunging response rates in tandem with escalating costs (Groves 2006; Brick and Williams 2013; Williams and Brick 2017). When considering the study of political identity groups, however, probability-based samples also present sizable limitations for the recruitment of rare and hidden populations (Heckathorn 1997; see also Kalton 2014). Because these groups are small, even a large general population sampling frame will contain a relatively small number of individuals from the target subpopulation. Similarly, if an identity group is both rare and hidden (meaning that group members cannot be readily identified from typical administrative data), then stratified designs involving oversampling may not be feasible either.

The increased use of online panels in survey research (regardless of whether members are recruited through probability or nonprobability methods) has resulted in a heightened capacity to identify and possibly recruit members of rare identity groups into surveys without the need to screen for such individuals from a general population sample. In some cases, oversampling of specific subgroups can facilitate identity-focused research. For example, Spence (2010) was interested in studying the effects of media frames on African Americans' attitudes toward HIV/AIDS, and relied on a probability-based sample of respondents from the Knowledge Networks (now Ipsos) online KnowledgePanel to obtain a nationally representative sample of 500 African Americans who self-identified during empanelment. Even if such data were collected from a nonprobability sample, the basic demographic characteristics of this group (African Americans) are known from official statistics. Poststratification weighting according to standard demographics (age, sex, education, region, and metropolitan area) might correct for sample biases if the variables used to poststratify are also related to the survey measures of interest that are collected (see Krueger and West 2014), though to date there is little empirical support for this. Geographical oversampling of areas with likely high concentrations of target group members presents another possibility when administrative or profile data do not provide relevant identity information (Lavelle et al. 2009).

For other subpopulations such as those that are generally hidden or unlikely to be geographically clustered, oversampling from a subpopulation stratum can prove difficult. Pew (2013)[1] attempted to study the hidden population of self-identified lesbian, gay, bisexual, or transgender (LGBT) individuals in the United States, again using the KnowledgePanel. According to the panel profile data, 3645 panel members (or 5.2% of all panelists) self-identified as LGBT. After Pew recruited respondents into the survey, 15% of selected respondents invited to participate indicated that they did not identify in one of these categories. Particularly troubling was that among those identified in the profile data as transgender, only 23% reidentified as such during interviewing.

Large online panels, the use of administrative data, and reliance upon geographical oversampling may enable a researcher to generate potentially large probability-based samples of small populations. But evaluating whether even a well-designed sample of an intersectional identity

1 http://www.pewsocialtrends.org/2013/06/13/a-survey-of-lgbt-americans/.

group is representative of that subpopulation as a whole with respect to a set of characteristics (e.g. for nonresponse adjustments, or poststratification) poses challenges. Population-level data collections (e.g. the US Census) only generate limited demographic data (namely, age, sex, race, and ethnicity) and official surveys generate reasonably precise estimates of only selected population distributions (e.g. estimated age distributions but not proportions of party identifiers); a lack of data may thus make it impossible to assess whether a probability-based sample of intersectional identity group members are descriptively representative of their subpopulation. Consider, e.g. Catholic LGBT individuals, which Pew Research Center (2013)[2] estimates to constitute 14% of the LGBT population. Given that Census questionnaires have historically omitted measures of religious identity and typically omit measures of sexual identity, it would be impossible to say whether any particular sample of Catholic LGBT respondents resembled the population of Catholic LGBT individuals with respect to demographics, behavior, or any other measure. And it would, of course, also be impossible to apply any postsurvey adjustments to correct for any bias.

21.2.1 Purposive Samples

Purposive sampling can be thought of as a subset of convenience sampling, in that respondents are chosen subjectively. However, purposive sampling differs from other forms of convenience sampling in which "expert judgment is not used to select a representative sample of elements," whereas in purposive sampling it is (Battaglia 2008b, p. 148). To reiterate, the use of purposive samples requires expert judgment about the suitability of the sample and a well-reasoned argument for why a particular set of respondents provides an adequate basis for survey-experimental inference. While convenience sampling necessarily relies upon untestable assumptions rather than probability-based sampling methods, we will argue that such methods may be more appropriate for survey-experimental research than for observational research. In particular, this requires explicit definition of one's research question and goal of inference in order to establish whether any given survey experiment on any given sample is appropriate.

When one's research objective is foremost to obtain sufficiently large samples of insufficiently large subpopulations, it can become necessary to use this alternative sampling strategy in order to achieve one's research goals. The issues with nonprobability sampling have been widely discussed so we do not repeat them here in depth (see, for a thorough overview, Tourangeau et al. 2014). But critically, purposive sampling does not rely upon a probability-based survey design to select respondents and thus has no design-based claim to representativeness with respect to a given population. This problematizes the use of such samples for many kinds of descriptive research (e.g. what is the modal Presidential vote choice of LGBT Catholics?). For the purposes of experimental research on small intersectional identity groups, however, many purposive samples may be fit for use because they trade off design-based representativeness against obtaining a sample size sufficiently large to powerfully estimate an experimental effect size.[3] Snowball sampling, respondent-driven sampling (Heckathorn 1997), and socially mediated Internet surveys (Cassese et al. 2013) all present methods of obtaining large numbers of respondents from small or hidden populations.[4] But what price in total survey error is

2 http://www.pewsocialtrends.org/2013/06/13/a-survey-of-lgbt-americans/.

3 This is in large part because they supply sufficient numbers of respondents to obtain minimal standards of experimental power where other sampling strategies would be too costly given the large amount of overcoverage in any situation where population members cannot be identified from administrative data alone.

4 For other small but potentially nonhidden populations (e.g. Hispanic women), statistical adjustments are possible but issues of sample size remain. Lacking pre-recorded demographic profile data, a common approach for sampling such populations would be surname sampling. Such methods can be successful but past research shows positive

paid by relying on these nontraditional, nonprobability methods of recruitment? While any purposive sampling method has limitations, those limitations mainly relate to coverage and representativeness. With respect to coverage: any sampling method that does not rely upon a well-defined sampling frame provides no guarantee of coverage, such that response rates and response biases are undefined. And even in some cases where panel profile data are available, we have seen from the Pew study just described that such cases can still suffer from high overcoverage and unknown undercoverage. We suggest that the potentially very high cost of reducing coverage errors for these rare subpopulations may be better spent reducing other sources of survey(-experimental) error.

21.2.2 The Question of Representativeness: An Answer in Replication

The issue of representativeness is particularly problematic for many kinds of research (such as those just discussed that have descriptive ambitions). For survey-experimental work, however, representativeness requires more nuance than simply pointing to sampling design or comparing demographics in a sample and its target population. Thus, for rare but commonly studied intersectional identity groups, neither the tools of design-based nor model-based approaches achieve the dual goals of large sample sizes (for experimental power) and population representativeness. For example, for experimental research, what does representativeness mean and is it an objective worth striving for at the expense of recruiting large but possibly unrepresentative samples? Further, are there any unique challenges to the use of purposive samples that might introduce other sources of survey error above and beyond the possible bias in experimental effect sizes?

Short of obtaining a large probability-based sample with overcoverage, our view is that it is not presently possible to adequately evaluate the representativeness of any sample of the subpopulations that intersectionality researchers are typically interested in. However, that limitation does not mean that research on these groups is futile, particularly if survey-experimental research is involved. Instead, we follow the advice of Shadish et al. (2001) on the fitness of different samples for experimental research. Their goal to obtain "causal generalization" emphasizes experimental replication across multiple data collections rather than a focus on the "surface similarities" of a given sample to a (often poorly defined) target population, setting, and context: "that similarity judgment is hampered by the fact that not everyone will agree about which components are prototypical. By itself, then, surface similarity rarely makes a persuasive case for external validity, despite its widespread use" (Shadish et al. 2001, p. 360). By replicating results across samples, contexts, forms of the treatment, and measures of the outcome, the focus is less on whether a given study provides unbiased estimates on a population effect but rather whether a literature as a whole provides evidence that points to a consistent or systematically varying pattern of results. Descriptive representativeness of a sample is therefore seen as something that can be traded off in favor of other aspects of research design.

To understand Shadish et al.'s perspective requires discussing the quantities of interest in survey-experimental as opposed to descriptive survey research. In the simplest survey experiment, individuals are randomly assigned to one of the two groups, each of which is exposed to a stimulus before an outcome is measured. For example, we may be interested in how Democratic parents respond to primes of their respective identities as a partisan or as a parent in terms of their opinions on criminal justice policy (Klar 2013). Random assignment of respondents to groups ensures that each group provides, on average, a counterfactual against which

---

predictive values (probability of selecting an eligible individual conditional on surname) is perhaps 80% for Asian Americans (Lauderdale and Kestenbaum 2000), 55–70% for Latinos (Pérez-Stable et al. 1992), etc. suggesting that it remains an imperfect approach.

the others can be compared. The groups assigned to the parental identity prime and to the Democratic identity prime are the same in expectation on all pretreatment covariates and in their potential outcomes in response to each treatment, making the difference in opinions after treatment attributable solely to the causal influence of the randomly assigned treatment. The primary experimental statistic is therefore the sample average treatment effect (SATE): a difference-in-means of opinions between the two experimental groups.

This counterfactual statistic is quite different from the population parameters typically estimated in descriptive survey research because its bias depends on the variance of individual-level (unobservable) treatment effects rather than on the observable, descriptive characteristics of the sample. While a probability-based sample will generate an unbiased SATE, a descriptively representative nonprobability sample may be biased or unbiased. In our view, the expert judgment that underlies purposive sampling is therefore uniquely useful in survey-experimental work because knowledge of possible sources of individual-level effect heterogeneity (e.g. from studies published in existing literature or knowledge of unpublished experimental analyses on the target population) might help to guide sample construction.

To elaborate, there are four (not wholly exclusive) ways in which the SATE might be an unbiased estimate of the population average treatment effect (PATE) for the target population (in the above example, the population of Democratic parents): (i) the sample might be probability-based and suffer no meaningful nonresponse bias; (ii) the sample might be reweighted to represent the population on *all* outcome-related covariates; (iii) the unobservable individual-level treatment effects for every population member might be identical (i.e. homogeneous effects); and/or (iv) the unobservable individual-level treatment effects might, by chance, have the same mean as the population (regardless of the shape of the sample or population effect distributions). If the individual-level effects are heterogeneous and the average effect in the purposive sample differs from the average effect in the population, then the SATE is biased. Thus the degree to which a purposive sample can provide an unbiased effect estimate hinges on the – often unobservable – amount of effect heterogeneity in both the sample and population. Expert judgment about patterns of effect heterogeneity – informed by both prior research in the accumulating literature and knowledge of the sample and target population – are therefore critical to evaluating any kind of sample other than a probability-based sample from a heavily overcovered sampling frame. Chapter 22 of this volume addresses this notion of effect heterogeneity in more depth.

What this discussion highlights is that the purpose of survey-*experimental* research is the identification of causal effects. Researchers are quick to draw a parallel between internal validity (whether some relationship is causally identified) and external validity (the generalizability of the causal effect across persons, settings, treatments, and outcomes), as if the two are opposite ends of a continuum with a heavily controlled laboratory at one end and population census on the other. The trade-off, however, is not so severe. The advantages for internal validity of experimental randomization make the resulting information useful even when a set of respondents are not representative of any population because the results highlight the *plausibility* of a causal relationship and contribute evidence about effects that is most useful in aggregate when compared against other experimental studies on different samples, settings, treatments, and outcomes. In our view, external validity thus derives not from features of a given study's sample (as, for example, Sears 1986 would argue), but instead across replications (Mullinix et al. 2015; Krupnikov and Levine 2014; Coppock and Green 2015; Klein et al. 2013; Open Science Collaboration 2015). This is not to say that researchers should actively avoid efforts to obtain a probability-based sample of an intersectional identity group, but that the inherent limitations of doing so may mean prioritizing other single-study features, such as those that maximize internal validity, and attempting to replicate studies across multiple samples.

For experimental studies of intersectional identity, then, purposive samples may well provide useful insights even when the set of respondents is not demographically representative of the population of interest. And, in many combinations of identities, assessment of representativeness may not even be possible.

21.2.3 Response Biases and Determination of Eligibility

One question that arises in experimental studies of intersectionality is how to best identify members of a particular group. We discuss two major issues here: (i) how to determine whether a respondent's demographics make them eligible for survey participation and (ii) what effects the process of eligibility determination might have on survey and experimental response behavior during interviewing.

First, how do researchers interested in intersectional identity groups identify eligible respondents? Were demographic characteristics and self-identification synonymous, this process would be quite simple, particularly when administrative data are available to construct a sampling frame. However, in situations where such data are unavailable (be it in general population surveys or in purposive samples), researchers must develop recruitment procedures that are likely to yield large numbers of eligible respondents and few ineligibles. Commonly used strategies for recruiting small populations involve geographically targeted sampling, surname sampling, or – in the digital age – socially mediated sampling. All of these strategies target individuals based on manifest traits that might align with specific identities (African Americans living in predominantly African American neighborhoods, Latinos with a Hispanic surname, participants in relevant online communities, respectively). And given possible disconnects between demographics and identities – e.g. Hispanic surnames for those who identify with other races, individuals who present as female but have other gender identifications – researchers must ultimately rely on respondents to self-identify in order to determine eligibility (but measuring self-identification can potentially introduce other biases, as we discuss below).

Purposive sampling techniques therefore are likely to substantially reduce the overcoverage characteristic of general population sampling but present undercoverage problems of unknown consequences. By limiting one's recruitment to a targeted geography or to a physical or digital environment, the question becomes: how do those group members recruited here differ from group members recruited elsewhere? Identity strength is a spectrum that is dependent on one's social surroundings (Huddy 2001). In both cases of geographical targeting and recruitment through community or digital networks, one can expect that individuals in areas with higher concentrations of group members will respond differently to experimental treatment than will those who fall outside of the targeted area. One might intuitively expect that group members living in close proximity to a large number of other in-group members are likely to have stronger identification with that group, but research in fact suggests that members of minority groups hold stronger group identifications than do members of majority groups (Brewer 1991; see also Huddy 2001 and Huddy 2013 for further discussion). As Huddy (2001) explains, existing work suggests that individuals with the strongest gradations of identity strength are least affected by context (p. 146). If we consider how this might impact the effects of experimental interventions related to identity, we should expect that studying the influence of an identity on individuals who are surrounded by relatively few in-group members will overstate the effect that one would see in a more concentrated area captured by purposive sampling.

Second, given the profound influence that identities hold over our expressed preferences, carefully worded and timed measurement of identity groups is pivotal to determining respondent eligibility while also not stimulating the identity that is being used to assess eligibility. There is indeed abundant evidence (particularly in the field of political psychology) to suggest that salient identities influence individual preferences (e.g. Transue 2007; Jackson 2011; Klar 2013). For example, Transue (2007) found that asking individuals about their attachment to the identity group of "American" increases the support that they subsequently express for a national tax increase, where asking them first about their attachment to an ethnic identity decreases their support. When measuring attitudes among identity groups, researchers must remain cognizant of how consequential demographic questions can be.

Klar (2013) employed a purposive sample to experimentally test the degree to which competing identities influence policy preference. The intersection upon which she chose to focus was individuals who identify as both Democrats (a liberal-leaning group) and parents (when it comes to issues of national security, a conservative-leaning group). With a team of 25 interviewers, Klar conducted a face-to-face survey of 701 adults who showed up at the polls to vote on Election Day of 2010 in Illinois's 9th Congressional District. The intent of the experiment was to measure the impact of priming identities – that is, by making them salient – on policy responses. Respondents were randomly assigned to be exposed to a specific combination of identity primes: one targeting the parental identity and one targeting the Democratic identity. Each "prime" was merely a survey question asking the respondent about their affiliation with the particular identity group. Some primes were designed to invoke threat against the group, others were designed to instill a sense of group efficacy, and a third type of prime was a mere mention of the identity group. Klar found that the identity targeted with threat was most likely to influence subsequent policy preference. For example, respondents who received a threatening prime toward their parental identity increased their support for national security measures even when a competing (nonthreatening) prime targeted their Democratic identity. Conversely, those who received a threatening prime toward their Democratic identity decreased their support for those same measures even when a competing (nonthreatening) prime targeted their parental identity.

Crucially, Klar did not know a priori who was a parent and who was a Democrat. However, she could not ask ahead of time, as the demographic question itself would contaminate the experimental design. That is, by merely asking individuals if they were part of a group, the researcher would prime that group. To avoid this contamination, Klar asked all respondents for their party identity and whether or not they had children at the very end of the survey, after the dependent variables had been measured. By targeting adults in a highly Democratic district, the percentage of respondents who fell at the intersection of the two targeted identities was high (61% identified as Democratic parents). Although the researcher was forced to exclude nearly 40% of the respondents, her use of a purposive sample increased the size of the usable sample. By contrast, only 15% of respondent in the 2012 American National Election Studies classified themselves as Democrats (including leaners) with children. Purposive sampling in this case ensured greater efficiency and, by asking respondents about their inclusion in these groups only at the end of the survey, the survey design was not contaminated.

21.3 How Limited Are Representative Samples for Intersectionality Research?

21.3.1 Example from TESS

To put our concerns about the viability of nationally representative samples for intersectionality research in perspective, we examine the demographic composition of a large set of research experimental samples. The source of these samples is the Time-Sharing Experiments in the

Social Sciences (TESS) program, a National Science Foundation-funded project to provide researchers with access to probability-based samples from GfK (through 2016) and NORC's AmeriSpeak Panel (from 2016 onward). We examine the complete set of 19 experiments implemented by TESS during 2014, the most recent year for which full data are publicly available. Sample sizes for these studies varied from 551 to 5835 with a cumulative sample size of $n = 37\,950$. Across all respondents (treating the data as an unweighted pool of respondents), gender demographics were balanced (50.8% female) and racial demographics were reflective of the general population: 73.3% were non-Hispanic whites, 9.4% were non-Hispanic blacks, 10.6% were Hispanic, 3.5% were another race, approximately 3.5% were from any other race, and 3.2% reported two or more races. Non-Christian religious groups were particularly small, with only 2.4% Jewish ($n = 907$), 0.4% Muslim ($n = 151$), 0.7% Buddhist ($n = 251$), 0.5% Hindu ($n = 179$), and 2.2% other non-Christian ($n = 850$). Nonreligious identifiers were 18.5% of all respondents ($n = 7001$). Partisan groups were also large, with more than 16 000 self-identified Republicans (43.0%) and more than 19 000 self-identified Democrats (51.4%).[5] TESS unfortunately does not report data for LGBT self-identification, but given the numbers reported by Pew above, we can imagine that these are low.

The demographic characteristics of the TESS respondents not only reveal both the ease of acquiring a modestly sized, one-off representative sample for survey-experimental research (conditional on costly recruitment into and maintenance of design-based online panels), but also the problem of acquiring large numbers of respondents from the kinds of intersectional identity groups in which researchers may be interested. Online panel vendors provide a finite set of demographic and identity variables that can be used for sample construction and are constrained in retaining large samples of intersectional identity group members by the sizes of the subpopulations and the costs of recruiting from them. Larger samples – even obtained from an online panel – tend to be more costly when targeting small groups. Thus, for example, Hispanics, who were more than 10% of the combined sample of respondents, still present sample size challenges because once intersectional considerations are taken into account, practical samples sizes are often much smaller. Hispanic women were a mere 5.4% of the total pooled sample and multiracial women were only 1.8%, with some TESS studies containing as few as 7 such individuals. Figure 21.1 puts these numbers in very clear perspective by showing the relative sizes of intersectional groups based on sex, race/ethnicity, and partisan self-identification using the TESS data.[6] While general population studies typically yield large numbers of white Democrats and Republicans, of either sex, they produce comparatively tiny samples of non-White partisans, particularly given the rarity of non-Whites' support for the Republican Party and their low propensity to identify as politically independent. When one considers the need – in experimental research – for multiple experimental groups, such sample sizes quickly become insufficient.

21.3.2 Example from a Community Association-Based LGBTQ Sample

We now turn to two examples of purposive samples employed for experimental studies of individuals at the intersection of multiple groups. The first is a sample collected by Bergersen et al. (2018) of the lesbian, gay, bisexual, transgendered, and queer (LGBTQ) community in Pima County, Arizona. To illustrate the difficulty of collecting data on this particular demographic intersection, consider the American National Election Studies. In 2012, the ANES interviewed 5914 respondents in total – among them, only 39 individuals identified as

5 These numbers include leaners.

6 The figure shows a hypothetical sample of 1000 respondents that reflects the demographic composition of all 37 950 respondents (one icon represents approximately 38 individual respondents from the data). Data are unweighted for this presentation.

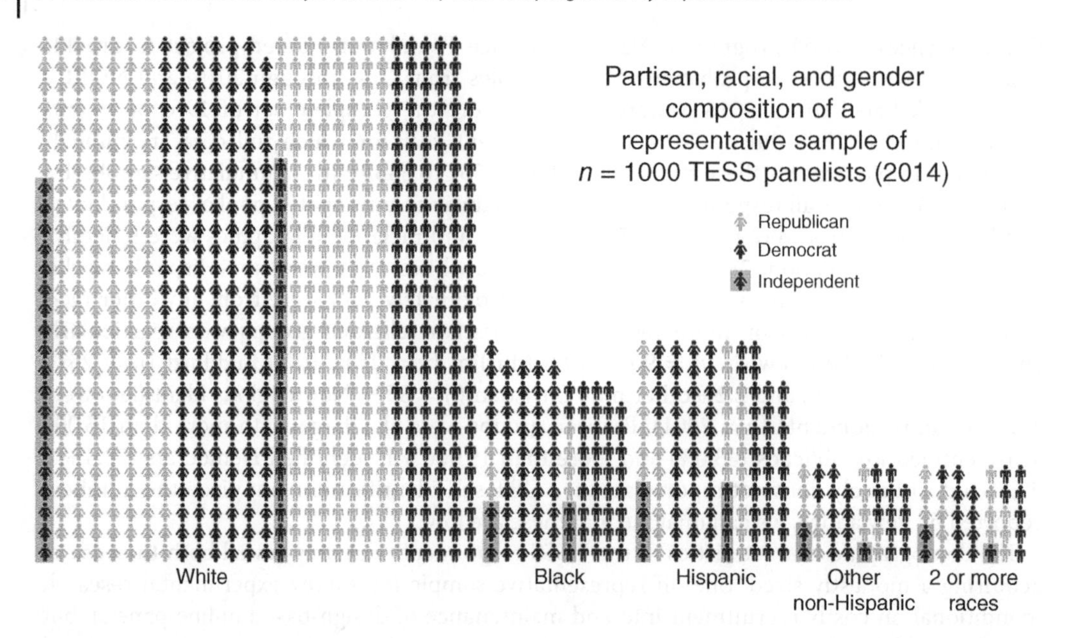

Figure 21.1 Partisan, racial, and gender composition of a hypothetical sample of $n = 1000$ TESS online panelists, based on unweighted data for $n = 37\,950$ respondents from GfK for all TESS studies conducted in 2014. Partisan groups include "leaners."

both nonheterosexual (i.e. homosexual or bisexual) and non-White. As this example suggests, collecting data on minority groups can be burdensome when large samples include only small proportions of the population of interest. Rather than analyzing only those relatively few respondents in a nationally representative sample who identify as LGBTQ, Bergersen et al. (2018) contacted 28 LGBTQ+ organizations throughout one county in the Southwestern region of the United States. The researchers asked if they could invite the members of each organization to participate in an anonymous survey regarding politics. Ten organizations agreed to share their membership list. Each participating organization contacted their respondents either through email or through their organizational Facebook page. Respondents were all provided with a link to the anonymous survey. After 21 days, the researchers closed the survey and analyzed the responses. In total, the researchers received a total of 200 responses; among them, 165 individuals identified as lesbian, gay, bisexual, transgendered or queer, and 35 identified as straight (or heterosexual). Although this sample is certainly limited by its geographic constraint (i.e. all respondents are located in the same Southwestern state and most within one particular county), it is in fact significantly more diverse than what we typically find in nationally representative surveys. To begin with, a third of this sample identified as non-White. The full demographics of this purposive sample are provided in Table 21.1.

The researchers' goal was to understand how governmental representation of a group to which an individual belongs (i.e. in-group) vs. governmental representation of a group that an individual views as outsiders (i.e. out-group) affects political engagement. To do so, they randomly assigned respondents to one of three groups. The first group read about congressional efforts to represent the interests of the LGBTQ+ community (the in-group), the second group read about congressional efforts to represent the Tea Party (a threatening out-group), and the third group read about congressional efforts to represent the African American community (a nonthreatening out-group). The results showed that learning about the Tea Party's

Table 21.1 Demographics of Pima County LGBTQ community purposive sample collected by Bergersen et al. ($n = 200$) (unweighted).

Gender identity[a]	Female	50%
	Male	33%
	Transgender[b]	18%
Sexual orientation[a]	Lesbian	14%
	Gay	30%
	Bisexual	21%
	Queer	10%
	Plus (includes asexual, other, nonhetero, and other write-in categories)	8%
	Heterosexual	18%
Race[c]	White	68%
	Hispanic or Latino	17%
	Asian	4%
	Black	3%
	Native American	1%
	Other	5%
Age	18–25	36%
	26–35	14%
	36–45	11%
	46–55	23%
	56–65	10%
	66 or older	5%
Party identification	Strong Democrat	28%
	Weak Democrat	23%
	Independent-Democrat	28%
	Independent	8%
	Independent-Republican	5%
	Weak Republican	1%
	Strong Republican	1%
	Other	7%

a) Gender identity and sexual orientation were open-ended questions; responses were subsequently coded for analyses.
b) "Transgender" refers to respondents who identified as transgender, agender, gender queer, and nonbinary.
c) Race categories do not sum to 100% because respondents could choose more than one.

representation in government significantly motivates LGBTQ+ individuals to elect representatives with shared demographic traits, as compared to learning about the efforts being done in government to represent their own group (the LGBTQ+ community).

Furthermore, because the sample of LGBTQ+ Americans was diverse on a series of demographic traits, Bergersen and colleagues were able to identify important differences within the community that a nationally representative survey (such as the ANES) would not allow. For example, LGBTQ+ individuals who identify as transgender are significantly less likely to feel that their vote matters. LGBTQ+ individuals who are not white are significantly less likely to vote, as are LGBTQ+ individuals who are not male. As Bergersen and colleagues explain, by collecting a purposive sample, it is possible to reveal important attitudinal distinctions within the pool of respondents.

21.3.3 Example from Hispanic Women Sample in Tucson, AZ

We now turn to a second purposive sample, collected by Klar et al. (2015) in November 2014. Interested in learning about political engagement among Hispanic women, the researchers collected a face-to-face purposive sample of residents in Tucson, Arizona. The Hispanic and Latino population constitutes approximately 17% of the total US population; women constitute approximately 50% of the population. It is thus no surprise that women who identify as "Spanish, Hispanic, or Latino" make up just 8.7% of the sample collected by the American National Election Studies – precisely the same proportion that we would expect to find in the population. Instead of pursuing a nationally representative sample, Klar and her colleagues collected a purposive sample in a Southwestern city whose population is 30% Hispanic. From one day of interviewing women throughout a diverse sample of neighborhoods in the city, the researchers collected a sample of 89 Hispanic women, each of whom participated in a five-minute face-to-face survey experiment.

The researchers were interested in studying the effects of interviewer race and gender on self-reported political engagement among Hispanic women. Interviewees were randomly assigned to an interviewer that was a Hispanic female, a Hispanic male, a white female, or a white male. Each interviewer administered the exact same survey, in which respondents reported how confident they are when they make political decisions and how interested they are in politics. Even with this small sample, the researchers found that Hispanic women express significantly more confidence and more interest when being interviewed by a Hispanic woman as compared to each of the other three types of interviewers (there were no significant differences among the other three). By conducting the interviews in one small geographical area (one county), the researchers did not have to combine respondents from distinct regions of the country to achieve a sufficient sample size and dismiss potentially important geographical variation. Whereas a nationally representative sample includes at most a handful of respondents from each geographic location across the country, concentrated purposive sampling ensures geographic homogeneity within the sampled population.

21.4 Conclusions and Discussion

Given the quickly changing demographics of the American population, it is important to consider the intersection of multiple identities as a determinant of attitudes and behaviors. Probability samples, while providing clear benefits for generalizing to broad populations, limit the degree to which researchers can select large samples of quite small identity groups that exist at the nexus of multiple identities or demographic characteristics. While online panels present the opportunity for stratified sampling, many rare populations of interest to intersectionality researchers are necessarily very small subsets of the respondents available in such panels. Because some groups are also hidden (e.g. non-White LGBT respondents), there is also a limited ability to use poststratification methods to reweight these groups to match population distributions on other characteristics. Thus population sampling, stratified sampling, and postsampling adjustments all have limited utility in the search for and study of intersectional identity groups.

We propose that purposive sampling may be the most viable strategy for obtaining large samples of respondents from these target groups and that the objectives of experimental research (i.e. the identification of group-specific causal effects across multiple study replications) makes such purposive samples an appropriate data base for identity research. When one considers that the objective of survey-experimental research is not always to obtain an estimate of the PATE, but instead to investigate the plausible influence of different putatively causal factors, the viability of nonprobability and perhaps nonrepresentative samples becomes clear. The costs of collecting large numbers of responses of many intersectional identity groups from general population samples is substantial given the sizes of such subgroups in the population. Even when recruiting participants from online panels or other sources where administrative or profile data provide demographic details about respondents, such demographic details alone do not reveal eligibility based on group *self-identification* rather than demographic membership.

While purposive samples tend to lack the descriptive representativeness that is typically associated with claims of unbiasedness is survey research, we have attempted to show that there is still – potentially considerable – value in experimental samples that lack this descriptive quality. Indeed, such samples may be able to generate unbiased and quite precise estimates of conditional average treatment effects and, when individual effects are relatively homogeneous, unbiased estimates of PATEs. Ultimately, the value of experimental research generated from any sample – be it representative or purposive – hinges on whether the findings replicate across settings, samples, treatments, and outcomes (Shadishet al. 2002). Researchers aiming to study intersectional identity groups should attempt, to the extent possible, to replicate their findings as well as identify the possible sources of treatment effect heterogeneity that might limit the value of any particular sample or setting. Such insight into possible sources of heterogeneity will make it easier for other researchers to design purposive samples that can replicate and possibly extend their findings.

Author Biographies

Samara Klar is an Assistant Professor of Political Science at the University of Arizona. Thomas Leeper is an Associate Professor in Political Behaviour at the London School of Economics and Political Science. Both received their PhDs in Political Science from Northwestern University in 2013. Their work on survey experiments and public opinion appears in over a dozen journals including *The American Political Science Review, The American Journal of Political Science, The Journal of Politics,* and *Public Opinion Quarterly*.

References

Battaglia, M. (2008a). Purposive sample. In: *Encyclopedia of Survey Research Methods* (ed. P.J. Lavrakas), 645. SAGE Publications.

Battaglia, M. (2008b). Convenience sample. In: *Encyclopedia of Survey Research Methods* (ed. P.J. Lavrakas), 645. SAGE Publications.

Baum, M.A. and Groeling, T. (2009). Shot by the messenger: partisan cues and public opinion regarding national security and war. *Political Behavior* 31 (2): 157–186.

Bergersen, M., Klar, S., and Schmitt, E. (2018). Intersectionality and engagement among the LGBTQ+ community. *Journal of Women, Politics, and Policy* 39 (2): 196–219.

Biemer, P.P. (2010). Total survey error: design, implementation, and evaluation. *Public Opinion Quarterly* 74 (5): 817–848.

Brewer, M.B. (1991). The social self on being the same and different at the same time. *Personality and Social Psychology Bulletin* 17 (5): 475–482.

Brick, J.M. and Williams, D. (2013). Explaining rising nonresponse rates in cross-sectional surveys. *The ANNALS of the American Academy of Political and Social Science* 645 (1): 36–59.

Caruso, E.M., Mead, N.L., and Balcetis, E. (2009). Political partisanship influences perception of biracial candidates' skin tone. *Proceedings of the National Academy of Sciences in the United States of America* 106 (48): 20168–20173.

Cassese, E., Huddy, L., Hartman, T. et al. (2013). Socially mediated Internet surveys: Recruiting participants for online experiments. *PS: Political Science & Politics* 46 (4): 775–784.

Citrin, J., Green, D.P., Mute, C., and Wong, C. (1997). Public opinion toward immigration reform: the role of economic motivations. *The Journal of Politics* 59 (3): 858–881.

Coppock, A. and Green, D.P. (2015). Assessing the correspondence between experimental results obtained in the lab and field: a review of recent social science research. *Political Science Research and Methods* 3 (1): 113–131.

Druckman, J.N. and Kam, C.D. (2011). Students as experimental participants: a defense of the "narrow data base". In: *The Cambridge Handbook of Experimental Political Science* (ed. J.N. Druckman, D.P. Green, J.H. Kuklinski and A. Lupia). Cambridge University Press.

Gaines, B.J., Kuklinski, J.H., Peyton, P.J.Q.B., and Verkuilen, J. (2007). Same facts, different interpretations: partisan motivation and opinion on Iraq. *The Journal of Politics* 69 (4): 957–974.

Groves, R.M. (2006). Nonresponse rates and nonresponse bias in household surveys. *Public Opinion Quarterly* 70 (5): 646–675.

Hancock, A.-M. (2007). When multiplication doesn't equal quick addition: examining intersectionality as a research paradigm. *Perspectives on Politics* 5 (1): 63–79.

Heckathorn, D.D. (1997). Respondent-driven sampling: a new approach to the study of hidden populations. *Social Problems* 44 (2): 174–199.

Herek, G. (2002). Gender gaps in public opinion about lesbians and gay men. *Public Opinion Quarterly* 66 (1): 40–66.

Huddy, L. (2001). From social to political identity: a critical examination of social identity theory. *Political Psychology* 22: 127–156. [Reprinted in Howard Lavine, ed., Political Psychology, Sage, 2010.

Huddy, L. (2013). Translating group identity into political cohesion and commitment. In: *Oxford Handbook of Political Psychology*, 2e (ed. L. Huddy, D.O. Sears and J. Levy), 737–773. New York: Oxford University Press.

Iyengar, S. and Bennett, L. (2008). A new era of minimal effects? The changing foundations of political communication. *Journal of Communication* 58 (4): 707–731.

Jackson, M. (2011). Priming the sleeping giant. *Political Psychology* 32 (4): 691–716.

Kalton, G. (2014). Probability sampling methods for hard-to-sample populations. In: *Hard to Sample Populations* (ed. R. Tourangeau, B. Edwards, T.P. Johnson, et al.), 401–423. New York: Cambridge.

Klar, S. (2013). The influence of competing identity primes on political preferences. *Journal of Politics* 75 (4): 1108–1124.

Klar, S., Bradshaw, S.C., Chavez, L., and Berman, Z. (2015). The influence of identity stereotypes on political engagement among Hispanic women. Presented at the Annual Meeting of the Midwest Political Science Association, Chicago, IL (16–19 April 2015).

Klein, R.A., Ratliff, K.A., Vianello, M. et al. (2013). Investigating variation in replicability: a 'many labs' replication project. *Social Psychology* 45 (3): 142–152.

Krupnikov, Y. and Levine, A.S. (2014). Cross-sample comparisons and external validity. *Journal of Experimental Political Science* 1 (March): 59–80.

Krueger, B.S. and West, B.T. (2014). Assessing the potential of paradata and other auxiliary data for nonresponse adjustments. *Public Opinion Quarterly* 78 (4): 795–831.

Lauderdale, D.S. and Kestenbaum, B. (2000). Asian American ethnic identification by surname. *Population Research and Policy Review* 19 (3): 283–300.

Lavelle, B., Larsen, M.D., and Gundersen, C. (2009). Research synthesis: strategies for surveys of American Indians. *Public Opinion Quarterly* 385–403.

Lewis-Beck, M.S., Jacoby, W.G., Norpoth, H., and Weisberg, H.F. (2008). *The American Voter Revisited*. Ann Arbor, MI: University of Michigan Press.

Leeper, T.J. and Slothuus, R. (2014). Political parties, motivated reasoning, and public opinion formation. *Advances in Political Psychology* 35 (S1): 129–156.

Mullinix, K.J., Leeper, T.J., Druckman, J.N., and Freese, J. (2015). The generalizability of survey experiments. *Journal of Experimental Political Science* 2 (2): 109–138.

Open Science Collaboration (2015). Estimating the reproducibility of psychological science. *Science* 349 (6251): aac4716–aac4716.

Mutz, D.C. (2011). *Population-Based Survey Experiments*. Princeton, NJ: Princeton University Press.

Pérez-Stable, E.J., Sabola, F., Otero-Sabogal, R. et al. (1992). Misconceptions about cancer among Latinos and Anglos. *Journal of the American Medical Association* 268 (22): 3219–3223.

Sears, D.O. (1986). College sophomores in the laboratory: influences of a narrow data base on social psychology's view of human nature. *Journal of Personality and Social Psychology* 51 (3): 515–530.

Shadish, W.R., Cook, T.D., and Campbell, D.T. (2001). *Experimental and Quasi-Experimental Designs for Generalized Causal Inference*, 2e. Cengage Learning.

Shadish, W.R., Cook, T.D., and Campbell, D.T. (2002). *Experimental and Quasi-Experimental Designs For Generalized Causal Inference*, 2002. Wadsworth Cengage Learning.

Spence, L.K. (2010). Episodic frames, HIV/AIDS, and African American public opinion. *Political Research Quarterly* 63 (2): 257–268.

Tamman, M. and Stephenson, E. (2015). Republicans come up short in search for diverse voters in 2016 election. Reuters. http://www.reuters.com/article/us-usa-election-republicans-idUSKBN0UD0X220151230 (retrieved 21 January 2016).

Transue, J.E. (2007). Identity salience, identity acceptance, and racial policy attitudes: American national identity as a uniting force. *American Journal of Political Science* 51 (1): 78–91.

Tourangeau, R., Edwards, B., Johnson, T.P. et al. (eds.) (2014). *Hard to Sample Populations*. New York: Cambridge University Press.

Williams, D. and Brick, J.M. (2017). Trends in U.S. face-to-face household survey nonresponse and level of effort. *Journal of Survey Statistics and Methodology*. https://doi.org/10.1093/jssam/smx019.

22

Designing Probability Samples to Study Treatment Effect Heterogeneity

Elizabeth Tipton[1,2], David S. Yeager[3], Ronaldo Iachan[4], and Barbara Schneider[5]

[1] *Human Development Department, Teachers College, Columbia University, New York, NY 10027, USA*
[2] *Statistics Department, Northwestern University, Evanston, IL 60201, USA*
[3] *Psychology Department, University of Texas at Austin, Austin, TX 78712, USA*
[4] *ICF International Fairfax, VA 22031, USA*
[5] *College of Education and Sociology Department, Michigan State University, East Lansing, MI 48824, USA*

22.1 Introduction

Population-based survey experiments are increasingly common in the social and behavioral sciences, particularly in psychology, political science, economics, and sociology. Much of this growth comes from the ability to randomize participants to different conditions via Internet-delivered surveys administered to panels of probability sampled adults (e.g. the time sharing experiments for the social sciences; TESS).[1] Survey experiments are particularly amenable to experimental hypotheses regarding individuals' attitudes, beliefs, or intended behaviors because questions or vignettes can be randomly assigned early in a survey and outcomes can be compared on later questions. Although less common, survey-administered manipulations have also been used to "nudge" individuals (Thaler and Sunstein 2008) or change "mindsets" (Dweck 2006) in ways that increase socially beneficial behaviors such as voting or college graduation, assessed via official registries like the validated vote file (e.g. Bryan et al. 2011, Studies 2 and 3), or the national student clearing house (e.g. Yeager et al. 2016b, Study 1).

Studies that nest a random-assignment experiment within a probability sample are well-positioned to estimate average treatment effects that generalize to a well-defined population.[2] This is in contrast to more typical experiments (e.g. psychology laboratory studies or field trials in education), which are often conducted with samples of convenience (see Olsen et al. 2013). Experiments conducted in convenience samples can provide initial evidence against the null hypothesis, but they are poorly positioned for assessing generalizability if there is any variation in treatment effects across groups – that is, *treatment effect heterogeneity*

1 See www.tessexperiments.org.
2 Throughout the chapter, we assume that nonresponse in probability samples is ignorable or addressable through poststratification. While our analysis does not depend on this assumption, there is some support for the idea that it is often defensible. Some find that nonresponse in surveys over the years has not led to lower accuracy in those surveys (e.g. Keeter et al. 2006; Yeager et al., 2011). However, those studies were used marginal distributions on nonweighted variables as a measure of validity; to our knowledge, there has been no evaluation of the effects of survey nonresponse on treatment effect sizes over time.

Experimental Methods in Survey Research: Techniques that Combine Random Sampling with Random Assignment, First Edition. Edited by Paul J. Lavrakas, Michael W. Traugott, Courtney Kennedy, Allyson L. Holbrook, Edith D. de Leeuw, and Brady T. West.

Companion Website: www.wiley.com/go/Lavrakas/survey-research

(see Tipton 2013). For example, if younger respondents with lower educational attainment are less responsive to a treatment, but a recruited sample overrepresents older, highly educated individuals, then the average effect estimated in the sample will be larger than that in the population.

Unfortunately, too little is known about treatment effect heterogeneity, and what is known has typically come from the analysis phase of research, not the design phase. Most experiments are designed to have power to detect *average* effects, and are only rarely well powered for detecting *differences* in treatment effects across subgroups (i.e. moderators). This is because moderator effects are statistically less precise than the estimate of the average effect, and this is exacerbated when subgroups are rare. As we will discuss, underpowered moderation analyses can result in high Type 1 *and* Type 2 errors (also see Gelman 2015).

We argue that probability sampling, as typically implemented, does not automatically produce samples that are suitable for studying treatment heterogeneity, even when samples are high quality and have low nonresponse bias. We propose that it is better to *design* the probability sample from the outset with the goal of understanding sources of treatment effect heterogeneity. By doing so, researchers are thus able to not only estimate an average effect but also identify for *whom, under what conditions, and why* a treatment may work best (see Bryk 2009; Shadish et al. 2001). This type of knowledge is central to the development of a theory of the causal mechanism – the highest goal of science (see Cook 1993).

In the remainder of this chapter, we explain a new approach that survey samplers can use when designing probability samples for survey experiments where there is a possibility of treatment heterogeneity. We begin by explaining why probability samples are preferred to nonprobability samples for estimating two quantities (or *estimands*): (i) population average treatment effects and (ii) treatment effects within subgroups. The chapter furthermore explains why typical probability sampling methods that optimize statistical power for the *average* effect in a population do not necessarily optimize statistical power for the *subgroup* effects of interest – especially when one's interest is in estimating effects within a rare subgroup.

Next, this chapter explains why even large, well-constructed, highly representative probability samples with randomized treatments can produce *confounded* analyses of differences across subgroups. Specifically, when one interacts a manipulated treatment with a measured, nonmanipulated variable (a moderator), then one faces the same set of confounds as any other observational study, despite randomization of the treatment (e.g. Morgan and Winship 2014; Shadish et al. 2001). For instance, two competing moderators (e.g. race and education) might be highly correlated with each other and with a host of other unmeasured, third-variable confounds (like neighborhood segregation); a moderation analysis, even with both in a model, can have difficulty disambiguating which (if any) has causal power. Furthermore, the off-diagonal cases that could be useful for disambiguating the two (e.g. people who are both minority and highly educated) might be strongly underrepresented, even in a typical probability sample. This basic insight – that moderators are often observational and therefore confounded, threatening moderation analysis – has been almost completely ignored in the experimental literature on subgroup analyses and interaction effects (though see Tucker-Drob 2011; Vanderweele 2015).

We recommend a focus on the development of sampling *designs* that are informed by a *theory of treatment effect heterogeneity*. Our recommended approach requires researchers to identify important subgroups and hypotheses regarding *to whom, where, and under what conditions* an intervention may work best *a priori*, rather than post hoc (see Gelman 2015; Rothman and Greenland 2005). In many cases, these potential moderators are not the usual demographics. Demographics are not always the most relevant to what enhances or precipitates a treatment effect. Instead, theoretically relevant moderators are often rich contextual factors, or a person's behavioral history. Sometimes measures of these moderators may need to be developed by combining existing measures from multiple sources of population data because direct measures are not available. These measures can then be used to stratify the population. Ultimately, the allocation of the sample to these strata can be based on power analyses for not only the average treatment effect but also subgroup impacts and comparisons between these subgroups.

We illustrate our proposed approach using an empirical case study of a survey-administered behavioral science intervention: The US National Study of Learning Mindsets (NSLM). This experiment evaluates the effects of an Internet-based, self-administered intervention on achievement-oriented behavior over time in a national probability sample of 76 US public high schools. The manipulation is delivered via survey, and the outcome data are collected from official school registrars at the end of the school year. By using a novel stratification method – which we show reduces confounding between two theoretically developed contextual factors and improves subgroup statistical power – it will be possible to test not only if the intervention has an effect on high school achievement on *average* but also if it has differential effects across school contexts. Answers to such questions directly inform causal, mechanistic theory. Although this case study has many unique features – like the use of high school students, the availability of rich auxiliary information for strata construction, and assessment of behavioral outcomes – the same basic logic could apply to more traditional survey experiments.

22.1.1 Probability Samples Facilitate Estimation of Average Treatment Effects

We start with a review of basic causal inference for the estimation of the average treatment impact in a population. We then expand to address treatment heterogeneity. As has been well documented (e.g. Morgan and Winship 2014; Shadish et al. 2001), as a result of random assignment to treatment, the baseline characteristics of the experimental and control conditions are equivalent, in expectation. Thus, in expectation, any differences in the outcomes of the two groups can be attributed to the treatment. This difference in average group outcomes in the sample is called the *sample average treatment effect* (SATE) and can be formally represented for units $i = 1, \ldots, n$ in the sample as the expected outcome Y when a treatment is present (1) minus the expected outcome when it is not (0), where the subscript S indicates that this is the average over the sample:

$$\text{SATE} = E_S[Y_i(1)] - E_S[Y_i(0)] \tag{22.1}$$

The estimate of the average treatment effect produced in an experiment is unbiased for the SATE (Holland 1986). Put another way, as a result of randomization to treatment, in a well-designed experiment with no attrition, there is no *treatment selection bias*.

Importantly, if treatment effects vary across units, this SATE may differ across different samples and populations (Imai et al. 2008; Olsen et al. 2013; Weiss et al. 2014). For this reason, researchers have increasingly conducted these types of experiments in national probability samples, since doing so allows for the direct estimation of the average treatment effect in the population (i.e. PATE; see Mutz 2011). The PATE has great value. For instance, policymakers may wish to implement a public health message for an entire nation, and they may want to know whether, on average, the message may improve people's health behaviors.

We can define the PATE formally based on units $i = 1, \ldots, N$ in an inference population with the subscript P indicating average over the population, as:

$$\text{PATE} = E_P[Y_i(1)] - E_P[Y_i(0)] \tag{22.2}$$

Unlike the SATE, estimation of the PATE based on a randomized experiment is only unbiased when in addition to randomization of treatment, the sample is selected from the inference population randomly and nonresponse is ignorable (i.e. uncorrelated with any source of treatment heterogeneity; Olsen et al. 2013; Tipton 2013). When probability sampling is not used, the estimate of the PATE may be biased as a result of *sample selection bias* (Allcott 2015; Tipton 2013).

Recently, researchers in education, medicine, and economics have become increasingly concerned about sample selection bias and its effect on ATEs (e.g. Allcott 2015; Cole and Stuart 2010; Tipton 2013). This stems from the fact that it is rare in the social and behavioral sciences for researchers to be certain that treatments have identical effects across all individuals in a population (e.g. Gelman et al. 2015; Green and Kern 2012; Rothman and Greenland 2005; Vanderweele 2015). As Gelman et al. (2015) state, it is "better to start with the admission of variation in the [treatment] effect and go from there" (p. 637). Indeed, effects vary in relation to implementation quality (e.g. Weiss et al. 2017), participant characteristics (e.g. gender or minority status in psychology experiments), and also contextual factors (Allcott 2015; Hulleman and Cordray 2009; Weiss et al. 2014).

One way to understand variation in treatment effects across subgroups in the estimation of PATEs is to decompose the overall effect into subgroup ATEs:

$$\text{PATE} = \text{CATE}_1\pi_1 + \text{CATE}_2\pi_2 + \cdots + \text{CATE}_H\pi_H = \sum \text{CATE}_h\pi_h \tag{22.3}$$

Here CATE_h is the conditional (i.e. subgroup) ATE for group $h = 1, \ldots, \text{H}$, and π_h is the proportion of the population in subgroup h (where $\sum \pi_h = 1$). The PATE is then the weighted sum of these subgroup ATEs.

Decomposing the PATE in this way (Eq. (22.3)) highlights two facts pertinent to survey sampling. First, when the CATE does not vary across subgroups – for instance, when investigating a basic cognitive or biological process that works in the same manner for all humans – then any sample's estimate of the ATE will match the PATE, within sampling error. In such cases, probability sampling is not needed. Second, if the CATE *does* vary across subgroups, but the achieved sample has the same proportions of individuals who are in each subgroup as in the population, then the ATE will match the population ATE, within sampling error (Olsen et al. 2013). A properly constructed and weighted probability sample with ignorable nonresponse ensures the latter, resulting in an unbiased estimate of the PATE, within sampling error.

If, however, a probability sample is not employed, improperly constructed, or has nonignorable nonresponse, then sample proportions in each of the H subgroups may differ from the population proportions (the π_h). When treatment impacts vary, then the resulting estimate based on the sample (the SATE) may be biased for the ATE for the population (the PATE).

In the last five years, this problem has received increased attention in the statistical and methodological literature, resulting in a new set of tools for making post hoc adjustments to the average treatment impact estimator (e.g. Stuart et al. 2011; Tipton 2013, 2014). In general, however, probability samples are preferred over nonprobability samples for making generalizations about effect sizes when there is heterogeneity (as we expect there nearly always is in typical social and behavioral research).

22.1.2 Treatment Effect Heterogeneity in Experiments

We next briefly review current approaches to studying this heterogeneity in the social and behavioral sciences. Identifying subgroups has tremendous practical utility. Policymakers who know for whom, and under what conditions, a treatment is effective can target a treatment to subgroups where it may be most useful, and avoid delivering it in settings where it may be harmful (Cook 1993; Solon et al. 2015). In addition, moderator analyses – which make comparisons between treatment impacts in subgroups – are critical for theory development because they point to causal mechanisms.

Consider several prominent experimental manipulations that showed different treatment effects across subgroups:

- a manipulation of questionnaire wording is weaker or stronger for individuals with high versus low levels of education (a means of testing satisficing theory; e.g. Krosnick 1999; Narayan and Krosnick 1996);
- A "nudge" to motivate people to decrease energy use is more effective in households with large square footage or swimming pools and less effective in households where even motivated individuals cannot decrease energy use (Allcott 2015);
- different policy framings are more or less compelling to members of different political parties or ideologies (Entman 2010); and
- motivational interventions delivered via Internet surveys change achievement-oriented behavior more or less for low-achieving students and/or racial/ethnic minority students (Paunesku et al. 2015; Yeager et al. 2016a).

Testifying to the importance of these moderator analyses for basic theoretical advances, our reanalysis of a random sample of psychology experiments published in premier journals in 2008 showed that 20% of all studies interacted a measured individual difference moderator with an experimental treatment or task (Open Science Collaboration 2015).

Despite this interest in moderators and treatment heterogeneity, social and behavioral scientists have lamented the state of subgroup analyses in experimental research for decades. For instance, Green and Kern (2012) state, "in practice, the investigation of treatment effect heterogeneity in survey experiments often seems ad hoc and unstructured." Such unstructured analysis can cause problems for cumulative science. For example, authors have noted the large number of cases in which moderation of a treatment effect by a demographic characteristic did not replicate upon further examination (Gelman 2015; Gelman et al. 2015; Rothman and Greenland 2005; Vanderweele 2015). In a famous clinical example, researchers discovered unexpected (and eventually unreplicated) moderation by gender for effects of aspirin in preventing stroke death, leading doctors to withhold aspirin from women for over a decade (see Rothman et al. 2005).

More broadly, Nosek and colleagues (Open Science Collaboration 2015) attempted to replicate 100 randomly sampled psychology experiments. In this analysis, only 1 out of 20 studies that examined differences in responsiveness to an experimental treatment across measured individual characteristics (e.g. race, gender, IQ) replicated a statistically significant interaction. By contrast, 46% of all other kinds of studies replicated a significant difference ($X^2(1) = 8.84$, $p = 0.003$).

In response to such pessimism, a growing amount of research has proposed novel analytic techniques for interrogating treatment effect heterogeneity, including regression trees (Imai and Strauss 2011), Bayesian additive regression trees (BART; Green and Kern 2012), generalized additive models (GAM; Feller and Holmes 2009), and semi-parametric differencing estimators (Horiuchi et al. 2007). These approaches are all carried out during the analysis phase, however, and do not require a strong theory when selecting potential moderators; instead, as data-driven approaches, they test nearly all possible moderators, with the focus on finding the best predictive model.

As statisticians consistently note, the key to strong inferences is not the model, but "design, design, design" (Bloom 2010). Surprisingly, almost no research has explained the implications of concerns about treatment effect heterogeneity tests for the *design* of survey samples or experiments from the start (though see Tipton 2014). We propose a design-oriented approach that is *theory*-driven, requiring researchers to design new studies based on hypotheses generated from prior work, thus resulting in a more cumulative model for science.

22.1.3 Estimation of Subgroup Treatment Effects

More formally, returning to Eq. (22.2), researchers may wish to estimate the conditional average treatment effect, or CATE, for subgroup 1 as well as the CATE for subgroup 2, and so on. They may then be interested in determining whether the differences between treatment effects in those subgroups is nonzero. Without planning carefully, estimates of subgroup impacts and comparisons between them can be both biased and imprecise.

22.1.3.1 Problems with Sample Selection Bias

One might imagine that as long as a sample had a sufficient number of individuals from each subgroup, then it would be possible to obtain a generalizable CATE effect size *for that subgroup*. That is, if there were no heterogeneity in effect sizes within the subgroups – if men are men, women are women, whites are whites – then it should not make a difference whether the subgroup in a sample was recruited through probability methods or not. This position has long been taken in psychological research. That is, researchers have compared arbitrary samples of subgroups in their responsiveness to experimental manipulations: western college students vs. East-Asian college students (Markus and Kitayama 2010), white college students vs. African-American college students (Steele and Aronson 1995), low-income people versus high income people (Mullainathan and Shafir 2013), and so on. This approach has become standard in political science as well, particularly as this field has migrated toward the use of national nonprobability samples for research on differential effects of messages within Republican and Democratic subgroups (Green and Kern 2012).

According to this logic, as long as one could obtain enough representatives from a given subgroup, through whatever means, then it is possible to estimate that subgroup's treatment effect. Alas, there can be significant heterogeneity in effect sizes *within* demographic subgroups and therefore methods for acquiring a subgroup can produce bias.

This can be seen theoretically by extending Eq. (22.2) to focus instead on a particular CATE for subgroup h,

$$\text{CATE}_h = \left[\text{CCATE}_{h1}\pi_{h1} + \text{CCATE}_{h2}\pi_{h2} + \cdots + \text{CCATE}_{hK}\pi_{hK} = \sum \text{CCATE}_{hk}\pi_{hk}\right] / \pi_h \tag{22.4}$$

where the subgroup CATE (e.g. for women) can itself be decomposed into $k = 1, \ldots, K$ subgroup ATEs (what we call conditional–conditional ATEs, CCATEs), with the proportion of the population in these doubly defined subgroups equal to π_{hk} (where now $\sum\pi_{hk} = \pi_h$). Equation (22.4) shows that just as when estimating the PATE (22.2), bias can result if the population and sample proportions (the π_{hk}) are not identical and treatment impacts are not constant within the defined subgroup. Thus, in general, simply having enough representatives *from* a given demographic subgroup does not ensure that a sample is representative *of* the demographic subgroup – if there is any heterogeneity in treatment impact within the group. This is one reason that probability sampling is important for experiments, since when implemented well (and with ignorable nonresponse), this bias is removed, in expectation.

22.1.3.2 Problems with Small Sample Sizes

Putting aside the issue of sample selection bias, the second issue in estimating subgroup ATEs has to do with small sample sizes. This concern arises regardless of the method used to recruit the sample (probability or nonprobability). Given recent innovations in statistical methodology regarding the detection of treatment effect heterogeneity (e.g. regression trees, Feller and Holmes 2009; Green and Kern 2012; Horiuchi et al. 2007; Imai and Strauss 2011) and the (often large) sample sizes often found in survey experiments, it is easy to overlook the importance of adequate subgroup sample sizes when understanding possible sources of treatment effect heterogeneity.

This problem becomes particularly likely as the number of subgroups studied (H) increases, thus resulting in smaller and smaller sample sizes in each group. For example, imagine if subgroup effects are desired not only in terms of racial or ethnic minority status but also by education level. This would mean estimating the subgroup CATE for African Americans with a four-year degree or higher; in the 2016 Current Population Survey this group accounted for only 2% of the US population,[3] or just 20 respondents in a typical 1000 person phone survey.

Small subgroup sample sizes can lead to two problems. First, small sample sizes typically lead to low statistical power, thus making it likely that one will erroneously conclude that a treatment was not effective for a subgroup (a Type II error). This is particularly likely given that most experiments are powered only for estimation of the SATE, not subgroup effects (Rothman and Greenland 2005). For instance, in the example above, the minimal detectable effect for highly educated African-Americans, with 80% power, would be $d = 1.16$, a massive effect as far as survey manipulations go. In practice, this lack of a statistically significant effect within a subgroup can be mistaken for ineffectiveness.

Second, somewhat ironically, small sample sizes – and the resulting low power – might lead researchers to search for subgroup effects across a variety of potential groupings, thus producing statistically significant but nonreplicable results (a Type I error) (Gelman 2015). This is the multiple comparisons problem (Gelman et al. 2015; Green 1999; Rothman and Greenland 2005; Simmons et al. 2011). A famous example of this is the ISIS-2 trial (testing aspirin vs. placebo in acute myocardial infarction). Aspirin was effective for patients born under the signs of Libra and Gemini, but not others (see Rothman and Greenland 2005). Results such as these could reasonably be obtained via post hoc flexibility in data analysis, or *p-hacking* (Simmons et al. 2011). As a result, even a probability sample can produce nonreplicable findings if the sample did not afford adequate power for a subgroup and therefore invited *p*-hacking (Gelman 2015).

The design-oriented approach that we develop in this chapter reduces these problems by requiring researchers to articulate, from the very beginning, potential subgroups for whom estimating a CATE might be of interest. Information on the distribution of the related variables (e.g. education level, racial composition) is then gathered in the population, and power analyses are conducted based upon this information. For example, if the sample will include 1000 individuals, but a CATE estimate is desired for a subgroup accounting for only 10% of this population, a power analysis would be conducted based on a sample of size 100 (=1000*0.10). In many cases, this will lead to a need for oversampling (particularly with rare subgroups). Importantly, as we will show in an example, this means that when the PATE estimate is also desired from the same sample, a poststratification adjustment will be necessary (otherwise, the simple, sample-based estimate will be biased).

22.1.4 Moderator Analyses for Understanding Causal Mechanisms

As we have stated, the goal of mechanistic, theory-driven science is to not only understand *if* an intervention changes outcomes for particular subgroups but also *why* or *how* (Cook 1993). If we begin from this position, the detection of subgroup and moderator effects becomes not simply *descriptive* – as occurs when researchers report results by demographic subgroups without theory, in an actuarial sense – but explanatory, leading to a theory of the *causal mechanism*.

For example, one may conclude that an intervention has larger effects for women than for men – thus eliciting the question: Why? Is it something inherent about gender that leads

3 Analyses conducted at http://www.census.gov/cps/data/cpstablecreator.html.

to these differential effects? This is where the benefits of identifying possible moderators in advance are particularly large, since very often the focus is on contextual and social factors that may be more theoretically relevant to the mechanism of the intervention. In this section, we address two concerns that occur when testing these moderator effects – confounder bias and statistical power.

22.1.4.1 Problems of Confounder Bias in Probability Samples

In a probability sample, the estimate of a subgroup CATE is unbiased. However, just because the treatment impacts are found to vary in relation to a particular measured variable (e.g. education level) does not mean that the moderator *causes* treatment effects to differ.

Consider that demographic characteristics are not, in and of themselves, causes of moderated treatment effects. Instead, they are proxies for unobserved material, social, or psychological realities that titrate a person's responsiveness to a given experimental manipulation (Vanderweele 2015). Recall the example of moderation of questionnaire design effects by educational attainment (Narayan and Krosnick 1996). Researchers did not have a theory that a *diploma* causes moderation of a questionnaire manipulation; instead, educational attainment covaries with qualities such as IQ, need for cognition, social class, openness to experience, and more, which can make people pay greater or less attention to novel information (Narayan and Krosnick 1996). Likewise, when assessing moderation by race and ethnicity, researchers rarely start an analysis with a theory about inherent qualities of skin pigment or of self-labels; instead, researchers may expect that race covaries with access to resources or experiences of discrimination or stereotyping over the lifespan (Spencer 2006), for instance those that affect trust in institutions (Smith 2010) or performance in school (Steele 1997).

The issues of confounding we are raising here have long been discussed and acknowledged in observational or quasi-experimental studies of the "X causes Y" sort (Morgan and Winship 2014; Shadish et al. 2001). Researchers would not say that education causes longer life solely on the basis of a correlation between the two. Instead, researchers routinely attempt to control or adjust for baseline differences between the groups they wish to compare, so as to isolate the effect of a category or status.

We argue that researchers need to address these same problems of confounding in the "M causes the effect of X on Y to be larger (or smaller)" case. This means not only defining a handful of relevant subgroups but also the development of a theory of treatment effect heterogeneity – that is a mechanism that potentially *causes* impacts to vary – and the development of statistical and design-oriented approaches to directly address concerns with identifying that mechanism.

22.1.4.2 Additional Confounder Bias Problems with Nonprobability Samples

In nonprobability samples, concerns with confounder bias in moderator analyses are even larger. Just as the estimates of subgroup effects can be biased in nonrandom samples, so too are *differences* in subgroup effects. This is because the correlation between a demographic characteristic and an unobserved moderator may vary across the population, and thus may differ in a nonrandom sample compared to in a larger population.

Recent audits have investigated this possibility, comparing the covariance between demographic variables and unobserved characteristics in seven nonprobability samples as compared to two probability samples and government benchmarks (Yeager et al. 2011). Yeager et al. (2011) showed that the probability samples were more demographically representative than the nonprobability samples and were more accurate when estimating nondemographic characteristics. Critically, when weights addressing demographic imbalances were applied, the accuracy on the variables not used in weighting was increased in the probability samples.

By contrast, in the nonprobability samples, accuracy on nonweighted characteristics did *not* improve with weighting; in some cases, accuracy was worse (Yeager et al. 2011). This directly illustrates that even well-constructed and properly weighted *non*probability samples do not necessarily accurately represent the latent, unobserved characteristics that typically covary with measurable demographics in a population (for another perspective, see Chapter 21). Hence, using data from large nonprobability surveys does not, in our view, yet address the issues of confounding and generalizability that we raise here.

22.1.4.3 Problems with Statistical Power

Until now, we have focused on issues of bias when *attributing* tests of moderation in the development of a theory of the causal mechanism. As with subgroup impact estimation, however, with moderator analyses another issue arises: statistical power. In moderation analyses, this concern with power is even larger than it is when estimating subgroup CATEs, since the resulting test involves comparisons between *estimated* quantities. Recall, for example, that in the simplest case,[4]

$$V(\text{CATE}_1 - \text{CATE}_2) = V(\text{CATE}_1) + V(\text{CATE}_2) \tag{22.5}$$

Equation (22.5) illustrates that the variance of a difference between subgroups is – in the best case – twice as large as the variance of a subgroup-specific CATE.

If the goal is to minimize the variance of this difference for a fixed total sample size (n) – thus increasing power – a well-known result (see Raudenbush and Liu 2000; Bloom and Spybrook 2017) is that the best strategy is to divide the sample evenly across these subgroups (i.e. $n/2$ to estimate CATE_1, $n/2$ to estimate CATE_2). If multiple comparisons are desired – as when all pair-wise subgroup comparisons are of interest – the end result is a strategy in which not only are some subgroups oversampled, but furthermore the sample size in each of the subgroups is the same. As noted previously, this uniform sampling across subgroups biases the estimate of the population ATE, which means that poststratification weighting will be required.

To see why poststratification weighting may be problematic, however, let us now examine the variance of such an estimator,[5]

$$V\left(\sum \pi_h \text{CATE}_h\right) = \sum \pi_h^{\,2} V(\text{CATE}_h) \approx \sum (\pi_h/p_h)\pi_h^{\,*} V(\text{PATE})$$

where π_h is the proportion of the population in subgroup h, p_h is the proportion of the sample, $V(\text{PATE})$ is the variance of the PATE estimate if no poststratification is required (i.e. $p_h = \pi_h$), and where the last approximation is due to Tipton (2013). In this estimator, the total sampling variance is thus a function of the ratio of population and sample proportions, leading to a larger variance whenever an allocation scheme other than proportional allocation is implemented. This particularly occurs if there are very small subgroups in the population (i.e. small π_h), but equal allocation is used across subgroups to ensure power for moderator tests (i.e. $p_h = 1/H$).

The end result is that power for detecting differences in subgroup treatment effects (i.e. having a significant interaction comparing two CATEs) is lower than for detecting a simple effect of the treatment within the subgroup itself (estimating a single CATE). Thus, the optimal strategy for powering for the detection of moderation can be far from optimal for estimating the population ATE. We return to this tension in the next section and in an example from a national experimental study.

---

4 Certainly, complex sampling will result in an additional covariance term here, due to possible covariances between the subgroups introduced by cluster sampling and/or combining PSUs for variance estimation purposes.

5 This variance estimator assumes the subgroups are independent. In more complex samples involving cluster sampling, they will not be and the variance formula will need to take these covariances into account. We focus on the simplest case to highlight the roles that p_h and π_h play.

22.1.5 Stratification for Studying Heterogeneity

In this section, we outline a general approach to sample design for survey experiments interested in estimating three parameters: the average treatment impact in the population (PATE), subgroup treatment effects (CATEs), and predefined differences in treatment impacts across subgroups. Throughout we focus on designs in which observations are nested within clusters of some type, with a focus on treatment effect heterogeneity defined at the cluster level. In the experiment used as a motivating example in this chapter, these clusters are schools. In other survey experiments, clusters might be regions, villages, counties, demographic subgroups, or firms.

22.1.5.1 Overview

The approach we develop builds off one proposed by Tipton (2014) and Tipton et al. (2014) for estimation of the average treatment impact in field experiments in education research. The approach is a version of stratified sampling, though the definition of the strata deviates from the traditional approach found in surveys.

To begin, an inference population must be clearly defined – for example, it might include all high schools in the United States – and information on these clusters in the population must be available. For example, in the education context, this information can come from the Common Core of Data and include information on school-level demographics. When participants are drawn from an existing probability sample panel, such as the GfK knowledge panel or the NORC Amerispeak panel, this may include information on person-level demographics, cognitive abilities, or sociodemographic questions answered on a profile survey, that can be aggregated to important subgroups (e.g. political party membership). When one's goal is to conduct experiments with adults who are not yet part of a panel, then in the future it may be possible to obtain richer microdata on people's past behavior or characteristics (for instance, whether they voted in past elections, and what their party identification was, from the validated voter file) and then invite people to participate within strata.

Strata are commonly used in surveys to reduce residual variation when the measures of interest vary substantially across strata. In most surveys, these measures of interest are the *outcomes*. For example, in education experiments, strata are typically defined by the percent of students in free or reduced lunch, a measure of socioeconomic status, which predicts schools' test scores (Battle and Lewis 2002). In this usage, strata are employed only for reducing variation in the sample, and, while taken into account in the variance calculation, are not the focus of the analysis.

In contrast to the typical approach, we argue that the covariates that matter for stratification in survey experiments are instead those that explain *variation* in *treatment impacts* (Tipton 2014). Since these cannot be known *a priori*, researchers can identify possible treatment effect moderators based upon both previous empirical research and a theoretical understanding of the mechanism of the intervention. In terms of practicality, however, not all potential moderators can be included, since these must be available in existing population level data. For example, demographic variables are often available, whereas variables related to attitudes and beliefs are not. Once defined, these variables are then used to stratify the population.

When there are multiple variables, stratifying is more difficult, and is limited by the total number of experimental units; for example, a study including 100 sites could have at most 100 strata (and likely, many fewer). Tipton (2014) shows that one approach is to reduce the dimensionality to a single variable through the use of k-means cluster analysis.[6] In this approach, k is the total

6 Tipton (2014) shows that k-means can be used with both continuous and categorical variables through choice of the distance metric.

number of strata that study resources can accommodate, and the k-means procedure results in k clusters (strata) that are maximally heterogeneous. Tipton et al. (2014) show that another approach, propensity scoring, can also be used and is particularly effective in "synthetic" generalizations. These are generalizations in which the population to whom researchers want to generalize cannot take part in the study; for example, this arises when an effect is desired for the population of schools currently using a program.

In this previous research by Tipton (2014) and Tipton et al. (2014), the goal of stratification is to gather a sample that is compositionally similar to the inference population so that the SATE estimated in the study is an unbiased estimate of the PATE. In this chapter, however, we shift focus to an alternative approach to stratification that adds a concern with *treatment contrasts*.

22.1.5.2 Operationalizing Moderators

After researchers have identified a set of potential moderators, a next question is how to operationalize these. For example, socioeconomic status may be of interest, in which case researchers must develop a measure that operationalizes this construct. This might involve the use of a *latent variable* approach, pooling data from several sources, for instance, through structural equation modeling. These sources might include multiple variables from the same data source, as well as variables gathered from several different sources of administrative data. For example, in a study evaluating a social welfare intervention, potential moderators were drawn from survey data from the Bureau of Labor Statistics, the American Community Survey, and the National Association of Counties (Tipton and Peck, 2017). In the context of the NSLM, this included data on school quality collected on a website aimed at parents (GreatSchools .org), as well as data on PSAT scores and AP course enrollments from the College Board, and state National Assessment of Education Progress (NAEP) scores. This approach is useful when the moderators of interest are contextual variables available in administrative data (e.g. from the government or existing surveys). This approach can also be useful when participants are members of panels of respondents who have completed profile surveys or other prior assessments (such as the participants available via TESS proposals).

22.1.5.3 Orthogonalizing Moderators

A second concern is the degree to which the hypothesized moderators covary, or, put another way, to what degree the effect of one moderator is *confounded* by another possible moderator. For example, in school data, the proportion of minority students in a school is highly correlated with the academic rigor and achievement level of a school. If the population is *not* stratified on these variables, then it is likely that in the resulting sample they will be so highly correlated as to make it difficult or impossible to test hypotheses regarding them separately. Being able to distinguish between these potential causes is important, since it allows for greater understanding of the causal mechanism. To orthogonalize moderators, it is helpful to have adequate sample not only from the marginal distribution cells but also from the off-diagonals.

22.1.5.4 Stratum Creation and Allocation

Population units should then be divided into strata based upon this set of moderators. The next step is to allocate the population to these H strata, noting which proportion of the population is found in each stratum (i.e. π_h, $h = 1, \ldots, H$). An important question in all stratified sample designs is how to allocate the n units in the sample to these H strata (i.e. the $p_h = n_h/n$).

If the goal is only to estimate the average treatment impact, the optimal strategy is to allocate the sample proportionally (i.e. $p_h = \pi_h$). This strategy, however, may not be optimal for testing contrasts between the treatment impacts in different strata. This approach is particularly problematic when the sample size n is small to moderate and some of the strata being tested contain only a very small proportion of the population. In fact, as shown above, if pairwise contrasts are desired between all strata, the optimal strategy for moderator analyses is to instead divide the sample evenly across strata (i.e. $p_h = 1/H$).

In practice, the competing goals of estimating the population ATE and testing comparisons between strata must be balanced. If the allocation is not proportional (i.e. $p_h \neq \pi_h$), however, this means that in order to estimate the average treatment impact, a reweighting approach will be needed (see Eq. (22.5)). While the technical details are beyond the scope of this chapter, in order to determine the "ideal" stratum allocation, these three separate parameters and their competing demands for power must be taken into account. In practice, this often means prioritizing hypotheses into "confirmatory and "exploratory," giving greater preference for allocation schemes for "confirmatory" analyses than others (Open Science Framework 2016).

22.2 Nesting a Randomized Treatment in a National Probability Sample: The NSLM

To illustrate the issues addressed in this chapter, we turn to an example based upon the NSLM study. So-called "mindset interventions" (e.g. Aronson et al. 2002; Paunesku et al. 2015) teach students new beliefs that can increase their motivation to learn – in particular, they teach that "smartness" is not a fixed quantity, but can be developed with effort over time (Dweck 2006; Yeager and Dweck 2012). Psychologists have advocated for their broader use in policy and practice (e.g. Rattan et al. 2015). The goal of the NSLM is to determine if brief mindset interventions can increase student motivation and improve learning outcomes for high school students throughout the US, especially during the difficult transition to high school (Yeager et al. 2016a). The procedures described here were validated in a successful pilot intervention conducted in a convenience sample, which raised low-achieving students' grades (Yeager et al. 2016a).

The NSLM provides a motivating example for this chapter for three reasons. First, questions regarding the generalizability of mindset interventions have received recent policy interest (e.g. see Executive Order No. 13707, 2015). Second, the NSLM is one of only a handful of educational and social welfare experiments to include both probability sampling and random assignment (see Olsen et al. 2013). Third, and most critically, the NSLM illustrates well the theoretical points raised in this chapter so far: why a convenience sample is inadequate for estimating treatment effects, why typical probability samples are inadequate, why typical moderation analyses are confounded, and how a theory of heterogeneity can inform better design that addresses these issues.

22.2.1 Design of the NSLM

22.2.1.1 Population Frame

The inference population was based on a school sampling frame created by ICF International that included both information from the Common Core of Data (CCD), a file of public schools obtained from the National Center for Education Statistics (NCES), and data from Market Data Retrieval (MDR). The MDR data files augment and update CCD data for public schools. This inference population was defined to include only schools serving grades 9–12 (e.g. excluding $K-12$ schools), since a focus was on the effect of a program for students transitioning to a new school environment. Only public "regular" schools were included, thus excluding private, charter, Bureau of Indian Affairs, Department of Defense, and other exceptional schools. This resulted in a frame that included over 12 000 schools.

22.2.1.2 Two-Stage Design

The sample included two stages of selection. First, more than 12 000 schools were divided into primary sampling units (PSUs). In some cases, these aligned with school district boundaries, while in other cases, they involved combining smaller districts. This created 4693 PSUs. Methods for probability proportional to size (PPS) sampling were then employed to first select PSUs and then schools. This was done to reduce the cost of recruitment (which was done face-to-face). Within most schools, a census of students was recruited. In some schools, where computer lab availability was limited, a random sample of students was recruited.

22.2.1.3 Nonresponse

While the intervention in the NSLM is brief, it requires school resources, including working computers and the ability of a school to get all ninth-grade students through the computer lab in adequate time, and the evaluation requires extraction of sensitive student records. In anticipation of school-level nonresponse, ICF selected a stratified random sample of 139 high schools, with the expectation that more than 70 would agree to participate in the study; in the end, 76 agreed to participate.

22.2.1.4 Experimental Design

The experiment begins at the beginning of high school, in the fall semester, and is delivered through a computer program. Within each of the schools, ninth-grade students were randomized to the treatment or control conditions. Randomization was conducted within the software itself. This person-level random assignment also allowed for separate treatment impacts to be estimated in each school.

22.2.1.5 Treatment

Students randomized to the treatment received a brief mindset intervention (less than two hours) (Yeager et al. 2016a); students randomized to the control condition received a brief, comparable nonmindset program focused on the transition to high schools (also less than two hours). The study involves two doses of the program, with at least two weeks between doses. "Intent-to-treat" analyses are conducted, which means that students are included in the analytic sample regardless of whether they received the second dose (see Bloom 1984).

22.2.1.6 Outcomes

Outcomes are survey responses regarding motivation (assessed during the second computerized session), as well as outcomes from administrative records gathered at the end of the academic year, such as average grade point average and proportion earning D or F averages in core classes.

22.2.2 Developing a Theory of Treatment Effect Heterogeneity

22.2.2.1 Goals

In addition to estimating a population ATE, the goal of the NSLM from the outset was to determine what kinds of schools show weaker or stronger treatment effects. This focus on treatment effect heterogeneity led to the considerations and design developed in this chapter, with a particular concern on issues of confounding in moderator analyses.

22.2.2.2 Potential Moderators

In order to address concerns with confounder bias in the moderator analyses, a list of possible moderators was developed at the study outset and, based upon theory, two moderators were selected: school achievement and school minority composition. More specifically, these were selected based upon the theory that a mindset message should boost motivation and achievement when (i) the mindset message counteracts some other message that is suppressing motivation and when (ii) students attend schools where motivation matters for learning. Hence, the mindset intervention, which teaches that "smartness" is not fixed but can be developed, might be more effective (i) for students whose intelligence may be impugned by negative stereotypes (Steele 2011), which in the US is often students of color (specifically black or Hispanic/Latino students) and (ii) in schools that reward at least some level of motivation (i.e. not the worst schools) but where motivation is not already maximal (i.e. not the best schools).

22.2.2.3 Operationalizing School Achievement Level

As noted, intervention effects might vary in relation to school achievement level, including features of the students (e.g. test scores, motivation), teachers, and schools (e.g. offering AP classes, quality). Information of this type, however, was not readily available in the CCD, which ICF used to create the sample selection plan. This meant that the team needed to develop such a measure.

The creation of a school achievement variable involved pooling data from several outside sources. The first of these sources was GreatSchools.org, which provides within-state information on schools (and rankings from 0 to 10) based upon school test scores, as well as other features (e.g. parent and community ratings, absenteeism, programs offered). The second source was high school average PSAT scores and Calculus AB and English (Literature and Language) AP participation rates, which were provided from the College Board. The third source was state proficiency levels for math and reading for eighth grade (gathered from the NAEP). Notably, while PSAT and NAEP scores all identify features of student achievement, the offering of AP courses speaks instead to decisions made at the institutional level in schools.

With these three data sources combined, a structural equation model was used to estimate a latent "school achievement" variable. The model and factor loadings are given in Figure 22.1

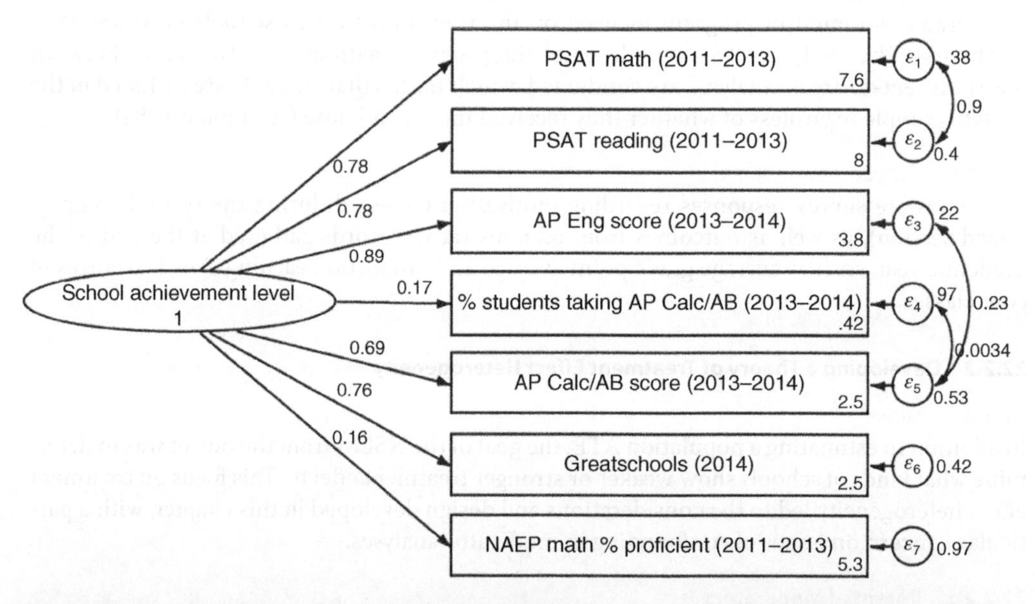

Figure 22.1 A latent variable model of school-level achievement, estimated from various indicators of test scores, advanced achievement, and community ratings. Measures coming from the same data source were allowed to be correlated to adjust for shared method variance and to improve model fit; RMSEA = 0.10 [0.095, 0.105]. The final measure included NAEP scores despite low factor loadings in order to provide a valid between-state test score adjustment.

below. Based on this variable and the associated loadings, a "school achievement" value was estimated for each school; this variable was then standardized. Finally, the school achievement variable was divided into strata. Based upon a theory that the treatment impact would be largest in "average" schools, three strata were created based on the 25th and 75th percentiles.

22.2.2.4 Orthogonalizing for Minority Composition

The obvious empirical challenge is that school minority composition is confounded with school average achievement. See Figure 22.2 that visually depicts this. The scatterplot in Figure 22.2a shows the sampling frame of more than 12 000 regular US public schools in 2014, and association between the % minority students (Black or Hispanic/Latino), on the y-axis, and, on the x-axis, the average school achievement level (variable described below in Figure 22.2; $r = -0.67$). A solid line in Figure 22.2a depicts the loess smoothing curve for the bivariate relation. Note the paucity of "off-diagonal" cases; very few schools have below-median minority composition and bottom 25% achievement level (the bottom left cell) and very few schools have above-median minority composition and top 25% achievement level (the top right cell).

Given this concern, minority composition (% black or Hispanic/Latino) was divided into two groups based upon the median value in the population (26%). The strata for minority composition were then crossed with those for school achievement. This resulted in six possible strata; as described in the next section, two strata were combined, resulting in five final strata. Hypothesized CATE results for each of the five strata are depicted in Figure 22.2b.

22.2.3 Developing the Final Design

22.2.3.1 Stratum Creation

In order to develop the final design, hypotheses regarding the CATEs in each stratum were specified. In Figure 22.2b, theorized control and treatment group levels of student achievement – that is, hypothesized CATEs – are provided for each stratum. Notice that, from left to right in Figure 22.2b, the control group rises regardless of minority composition, because higher-achieving schools are expected to have higher achievement. And yet the size of the treatment *contrast* is thought to be largest in the middle range of school achievement, because of our hypothesis that in the lowest-achieving schools in the US motivation is swamped by more basic concerns (such as safety or curricular resources), while for the highest-achieving schools, all students may be nearly maximally motivated.

In Figure 22.2b, when comparing CATEs at the top to those at the bottom, control condition levels are expected to be weaker in high-minority schools due to stereotype threat, which is the worry about contending with negative stereotypes about one's group (Steele 2011). However, because the growth mindset intervention presumably alleviates a portion of the consequences of stereotype threat (Aronson et al. 2002), then the treatment *contrast* is hypothesized to be larger in high-minority schools versus low-minority schools. Adding these two moderation effects together, the largest treatment contrast was expected for high-minority, medium-quality schools (Stratum 3 in Figure 22.2a,b).

Based on this theory, it was decided that two of the strata could be combined – the "high" and "low" minority by "low" context strata – since it was hypothesized that effects would be similar in both and since testing these differences were not a high priority. This resulted in five strata, given in Table 22.1.

22.2.3.2 Stratum Allocation

The first row of Table 22.1 provides the proportion of the population of schools in each of the five strata. Importantly, while these are mostly evenly distributed, the fifth stratum – containing "high" minority "high" context schools – included only a very small

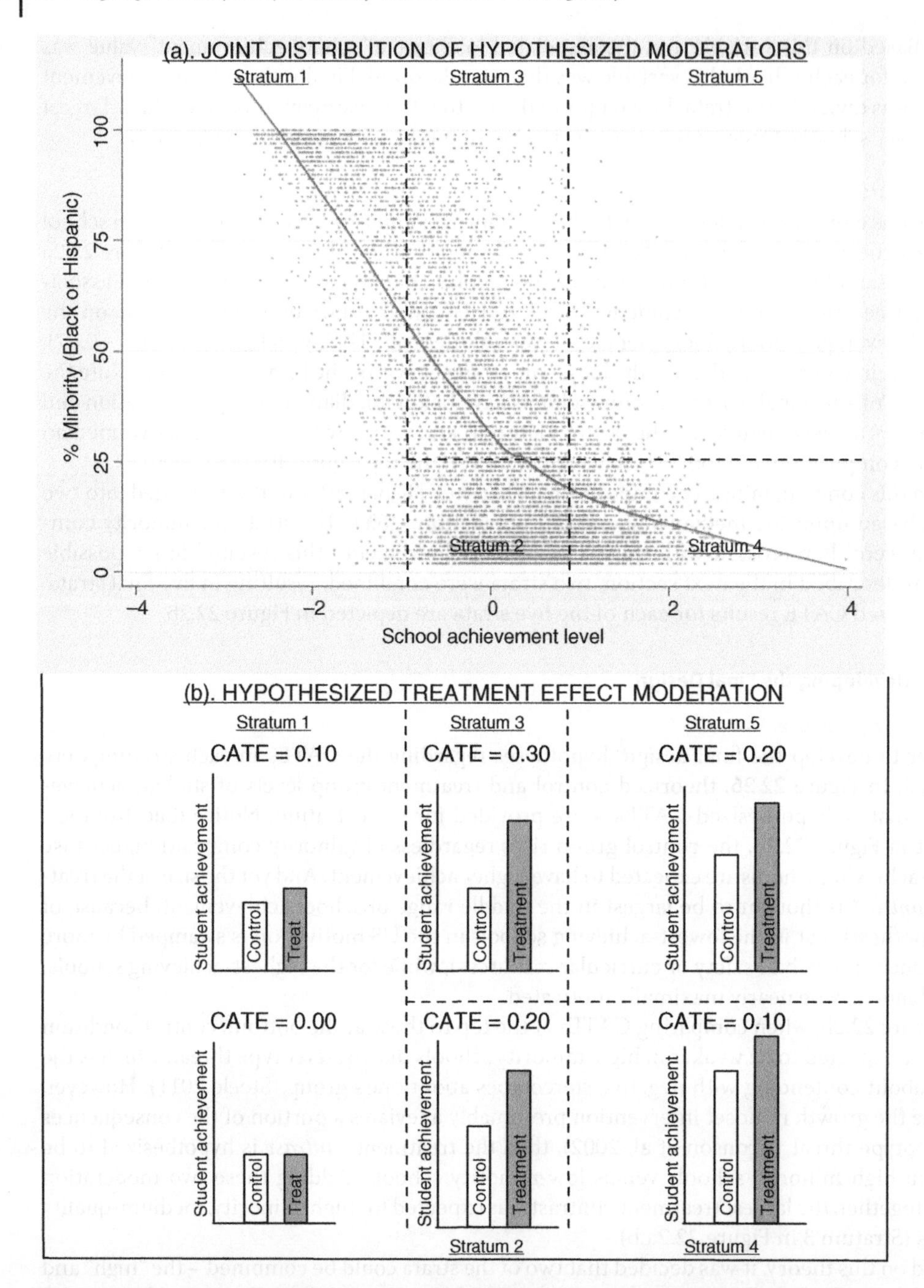

Figure 22.2 Scatterplot showing the confounding between the hypothesized moderators (school achievement and % minority), with strata cutpoints (a) and hypothesized pattern of CATEs, by strata (b).

proportion of the population (approximately 5%; also recall Figure 22.2a). A typical probability sample could easily have included far too few of these schools to reliably test hypotheses.

In the second row of the table, the number of schools (out of 140) is given that would have been included using proportional allocation. Given the goals of pair-wise comparisons between strata, the third row indicates the allocation that would have been best for stratum contrasts

Table 22.1 Strata definitions and allocations in NSLM.

School achievement level Minority %	Low Low/high	Med Low	Med High	High Low	High High
Population of schools (100%)	25%	27%	23%	20%	5%
Proportional allocation ($n = 140$)	35 (25%)	39 (27%)	32 (23%)	28 (20%)	7 (5%)
Optimal for strata contrasts ($n = 140$)	28 (20%)	28 (20%)	28 (20%)	28 (20%)	28 (20%)
Final sample ($n = 139$)	28 (20%)	34 (24%)	32 (23%)	19 (14%)	26 (19%)
Final respondent counts ($n = 76$)	13 (17%)	25 (33%)	18 (24%)	9 (12%)	11 (15%)

Note: The final sample involved 139 schools, not 140, because one school was determined ineligible after drawing the sample.

(equal proportions across strata). Here the differences are largest with respect to the fifth stratum again, which moved from 7 to 28 schools, a fourfold increase. The fourth row indicates the final allocation used for sample selection. This allocation offers a compromise between the previous two, taking into account expected CATEs in each stratum (which were higher, based on theory, in some strata than others).

22.2.3.3 Implementing Sample Selection

Implementing the stratified selection design was not straightforward because the strata were defined based upon *school* characteristics, but the sample selection process involved first the selection of geographic PSUs. The final design included the selection of 70 sample PSUs with PPS at the first stage, using a composite measure of size.[7] At the second stage, 140 sample schools were randomly selected from the PSUs according to the final sample allocation defined in Table 22.1.

22.2.3.4 Final Sample

Sample selection for the study took place over the Summer and Fall of 2016. Overall, 76 schools agreed to take part. The final numbers of schools per stratum can be located in the bottom row of Table 22.1. Even with nonresponse, this increased the proportion of the sample in the High–High stratum (the most rare subgroup) threefold. The analysis of the study results is currently in process.

22.3 Discussion and Conclusions

The NSLM is unique in several ways compared to other survey experiments. However, we argue that the focus on treatment effect heterogeneity and the design-based perspective are important in *all* survey experiments. This is particularly important as behavioral experiments are rapidly increasing in their prevalence in public policy, economics, and political science (Wilson and Juarez 2015). As we conclude this chapter, we therefore focus on five discussion points for researchers designing survey experiments.

7 The measure of size for each PSU was computed such that MOS = F1*M1 + F2*M2 + F3*M3 + F4*M4 + F5*M5, where F(h) is the sampling rate for stratum-h and M(h) is the number of schools for that stratum in the PSU. The sampling rates that best produced the desired allocations were constructed using 200 simulation samples. In the end, the final MOS for a PSU was MOS = 0.01*M1 + 0.0076*M2 + 0.0089*M3 + 0.011*M4 + 0.07*M5, for strata 1–5, respectively.

22.3.1 Estimands

Survey experiments typically focus on estimation of the ATE in a population. However, as we have argued throughout, if treatment impacts do in fact vary, then the average effect is not adequate for answering many questions regarding an intervention. Instead, testing hypotheses regarding this variation offers greater insight into *how* and *why* an intervention changes outcomes.

22.3.2 Contextual Effects Matter

In much of the survey experiment literature, the focus is on *individual* behaviors and treatment heterogeneity related to these features. However, we argue that in these situations – just as in education – context matters and should be studied. Individual behaviors are often affected by social conditions, including local culture, region, and organizational affiliations. Additionally, very often the individual is not the decision maker when implementing an intervention outside a study. For example, in political science studies focused on Get-Out-the-Vote campaigns, resources may be allocated at the local, village, regional, or state level, and as such moderators associated with these contexts matter.

22.3.3 Plan for Moderators

Much of recent innovations regarding treatment effect heterogeneity and generalizability in both experiments conducted in probability samples and convenience samples have focused on post hoc analysis methods. In contrast, the approach developed here – in following recent work on generalization more broadly (e.g. Tipton 2014; Tipton et al. 2014; Tipton and Miller 2015) – asks researchers to begin their study with a discussion of these moderators. There are three benefits to this. First, it is possible that even in large studies, there are small subgroups that while *representing* a very small fraction of the population are important for *testing* hypotheses regarding moderator effects. Second, by identifying these at the beginning of the study, better pretreatment data can be gathered both from other sources, as well as *from the study sample* itself. Third, by shifting to "confirmatory" and "exploratory" hypotheses, problems from multiple testing and data fishing are circumvented. This follows best practice in the design of experiments more generally, particularly in clinical trials, and speaks also to reproducibility issues in social science studies.

22.3.4 Designing Strata

While standard in survey design, strata are typically included with the goal of reducing variance of the estimators; these are typically related to measures of geography or basic demographics. In this chapter, we argued that in survey experiments, strata should be designed instead with respect to variables that potentially explain variation in treatment impacts. Creating these strata can take work – and may require the collection or combination of new sources of data.

Importantly, in some cases, scientific hypotheses are highly correlated, and in these cases, stratifying on multiple covariates can greatly increase the ability to unconfound the effects. In the NSLM example, unplanned moderation analysis of either of racial composition or school achievement would have been confounded in critical ways. Indeed, although racial-composition and school achievement level are positively correlated, the direction of the moderation, at least in the top 75% of schools, is opposite (see Figure 22.2b). Without understanding one or the other, then the two relations might have cancelled out.

22.3.5 Power Concerns

In this chapter, we argue that when designing a survey experiment to study treatment effect heterogeneity, compromises will be required in order to accommodate the multiple estimands of focus. This is because the design that is optimal for one estimand (e.g. the PATE) is not always optimal for another estimand (e.g. moderators). As we showed in the NSLM, with only 76 clusters, moderator analyses would have been greatly underpowered had prior planning not taken these into account in the study design.

In this chapter, we have argued for a new design-based approach to studying treatment effect heterogeneity and causal mechanisms in survey experiments. The focus of this approach is on the development of a sample selection procedure that enables estimation of three parameters – the population average treatment effect, subgroup impacts, and contrasts between subgroups. This approach was developed in relation to a real example – that of the NSLM – that included 76 public high schools throughout the United States. Our hope is that future survey experiments will learn from the NSLM design, leading researchers to develop not only generalizable estimates of average treatment effects but also better theory regarding the mechanisms behind these effects.

Acknowledgments

Writing of this chapter was supported by the Spencer Foundation and, for the second author, by the Raikes Foundation, the William T. Grant Foundation, the National Institute of Child Health and Human Development under award number R01HD084772, and a fellowship from the Center for Advanced Study in the Behavioral Sciences (CASBS). The National Study of Learning Mindsets (PI: Yeager; Co-Is: Dweck, Walton, Crosnoe, Schneider, and Muller) was made possible through methods and data systems created by the Project for Education Research That Scales (PERTS), data collection carried out by ICF International, and received support from the Raikes Foundation, the William T. Grant Foundation, the Spencer Foundation, the Bezos Family Foundation, the Character Lab, the Houston Endowment, Angela Duckworth (personal gift), and the President and Dean of Humanities and Social Sciences at Stanford University.

References

Allcott, H. (2015). Site selection bias in program evaluation. *The Quarterly Journal of Economics* 130 (3): 1117–1165. https://doi.org/10.1093/qje/qjv015.

Aronson, J.M., Fried, C.B., and Good, C. (2002). Reducing the effects of stereotype threat on African American college students by shaping theories of intelligence. *Journal of Experimental Social Psychology* 38 (2): 113–125. https://doi.org/10.1006/jesp.2001.1491.

Battle, J. and Lewis, M. (2002). The increasing significance of class: the relative effects of race and socioeconomic status on academic achievement. *Journal of Poverty* 6 (2): 21–35.

Bloom, H.S. (1984). Accounting for no-shows in experimental evaluation designs. *Evaluation Review* 8 (2): 225–246.

Bloom, H.S. (2010). Nine lessons about doing evaluation research: Remarks on accepting the Peter H. Rossi Award. Retrieved from http://dev.mdrc.org/sites/default/files/presentation.pdf.

Bryan, C.J., Walton, G.M., Rogers, T., and Dweck, C.S. (2011). Motivating voter turnout by invoking the self. *Proceedings of the National Academy of Sciences* 108 (31): 12653–12656. https://doi.org/10.1073/pnas.1103343108.

Bryk, A.S. (2009). Support a science of performance improvement. *Phi Delta Kappan* 90: 597–600.

Cole, S.R. and Stuart, E.A. (2010). Generalizing evidence from randomized clinical trials to target populations: the ACTG 320 trial. *American Journal of Epidemiology* 172 (1): 107–115. https://doi.org/10.1093/aje/kwq084.

Cook, T.D. (1993). A quasi-sampling theory of the generalization of causal relationships. *New Directions for Program Evaluation* 1993 (57): 39–82. https://doi.org/10.1002/ev.1638.

Dweck, C.S. (2006). *Mindset: The New Psychology of Success.* New York, NY: Random House.

Entman, R.M. (2010). Media framing biases and political power: explaining slant in news of Campaign 2008. *Journalism* 11 (4): 389–408. https://doi.org/10.1177/1464884910367587.

Feller, A. and Holmes, C.C. (2009). Beyond toplines: Heterogeneous treatment effects in randomized experiments. *Unpublished Manuscript*, Oxford University. Retrieved from http://www.stat.columbia.edu/~gelman/stuff_for_blog/feller.pdf.

Gelman, A. (2015). The connection between varying treatment effects and the crisis of unreplicable research: A Bayesian perspective. *Journal of Management* 41 (2): 632–643.

Gelman, A., Su, Y.-S., Yajima, M. et al. (2015). Package "arm." Data Analysis Using Regression and Multilevel/Hierarchical Models.

Green, C.L. (1999). Ethnic evaluations of advertising: interaction effects of strength of ethnic identification, media placement, and degree of racial composition. *Journal of Advertising* 49–64.

Green, D.P. and Kern, H.L. (2012). Modeling heterogeneous treatment effects in survey experiments with Bayesian additive regression trees. *Public Opinion Quarterly* 76 (3): 491–511. https://doi.org/10.1093/poq/nfs036.

Holland, P.W. (1986). Statistics and causal inference. *Journal of the American Statistical Association* 81 (396): 945–960. https://doi.org/10.2307/2289064.

Horiuchi, Y., Imai, K., and Taniguchi, N. (2007). Designing and analyzing randomized experiments: application to a Japanese election survey experiment. *American Journal of Political Science* 51 (3): 669–687.

Hulleman, C.S. and Cordray, D.S. (2009). Moving from the lab to the field: the role of fidelity and achieved relative intervention strength. *Journal of Research on Educational Effectiveness* 2 (1): 88–110. https://doi.org/10.1080/19345740802539325.

Imai, K., King, G., and Stuart, E.A. (2008). Misunderstandings between experimentalists and observationalists about causal inference. *Journal of the Royal Statistical Society: Series A (Statistics in Society)* 171 (2): 481–502.

Imai, K. and Strauss, A. (2011). Estimation of heterogeneous treatment effects from randomized experiments, with application to the optimal planning of the get-out-the-vote campaign. *Political Analysis* 19 (1): 1–19. https://doi.org/10.1093/pan/mpq035.

Keeter, S., Kennedy, C., Dimock, M. et al. (2006). Gauging the impact of growing nonresponse on estimates from a national RDD telephone survey. *Public Opinion Quarterly* 70 (5): 759–779. https://doi.org/10.1093/poq/nfl035.

Krosnick, J.A. (1999). Survey research. *Annual Review of Psychology* 50 (1): 537–567. https://doi.org/10.1146/annurev.psych.50.1.537.

Markus, H.R. and Kitayama, S. (2010). Cultures and selves: a cycle of mutual constitution. *Perspectives on Psychological Science* 5 (4): 420–430. https://doi.org/10.1177/1745691610375557.

Morgan, S.L. and Winship, C. (2014). *Counterfactuals and Causal Inference.* Cambridge University Press.

Mullainathan, S. and Shafir, E. (2013). *Scarcity: Why Having Too Little Means So Much.* Macmillan.

Mutz, D.C. (2011). *Population-Based Survey Experiments.* Princeton, NJ: Princeton University Press.

Narayan, S. and Krosnick, J.A. (1996). Education moderates some response effects in attitude measurement. *Public Opinion Quarterly* 60 (1): 58–88. https://doi.org/10.1086/297739.

Olsen, R.B., Orr, L.L., Bell, S.H., and Stuart, E.A. (2013). External validity in policy evaluations that choose sites purposively. *Journal of Policy Analysis and Management* 32 (1): 107–121. https://doi.org/10.1002/pam.21660.

Open Science Collaboration (2015). Estimating the reproducibility of psychological science. *Science* 349 (6251): aac4716-1–aac4716-8. https://doi.org/10.1126/science.aac4716.

Open Science Framework (2016). The $1 million preregistration challenge by the center for open science. https://osf.io/peut2/ (accessed 17 January 2019).

Paunesku, D., Walton, G.M., Romero, C. et al. (2015). Mindset interventions are a scalable treatment for academic underachievement. *Psychological Science* 26 (6): 284–293. https://doi.org/10.1177/0956797615571017.

Rattan, A., Savani, K., Chugh, D., and Dweck, C.S. (2015). Leveraging mindsets to promote academic achievement policy recommendations. *Perspectives on Psychological Science* 10 (6): 721–726.

Raudenbush, S.W. and Liu, X. (2000). Statistical power and optimal design for multisite randomized trials. *Psychological Methods* 5 (2): 199.

Rothman, K. and Greenland, S. (2005). Causation and causal inference in epidemiology. *American Journal of Public Health* 95 (S1): S144–S150.

Rothman, R.L., Malone, R., Bryant, B. et al. (2005). A randomized trial of a primary care-based disease management program to improve cardiovascular risk factors and glycated hemoglobin levels in patients with diabetes. *The American Journal of Medicine* 118 (3): 276–284. https://doi.org/10.1016/j.amjmed.2004.09.017.

Shadish, W.R., Cook, T.D., and Campbell, T.D. (2001). *Experimental and Quasi-Experimental Designs for Generalized Causal Inference*, 2e. Boston, MA: Wadsworth Publishing.

Simmons, J.P., Nelson, L.D., and Simonsohn, U. (2011). False-positive psychology: undisclosed flexibility in data collection and analysis allows presenting anything as significant. *Psychological Science* 22 (11): 1359–1366. https://doi.org/10.1177/0956797611417632.

Smith, S.S. (2010). Race and trust. *Annual Review of Sociology* 36 (1): 453–475. https://doi.org/10.1146/annurev.soc.012809.102526.

Solon, G., Haider, S.J., and Wooldridge, J.M. (2015). What are we weighting for? *Journal of Human Resources* 50 (2): 301–316.

Spencer, M.B. (2006). Phenomenology and ecological systems theory: development of diverse groups. In: *Handbook of Child Psychology, Vol. 1: Theoretical Models of Human Development*, 6e (ed. W. Damon and R.M. Lerner). New York: Wiley.

Bloom, H.S. and Spybrook, J. (2017). Assessing the precision of multisite trials for estimating the parameters of a cross-site population distribution of program effects. *Journal of Research on Educational Effectiveness* 10 (4): 877–902.

Steele, C.M. (1997). A threat in the air: how stereotypes shape intellectual identity and performance. *American Psychologist* 52 (6): 613–629. https://doi.org/10.1037/0003-066X.52.6.613.

Steele, C.M. (2011). *Whistling Vivaldi: How Stereotypes Affect Us and What We Can Do*. New York, NY: W. W. Norton & Company.

Steele, C.M. and Aronson, J. (1995). Stereotype threat and the intellectual test performance of African Americans. *Journal of Personality and Social Psychology* 69 (5): 797–811. https://doi.org/10.1037/0022-3514.69.5.797.

Stuart, E.A., Cole, S.R., Bradshaw, C.P., and Leaf, P.J. (2011). The use of propensity scores to assess the generalizability of results from randomized trials. *Journal of the Royal Statistical Society: Series A (Statistics in Society)* 174 (2): 369–386. https://doi.org/10.1111/j.1467-985X.2010.00673.x.

Thaler, R.H. and Sunstein, C.R. (2008). *Nudge: Improving Decisions About Health, Wealth, and Happiness.* Yale University Press.

Tipton, E. (2013). Improving generalizations from experiments using propensity score subclassification: assumptions, properties, and contexts. *Journal of Educational and Behavioral Statistics* 38 (3): 239–266. https://doi.org/10.3102/1076998612441947.

Tipton, E. (2014). How generalizable is your experiment? An index for comparing experimental samples and populations. *Journal of Educational and Behavioral Statistics* 39 (6): 478–501. https://doi.org/10.3102/1076998614558486.

Tipton, E., Hedges, L., Vaden-Kiernan, M. et al. (2014). Sample selection in randomized experiments: a new method using propensity score stratified sampling. *Journal of Research on Educational Effectiveness* 7 (1): 114–135. https://doi.org/10.1080/19345747.2013.831154.

Tipton, E. and Miller, K. (2015). Generalizer [Web-tool]. http://www.thegeneralizer.org (accessed 24 January 2019).

Tipton, E. and Peck, L.R. (2017). A design-based approach to improve external validity in welfare policy evaluations. *Evaluation Review* 41 (4): 326–356.

Tucker-Drob, E.M. (2011). Individual differences methods for randomized experiments. *Psychological Methods* 16 (3): 298–318. https://doi.org/10.1037/a0023349.

Vanderweele, T.J. (2015). *Explanation in Causal Inference: Methods for Mediation and Interaction.* Oxford, UK: Oxford University Press.

Weiss, M.J., Bloom, H.S., and Brock, T. (2014). A conceptual framework for studying the sources of variation in program effects. *Journal of Policy Analysis and Management* 33 (3): 778–808. https://doi.org/10.1002/pam.21760.

Weiss, M.J., Bloom, H.S., Verbitsky Savitz, N. et al. (2017). How much do the effects of education and training programs vary across sites? Evidence from past multisite randomized trials. *Journal of Research on Educational Effectiveness* 10 (4): 843–876.

Wilson, T.D. and Juarez, L.P. (2015). Intuition is not evidence: prescriptions for behavioral interventions from social psychology. *Behavioral Science & Policy* 1 (1): 13–20.

Yeager, D.S. and Dweck, C.S. (2012). Mindsets that promote resilience: when students believe that personal characteristics can be developed. *Educational Psychologist* 47 (4): 302–314. https://doi.org/10.1080/00461520.2012.722805.

Yeager, D.S., Krosnick, J.A., Chang, L. et al. (2011). Comparing the accuracy of RDD telephone surveys and internet surveys conducted with probability and non-probability samples. *Public Opinion Quarterly* 1–39. https://doi.org/10.1093/poq/nfr020.

Yeager, D.S., Romero, C., Paunesku, D. et al. (2016a). Using design thinking to improve psychological interventions: the case of the growth mindset during the transition to high school. *Journal of Educational Psychology* 108 (3): 374–391. https://doi.org/10.1037/edu0000098.

Yeager, D.S., Walton, G.M., Brady, S.T. et al. (2016b). Teaching a lay theory before college narrows achievement gaps at scale. *Proceedings of the National Academy of Sciences* 113 (24): E3341–E3348. https://doi.org/10.1073/pnas.1524360113.

23

Design-Based Analysis of Experiments Embedded in Probability Samples

Jan A. van den Brakel

Department of Statistical Methods, Statistics Netherlands, Heerlen, the Netherlands
Department of Quantitative Economics, Maastricht University School of Business and Economics, Maastricht, the Netherlands

23.1 Introduction

Randomized experiments embedded in probability samples typically find their applications in survey methodology to test the effects of alternative survey implementations on the outcomes of a sample survey. The purpose of such empirical research is to improve the quality and efficiency of the underlying survey processes or to obtain more quantitative insight into the various sources of nonsampling errors. Many experiments conducted in this context are small scaled or conducted with specific groups. The value of empirical research into survey methods is strengthened as conclusions can be generalized to populations larger than the sample that is included in the experiment. This can be achieved by selecting experimental units randomly from a larger target population and naturally leads to randomized experiments embedded in probability samples.

The fields of randomized experiments and probability sampling are traditionally two separated domains of applied statistics. Both fields share one similarity, which makes them unique from other areas of statistics: design plays a crucial role in this type of empirical research. While design of randomized experiments is traditionally focused on establishing the causality between differences in treatments and observed effects (internal validity), design of probability samples is focused on generalizing results observed in a small sample to an intended target population (external validity). Designing experiments that are embedded in probability samples, which are drawn from a finite target population, results in experiments that potentially combine the strong internal validity from randomized experiments with the strong external validity of probability sampling. This enables the generalization of conclusions observed in an experiment to larger target populations and is particularly important if experiments are conducted to improve survey methods or to obtain quantitative insight into different sources of nonsampling errors in survey research.

An important class of applications is experiments conducted to quantify the effect of a redesign of a repeatedly conducted survey (Van den Brakel et al. 2008). An important issue in the analysis of such a redesign, but also in the case of experiments aimed to improve survey strategies, is to find the right inferential framework. Inference in survey sampling is traditionally design-based, whereas inference in randomized experiments is traditionally model-based.

Experimental Methods in Survey Research: Techniques that Combine Random Sampling with Random Assignment, First Edition.
Edited by Paul J. Lavrakas, Michael W. Traugott, Courtney Kennedy, Allyson L. Holbrook, Edith D. de Leeuw, and Brady T. West.

Companion Website: www.wiley.com/go/Lavrakas/survey-research

In the design-based and model-assisted approach, the statistical inference is based on the stochastic structure induced by the sampling design. Parameter and variance estimators are derived under the concept of repeatedly drawing samples from a finite population according to the same sampling design, keeping all other population parameters fixed. Statistical modeling plays a minor role. This is the traditional approach of survey sampling theory, followed by authors like Hansen et al. (1953), Kish (1965), Cochran (1977), and Särndal et al. (1992). In the model-based context, the probability structure of the sampling design plays a less pronounced role, since the inference is based on the probability structure of an assumed statistical model. This approach is predominantly followed in the analysis of randomized experiments by authors like Scheffé (1959) and Searle (1971). The observations obtained in the experiment are assumed to be the realization of a linear model. To test hypotheses about treatment effects, F-tests are derived under the assumption of normally and independently distributed observations.

The use of (approximately) design-unbiased estimators for unknown population parameters naturally fits with the purpose of probability sampling to generalize conclusions to larger target populations. If experiments are embedded in probability samples with the purpose to generalize conclusions to larger target populations, a design-based inference framework might be more appropriate than the model-based approach traditionally used in the analysis of randomized experiments. This requires an analysis procedure that accounts for the sample design used to select a random sample from a target population as well as the experimental design used to assign the sampling units to the different treatments of the experiment. The purpose of this chapter is to present a general design-based framework for the design and analysis of experiments embedded in probability samples.

This chapter is organized as follows. Design considerations for experiments embedded in probability samples are discussed in Section 23.2. A design-based inference approach for single-factor experiments where sampling units are also the experimental units is given in Section 23.3. In Section 23.4, results are generalized to experiments where clusters of sampling units are the experimental units in randomized experiments. The results of Sections 23.3 and 23.4, for single-factor experiments, are generalized to factorial designs in Section 23.5. An experiment with different data collection modes in the Dutch Crime Victimization Survey is used as an illustrative example in Section 23.6. This section also contains a brief description of a software package developed for the design-based inference approach proposed in this chapter. This chapter concludes with a discussion in Section 23.7.

23.2 Design of Embedded Experiments

In an embedded experiment, a probability sample is drawn from a finite target population. This sample is then randomly divided into two or more subsamples according to a randomized experiment. In the survey literature, such experiments are also referred to as split-ballot designs or interpenetrating subsampling and date back to Mahalanobis (1946), but see also Cochran (1977, section 13.15), Hartley and Rao (1978), and Fienberg and Tanur (1987, 1988, 1989).

The design of embedded experiments starts with a clear specification of definitions and decisions about the type and number of treatment factors and levels to be tested in the experiment. Based on this, hypotheses about main effects and interactions can be specified for a prespecified set of target variables or dependent variables.

The most straightforward approach is to split the sample into subsamples by means of a completely randomized design (CRD). Generally this is not the most efficient design available. The power of an experiment might be improved by using sampling structures such as strata, clusters, or interviewers as block variables in a randomized block design (RBD) (Fienberg and Tanur 1987, 1988). Unrestricted randomization by means of a CRD might also result in practical complications, like long traveling distances for interviewers. This can be avoided by using small geographical regions as blocking variables.

It might be impractical to randomize respondents over the treatments. There might be practical objections to assigning respondents belonging to the same household or interviewed by the same interviewer to different treatments in the experiment. In such situations, we can consider randomizing *clusters* of sampling units over the treatments, at the cost of reduced power (Van den Brakel 2008).

The field staff, responsible for data collection, also requires special attention in the planning and design stage of an experiment. To draw conclusions that can be generalized to a situation where the new approach is implemented as a standard, it is advisable to use either the entire field staff or a representative sample of the field staff. Newly recruited staff, on the other hand, might be precluded for this reason. It is also advisable to provide sufficient training, to ensure that the field staff has sufficient experience with the data collection under the new approach. One might also anticipate that the data collected under the new approach in the first period of the experiment cannot be used in the analysis, since the interviewers must adapt to or gain sufficient experience with the new methods.

From a statistical point of view, it is efficient to use interviewers as the block variable in an RBD, since this removes the interviewer variance component from the analysis of the experiment. A major drawback is that this implies that each interviewer has to collect data under both the regular and the new methodology, which might give rise to confusion; Lavrakas et al. discuss additional negative implications of such a design elsewhere in this volume. If it is decided that interviewers are assigned to one treatment only, then this must be done randomly to avoid the possibility that treatment effects are confounded with other interviewer characteristics, such as experience. Another issue is whether the interviewers should be informed that they are participating in an experiment or not. The advantage of keeping interviewers uninformed is that they do not adjust their behavior because they are aware that their performance is being supervised in an experiment. It depends on the treatments whether it is possible to keep interviewers uninformed.

Whether it is possible to use interviewers as blocks depends on the number and type of treatments and the field staff's experience with collecting data under different treatments simultaneously. Assigning interviewers to one treatment only can be accomplished as follows in a survey where data collection is based on computer-assisted telephone interviewing (CATI). First interviewers as well as sampling units are randomly assigned to the different treatments. Next sampling units are assigned randomly to interviewers within each subsample or treatment. In a survey where data collection is based on computer-assisted personal interviewing (CAPI) and interviewers are working on the data collection in relatively small areas around their own place of residence, unrestricted randomization of sampling units and interviewers over the treatments is often not feasible. This randomization mechanism might result in an unacceptable increase in the travel distance for interviewers, particularly if the subsample sizes of the alternative treatments are small. An alternative is to assign sampling units to interviewers. Subsequently, the interviewers with their clusters of sampling units are randomized over the treatments of the experiment. Here, however, the interviewers are the experimental units instead of the sampling units, which decreases the effective sample size for variance estimation and power for testing hypotheses. One compromise is to use geographical regions defined by linked adjacent interviewer regions as blocks. The sampling units within each block are randomized over the treatment combinations, and each interviewer within each block is randomly assigned to one of the treatment combinations. This implies that the number of interviewers in each block must be equal to the number of treatments. Subsequently, each interviewer visits the sampling units assigned to his or her treatment combination. This results in a relatively small increase in the travel distance for the interviewers and no increase in variance, since the sampling units are the experimental units in this design. See Van den Brakel and Renssen (1998) and Van den Brakel (2008) for more details about issues concerning the field staff in embedded experiments.

Another consideration is the minimum required sample size. The researcher must give an indication about the minimum size of the treatment effects that should at least reject the null hypothesis at prespecified levels of significance and power. Based on these, the minimum subsample sizes can be determined by an appropriate power calculation; see, e.g. Montgomery (2001). A general framework and practical guidelines for planning and conducting experiments is provided by Robinson (2000). More specific details on design issues for embedded experiments in probability samples can be found in Van den Brakel and Renssen (1998), Van den Brakel et al. (2008), and Van den Brakel (2008).

23.3 Design-Based Inference for Embedded Experiments with One Treatment Factor

The purpose of this section is to develop a theoretical framework to test hypotheses about differences between finite population parameter estimates observed under different survey implementations or treatments, based on a field experiment embedded in a probability sample. Testing hypotheses about systematic differences between a finite population parameter observed under different treatments implies the occurrence of measurement bias. Explaining systematic differences between a target variable observed under different treatments or survey implementations requires a measurement error model, which is developed in Section 23.3.1. This enables us to specify sensible hypotheses about treatment effects on finite population means (Section 23.3.1) and a design-based analysis procedure (Sections 23.3.2 and 23.3.3). In Section 23.3.4, special cases are considered where the proposed procedure coincides with the more familiar model-based analysis such as ANOVA F-tests or two sample t-tests. Finally, some extensions to population parameters defined as totals or ratios are provided in Section 23.3.5.

23.3.1 Measurement Error Model and Hypothesis Testing

Consider a finite population of N units. Let u_i denote the true but not directly observable target variable of the ith population unit ($i = 1, \ldots, N$). Consider an experiment conducted to test systematic differences between a finite population mean, $\overline{U} = 1/N \sum_{i=1}^{N} u_i$, observed under $K \geq 2$ treatment levels of one factor. Let y_{iqk} denote an observation obtained from the ith sampling unit that is assigned to the kth treatment and qth interviewer. These observations are assumed to be a realization of the following measurement error model:

$$y_{iqk} = u_i + b_k + \gamma_q + \varepsilon_{ik} \tag{23.1}$$

In (23.1), b_k is the effect of the kth treatment or survey implementation ($k = 1, \ldots, K$), γ_q is the effect of the qth interviewer ($q = 1, \ldots, Q$), and ε_{ik} is a random measurement error if the target variable of the ith population unit is measured under the kth treatment. We allow for mixed interviewer effects, i.e. $\gamma_q = \psi + \xi_q$ with ψ a fixed effect and ξ_q a random interviewer effect. Let E_m and Cov_m denote the expectation and (co)variance with respect to the measurement error model. It is assumed that $E_m(\varepsilon_{ik}) = 0$, $\text{Var}_m(\varepsilon_{ik}) = \sigma_{ik}^2$, $\text{Cov}_m(\varepsilon_{ik}\, \varepsilon_{i'k}) = 0$, $\text{Cov}_m(\varepsilon_{ik}\, \varepsilon_{ik'}) = \sigma_{ikk'}$, $E_m(\xi_q) = 0$, $\text{Var}_m(\xi_q) = \tau_q^2$, $\text{Cov}_m(\xi_q\, \xi_{q'}) = 0$, and $\text{Cov}_m(\varepsilon_{ik}\, \xi_q) = 0$. No parametric assumptions about the distributions are made since large sample theory is applied to derive a limit distribution for the test statistics in Section 23.3.3. From these assumptions it follows that measurement errors of different population units are independent but that units observed by the same interviewer can have correlated responses through the interviewer effects. Let $\overline{Y}_k$ denote the population mean observed under the kth treatment. From the measurement error model, it follows that $\overline{Y}_k = \overline{U} + b_k + \overline{\gamma} + \overline{\varepsilon}_k$, with $\overline{\gamma}$ and $\overline{\varepsilon}_k$ the finite population means of the interviewer effects and random measurement errors. Then $\overline{\boldsymbol{Y}} = (\overline{Y}_1, \ldots, \overline{Y}_K)^t$ denotes the K dimensional vector with population means observed under the different treatments of the experiment.

A linear measurement error model (23.1) is appropriate for quantitative variables. In many applications, however, the target variables are binary or categorical. In such cases, a logistic model or multinomial model might be more appropriate. The linear model defined in (23.1) is nevertheless applied in such situations. This might appear rigid at first sight, but similar linear models are used to motivate the generalized regression (GREG) estimator that is generally used in survey sampling to estimate sample means or totals of binary or categorical variables. Model (23.1) is also required for the use of the GREG estimator in the design-based analysis procedure proposed in the next subsection. In the case of binary variables, the population means are interpreted as the fraction of persons meeting a specific requirement. The treatment effects b_k in (23.1) can still be interpreted as the average effect on this fraction if this finite population parameter is measured under the kth treatment. Model (23.1) is still useful to quantify systematic differences between a finite population parameter that is observed under different survey implementations. Examples where model (23.1) is applied to binary data can be found in Van den Brakel and Van Berkel (2002), Van den Brakel (2008), and Van den Brakel et al. (2008). Applications of linear models to binary data in the context of small area estimation can be found in, for example, Boonstra et al. (2008) and Boonstra (2014).

The purpose of the experiment is to test the null hypothesis that the population means observed under the different treatments are equal against the alternative that at least one pair is significantly different. Only systematic differences between the treatments, reflected by b_k, should lead to a rejection of the null hypothesis. Random deviations due to measurement errors and interviewer effects should not lead to significant differences in the analysis. This is accomplished by formulating hypotheses about $\overline{\boldsymbol{Y}}$ in expectation over the measurement error model, i.e.

$$\begin{aligned} H_0 &: \boldsymbol{C}E_m\overline{\boldsymbol{Y}} = \boldsymbol{0} \\ H_1 &: \boldsymbol{C}E_m\overline{\boldsymbol{Y}} \neq \boldsymbol{0} \end{aligned} \tag{23.2}$$

where $\boldsymbol{C} = (\boldsymbol{j} \mid -\boldsymbol{I})$ denotes a $(K-1)\times K$ matrix with the $(K-1)$ contrasts or treatment effects, $\boldsymbol{0}$ and $\boldsymbol{j}$ vectors of order $(K-1)$ with each element equal to zero and one, respectively, and $\boldsymbol{I}$ an identity matrix of order $(K-1)$. Since $E_m\overline{Y}_k = \overline{U} + b_k + \psi$, it follows that $\boldsymbol{C}E_m\overline{\boldsymbol{Y}} = (b_1 - b_2, \ldots, b_1 - b_K)^t$ exactly corresponds to the observable treatment effects. Let $\hat{\overline{\boldsymbol{Y}}}$ denote a design-unbiased estimator for $E_m\overline{\boldsymbol{Y}}$ and $V(\boldsymbol{C}\hat{\overline{\boldsymbol{Y}}})$ the design-based covariance matrix of the contrasts between $\hat{\overline{\boldsymbol{Y}}}$. Both estimators account for the sample design used to draw a random sample form the finite target population and the experimental design used to randomize the sampling units over the K treatments. Expressions are derived in the next subsection. Now hypothesis (23.2) can be tested with the Wald statistic

$$W = \hat{\overline{\boldsymbol{Y}}}^t\boldsymbol{C}^t V(\boldsymbol{C}\hat{\overline{\boldsymbol{Y}}})^{-1}\boldsymbol{C}\hat{\overline{\boldsymbol{Y}}} \tag{23.3}$$

Section 23.3.3 shows that under the null hypothesis, W is approximately chi-squared distributed with $K-1$ degrees of freedom.

23.3.2 Point and Variance Estimation

A design-based inference procedure for the analysis of embedded experiments is obtained by constructing GREG estimators for the population means in $\overline{Y}$ and the covariance matrix of the contrasts. Such a procedure starts with deriving the first-order inclusion probabilities that a sampling unit is drawn in the sample and assigned to one of the K treatments. The GREG estimator is widely applied in survey sampling and uses a priori knowledge about the finite population available from registers (Särndal et al. 1992). In the case that no register information is available, the ratio estimator for a population mean, proposed by Hájek (1971), should be applied. This estimator follows as a special case from the GREG estimator with a weighting model that only uses the population size as auxiliary information, which will be known in most applications.

Consider a sample s of size n drawn by a complex sample design that can be described with first- and second-order inclusion probabilities π_i and $\pi_{ii'}$ of the ith and i, i'th sampling unit(s), respectively. In the case of a CRD, s is randomly divided into K subsamples s_k of size n_k. Conditional on the realization of s, the probability that the ith sampling unit is assigned to subsample s_k equals n_k/n, and the unconditional inclusion probability that the ith sampling unit is included in s_k is $\pi_i^* = \pi_i(n_k/n)$. In the case of an RBD, s is deterministically divided into B blocks of size n_b ($b = 1, \dots, B$). Subsequently, within each block, n_{bk} sampling units are randomly assigned to subsample s_k. As a result, the conditional probability that the ith sampling unit from block b is assigned to subsample s_k equals n_{bk}/n_b and the unconditional inclusion probability that the ith sampling unit is included in s_k is $\pi_i^* = \pi_i(n_{bk}/n_b)$.

For notational convenience, the subscript q will be omitted in y_{iqkl} since there is no need to sum explicitly over the interviewer subscript in most of the formulas developed in the rest of this chapter. To apply the model-assisted inference approach to the analysis of embedded experiments, it is assumed for each unit in the population that the intrinsic value u_i in measurement error model (23.1) is an independent realization of the following linear regression model:

$$u_i = \boldsymbol{\beta}^t \boldsymbol{x}_i + e_i \tag{23.4}$$

where $\boldsymbol{x}_i$ denotes a H-vector with auxiliary information, $\boldsymbol{\beta}$ a H-vector with the regression coefficients, and e_i the residuals, which are independent random variables with variance ω_i^2. It is required that all ω_i^2 are known up to a common scale factor, that is $\omega_i^2 = \omega^2 v_i$, with v_i known. It is assumed that these auxiliary variables are intrinsic variables observed without measurement error and are not affected by the treatments. Data collected in the K subsamples can be used to estimate the population means $\overline{Y}_k$ with the GREG estimator, which is defined as

$$\widehat{\overline{Y}}_{k;r} = \widehat{\overline{Y}}_k + \widehat{\boldsymbol{\beta}}_k^t(\overline{\boldsymbol{X}} - \widehat{\overline{\boldsymbol{X}}}_k), \quad k = 1, \dots, K \tag{23.5}$$

with $\overline{\boldsymbol{X}}$ a H-vector containing the finite population means of the auxiliary variables $\boldsymbol{x}$,

$$\widehat{\overline{Y}}_k = \frac{1}{N} \sum_{i=1}^{n_k} \frac{y_{ik}}{\pi_i^*} \quad \text{and} \quad \widehat{\overline{\boldsymbol{X}}}_k = \frac{1}{N} \sum_{i=1}^{n_k} \frac{\boldsymbol{x}_i}{\pi_i^*}$$

the Horvitz–Thompson (HT) estimators for $\overline{Y}_k$ and $\overline{\boldsymbol{X}}$, and

$$\widehat{\boldsymbol{\beta}}_k = \left(\sum_{i=1}^{n_k} \frac{\boldsymbol{x}_i \boldsymbol{x}_i^t}{\omega_i^2 \pi_i^*} \right)^{-1} \sum_{i=1}^{n_k} \frac{\boldsymbol{x}_i y_{ik}}{\omega_i^2 \pi_i^*} \tag{23.6}$$

a HT-type estimator for the regression coefficients $\boldsymbol{\beta}$ in (23.4). Now $\widehat{\overline{\boldsymbol{Y}}}_r = (\widehat{\overline{Y}}_{1;r}, \dots, \widehat{\overline{Y}}_{K;r})^t$ is an approximately design-unbiased estimator for $\overline{\boldsymbol{Y}}$ and also for $E_m\overline{\boldsymbol{Y}}$ by definition. An estimator for the covariance matrix of the contrasts between the elements of $\hat{\hat{\boldsymbol{Y}}}_r$, where the covariance is taken over the sampling design, the experimental design, and the measurement error model, is given by

$$\widehat{\text{Cov}}(\boldsymbol{C}\hat{\hat{\boldsymbol{Y}}}_r) = \boldsymbol{C}\hat{\boldsymbol{D}}\boldsymbol{C}^t \tag{23.7}$$

In the case of an RBD, $\hat{\boldsymbol{D}}$ is a $K \times K$ diagonal matrix with elements

$$\hat{d}_k = \sum_{b=1}^{B} \frac{1}{n_{bk}(n_{bk}-1)} \sum_{i=1}^{n_{bk}} \left(\frac{n_b \hat{e}_{ik}}{N\pi_i} - \frac{1}{n_{bk}} \sum_{i'=1}^{n_{bk}} \frac{n_b \hat{e}_{i'k}}{N\pi_{i'}} \right)^2 \equiv \sum_{b=1}^{B} \frac{\hat{S}^2_{E_{bk}}}{n_{bk}} \tag{23.8}$$

with $\hat{e}_{ik} = y_{ik} - \hat{\boldsymbol{\beta}}_k^t \boldsymbol{x}_i$. The diagonal elements for a CRD follow as a special case from (23.8) with $B = 1$, $n_b = n$, and $n_{bk} = n_k$. A full proof of this result is given by Van den Brakel and Renssen (2005), under the assumption of a weighting model for the GREG estimator for which it holds that a constant H-vector $\boldsymbol{a}$ exists such that $\boldsymbol{a}^t\boldsymbol{x}_i = 1$ for all elements in the population. This is a relatively weak condition, since it assumes that the size of the finite target population is known and is used as auxiliary information in the GREG estimator. As an alternative, the residuals $\hat{e}_{ik}$ can be multiplied with correction weights of the GREG estimator (also called g-weights; Särndal et al. (1992, result 6.6.1)).

The derived point and variance estimators for the treatment effects are valid for CRDs and RBDs embedded in general complex sample designs, since the sample design is specified in general terms by first- and second-order inclusion probabilities. The variance estimator has an appealingly simple structure as if the K subsamples are drawn independently from each other, where sampling units are selected with unequal selection probabilities π_i/n with replacement in the case of a CRD and π_i/n_b with replacement within each block in the case of an RBD; compare (23.8) with Cochran (1977, equation (9A.16)). No joint inclusion probabilities or design covariances between the different subsamples are required, which simplifies the analysis considerably. This is the result of (i) the super imposition of the experimental design on the sample design in combination with the fact that variances about contrasts between subsample estimates are calculated, (ii) a weighting model is used that meets the condition that there is a constant H-vector $\boldsymbol{a}$ such that $\boldsymbol{a}^t\boldsymbol{x}_i = 1$ for all elements in the population, and (iii) the assumption that measurement errors between sampling units are independent. See Van den Brakel and Renssen (2005) for a more detailed discussion and interpretation of this result.

The covariance structure in (23.7) and (23.8) illustrates the importance of blocking on sampling structures such as strata, clusters, or interviewers. Note, for example, that the variance reduction of stratified sampling is only preserved in the variance of the treatment effects if the strata of the sample design are used as block variables in the experiment. In the case of unrestricted randomization of a CRD, the between stratum variance will be reintroduced in the variance of the treatment effects. Using clusters or primary sampling units (PSUs) as block variables excludes the between cluster variation from the variance of the treatment effects in a similar way. Randomizing clusters or PSUs over the treatments results in experiments where clusters instead of respondents within clusters are the experimental units. This reduces the effective sample size and the power of the experiment; see Section 23.4.

Ignoring sampling structures in the design of an embedded experiment reduces the power and results in a less efficient but still valid experiment. Using auxiliary information through the GREG estimator improves the precision of the estimated treatment effects and can correct, at least partially, for selective nonresponse. To account for the complexity of the sample design, it is sufficient to incorporate the first-order inclusion probabilities in the point and variance estimators as specified above. Ignoring these features of the sample design in the analysis of the experiment might result in biased estimates for point and variance estimates. Van den Brakel and Van Berkel (2002) analyze an experiment embedded in the Dutch Labor Force Survey. In this survey addresses in the Employment Exchange have selection probabilities that are three times larger than other addresses. Ignoring these unequal inclusion probabilities results in an overestimation of the treatment effects in the Unemployed Labor Force. Van den Brakel and Renssen (2005) conducted a simulation and showed that ignoring inclusion probabilities chosen proportional to the value of the target parameter results in biased estimates for the treatment effects including their standard errors.

The condition that a constant H-vector $\boldsymbol{a}$ exists such that $\boldsymbol{a}^t\boldsymbol{x}_i = 1$ for all elements in the population precludes the ratio estimator and the HT estimator for the proposed design-based inference procedure. As an alternative for the HT estimator, a GREG estimator with weighting model $\boldsymbol{x}_i = (1)$ and $\omega_i^2 = \omega^2$ for all elements in the population can be used. This weighting model is known as the common mean model and only uses the size of the finite population as a priori knowledge (Särndal et al. 1992, section 7.4). Under this weighting model, it follows that

$$\hat{\bar{Y}}_{k;r} = \left(\sum_{i=1}^{n_k} \frac{1}{\pi_i^*}\right)^{-1} \left(\sum_{i=1}^{n_k} \frac{y_{ik}}{\pi_i^*}\right) \equiv \tilde{y}_k \tag{23.9}$$

which can be recognized as the ratio estimator for a population mean, originally proposed by Hájek (1971). In this case, the covariance matrix can be estimated by (23.7) and (23.8) with $\hat{\boldsymbol{\beta}}_k^t \boldsymbol{x}_i = \tilde{y}_k$. This estimator is preferable since it avoids the extreme estimates sometimes obtained with the HT estimator and it has relatively simple approximately design-unbiased estimates for the variance of the treatment effects. Variance expressions for the HT estimator are more complex for sample designs where $\sum_{i=1}^{n_k} 1/\pi_i^* \neq N$, and are given by Van den Brakel (2001).

23.3.3 Wald Test

The design-unbiased estimators for the subsample means and the covariance matrix give rise to the following Wald statistic:

$$W = \hat{\bar{\boldsymbol{Y}}}_r^t \boldsymbol{C}^t (\boldsymbol{C}\hat{\boldsymbol{D}}\boldsymbol{C}^t)^{-1} \boldsymbol{C}\hat{\bar{\boldsymbol{Y}}}_r \tag{23.10}$$

Van den Brakel and Renssen (2005) show that this expression can be simplified to

$$W = \sum_{k=1}^{K} \frac{\hat{\bar{Y}}_{k,r}^2}{\hat{d}_k} - \left(\sum_{k=1}^{K} \frac{1}{\hat{d}_k}\right)^{-1} \left(\sum_{k=1}^{K} \frac{\hat{\bar{Y}}_{k,r}}{\hat{d}_k}\right)^2 \tag{23.11}$$

Under general complex sampling designs, it can be conjectured that the limit distribution of $\boldsymbol{C}\hat{\bar{\boldsymbol{Y}}}_r$ is a $K-1$ dimensional multivariate normal distribution, i.e. $\boldsymbol{C}\hat{\bar{\boldsymbol{Y}}}_r \to N(\boldsymbol{C}E_m\bar{\boldsymbol{Y}}, V(\boldsymbol{C}\hat{\bar{\boldsymbol{Y}}}))$. Then it can be shown that W is asymptotically chi-squared distributed with $(K-1)$ degrees of freedom (Searle 1971, theorem 2, Ch. 2).

23.3.4 Special Cases

It follows from (23.8) that under an RBD, BK separate population variances, $\hat{S}_{E_{bk}}^2$, have to be estimated. It might be efficient to consider a pooled estimator within each block,

$$\hat{S}_{E_b p}^2 = \frac{1}{(n_b - K)} \sum_{k=1}^{K} \sum_{i=1}^{n_{bk}} \left(\frac{n_b \hat{e}_{ik}}{N\pi_i} - \frac{1}{n_{bk}} \sum_{i'=1}^{n_{bk}} \frac{n_b \hat{e}_{i'k}}{N\pi_{i'}} \right)^2 \tag{23.12}$$

as an alternative for $\hat{S}_{E_{bk}}^2$ in (23.8).

Van den Brakel and Renssen (2005) show that under (i) a self-weighted sample design where sampling units are allocated proportionally to the treatments over the blocks (i.e. $n_{bk}/n_b = n_{b'k}/n_{b'}$ for all b, b'), (ii) the use of the ratio estimator for a population mean defined by (23.9), and (iii) the pooled variance estimator defined by (23.12), it can be shown that $W/(K-1)$ is equal to the F-statistic of an ANOVA for a two-way layout with an interaction. In the case of a CRD, they show that $W/(K-1)$ is equal to the F-statistic of an ANOVA for a one-way layout.

In the case of a two-treatment experiment (designed as a CRD or RBD), the analysis can be based on a design-based t-type statistic, i.e.

$$t = \frac{\hat{\overline{Y}}_{1,r} - \hat{\overline{Y}}_{2,r}}{\sqrt{\hat{d}_1 + \hat{d}_2}} \tag{23.13}$$

This enables the testing of specified and unspecified alternative hypotheses. Estimators for the sample means and variances follow from (23.5) and (23.8). If it is conjectured that both sample means are asymptotically normally distributed, then t is asymptotically a standard normal distributed variable. In the case of a self-weighted sample design, a CRD, and the use of estimator (23.9), it can be shown that (23.13) equals Welch's t-statistic (Miller 1986). If in addition the pooled variance estimator (23.12) is used, then it follows that (23.13) is equal to the standard t-statistic. See Van den Brakel (2001, Ch. 5.4) for details.

23.3.5 Hypotheses About Ratios and Totals

In many surveys, target parameters are defined as the ratio of two population means or totals. Testing hypotheses about ratios of two survey estimates requires different point and variance estimators for the Wald statistic. Let $R_k = \overline{Y}_k/\overline{Z}_k$ denote the ratio of two population means observed under treatment $k = 1, \ldots, K$. Then $\boldsymbol{R} = (R_1, \ldots, R_K)^t$ denotes the K dimensional vector containing the population ratios observed under the K different treatments of the experiment. Analogously to (23.2), the hypothesis of no treatment effects is formulated about the ratios where the numerator and denominator both denote the population mean in expectation over the measurement error model. This can be tested with a Wald statistic $W = \hat{\boldsymbol{R}}^t \boldsymbol{C}^t V(\boldsymbol{C}\hat{\boldsymbol{R}})^{-1}\boldsymbol{C}\hat{\boldsymbol{R}}$, where $\hat{\boldsymbol{R}}$ denotes a design-based estimator for $\boldsymbol{R}$.

Let y_{iqk} and z_{iqk} denote the observations for the parameter in the numerator and denominator, respectively, for the ith sampling unit assigned to the kth treatment and qth interviewer. It is assumed that both observations are a realization of the same type of measurement error model defined by (23.1). The GREG estimator for R_k is defined as $\hat{R}_{k;r} = \hat{\overline{Y}}_{k;r}/\hat{\overline{Z}}_{k;r}$, where $\hat{\overline{Z}}_{k;r}$ is the GREG estimator for $\overline{Z}_k$ defined analogously to expression (23.5). The K GREG estimates $\hat{R}_{k;r}$ can be combined in the vector $\hat{\boldsymbol{R}}_r = (\hat{R}_{1;r}, \ldots, \hat{R}_{K;r})^t$ as the GREG estimator for $\boldsymbol{R}$. An approximately design-unbiased estimator for the covariance matrix of the $K-1$ contrasts between $\hat{\boldsymbol{R}}_r$ is given by (23.7), where $\hat{\boldsymbol{D}}$ is replaced by $\hat{\boldsymbol{D}}^{(R)}$ with diagonal elements:

$$\hat{d}_k^{(R)} = \frac{1}{\hat{\overline{Z}}_{k;r}^2} \sum_{b=1}^{B} \frac{1}{n_{bk}(n_{bk}-1)} \sum_{i=1}^{n_{bk}} \left(\frac{n_b \hat{e}_{ik}}{N\pi_i} - \frac{1}{n_{bk}} \sum_{i'=1}^{n_{bk}} \frac{n_b \hat{e}_{i'k}}{N\pi_{i'}} \right)^2 \tag{23.14}$$

with $\hat{e}_{ik} = (y_{ik} - \hat{\boldsymbol{\beta}}_k^{y^t} \boldsymbol{x}_i) - \hat{R}_{k;r}(z_{ik} - \hat{\boldsymbol{\beta}}_k^{z^t} \boldsymbol{x}_i)$. Here $\hat{\boldsymbol{\beta}}_k^y$ is defined by (23.6), and $\hat{\boldsymbol{\beta}}_k^z$ denotes the H-dimensional vector with the HT-type estimator for the regression coefficients of the regression function of z_{ik} on $\boldsymbol{x}_i$, and is defined in a similar way as (23.6). The diagonal elements for a CRD follow as a special case from (23.14) with $B = 1$, $n_b = n$, and $n_{bk} = n_k$. A full proof of this result is given by Van den Brakel (2008). The hypothesis of no treatment effects can be tested with Wald statistic (23.11), where $\hat{\overline{Y}}_{k,r}$ is replaced by $\hat{R}_{k;r}$ and $\hat{d}_k$ is replaced by $\hat{d}_k^{(R)}$. In the case of two-treatment experiments, the design-based t-type statistic (23.13) can be used where $\hat{\overline{Y}}_{k,r}$ is replaced by $\hat{R}_{k;r}$ and $\hat{d}_k$ is replaced by $\hat{d}_k^{(R)}$.

Expressions for the Hájek estimator are now obtained in a straightforward way by taking $\tilde{R}_k = \tilde{y}_k / \tilde{z}_k$, where $\tilde{y}_k$ is defined by (23.9) and $\tilde{z}_k$ is defined analogously. An approximation of the covariance matrix of the contrasts between the subsample estimates is defined by (23.14) with residuals $\hat{e}_{ik} = (y_{ik} - \tilde{y}_k) - \hat{R}_{k;r}(z_{ik} - \tilde{z}_k)$.

Wald and t-statistics for testing hypotheses about population totals follow in a straightforward manner from the results obtained for means, by multiplying the parameter and variance estimators by N and N^2, respectively. The test statistics for population totals are equivalent to the test statistics for population means since they are invariant under scale transformations with a constant like the population size.

23.4 Analysis of Experiments with Clusters of Sampling Units as Experimental Units

As mentioned in Section 23.2, there can be practical reasons to randomize clusters of respondents or sampling elements over the different treatments, resulting in experiments where the sampling elements and experimental units are at different levels. This section explains the analysis procedure for an experiment embedded in a two-stage sample design where PSUs are randomized over K different treatments. Consider a finite population that consists of M PSUs. The jth PSU consists of N_j sampling elements or secondary sampling units (SSUs) in the case of a two-stage sample design. The population size equals $N = \sum_{j=1}^{M} N_j$. Let y_{ijqk} denote the response obtained from the ith SSU belonging to the jth PSU that is assigned to the qth interviewer and kth treatment. To allow for correlated response between SSUs belonging to the same PSU, measurement error model (23.1) is extended by

$$y_{ijqk} = u_{ij} + b_k + \gamma_q + \varepsilon_{ijk} \tag{23.15}$$

with u_{ij} the true intrinsic value of sampling unit (i,j) and ε_{ijk} its measurement error under treatment k. In addition to the assumptions of measurement error model (23.1), it is assumed that $E_m \varepsilon_{ijk} = 0$ and

$$\mathrm{Cov}_m(\varepsilon_{ijk}, \varepsilon_{i'j'k'}) = \begin{cases} \sigma_{ijk}^2 + \sigma_{jk}^2 & \text{if } i = i', j = j', \text{and } k = k' \\ \sigma_{jk}^2 & \text{if } i \neq i', j = j', \text{and } k = k' \\ 0 & \text{if } i \neq i', j \neq j', \text{and } k = k' \end{cases} \tag{23.16}$$

Recall that sampling units (i,j) assigned to the same interviewer also have correlated response through the interviewer effects $\gamma_q = \psi + \xi_q$ with assumptions $E_m(\xi_q) = 0$, $\mathrm{Var}_m(\xi_q) = \tau_q^2$, $\mathrm{Cov}_m(\xi_q\, \xi_{q'}) = 0$, and $\mathrm{Cov}_m(\varepsilon_{ijk}\, \xi_q) = 0$. The purpose of the experiment is to test the hypothesis that the population means $\overline{Y}_k = \overline{U} + b_k + \overline{\gamma} + \overline{\varepsilon}_k$ are equal against the alternative that at least one pair of means is significantly different. Hypotheses are formulated in expectation over the measurement error model to avoid the possibility that random measurement errors and random interviewer effects lead to significant differences and are given by (23.2). To test this hypothesis, a design-based Wald statistic is derived analogous to the approach followed in Sections 23.3.2 and 23.3.2.

Consider a general complex two-stage sample design. In the first stage, m PSUs are selected, where π_j^I denote the first-order inclusion expectation of the jth PSU in the first stage of the sampling design. In the second stage, n_j SSUs are selected from each of these m selected PSUs. Let $\pi^{\mathrm{II}}_{i|j}$ denote the first-order inclusion expectation of the ith SSU in the second stage conditionally on the realization of the first-stage sample. This results in a sample of $n = \sum_{j=1}^{m} n_j$ SSUs. In a CRD, the m PSUs are randomized over K subsamples s_k of size m_k. Now m_k/m is the conditional probability that the jth PSU is assigned to subsample s_k, given the realization of the first-stage sample. In the case of an RBD, the PSUs are deterministically divided into B blocks of size m_b. In the case of multistage sampling designs, strata are potential block variables. Within each block, the PSUs are randomized over the K different treatments or subsamples. Let m_{bk} denote the number of PSUs in block b that are assigned to treatment k. Now m_{bk}/m_b is the conditional probability that the jth PSU is assigned to subsample s_k, given the realization of the first-stage sample and that PSU j is an element of block b. It follows that the first-order inclusion probability of the jth PSU in the first stage of subsample s_k equals $\pi^{*\mathrm{I}}_j = (m_k/m)\pi^{\mathrm{I}}_j$ in a CRD and $\pi^{*\mathrm{I}}_j = (m_{bk}/m_b)\pi^{\mathrm{I}}_j$ in an RBD. Finally, the first-order inclusion probability for the ith SSU in subsample s_k equals $\pi^{*\mathrm{I}}_j\pi^{\mathrm{II}}_{i|j}$.

After having derived first-order inclusion probabilities for the K subsamples, the GREG estimator can be used to obtain estimates for the population parameter under the K different treatments. The GREG estimator $\widehat{\overline{Y}}_{k;r}$ for $\overline{Y}_k$ is defined by (23.5) with

$$\widehat{\overline{Y}}_k = \frac{1}{N}\sum_{j=1}^{m_k}\sum_{i=1}^{n_j}\frac{y_{ijk}}{\pi^{*\mathrm{I}}_j\pi^{\mathrm{II}}_{i|j}}, \quad \widehat{\overline{X}}_k = \frac{1}{N}\sum_{j=1}^{m_k}\sum_{i=1}^{n_j}\frac{\boldsymbol{x}_{ij}}{\pi^{*\mathrm{I}}_j\pi^{\mathrm{II}}_{i|j}}, \text{ and}$$

$$\hat{\boldsymbol{\beta}}_k = \left(\sum_{j=1}^{m_k}\sum_{i=1}^{n_j}\frac{\boldsymbol{x}_{ij}\boldsymbol{x}^t_{ij}}{\omega^2_i\pi^{*\mathrm{I}}_j\pi^{\mathrm{II}}_{i|j}}\right)^{-1}\sum_{j=1}^{m_k}\sum_{i=1}^{n_j}\frac{\boldsymbol{x}_{ij}y_{ijk}}{\omega^2_i\pi^{*\mathrm{I}}_j\pi^{\mathrm{II}}_{i|j}}$$

the HT estimators for $\overline{Y}_k$, $\overline{\boldsymbol{X}}$ and the vector with regression coefficients, respectively. Here $\boldsymbol{x}_{ij}$ denotes an H-vector with auxiliary information for sampling unit (i,j). The GREG estimators for the population mean observed under the K treatments can be collected in a K-vector $\widehat{\overline{\boldsymbol{Y}}}_r = (\widehat{\overline{Y}}_{1;r}, \ldots, \widehat{\overline{Y}}_{K;r})^t$. An approximately design-unbiased estimator for the covariance matrix of the $K-1$ contrasts between $\widehat{\overline{\boldsymbol{Y}}}_r$ is given by (23.7). In the case of an RBD, $\hat{\boldsymbol{D}}$ is a $K \times K$ diagonal matrix with elements

$$\hat{d}_k = \sum_{b=1}^{B}\frac{1}{m_{bk}(m_{bk}-1)}\sum_{j=1}^{m_{bk}}\left(\frac{m_b\hat{e}_{jk}}{N\pi^{\mathrm{I}}_j} - \frac{1}{m_{bk}}\sum_{j'=1}^{m_{bk}}\frac{m_b\hat{e}_{j'k}}{N\pi^{\mathrm{I}}_{j'}}\right)^2 \tag{23.17}$$

with

$$\hat{e}_{jk} = \sum_{i=1}^{n_j}\frac{y_{ijk} - \hat{\boldsymbol{\beta}}^t_k\boldsymbol{x}_{ij}}{\pi^{II}_{i|j}}.$$

The diagonal elements for a CRD follow as a special case from (23.17) with $B = 1$, $m_b = m$, and $m_{bk} = m_k$. A full proof of this result is given by Van den Brakel (2008).

The minimum use of information required to meet the condition that a constant H-vector $\boldsymbol{a}$ exists such that $\boldsymbol{a}^t\boldsymbol{x}_{ij} = 1$ for all elements in the population is a GREG estimator with weighting model $\boldsymbol{x}_{ij} = (1)$ and $\omega^2_{ij} = \omega^2$ for all elements in the population. Analogously to (23.9), it follows under this weighting model that the GREG estimator is equal to Hájek's ratio estimator for a population mean:

$$\widehat{\overline{Y}}_r = \left(\sum_{j=1}^{m_k}\sum_{i=1}^{n_j}\frac{1}{\pi^{*\mathrm{I}}_j\pi^{\mathrm{II}}_{i|j}}\right)^{-1}\sum_{j=1}^{m_k}\sum_{i=1}^{n_j}\frac{y_{ijk}}{\pi^{*\mathrm{I}}_j\pi^{\mathrm{II}}_{i|j}} \equiv \tilde{y}_k \tag{23.18}$$

It also follows that $\widehat{\beta}_k = \widetilde{y}_k$. An approximately design-unbiased estimator for the covariance matrix of the $K-1$ contrasts is given by (23.17) with

$$\widehat{e}_{jk} = \sum_{i=1}^{n_j} \frac{y_{ijk} - \widetilde{y}_k}{\pi^{\mathrm{II}}_{i|j}} \equiv \widehat{y}_{jk} - \widetilde{y}_k \widehat{N}_j$$

If the number of experimental units within each block is small, the variance estimation procedure might be improved by pooling the variance estimators for the separate subsamples, i.e.

$$\widehat{d}^{p}_{k} = \sum_{b=1}^{B} \frac{1}{m_{bk}(m_{bk}-K)} \sum_{k'=1}^{K} \sum_{j=1}^{m_{bk'}} \left(\frac{m_b \widehat{e}_{jk'}}{N\pi^{\mathrm{I}}_{j}} - \frac{1}{m_{bk'}} \sum_{j'=1}^{m_{bk'}} \frac{m_b \widehat{e}_{j'k'}}{N\pi^{\mathrm{I}}_{j'}} \right)^2 \tag{23.19}$$

and $\widehat{e}_{jk'}$ defined similarly as in (23.17).

The hypothesis of no treatment effects (23.2) is tested with a Wald statistic (23.10) using GREG estimates derived in this subsection with variance estimators (23.18) or (23.19). In the case of two-treatment experiments, the design-based t-type statistic (23.13) can be used with the subsample and variance estimators derived in this subsection.

Now consider an experiment where clusters of sampling units that are assigned to the same interviewer are randomized over the treatments (i.e. all cases worked by one interviewer, regardless of the areas in which they are worked, are assigned to the same treatment). Examples of this design include the stratified two-stage samples in the Netherlands with CAPI data collection. Interviewers work in multiple areas around their place of residence that do not coincide with the PSUs of the sample design, and these areas might be assigned to different treatments. The analysis of this type of experiment can be conducted with the procedure proposed in this subsection by taking $\pi^{\mathrm{I}}_{j} = 1$ for all j and considering $\pi^{\mathrm{II}}_{i|j} = \pi_i$ as the first-order inclusion probabilities of the sampling design. Furthermore, m_{bk} denotes the number of interviewers in block b who are assigned to treatment k, m_b the number of interviewers in block b, m_k the number of interviewers assigned to treatment k, and m the total number of interviewers in the experiment. In the example above, blocks could be the strata of the sample design if they contain multiple interviewer areas. This result is obtained by conceptually dividing the target population into M subpopulations, with M the number of interviewers available for the data collection. Each subpopulation consists of the sampling units that are interviewed by the same interviewer if they are included in the sample. These M subpopulations are included in the first stage of the sample and randomized over the treatments.

Results for ratios follow analogously from the preceding results. The ratio under each treatment is estimated as the ratio of two GREG estimators derived in this subsection. The variance components are estimated as

$$\widehat{d}^{(R)}_{k} = \frac{1}{\widehat{Z}^2_{k;r}} \sum_{b=1}^{B} \frac{1}{m_{bk}(m_{bk}-1)} \sum_{j=1}^{m_{bk}} \left(\frac{m_b \widehat{e}_{jk}}{N\pi^{\mathrm{I}}_{j}} - \frac{1}{m_{bk}} \sum_{j'=1}^{m_{bk}} \frac{m_b \widehat{e}_{j'k}}{N\pi^{\mathrm{I}}_{j'}} \right)^2 \tag{23.20}$$

with

$$\widehat{e}_{jk} = \sum_{i=1}^{n_j} \frac{(y_{ijk} - \widehat{\beta}^{y^t}_{k} \boldsymbol{x}_{ij}) - \widehat{R}_{k;r}(z_{ijk} - \widehat{\beta}^{z^t}_{k} \boldsymbol{x}_{ij})}{\pi^{\mathrm{II}}_{i|j}}.$$

23.5 Factorial Designs

23.5.1 Designing Embedded $K \times L$ Factorial Designs

So far, we have considered single-factor experiments. If two or more factors are investigated, it is generally efficient to combine them in one factorial design instead of conducting separate single-factor experiments since fewer experimental units are required to test main effects of the treatment factors and interactions between the treatment factors can be analyzed. Another advantage is that the validity of the results is extended, since the treatment effects are observed under a wider range of conditions (Hinkelmann and Kempthorne 1994).

In this section, the theory for factorial designs is explained for experiments where the effects of two factors are tested simultaneously. The first factor, denoted A, contains $K \geq 2$ levels. The second factor, denoted B, contains $L \geq 2$ levels. In a $K \times L$ factorial design, the K levels of factor A are crossed with the L levels of factor B. A probability sample s of size n is drawn from a finite target population U of size N, where π_i denote the first-order inclusion probabilities of the sample design. This sample is randomly divided into KL subsamples according to a randomized experiment. Each subsample is assigned to one of the KL treatment combinations.

As in the case of single-factor experiments, the most straightforward approach is a CRD, where the n sampling units are randomized over KL subsamples, say s_{kl} of size n_{kl}. In the case of an RBD, the sample s is first deterministically divided into B blocks of size n_b. Subsequently, n_{bkl} sampling units within each block are randomly assigned to each of the KL treatment combinations or subsamples. Following similar arguments as in Section 23.3.2, it follows that $\pi_i^* = (n_{kl}/n)\pi_i$ denotes the first-order inclusion probability for the elements in subsample s_{kl} in the case of a CRD. In the case of an RBD, the first-order inclusion probability is $\pi_i^* = (n_{bkl}/n_b)\pi_i$.

23.5.2 Testing Hypotheses About Main Effects and Interactions in $K \times L$ Embedded Factorial Designs

Let y_{iqkl} denote the observation obtained from the ith sampling unit that has been assigned to the klth treatment combination and the qth interviewer. The measurement error model, introduced in Section 23.3.1, is now extended for a factorial design to

$$y_{iqkl} = u_i + b_{kl} + \gamma_q + \varepsilon_{ikl} \tag{23.21}$$

In (23.21), u_i is the intrinsic value of the ith respondent. The mixed interviewer effect, γ_q, and the random measurement error, ε_{ikl}, are based on the same assumptions as in (23.1). Finally, b_{kl} is the treatment effect of the klth treatment combination ($k = 1, \ldots, K$ and $l = 1, \ldots, L$). The treatment effects can be decomposed into main and interaction effects in the traditional way of an analysis of variance for a two-way layout:

$$b_{kl} = A_k + B_l + AB_{kl} \tag{23.22}$$

with A_k and B_l the main effects of factor A and B and AB_{kl} the interactions. Note that (23.22) does not have an overall mean since the treatment effects b_{kl} are defined as fixed deviations from the true intrinsic values u_i in (23.21). In fact, the population mean of u_i in (23.21) replaces the overall mean in (23.22). The following restrictions are required to identify (23.22):

$$\sum_{k=1}^{K} A_k = 0,\ \sum_{l=1}^{L} B_l = 0,\ \sum_{k=1}^{K} AB_{kl} = 0, \quad l = 1, \ldots, L, \text{ and } \sum_{l=1}^{L} AB_{kl} = 0, \quad k = 1, \ldots, K \tag{23.23}$$

Population means observed under the KL different treatments or survey implementations are defined as $\overline{Y}_{kl} = \overline{U} + b_{kl} + \overline{\gamma} + \overline{\varepsilon}_{kl}$ and can be collected in a KL vector $\overline{\boldsymbol{Y}} = (\overline{Y}_{11}, \ldots, \overline{Y}_{1L}, \ldots, \overline{Y}_{K1}, \ldots, \overline{Y}_{KL})^t$. It is important to note that the subscript of factor B runs within factor A, since this determines the order of the elements in the vector $\overline{\boldsymbol{Y}}$. Hypotheses of contrasts between the population means are formulated in expectation over the measurement error model, as stated by (23.2), where $\boldsymbol{C}$ denotes an appropriate contrast matrix, depending on the type of hypothesis to be tested. In the case of $K \times L$ factorial designs, hypotheses of interest are about the main effects of factors A and B and the interactions between both factors.

The hypothesis about the main effects of factor A is defined as the $K-1$ contrasts between the K levels of factor A, averaged over the L levels of factor B and is obtained by the following contrast matrix:

$$C_A = \frac{1}{L}(j_{(K-1)} \mid -I_{(K-1)}) \otimes j^t_{(L)} \equiv \widetilde{C}_A \otimes j^t_{(L)} \tag{23.24}$$

where $j_{(p)}$ denotes a p-vector with each element equal to 1, $I_{(p)}$ the identity matrix of order p, and $\otimes$ the Kronecker product. Under the measurement error model defined in (23.21)–(23.23), it follows that $C_A E_m \overline{Y} = (A_1 - A_2, A_1 - A_3, \dots, A_1 - A_K)^t$ and thus exactly corresponds to the contrasts between the main effects of the first factor. The contrast matrix for the hypothesis about the main effects of factor B is defined as $L-1$ contrasts between the L levels of factor B, averaged over the K levels of factor A:

$$C_B = \frac{1}{K} j^t_{(K)} \otimes (j_{(L-1)} \mid -I_{(L-1)}) \equiv j^t_{(K)} \otimes \widetilde{C}_B \tag{23.25}$$

Under the measurement error model defined in (23.21)–(23.23), it follows that $C_B E_m \overline{Y} = (B_1 - B_2, B_1 - B_3, \dots, B_1 - B_L)^t$. Interactions between the two-treatment factors are defined as the $L-1$ contrasts of factor B between the $K-1$ contrast of factor A, or vice versa (Hinkelmann and Kempthorne 1994, Ch. 11). Therefore, the contrast matrix for the $(K-1) \times (L-1)$ interactions between factors A and B is defined as:

$$C_B = (j_{(K-1)} \mid -I_{(K-1)}) \otimes (j_{(L-1)} \mid -I_{(L-1)}) = \widetilde{C}_A \otimes \widetilde{C}_B \tag{23.26}$$

Similar to the main effects it follows that under the measurement error model the contrasts between the population parameter exactly correspond to the interactions between the first and second factor, since

$$\begin{aligned} C_{AB} E_m \overline{Y} = (&AB_{11} - AB_{12} - AB_{21} + AB_{22}, \dots, AB_{11} - AB_{1L} - AB_{21} + AB_{2L}, \dots, \\ &AB_{11} - AB_{12} - AB_{K1} + AB_{K2}, \dots, AB_{11} - AB_{1L} - AB_{K1} + AB_{KL}) \end{aligned}$$

Similar to the approach followed in the preceding sections, the Wald test can be used to test hypotheses about main effects and interaction effects. To this end, the Wald statistic (23.3) is used, where C is replaced by (23.24)–(23.26). The data obtained from the n_{kl} sampling units in subsample s_{kl} can be used to construct KL GREG estimators $\widehat{\overline{Y}}_{kl;r}$ for the population means $\overline{Y}_{kl}$, for $k = 1, \dots, K$ and $l = 1, \dots, L$, and are defined analogously to (23.5) using the inclusion probabilities derived for the units included in the subsamples s_{kl}. Now $\widehat{\overline{Y}}_r = (\widehat{\overline{Y}}_{11;r}, \dots, \widehat{\overline{Y}}_{kl;r}, \dots, \widehat{\overline{Y}}_{KL;r})^t$ is an approximately design-unbiased estimator for $\overline{Y}$ and thus also for $E_m \overline{Y}$. An estimator for the covariance matrix of the contrasts is defined by (23.7) where C is replaced by (23.24), (23.25) or (23.26), and (23.8) is based on the n_{kl} observations obtained in subsample s_{kl}. Furthermore, n_{bk} is replaced by n_{bkl} and $\widehat{e}_{ik}$ by $\widehat{e}_{ikl} = y_{ikl} - \widehat{\beta}^t_{kl} x_i$. This gives rise to a Wald statistic defined by (23.10). For the test of main effects, it can be shown that (23.10) can be simplified to (23.11) with $\widehat{\overline{Y}}_{k;r}$ and $\widehat{d}_k$ replaced by

$$\widehat{\overline{Y}}_{k.;r} = \frac{1}{L}\sum_{l=1}^{L} \widehat{\overline{Y}}_{kl;r} \quad \text{and} \quad \widehat{d}_{k.} = \frac{1}{L^2}\sum_{l=1}^{L} \widehat{d}_{kl}$$

in the case of the test of the main effects of factor A or

$$\widehat{\overline{Y}}_{.l;r} = \frac{1}{K}\sum_{k=1}^{K} \widehat{\overline{Y}}_{kl;r} \quad \text{and} \quad \widehat{d}_{.l} = \frac{1}{K^2}\sum_{k=1}^{K} \widehat{d}_{kl}$$

in the case of the test of the main effects of factor B. For the test of the interaction, expression (23.10) is required. Let $\chi^2_{[p]}$ denote the (central) chi-squared distribution with p degrees of freedom. If it is conjectured that $\widehat{C\boldsymbol{Y}}_r$ is multivariate normally distributed, then it follows under the null hypothesis for the Wald statistic that $W \rightarrow \chi^2_{[K-1]}$ for the test about the main effects of factor A, $W \rightarrow \chi^2_{[L-1]}$ for the test about the main effects of factor B, and $W \rightarrow \chi^2_{[(K-1)(L-1)]}$ for the test about the interaction between factor A and B.

23.5.3 Special Cases

The Hájek estimator that only uses the population size as auxiliary information is defined analogously to (23.9), where n_k is replaced by n_{kl} and y_{ik} by y_{ikl}. In the case of small sample sizes, the variance estimation procedure for an RBD can be stabilized by applying a pooled variance estimator:

$$\hat{d}^p_{kl} = \sum_{b=1}^{B} \frac{1}{n_{bkl}(n_b - KL)} \sum_{k'=1}^{K} \sum_{l'=1}^{L} \sum_{i=1}^{n_{bk'l'}} \left(\frac{n_b \hat{e}_{ik'l'}}{N\pi_i} - \frac{1}{n_{bk'l'}} \sum_{i'=1}^{n_{bk'l'}} \frac{n_b \hat{e}_{i'k'l'}}{N\pi_{i'}} \right)^2$$

A pooled estimator for a CRD follows as a special case by taking $B = 1$, $n_{bkl} = n_{kl}$, and $n_b = n$.

In the case of a CRD embedded in a self-weighted sample design with equal subsample sizes under Hájek's estimator, Van den Brakel (2013) showed that the Wald statistic for the test of the null hypothesis of the main effects of factors A and B in a $K \times L$ factorial design is equal to the F-statistic for the main effects of an analysis of variance in a two-way layout. In the case of an RBD under the same conditions, Van den Brakel (2013) showed that the Wald statistic for the test of the null hypothesis of the main effects of the factors A and B in a $K \times L$ factorial design is equal to the F-statistic for the main effects of an analysis of variance in a three-way layout. These results correspond with what would be expected intuitively.

23.5.4 Generalizations

The generalization to factorial designs with more than two factors is relative straightforward. The point and variance estimation procedure is similar for one- and two-factor experiments. The only complication is that more hypotheses can be tested. The specification of the corresponding contrast matrices requires a more complicated notation. In this subsection, the contrast matrices for a three-factor experiment are given as an example. The specification of the contrast matrices for the general case of an experiment with, say, G factors is spelled out in detail in Van den Brakel (2013).

Consider an experiment where three factors are tested. Similarly to Section 23.5.1, the first two factors A and B are tested at K and L levels, respectively. The third factor, say C, is tested at $M \geq 2$ levels. The estimates of the population means obtained under treatment combination k, l, and m are denoted as $\widehat{Y}_{klm;r}$ and are combined in KLM dimensional vector $\widehat{\boldsymbol{Y}}_r$ where the order of the elements is determined by running index $m = 1, \ldots, M$ within each klth combination and $l = 1, \ldots, L$ within each level k. Let $\tilde{\boldsymbol{C}}_C = (\boldsymbol{j}_{(M-1)} \mid -\boldsymbol{I}_{(M-1)})$. Now the contrast matrix for the main effects for factor A is defined as the $K-1$ contrasts between the K levels of A, averaged over the L levels of factor B and M levels of factor C and is defined as $\boldsymbol{C}_A = (LM)^{-1}\,\tilde{\boldsymbol{C}}_A \otimes \boldsymbol{j}^t_{[LM]}$. In a similar way, the matrix for the $L-1$ contrasts of factor B is defined as $\boldsymbol{C}_B = (KM)^{-1}\boldsymbol{j}^t_{[K]} \otimes \tilde{\boldsymbol{C}}_B \otimes \boldsymbol{j}^t_{[M]}$ and the $M-1$ contrasts of factor C as $\boldsymbol{C}_C = (KL)^{-1}\boldsymbol{j}^t_{[KL]} \otimes \tilde{\boldsymbol{C}}_C$. Second-order interactions between factors A and B are defined as the $(K-1)$ contrasts of A between the $L-1$ contrasts of B, averaged over the M levels of C, resulting in $(K-1)(L-1)$ contrasts that can be defined with the matrix $\boldsymbol{C}_{AB} = M^{-1}\,\widetilde{\boldsymbol{C}}_A \otimes \widetilde{\boldsymbol{C}}_B \otimes \boldsymbol{j}_{[M]}^t$. Similarly, the $(K-1)(M-1)$ contrasts of the second-order interactions between factors A and C can be defined as $\boldsymbol{C}_{AC} = L^{-1}\,\widetilde{\boldsymbol{C}}_A \otimes \boldsymbol{j}_{[L]}^t \otimes \widetilde{\boldsymbol{C}}_C$ and the $(L-1)(M-1)$ contrasts of the second-order interactions between factors B and C as $\boldsymbol{C}_{BC} = K^{-1}\,\boldsymbol{j}_{[K]}^t \otimes \widetilde{\boldsymbol{C}}_B \otimes \widetilde{\boldsymbol{C}}_C$. Finally, the third-order interactions between factors A, B, and C are defined by the $M-1$ contrasts of C between the second-order interactions between A and B. This results in $(K-1)(L-1)(M-1)$ contrasts that are defined by the contrast matrix $\boldsymbol{C}_{ABC} = \widetilde{\boldsymbol{C}}_A \otimes \widetilde{\boldsymbol{C}}_B \otimes \widetilde{\boldsymbol{C}}_C$.

The different hypotheses can be tested with Wald statistic (23.10). Under the null hypothesis, this Wald statistic is a chi-squared distributed random variable where the number of degrees of freedom is equal to the number of contrasts specified in the contrast matrix.

Testing hypotheses about parameters that are defined as the ratio of two population means proceeds by applying the point and variance estimators described in Section 23.3.5 to each subsample in a factorial setup. Subsequently, hypotheses are tested with the contrasts matrices derived in Sections 23.5.2 and 23.5.4 for main and interaction effects in factorial designs using Wald statistic (23.10).

To analyze factorial designs where clusters of sampling elements are randomized over the treatment combinations of an experiment, the point and variance estimators described in Section 23.4 must be applied to each subsample, i.e. the group of sampling elements assigned to each specific treatment combination. Hypotheses are tested with the contrast matrices from Sections 23.5.2 and 23.5.4 for main and interaction effects in factorial designs using the Wald statistic (23.10).

23.6 A Mixed-Mode Experiment in the Dutch Crime Victimization Survey

23.6.1 Introduction

Information on crime victimization, public safety, and satisfaction with police performance in the Netherlands is obtained by the Dutch Crime Victimization Survey (CVS), which is conducted by Statistics Netherlands. This is an annual survey designed to produce sufficiently precise estimates at the national level and the level of the Netherlands' 25 police districts. Before 2008, data collection was based on a mixed-mode design via CAPI and CATI. Persons for whom a telephone number is available are interviewed by CATI, and the remaining persons are interviewed by CAPI.

To reduce administration costs, a sequential mixed-mode design has been considered. Under this mixed-mode design, all persons included in the sample receive an advance letter with the request to complete the questionnaire on the web. After two reminders, nonrespondents are followed up with CATI if a telephone number is available or CAPI otherwise. Changes in data collection modes generally affect response rates and, therefore, the selection bias in the outcomes of a survey. In addition, a different data collection mode results in different amounts of measurement bias in the answers of respondents; see, e.g. De Leeuw (2005). To obtain quantitative insight into the effect of the introduction of a sequential mixed-mode design, an experiment embedded in the regular survey of the CVS was conducted in 2006. This experiment is used as an illustration of the methods described in the preceding sections.

23.6.2 Survey Design and Experimental Design

The CVS is based on a stratified simple random sample of people aged 15 years or older residing in the Netherlands. The sampling frame is the Municipal Basis Administration, which is the Dutch government's registry of all residents in the country. The 25 police districts are used as strata in the sample design. In a regular yearly sample, about 750 respondents are observed in each police district, resulting in a total net sample size of about 19 000 respondents. With a response rate of about 62%, this requires a gross sample size of about 30 500 persons. Since police districts have unequal population sizes, inclusion probabilities vary between police districts. The estimation procedure is based on the GREG estimator where the weighting model contains sociodemographic categorical variables such as gender (2), age class (11), marital status (4), urbanization level (5), police region (25), and household size (5), where the number of categories is specified in parentheses.

An important step in the design of an experiment is to decide in advance which hypotheses will be tested and which treatment effect must be observed at a prespecified significance and power level. Based on such considerations, the minimum required sample size of an experiment can be derived. To test the effect of the aforementioned sequential mixed-mode design, there was budget to increase the gross annual sample in 2006 by 3750 persons. This gave rise to an experiment embedded in the CVS where the regular CVS used for official publication purposes was used as the control group with an expected sample size of 19 000 respondents and the experimental group had an expected sample size of 2350 respondents. Instead of calculating the minimum sample size, we calculated which differences could be minimally observed at a prespecified significance and power level, which was useful to obtain insight about what could be achieved with this experiment. See Table 23.1 for an overview of the key parameters selected to test hypotheses about treatment effects.

Table 23.2 contains an overview of the differences that can be minimally detected at a 5% significance level and a power of 50%, 80%, and 90%. In columns two, three, and four, these calculations are made for the applied design, i.e. a control group of size 19 000 and an experimental group of size 2350. In columns five, six, and seven, the minimal observable differences are specified if the total sample size of $19\,000 + 2350 = 21\,350$ is equally divided over the two subsamples, i.e. a balanced design with an equal subsample size of 10 675 respondents.

Embedding an experiment in an ongoing survey is efficient in the sense that the regular survey conducted for official publication purposes is simultaneously used as the control group. On the other hand, it should be realized that this type of experiment combines two competing purposes. The purpose of the regular survey is to estimate population parameters as precisely as possible, which is achieved if as much sample size as possible was allocated to the regular survey. The purpose of the experiment, by contrast, is to estimate treatment effects as precisely as possible, which can be achieved with balanced designs where both subsamples are equally sized as illustrated in Table 23.2.

Finally, the experiment was designed as an RBD where the stratification variable of the sampling design was used as a block variable. Within each stratum, a sample of 1370 persons was

Table 23.1 Descriptions of key parameters, with labels used for reference.

Label	Parameter description
Satispol	Percentage of people satisfied with police performance during their last contact (if there was contact in the last 12 months)
Nuisance	Mean perceived amount of irritation due to antisocial behavior by drunk people, neighbors, or groups of youngsters and harassment and drug-related problems measured on a 10-point Likert scale in the last 12 months
Propvic	Percentage of people said to be victim of a property crime in the last 12 months
Violvic	Percentage of people said to be victim of a violent crime or assault in the last 12 months
Repvic	Overall victimization rate of reported crimes over the last 12 months

Table 23.2 Minimal size of treatment effects that can be detected at a 5% significance level and a power of 50%, 80%, or 90% under the applied allocation and a balanced design.

Estimate	Power level					
	50%	80%	90%	50%	80%	90%
	Δ applied design			Δ balanced design $n_k = 10\,675$		
Satispol	4.55	6.49	7.51	2.85	4.07	4.70
Nuisance	0.06	0.09	1.00	0.04	0.05	0.06
Propvic	2.06	2.94	3.41	1.29	1.84	2.13
Violvic	2.30	3.29	3.81	1.44	2.06	2.38
Repvic	4.49	6.41	7.41	2.81	4.01	4.64

drawn. We randomly assigned a fraction of 0.9 to the regular survey and 0.1 to the experimental group. With an expected response rate of 62%, 750 and 85 responses within each stratum were expected under the control group and treatment group, respectively.

23.6.3 Software

The design-based analysis procedures proposed for single-factor experiments were implemented in a software package called X-tool. This package is available as a component of the Blaise survey processing software package, developed by Statistics Netherlands (Statistics Netherlands 2002). X-tool is a software package to test hypotheses about differences between population parameter estimates observed under different survey implementations in randomized experiments embedded in complex probability samples. X-tool handles experiments designed as CRDs and RBDs. It is possible to analyze experiments where the sampling elements as well as clusters of sampling elements are randomized over the different treatments. Subsample estimates for means, totals, and ratios are based on the Hájek estimator or the GREG estimator. The integrated method of Lemaître and Dufour (1987) for weighting individuals and households can be applied under the GREG estimator to obtain equal weights for individuals belonging to the same household. Also a bounding algorithm based on Huang and Fuller (1978) can be applied to avoid negative correction weights. See Van den Brakel (2008) for more details on the functionality of X-tool. Access to X-tool requires a developer's license of Blaise 4. This software is distributed by Westat USA for North and South America and Statistics Netherlands for countries elsewhere. See http://blaise.com/ or https://www.westat.com/ for details.

23.6.4 Results

The first step in the analysis of this field experiment was to analyze the field work in both treatment groups. Table 23.3 contains an overview of the fieldwork results in the control group (the regular CVS) and the experimental group. In the experimental group, 1002 responses were obtained through the web after sending two reminders. From the remaining nonrespondents, 1958 persons were contacted by telephone (CATI), and 678 persons are approached at home by an interviewer (CAPI). From these two groups, an additional 91 responses were obtained through the web at this latter stage of the field period, 60 through CATI and 31 through CAPI. As a result, a total of 1093 complete web responses were obtained.

Table 23.3 Overview of fieldwork results for the CVS experiment.

	Web		CATI		CAPI		Total	
	Number	%[a)]	Number	%[b)]	Number	%[c)]	Numbers	%[d)]
Control group								
Gross sample			22 977		7 510		30 487	
Frame error[e)]			733		473		1 206	
Approached persons			22 244	100.0	7 037	100.0	29 281	100.0
Partial response			158	0.7	23	0.3	181	0.6
Refusal			2 816	12.7	1 204	17.1	4 020	13.7
No contact			1 209	5.4	464	6.6	1 673	5.7
Not approached			114	0.5	86	1.2	200	0.7
Rest			1 660	7.5	682	9.7	2 342	8.0
Complete response			16 287	73.2	4 578	65.1	20 865	71.3
Experimental group								
Gross sample	3 750		1 958		678		3 750	
Frame error[e)]			68		67		135	
Responded through web			60		31			
Approached persons	3 750	100.0	1 830	100.0	580	100.0	3 615	100.0
Partial response	0	0.0	0	0.0	0	0.0	0	0.0
Refusal	85	2.3	510	27.9	159	27.4	754	20.9
No contact	0	0.0	149	8.1	72	12.4	221	6.1
Not approached	0	0.0	9	0.5	9	1.6	18	0.5
Rest	2 572	68.6	172	9.4	77	13.3	284	7.9
Complete response	1 093	29.1	990	54.1	263	45.3	2 338	64.7

a) Distribution of the persons approached (excluding frame errors) under web.
b) Distribution of the persons approached (excluding frame errors) under CATI.
c) Distribution of the persons approached (excluding frame errors) under CAPI.
d) Distribution of the persons approached under all modes (CATI + CAPI or WEB + CATI + CAPI).
e) Frame error contains: respondent died, moved to another address, or not known at this address or telephone disconnected (CAPI only).

Table 23.3 illustrates that the response rate in the experimental group with the sequential mixed-mode design was about 7% lower and the refusal rate about 7% higher compared to the regular mixed-mode approach based on CATI and CAPI only. A sequential mixed-mode design starting with web clearly increased the refusal rate since the first contact did not involve a personal contact with an interviewer, which made it easier for sampled persons to refuse participation with the survey. The response rates under CATI and CAPI after web were indeed dramatically lower compared to the CATI and CAPI response rates in the regular CVS. For both modes, the response rates dropped by about 20 percentage points if the persons are first asked through a letter to respond through the Internet. This can be expected since the easy-to-reach respondents already participated with the web mode, but the net effect was a lower total response rate under the sequential mixed-mode design.

Overall, 20 865 responses were obtained under the regular survey and 2338 responses in the experimental treatment. With both subsamples, GREG estimates were calculated for the five key parameters under the CAPI–CATI mixed-mode design and the sequential mixed-mode design with web, CAPI, and CATI. Inclusion probabilities are based on the stratified sample design of the CVS and the RBD used to divide the sample into two subsamples. The GREG estimator was based on the following model: gender(2) × ageclass(11) + marital status(4) + urbanization level(5) + police region(25) + household size(5). Due to the relatively small sample size of the experimental group, the model contained mostly main effects and only one second-order interaction term. Following the procedure explained in Section 23.3.2, the design-based t-statistic (23.13) was used to test hypotheses about differences between the population estimates observed under the different data collection modes. Results are summarized in Table 23.4.

For nuisance, a significantly higher mean score was observed under the sequential mixed-mode design. For the other four variables, no significant differences were found. These differences would also not lead to a rejection of the null hypotheses if they were observed in an experiment where the regular and experimental approach were both conducted at the full sample size of 19 000 respondents. In Table 23.5, the estimates for the variables under the separate modes are given. The differences between these estimates are the net result of the different subpopulations that responded under the different modes and the mode-dependent measurement bias. Separation of mode-dependent selection and measurement bias is not possible with this experimental design. To separate mode-dependent measurement bias and mode-dependent selection bias, Schouten et al. (2013) proposed an experiment with repeated measurements.

Table 23.4 Differences between key CVS estimates under the regular mixed-mode data collection and the experimental sequential mixed-mode design.

Variable	Regular/control group		Experimental group		Difference		
	Estimate	$\sqrt{\hat{d}_1}$	Estimate	$\sqrt{\hat{d}_2}$	Treatment effect	t-Statistic	p Value
Satispol	55.46	(0.75)	53.95	(2.35)	1.51	0.610	0.542
Nuisance	2.94	(0.01)	3.19	(0.03)	−0.25	−6.958	0.000
Propvic	16.02	(0.38)	14.47	(1.06)	1.55	1.377	0.169
Violvic	8.47	(0.34)	8.71	(1.00)	−0.23	−0.220	0.826
Repvic	35.95	(0.74)	34.36	(2.20)	1.60	0.492	0.689

Table 23.5 Key CVS estimates under the different data collection modes of the regular and experimental group.

	Regular/control group				Experimental group					
	CAPI		CATI		CAPI		CATI		Web	
	Estimate	S.E.	Estimate	S.E.	Estimate	S.E.	Estimate	S.E.	Estimate	S.E.
Satispol	50.91	(1.46)	58.06	(0.83)	52.28	(6.65)	62.79	(3.50)	48.77	(3.31)
Nuisance	3.20	(0.02)	2.81	(0.01)	3.04	(0.11)	2.70	(0.05)	3.63	(0.05)
Propvic	23.34	(0.96)	12.61	(0.34)	10.29	(2.91)	11.13	(1.31)	18.62	(1.74)
Violvic	13.94	(0.86)	5.94	(0.29)	7.48	(2.42)	7.73	(1.23)	9.92	(1.73)
Repvic	33.75	(1.29)	37.80	(0.84)	36.54	(6.48)	43.92	(3.61)	29.62	(2.92)

S.E.: standard error for the subdomain estimate that responded through CATI, CAPI, or web.

This experiment showed that with the introduction of a sequential mixed-mode data collection with web, CAPI, and CATI, the administration costs per complete response can be reduced by about 25%. The experiment also showed that this results in a reduction of the response rate by about 7%. For the most important key parameters, the mean nuisance was estimated to be significantly higher. For the other variables, there were no indications that the introduction of a sequential mixed-mode design with web resulted in significantly different estimates. Since nuisance was measured on a Likert scale based on 10 underlying questions, the standard error of the mean was much smaller compared to the other variables. Therefore, a small difference resulted in a rejection of the null hypothesis for nuisance.

23.7 Discussion

This chapter presents a general framework for the design and analysis of experiments embedded in sample surveys. Embedding randomized experiments in probability samples results in experiments that potentially have strong internal and external validity. Principles from the theory of randomized experiments such as randomization, replication, and blocking are typically intended to guarantee the causal relationship between the treatments of the experiment and the observed effects. In the design stage of an embedded experiment, the sampling design of the probability sample provides a useful framework for the design of an experiment. Sampling structures such as strata, clusters, PSUs, and sampling units assigned to the same interviewer can be used as block variables in an RBD to improve the precision of the experiment. Randomized sampling, on the other hand, is intended to generalize results obtained in a sample to a larger target population from which the sample is drawn. In sampling theory, this is achieved by constructing (approximately) design-unbiased estimators for the unknown target parameters. If the experimental units in an embedded experiment are selected by means of a probability sample with the purpose to generalize results to an intended target population from which the experimental units are drawn, then a design-based inference framework, similar to the approach followed in sampling theory, is required. In this chapter, such a design-based approach is proposed for the analysis of embedded experiments that covers a broad set of situations. A software package, called X-tool, is available as a component of the Blaise suite to perform the analyses proposed in this chapter.

The methods are illustrated with a two-sample experiment embedded in the Dutch CVS to test the effect of different data collection modes. Another application can be found in Van den Brakel and Van Berkel (2002) where a two-sample experiment is described to test the effect of an alternative questionnaire in the Dutch Labor Force Survey (LFS). An application of an embedded RBD to test four different incentives in the LFS is described in Van den Brakel (2008). An example of a 2×2 factorial design embedded in the Dutch Family and Fertility Survey to test two different data collection modes and two different incentives is described in Van den Brakel et al. (2006). Finally, a 2×3 factorial design embedded in the LFS to test different forms of advance letters is described in Van den Brakel (2013).

Several directions for further research in this area can be identified. First of all, an implementation in an *R* package would make the methodology proposed in this chapter more accessible for empirical researchers. The design-based procedures can be further extended to more advanced experimental designs. The number of treatment combinations in full-factorial designs increases rapidly with the number of factors, which might hamper the implementation in the fieldwork. To reduce the number of treatment combinations, design-based analysis procedures for incomplete block designs and fractional factorial designs are required. These are more advanced experimental designs where the number of treatment combinations within a block or in the entire experiment is reduced (Hinkelmann and Kempthorne 2005). This requires that certain treatment effects, generally higher-order interactions, are indistinguishable from blocks (in the case of incomplete block designs) or from each other (in the case of fractional factorial designs).

Another important research area in mixed-mode designs is separating measurement bias from selection bias using repeated measurement experiments (Schouten et al. 2013). Extending the design-based analysis procedures for these types of cross-over designs is relevant for mixed-mode research.

Finally, there is a link with small area estimation. Sometimes interactions between treatment effects and important publication domains are relevant. Sample sizes of embedded experiments are often not sufficiently large to produce reliable estimates for the treatment effects for such domains. Van den Brakel et al. (2016) proposed a Fay–Herriot model to obtain more precise domain predictions for the treatment effects for the situation where outcomes under a regular survey obtained with a large sample size are compared with estimates obtained under an alternative approach observed under a much smaller sample size. Further extensions of small area estimation procedures for obtaining more precise predictions for treatment effects are relevant, since it improves the power of experiments in studies lacking sufficient sample sizes in small domains of interest.

Acknowledgments

The views expressed in this chapter are those of the author and do not necessarily reflect the policy of Statistics Netherlands. The author is grateful to the editors for careful reading of former drafts of this chapter and giving constructive comments, which proved to be very helpful for improving this chapter.

References

Boonstra, H.J. (2014). Time-series small area estimation for unemployment based on a rotating panel survey. Technical report 201417, Statistics Netherlands.

Boonstra, H.J., van den Brakel, J., Buelens, B. et al. (2008). Towards small area estimation at Statistics Netherlands. *Metron* LXVI (1): 21–49.

Cochran, W.G. (1977). *Sampling Techniques.* New York: Wiley.

De Leeuw, E. (2005). To mix or not to mix? Data collection modes in surveys. *Journal of Official Statistics* 21: 1–23.

Fienberg, S.E. and Tanur, J.M. (1987). Experimental and sampling structures: parallels diverging and meeting. *International Statistical Review* 55: 75–96.

Fienberg, S.E. and Tanur, J.M. (1988). From the inside out and the outside in: combining experimental and sampling structures. *Canadian Journal of Statistics* 16: 135–151.

Fienberg, S.E. and Tanur, J.M. (1989). Combining cognitive and statistical approaches to survey design. *Science* 243: 1017–1022.

Hájek, J. (1971). Comment on a paper by D. Basu. In: *Foundations of Statistical Inference* (ed. V.P. Godambe and D.A. Sprott), 236. Toronto: Holt, Rinehart and Winston.

Hansen, M.H., Hurwitz, W.N., and Madow, W.G. (1953). *Sample Survey Methods and Theory, Vol I and II*. New York: Wiley.

Hartley, H.O. and Rao, J.N.K. (1978). Estimation of nonsampling variance components in sample surveys. In: *Survey Sampling and Measurement* (ed. N.K. Namboodiri), 35–43. New York: Academic Press.
Hinkelmann, K. and Kempthorne, O. (1994). *Design and Analysis of Experiments, Volume 1: Introduction to Experimental Design*. New York: Wiley.
Hinkelmann, K. and Kempthorne, O. (2005). *Design and Analysis of Experiments, Volume 2: Advanced Experimental Design*. New York: Wiley.
Huang, E.T. and Fuller, W.A. (1978). *Nonnegative Regression Estimation for Survey Data*. Proceedings of the Social Statistics Session, 300–305. American Statistical Association.
Kish, L. (1965). *Survey Sampling*. New York: Wiley.
Lemaître, G. and Dufour, J. (1987). An Integrated Method for Weighting Persons and Families. *Survey Methodology* 13: 199–207.
Mahalanobis, P.C. (1946). Recent experiments in statistical sampling in the Indian statistical institute. *Journal of the Royal Statistical Society* 109: 325–370.
Miller, R.G. (1986). Beyond ANOVA,). *Basics of Applied Statistics*. New York: Wiley.
Montgomery, D.C. (2001). *Design and Analysis of Experiments*. New York: Wiley.
Robinson, G.K. (2000). *Practical strategies for experimenting*. New York: Sons.
Särndal, C.E., Swensson, B., and Wretman, J.H. (1992). *Model Assisted Survey Sampling*. New York: Springer Verlag.
Scheffé, H. (1959). *The Analysis of Variance*. New York: Wiley.
Schouten, B., Van den Brakel, J.A., Buelens, B. et al. (2013). Disentangling mode-specific selection and measurement bias in social surveys. *Social Science Research* 42: 1555–1570.
Searle, S.R. (1971). *Linear Models*. New York: Wiley.
Statistics Netherlands (2002). *Blaise Developer's Guide*. Heerlen: Statistics Netherlands. (Available from http://blaise.com/).
Van den Brakel, J.A. (2001). Design and analysis of experiments embedded in complex sample designs. PhD thesis. Rotterdam: Erasmus University of Rotterdam.
Van den Brakel, J.A. (2008). Design-based analysis of embedded experiments with applications in the Dutch Labour Force Survey. *Journal of the Royal Statistical Society, Series A* 171: 581–613.
Van den Brakel, J.A. (2013). Design-based analysis of factorial designs embedded in probability samples. *Survey Methodology* 39: 323–349.
Van den Brakel, J.A., Buelens, B., and Boonstra, H.J. (2016). Small area estimation to quantify discontinuities in repeated sample surveys. *Journal of the Royal Statistical Society, Series A* 179: 229–250.
Van den Brakel, J.A., Smith, P.A., and Compton, S. (2008). Quality procedures for survey transitions – experiments, time series and discontinuities. *Survey Research Methods* 2: 123–141.
Van den Brakel, J.A. and Renssen, R.H. (2005). Analysis of experiments embedded in complex sampling designs. *Survey Methodology* 31: 23–40.
Van den Brakel, J.A. and Van Berkel, C.A.M. (2002). A design-based analysis procedure for two-treatment experiments embedded in sample surveys. *Journal of Official Statistics* 18: 217–231.
Van den Brakel, J.A. and Renssen, R.H. (1998). Design and analysis of experiments embedded in sample surveys. *Journal of Official Statistics* 14: 277–295.
Van den Brakel, J.A., Vis-Visschers, R., and Schmeets, H. (2006). An experiment with data collection modes and incentives in the Dutch family and fertility survey for young Moroccans and Turks. *Field Methods* 18: 321–334.

24

Extending the Within-Persons Experimental Design: The Multitrait-Multierror (MTME) Approach

Alexandru Cernat[1] and Daniel L. Oberski[2]

[1] *Social Statistics Department, University of Manchester, Manchester M13 9PL, UK*
[2] *Department of Methodology & Statistics, Utrecht University, Utrecht, 3584 CH, the Netherlands*

24.1 Introduction

The typical aim of surveys is analytical as they are used to help investigate relationships between variables. To this end, methodologists have strived to estimate and minimize errors that can bias such estimates. Examples of possible measurement errors that could bias these results are random error, interviewer effects, response styles, and processing errors. Typically, these can also vary over respondents and are therefore stochastic, with the known effect of biasing estimates of relationships (Fuller 1987). For example, random errors, method variance, and interviewer effects may severely distort the analytic goals of the survey researcher (Beullens and Loosveldt 2014; Krosnick 1999; Saris and Gallhofer 2007). While research has been carried out separately on each of these different types of measurement errors, the need to estimate multiple stochastic errors concurrently has also been highlighted in the Total Survey Error (TSE) framework (Groves and Lyberg 2010).

If these biasing effects are to be evaluated, corrected for, or minimized by design, we first need to know how strong they are. Quantifying the extent to which such errors are present in survey answers is thus an essential prerequisite to attaining the analytic goals of surveys. Additionally, doing so concurrently for multiple error sources is paramount if we are to understand the mechanisms that cause them and the tradeoffs they imply.

Many approaches to estimating the extent of stochastic errors have been put forward in the literature. Suggested designs are "validation" data or "record check" studies (e.g. Katosh and Traugott 1981), interview–reinterview (test–retest) designs (Battese et al., 1976), as well as psychological scale evaluations, multitrait-multimethod (MTMM) (Saris and Gallhofer 2007), quasi-simplex (Alwin 2007), latent class (Biemer 2011), and other latent variable model approaches to estimate acquiescence, "yea-saying", extreme response style, or social desirability variance (Billiet and Davidov 2008; Billiet and McClendon 2000; Moors 2003; Moors et al., 2014; Oberski et al., 2012). Of these, the record check design is the strongest, if it can be assumed that the "validation" data are completely error-free themselves. Unfortunately, however, validation data are difficult and sometimes impossible to obtain, and the assumption of no measurement error is often questionable (Ansolabehere and Hersh 2010; Groen 2012). Although these approaches may appear very different, they are similar in that they all require

Experimental Methods in Survey Research: Techniques that Combine Random Sampling with Random Assignment, First Edition.
Edited by Paul J. Lavrakas, Michael W. Traugott, Courtney Kennedy, Allyson L. Holbrook, Edith D. de Leeuw, and Brady T. West.

Companion Website: www.wiley.com/go/Lavrakas/survey-research

some form of repeated data collection. From the perspective taken in this chapter, all methods to estimate the extent of stochastic error are a form of within-person design.

The within-persons approach necessary to quantify stochastic errors has three important drawbacks. First, it tends to allow for only one source of stochastic error, such as random error (Alwin 2007; Battese et al., 1976; Biemer 2011; Katosh and Traugott 1981) or two sources, such as random error and acquiescence response style (Billiet and McClendon 2000). Each of these approaches is therefore designed to model only one or two forms of stochastic error, assuming that the other forms are absent. Second, the designs often lack randomization of the order of different measurements (Alwin 2011), which means that carryover (Myers 1972) or test effects (Campbell and Stanley 1963) might bias the estimates. Third, respondents may remember their answer at the previous occasion in test–retest type designs such as the interview–reinterview, quasi-simplex, and MTMM designs. Although Saris and Andrews (1991) have suggested that more than 20 minutes between forms removes this effect, Alwin (2011) questioned this conclusion.

Due to the difficulties of the within-persons designs outlined above, Krosnick (2011) suggested abandoning them altogether. Unfortunately, if one abandons within-persons designs, the only alternative for evaluating stochastic errors is relying on validation data, which as noted above are generally nonexistent. Between-persons experiments indeed do not suffer from the above problems, but neither are they informative enough to estimate the stochastic components of TSE. Therefore, the view taken in this chapter is that we should not do away with within-persons experiments but strive to improve them.

This chapter introduces a general framework that uses a within-persons experimental design but deals with two of the three problems highlighted above: the assumption of only one error type and the assumption of zero test effects. The "multitrait-multierror" (MTME) framework does this by applying a simple idea: extend the within-persons design to vary several error sources at a time and randomize methodological variation such as question order. This design then enables researchers to concurrently estimate multiple sources of stochastic measurement errors from the TSE framework, allowing for the improvement of question design and removal of biasing effects from analyses.

In the next section, we will present the MTME framework. We then give practical advice on how to design and implement such experiments in surveys and on how to estimate the effects of the experimental treatments. The approach is illustrated using the Understanding Society Innovation Panel in the United Kingdom (UKHLS-IP). Finally, we comment on future research needed to improve within-persons designs, including the remaining problem of memory effects.

24.2 The Multitrait-Multierror (MTME) Framework

Our framework starts with the observation that any type of repeated observation on the same respondent can be viewed as a within-persons design. We will therefore use the term "within-persons design" to mean any situation in which multiple measures of the same variable have been obtained on the same respondent; we will call such designs "experiments" when the precise method of measurement and its timing is under the control of the researcher. Preferably, but not necessarily, such choices undergo randomization.

For instance, a record check study that observes respondents' answers to the question "have you voted in the last Presidential election?" together with an official vote record is a within-persons design, but it is not an experiment because the researcher does not control data collection in the administrative record. On the other hand, a survey company that calls back some of its respondents after a face-to-face interview to reinterview them is performing an experiment in our terms, because the company could have chosen to phone first, or reinterview them face-to-face. A *randomized* within-persons experiment might have randomized these choices among subgroups of respondents.

24.2.1 A Simple but Defective Design: The Interview–Reinterview from the MTME Perspective

In the familiar notation of Shadish et al. (2002), a possible interview–reinterview design to estimate mode effects could be depicted as:

Condition 1:	X_{CAPI}	O	X_{CATI}	O
Condition 2:	X_{CAPI}	O		

In this notation, the "X's" indicate a "treatment," and the "O's" an observation. The subscripts "CAPI" and "CATI" have been used to indicate that the treatments correspond to these data collection modes; other aspects of the treatments may also be relevant, however. Here, condition 1 is "exposed" to a CAPI interview and observed (i.e. interviewed). At a later date, this group of people is exposed to CATI and observed. In contrast, condition 2 has only the CAPI interview. To clarify these conditions, the data obtained from such a design can be coded in a design matrix, as shown in Table 24.1. In this design matrix, different columns encode factors that may be of interest, or that may cause methodological variation.

Table 24.1 represents the long form data where each row is a person/time record. As can be seen from the first column, some individuals have two rows, meaning that they were observed twice, while others have only one. This can be due to either being in condition 2 (who were not reinterviewed) or to being a nonrespondent in the second observation of condition 1. The second column shows an important component of TSE: random response error. This error varies every time we have a new measure. Thus, here we have two instances of random error (1 and 2) for each measurement made within a person. The third column recognizes that the number of visits may change over time. The "Repetition" column in Table 24.1 encodes the possibility of test effects, such as respondent fatigue leading to satisficing (Krosnick 2011). The data collection mode is known to have an effect on answers as well, so that the number of remembered visits may differ over the telephone versus face-to-face. The topic is a constant: all measures obtained are about the number of doctor visits, meaning we can only generalize the results to the respondents' doctor visits, which is intended in this case.

Table 24.1 The standard interview–reinterview design in which random error, true change, repetition, and data collection mode have been confounded.

Person ID	Random error	Change	Repetition	Mode	Topic
1	1	No	No	CAPI	Doctor's visits
1	2	Yes	Yes	CATI	Doctor's visits
2	1	No	No	CAPI	Doctor's visits
3	1	No	No	CAPI	Doctor's visits
3	2	Yes	Yes	CATI	Doctor's visits
4	1	No	No	CAPI	Doctor's visits
...	...	...	...	...	...

It should be clear from the design matrix in Table 24.1 that the standard interview–reinterview design leaves a lot to be desired. Inadvertently or not, a number of factors have been confounded with one another, so that random error, change, test effects, and data collection mode all vary together and cannot be disentangled. Another key point to note is that any Topic × Person interaction can be interpreted as the person's score for that topic, i.e. as their "true score" or "trait." In the above design, the Topic × Person interactions are the same as the Person main effects since Topic is constant. In other designs, with multiple topics, this will not be the case.

When the survey goal is to estimate relationships among variables, these relationships are primarily biased by *variation across persons* in the methodological factors, i.e. in the Factor × Person interactions and their correlations (Carroll et al. 2006; Fuller 1987). The Error × Person variance is one such example. It is known as the variance of the random errors, the error variance, random error or unreliable variance, and it is well-known to cause bias when estimating relationships between variables. From this point, we will refer to it as random variance. For example, Alwin (2007) found random variance to be around 35% of the total variation in self-reported questions in a number of surveys. This can attenuate correlations between variables, but the bias in estimates of relationships can be in either direction when we conduct multivariate analyses such as multiple regression. So, if we are interested in the relationship between the number of doctor visits and another self-report, such as physical health, with 35% of its total variance arising from random variance, a true correlation of 0.8 would be estimated as 0.5 using the observed data. From an experimental design perspective (e.g. Cox and Reid 2000, ch. 6), the main issue is the confounding among the Topic × Person and Error × Person design columns, so that error and true variance cannot be separated without a full-rank design. A full-rank design in this case would randomize the respondents to all possible combinations of the levels of our design factors.

In short, if a design can be developed so that the methodological factors' interactions with the Person factor are estimable, then the amount of bias in relationship estimates caused by these factors can be estimated. One goal in the above example is therefore to estimate how strong the effect of random error is relative to the overall variation in doctor's visits: expressed as a proportion of variance, this corresponds to the "(un)reliability" of the question. After collecting the data, the following model might be fitted to the jth observation on the ith person to estimate this effect:

$$y_{ij} = \beta_0 + \beta_1 Change_j + \beta_2 Error_j + \beta_3 Repetition_j + \beta_4 Mode_j + \beta_5 Person_i \\ + \beta_{0,5,i}(Topic_j)(Person_i) + \beta_{1,5,i}(Error_j)(Person_i)$$

where the coding of the categorical variables Time, Repetition, Mode, and Person has been omitted for clarity. In this case, the variables are dummies (two categories), but the model can be easily extended to variables with multiple categories. Since the Topic is constant across conditions, it has been absorbed into the other terms. Note that the usual residual error term has been replaced here by an Error × Person interaction. This is an equivalent formulation, i.e. $\beta_{1,5,i}$ plays the role of a residual here. As mentioned previously, in the interview–reinterview design, Repetition, Error, and Mode are all confounded with one another, and the Person main effect is confounded with the Topic × Person effect. This shows that for the classic interview–reinterview design, the following assumptions are necessary to identify the Error × Person interaction:

- There are no test, repetition, or mode effects. This implies that the effects of $Change_j$, $Error_j$, and $Repetition_j$ are all 0 (i.e. $\beta_2 = \beta_3 = \beta_4 = 0$).
- There is no $Person_j$ main effect ("style factor") beyond the person's true opinion (i.e. $\beta_5 = 0$).

- There are no further interactions with $Person_j$, e.g. the effects of $Change_j \times Person_j$, $Mode_j \times Person_j$, etc. are zero, as are any higher-order interactions. This would not be true if some people get fatigued faster than others, for example.

This leads to the model:

$$y_{ij} = \beta_0 + \beta_1 Change_j + \beta_{0,5,i}(Topic_j)(Person_i) + \beta_{1,5,i}(Error_j)(Person_i)$$

Assuming that all factors have been coded to sum to zero ("effect-coding"), β_0 is the mean number of doctor visits over people and repetitions, β_1 is a deviation from that mean depending on the first or second occasion of asking the question, the $\beta_{0,5,i}$ are person scores on doctor visits, and the $\beta_{1,5,i}$ are random error scores. Finally, taking $Person_j$ to be a random factor, the model can be estimated using linear random-effects modeling or, equivalently, confirmatory factor analysis. Either of these techniques will allow estimation of var($\beta_{1,5,i}$), the error variance of interest and the corresponding reliability, $1 - \frac{var(\beta_{1,5,i})}{var(y_i)}$.

To conclude this example, the classic interview–reinterview design can be seen as a within-persons design. However, for it to yield the error variance (reliability) of interest, a number of strong assumptions are necessary. Moreover, the discussion above has only factored in a few of the possible components of TSE, allowing for none of them besides random error in the model. This design, therefore, is unrealistic in light of the previous empirical findings in survey methodology and from the theoretical considerations of the TSE framework.

24.2.2 Designs Estimating Stochastic Survey Errors

With the limitations of the basic interview–reinterview approach now spelled out as a within-persons design, a partial solution also presents itself. In theory, if we can create a within-persons factorial experiment that varied within the design, we will be able to account for all of the effects assumed to be 0 so far. This suggests the following procedure:

1. Define the main types of stochastic (i.e. varying between people) measurement error sources whose influence is to be estimated;
2. Manipulate the survey questions to vary these error sources' influence;
3. Collect data using a random probability sample of persons; and
4. Estimate an appropriate model, taking *Person* to be a random factor.

In practice, this is possible for many factors, but not for all of them. In particular, any Person × Repetition interactions will remain confounded with random error, so this approach can solve all but the memory effect problem.

Examples of this approach are some MTMM designs (Andrews 1984; W. Saris and Gallhofer 2007). In these designs, the errors defined to be of interest are (i.) random error, (ii.) "method effects," defined as *Person* × "Question formulation" interaction effects, and sometimes also (iii.) order effects. In addition, the design varies the topic ("trait") of the question and allows for differences in errors over the different traits and question formulations. In this context, we would be able to disentangle the "method effect," the impact of the wording of the question or of the response scale, from the "trait." An MTMM design matrix is shown in Table 24.2.

As can be seen in Table 24.2, the MTMM design still confounds *Change, Repetition,* and *Method.* However, the MTMM design is a considerable improvement over the interview–reinterview design in other respects. First, the crossing with the *Topic* factor allows the *Error* × *Person* interaction to vary by *Topic* and *Repetition/Method.* In other words, it is no longer necessary to assume that the error variance is equal between the two time

Table 24.2 Typical survey MTMM design.

Person ID	Error × Topic	Change	Repetition	Mode	Topic
1	E_1	No	No	CAPI	Doctor's visits
1	E_2	Yes	Yes	CAPI	Doctor's visits
1	E_3	No	No	CAPI	Smoking behavior
1	E_4	Yes	Yes	CAPI	Smoking behavior
1	E_5	No	No	CAPI	General health
1	E_6	Yes	Yes	CAPI	General health
2	E_1	No	No	CAPI	Doctor's visits
…	…	…	…	…	…

points. Second, under the assumption that there are no *Change* × *Person* or *Repetition* × *Person* (e.g. memory) effects, the *Method* × *Person* interactions ("method effects") are now identifiable, allowing the researcher to study another source of stochastic TSE. Third, mode is now constant, so that this factor is no longer confounded with *Repetition/Change/Method.*

A possible linear model for the design in Table 24.2 is, given observation y_{ijt}, where i indexes the person, j the repetition, and t the topic:

$$y_{ijt} = \beta_{0,jt}(Method_j)(Topic_t) + \beta_{2,it}(Topic_t)(Person_i) + \beta_{3,ij}(Method_j)(Person_i) + \beta_{4,ijt}(Error_j)(Topic_t)(Person_i)$$

Here $\beta_{0,jt}$ represents the expected mean for each question and method, the $\beta_{2,it}$ represents impact of the *Topic* on the expected score (i.e. "trait") while $\beta_{3,ij}$ is the impact of method on the answers (i.e. "method" effects). Finally, $\beta_{4,ijt}$ represents deviations from the expected score given the method and the topic and it can be considered an estimate of random error.

If we take *Person* to be a random factor, the first equation can be rewritten as a confirmatory factor analysis (CFA) model.

$$y_{ijt} = \lambda_{it} T_j + \lambda_{ij} M_k + E_{ijt}$$

Thus, $(Topic_t)(Person_i)$ can be estimated as a latent variable coded as T_j (from "trait"), while the $(Method_j)(Person_i)$ quantity can be estimated using another latent variable M_k (from "method"). Finally, the random error can be estimated for each repetition and topic. The variances of these factors represent the stochastic effects we want to estimate. In a simple CFA model that just estimates the stochastic error, we can ignore the mean structure in the data and $(Method_j)(Topic_t)$ will be fixed to 0. Of course these can also be estimated if needed. The coefficients that link T_j and M_k to the observed scores (y_{ijt}), the λ's, are also known as loadings in CFA. In order to estimate the stochastic errors, these loadings should be fixed based on the experimental design. Examples of how to fix these will be shown below.

To sum up the MTME perspective on MTMM experiments, some of the disadvantages of the standard interview–reinterview design could be solved by varying additional factors in the within-persons design: this is what leads to the MTMM design (Campbell and Fiske 1959). However, the MTMM design has its own shortcomings. It does not recognize the possibility of an overall acquiescence random effect, for example, or of social desirability variance. The MTME approach suggests that such issues can be accounted for by continuing the same line of thought: those factors that are unaccounted for should be varied in the within-persons design. The resulting data can then be analyzed using CFA-type models. The following sections explain this extended within-persons approach in more detail.

24.3 Designing the MTME Experiment

Designing, implementing, and analyzing data from an MTME experiment can be daunting. As such, we put forward a list of questions researchers and practitioners need to consider when using the MTME approach. This is divided into two stages: designing the experiment and estimating the statistical model.

There are five essential questions researchers need to answer when implementing the MTME design.

24.3.1 What are the main types of measurement errors that should be estimated?

Before planning the MTME design, researchers should decide what types of stochastic errors are known/expected to have an impact on the measures of interest. For some types of survey questions, we can expect small amounts of error. Examples of these are sociodemographic questions, such as sex or age or other factual information, such as number of household members. For some other types of questions, we can expect specific types of stochastic errors. For example, social desirability can have an important impact on sensitive topics such as sexual behavior or income (DeMaio 1984; Tourangeau et al., 2000). Other types of survey questions might be influenced by other biases such as acquiescence, method effects, or extreme response styles. The researchers have to decide which types of stochastic errors are the most important for the key measures of interest.

In order to illustrate these points, we will assume here that researchers are interested in estimating method effects, acquiescence, and random variance in their variables of interest.

24.3.2 How can the questions be manipulated in order to estimate these types of error?

Once the researcher has decided on the types of systematic errors of interest, they must consider if it is possible to manipulate the survey attributes, the questions, or the response categories in order to impact these stochastic errors. For example, MTMM models often manipulate the response scale (e.g. number of response categories, labeling of the categories, their numbering) in order to estimate method effects. Similarly, acquiescence can be manipulated by changing the ordering of the labels used for the response categories. For example, instead of using Agree–Disagree response categories, they can be reversed, leading to a Disagree–Agree formulation. Social desirability can be manipulated in a number of ways. For example, the mode of the questions can be changed, as some modes (e.g. self-administered ones) are better at minimizing this type of systematic error. This approach has the disadvantage of confounding mode and social desirability. Another way could be to change the question wording or present vignettes that imply what is the socially desirable answer. Thus, people could be primed with knowledge about what the majority supports or does.

Once the researchers decide on the types of systematic errors and how to manipulate them, they have to decide on the number of levels for each treatment. For example, if researchers want to estimate acquiescence and method effects, they have to decide on the level and the types of treatments they want to apply. As such, in the case of acquiescence, they can have two levels of the treatment: Disagree–Agree response categories and Agree–Disagree categories. Any "acquiescence" effect would then presumably positively bias the first form while having the reverse effect on the second form. An additional form for which acquiescence effects could be assumed zero, such as item-specific scales, is also possible (Saris et al., 2010), but not considered in this example. For the method effect, there are a number of options depending on the number of response categories, the amount of labeling, and the numbering of the categories. Let us

Table 24.3 Four question "forms" as a result of combining two "methods" and two response scale orders.

Form	Method	Acquiescence
F_1	5 point	Agree/disagree
F_2	5 point	Disagree/agree
F_3	10 point	Agree/disagree
F_4	10 point	Disagree/agree

assume researchers choose two types of methods: a fully labeled five-point scale and a ten-point scale with only the extreme categories labeled.

The combination of the two acquiescence manipulations and two methods leads to four different "forms" of the questions (Table 24.3). Implementing the forms in the split-ballot design (Saris and Gallhofer 2007; Saris et al., 2004) leads to four combinations of two $\left(\binom{4}{2} = 6\right)$ forms. This implies that six different pairs of forms must be administered to the respondents. Such a design can be implemented by giving a pair of forms to one of six randomized groups.

24.3.3 Is it possible to manipulate the form pair order?

After deciding on the types of errors to be estimated, the treatments, and the form pairs, the researchers must decide if the order of the forms will be randomized. The advantage of implementing such an approach is that it tackles some of the possible carryover effects that can appear due to the lack of independence of the two within-person measurements. On the other hand, this can increase the amount of groups to be created and analyzed. While this does not have an effect on respondent burden, as each individual still receives two measures, it does have an effect on the resources needed by the data collection agency and the analysts. If the researchers decide to randomize the order in our hypothetical example, this results in 12 form combinations (6×2). These are presented in Table 24.4.

In conclusion, to implement this MTME design that makes it possible to estimate method effects, acquiescence and random variance while controlling for order effects, the researcher must create 12 random groups, each of which will receive a combination of two formats of the questions.

24.3.4 Is there enough power to estimate the model?

When dividing the sample into such a large number of groups, the power of the analysis has to be taken into consideration. The first advantage of using the latent variable modeling framework

Table 24.4 Six form combinations with randomized order.

Number	Time 1	Time 2	Number	Time 1	Time 2
1	F_1	F_2	7	F_2	F_1
2	F_1	F_3	8	F_3	F_1
3	F_1	F_4	9	F_4	F_1
4	F_2	F_3	10	F_3	F_2
5	F_2	F_4	11	F_4	F_2
6	F_3	F_4	12	F_4	F_3

is the ease of implementing maximum likelihood methods for dealing with missing data. This approach uses all the available information in the analysis, maximizing power. Additionally, as the groups are randomized, no bias will be introduced as missing information is missing completely at random (see for an overview Enders 2010).

Nevertheless, even with this statistical method of dealing with missing data, enough information must be present to estimate each parameter. It is good practice to first implement a simulation study to consider the power of the design under different (conservative) nonresponse rates.

24.3.5 How can data collection minimize memory effects?

As mentioned in the previous section, memory effects (or other carryover effects) are an important threat to the validity of within-persons designs. That is also true for MTME experiments. Nevertheless, researchers can adopt a number of strategies in order to minimize the possibility of such bias.

One approach is to minimize memory effects by design. This can be done in a number of ways, for example, by having a minimum period of time between the two measurements, such as the 20 minutes proposed by Saris and Andrews (1991). If it is possible to collect data again, at a different point in time, researchers have to take into consideration two different aspects: memory and change. The ideal period for collecting the second measure in a within-persons experimental design is one that minimizes both any memory effects and change in the true score. The nature of these two dimensions depends on the topic of the questions used in the experiment. If the second point is in the same interview, then the distance should be maximized with the first measure being implemented as early as possible and the second one toward the end of the questionnaire.

Identifying the optimal time within the interviews to collect the second measurement is often challenging. To aid in this task, researchers can collect paradata to facilitate sensitivity analyses. One class of paradata that can be collected is time stamps or time latencies between the first measurement and the second one. These can now be easily collected in most computer-assisted data collection software. Similarly, researchers can collect information regarding individuals' memory capabilities. These two measures can be used after data collection for sensitivity analysis by estimating their effect on the MTME coefficients. It should be noted that these approaches are not ideal, as possible confounds exist in such observational designs. Nevertheless, such sensitivity analyses might prove useful and insightful by providing evidence regarding the presence or absence of memory effects.

24.4 Statistical Estimation for the MTME Approach

The statistical estimation of the MTME model is closely linked to the latent variable modeling tradition of MTMM (Andrews 1984; Campbell and Fiske 1959; Eid 2000; Saris et al., 2004). As such, each observed item is a combination of the true/trait score (*Person* × *Topic*) and stochastic error (*Person* × *Error* source). The contribution of the MTME approach is the possibility of experimentally manipulating multiple types of systematic errors concurrently while explicitly controlling for order effects.

In the MTME experiment proposed previously, we included both method and acquiescence as treatments. The model can be written as:

$$y_{jkl} = \lambda_{Tjkl}T_j + \lambda_{Mjkl}M_k + \lambda_{Ajkl}A_l + E_{jkl}$$

Where the observed items Y_{jkl} are measured by trait (question) j, method k with acquiescence effect l and decomposed into trait T_j, method M_k, acquiescence A_l and a specific residual

Table 24.5 Using the MTME design matrix to inform the model constraints needed for estimation.

Form	Method	Acquiescence
F_1	0	+1
F_2	0	−1
F_3	1	+1
F_4	1	−1

component, E_{jkl}. Additionally, the $\lambda\, T_{jkl}$ are the trait loadings, the $\lambda\, M_{jkl}$ are the method loadings, and the $\lambda\, A_{jkl}$ are the acquiescence loadings.

The design matrix (Table 24.5) can be used as a guide to the constraints needed for estimation. For the method effect, we use "dummy" (0/1) coding, which corresponds to the "MTM(M-1)" coding proposed for such models proposed by Eid (2000) and Eid et al. (2003). Thus, only one method is represented by a latent variable while the other is considered as a reference. Here we estimate the effect of method 2 (10-point scale) by fixing loadings of the items measured using forms 3 and 4 to +1. We estimate acquiescence as explained by Billiet and McClendon (2000) and Billiet and Davidov (2008), using one latent variable. The questions measured with forms 1 and 3 should have the loadings fixed to +1 as the Disagree-Agree response scales suffer this directional bias, while questions measured with forms 2 and 4 should have the loadings fixed to −1, as the Agree–Disagree wording is thought to reverse acquiescence effects relative to Disagree–Agree.

We can also use a graphical representation as another way to understand the statistical model needed to estimate the different types of errors. Figure 24.1 presents the equivalent of the formula above in the visual form. Here large circles represent latent variables/random effects, while squares represent observed variables. For example, the four squares are the same question, Q_1, measured using the 4 different forms, F_1–F_4. Each arrow can be conceived as a regression

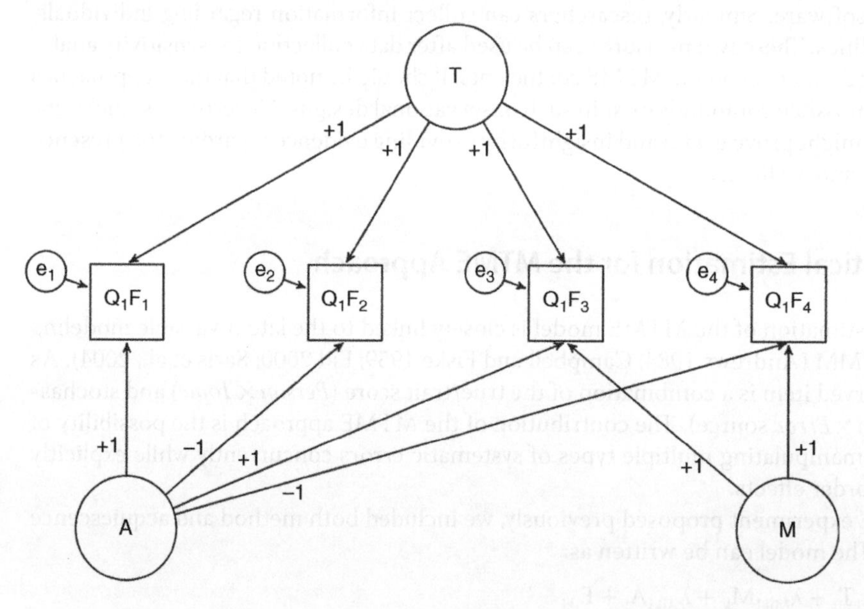

Figure 24.1 Visual representation of the Structural Equation Model for MTME with two stochastic errors.

coefficient. They can be either unrestricted, meaning they have to be estimated, or they can be restricted. For example, we can see that the arrows from A, the estimate of acquiescence, have been coded as shown in Table 24.5. Forms 1 and 3 should have higher acquiescence so we expect a positive relationship, while forms 2 and 4 should have less of it, thus we fix them to −1. We do the same for the Method effect, M. In the graphical representation of the SEM, any missing arrows between objects assume that there is no relationship, i.e. the coefficient is 0. Thus, the lack of arrows from M to the question measured in the first two forms is equivalent to having arrows restricted to 0, and is in accordance with Table 24.5. The small circles (coded e_1–e_4) are the residuals, i.e. what remains unexplained after we control for the other three sources of variance. It can be considered an estimate of unreliability/random error. The T in the figure represents an estimation of the "true" score or "trait" after controlling for the other influences. It is our estimate of what people would have answered to question 1 if we control for acquiescence (A), method effect (M), and random error (e_1–e_4).

24.5 Measurement Error in Attitudes Toward Migrants in the UK

In this section, we will give an example of an MTME experiment implemented in the UK Household Longitudinal Study – Innovation Panel (UKHLS-IP)[1]. The measures of interest are attitudes toward immigrants (Table 24.6) and have been previously used in other surveys such as the European Social Survey. In the first subsection, we will go through the five points discussed above in order to highlight the design process and how we implement it. In the second part, we will present the first results from the analysis. This is just one possible design. Researchers should adapt the MTME approach and analysis to best fit their needs.

24.5.1 Estimating Four Stochastic Error Variances Using MTME

24.5.1.1 What are the main types of measurement errors that should be estimated?

Attitudinal questions are notoriously hard to measure as they are less stable than values and much more subjective and prone to misunderstanding than factual questions. Because of their ephemeral nature, they can be easily influenced by response scale formatting. This influence

Table 24.6 Six questions ("traits") measuring attitudes toward immigrants in the UKHLS-IP.

Trait	Wording
Q_1	The UK should allow **more** people of the same race or ethnic group as most British people to come and live here
Q_2	UK should allow **more** people of a different race or ethnic group from most British people to come and live here
Q_3	UK should allow **more** people from the poorer countries outside Europe to come and live here
Q_4	It is generally **good** for UK's economy that people come to live here from other countries
Q_5	UK's cultural life is generally **enriched** by people coming to live here from other countries
Q_6	UK is made a **better** place to live by people coming to live here from other countries

1 In this study, we ignore weights in order to facilitate the understanding of the model and results. The aim of the analysis here is for illustration and not finite population inferences regarding these relationships. Weights can be easily included in structural equation modeling (Oberski 2014).

can appear in multiple forms, from method effects to acquiescence or extreme response styles. Additionally, some topics can be considered sensitive, thus increasing threats to validity. As such, we believe that multiple sources of errors must be taken into consideration to validly measure some attitudinal scales.

Here we decided to manipulate two formatting characteristics that might bias answers to attitudinal questions: method and acquiescence. We also believe that questions regarding migration are prone to social desirability bias as there are important cultural norms and debates around this topic. As such, we also want to estimate the random effects of social desirability. Finally, random variance can also play an important role when collecting data about attitudes. This source of variation will be estimated by taking advantage of the within-persons nature of the MTME experimental design.

24.5.1.2 How can the questions be manipulated in order to estimate these types of errors?

To estimate the three types of systematic errors, two treatment levels were manipulated for each one:

- **Method**: Number of scale points (2 point vs. 11-point scale);
- **Acquiescence**: Agree–Disagree vs. Disagree–Agree response scale;
- **Social desirability**: positively vs. negatively formulated item on immigration.

This yields $2 \times 2 \times 2 = 8$ possible item wordings (forms) for each of the six items (traits). By combining them, there are ($\binom{8}{2} = 28$) possible pairs of question formats to be applied in the split-ballot MTME experiment. In this application, we randomized the order of the form pairs, thus leading to a total of 56 experimental groups (28×2). For example, we have implemented separately both a design that uses F_1 at the start of the survey and F_2 at the end as well as the reverse of this (F_2 at the start and F_1 at the end). This was done in order to minimize any potential carryover effects (Table 24.7).

24.5.1.3 Is it possible to manipulate the form pair order?

In this application, we have decided to randomize the order of the form pairs, thus leading to a total of 56 experimental groups (28×2). This was done in order to minimize any potential carryover effects.

24.5.1.4 Is there enough power to estimate the model?

While this design seems to define a large number of groups (i.e. small sample size per group), it should be kept in mind that the observed correlations are to be projected into a much smaller

Table 24.7 Eight forms to measure three types of systematic error in UKHLS-IP.

Form	Scale points	Agree–disagree	Social desirability	Wording
F_1	2	AD	Higher	Negative
F_2	2	AD	Lower	Positive
F_3	11	AD	Higher	Negative
F_4	11	AD	Lower	Positive
F_5	2	DA	Higher	Positive
F_6	2	DA	Lower	Negative
F_7	11	DA	Higher	Positive
F_8	11	DA	Lower	Negative

parameter space based on an SEM model. In the most complex version of our model, including all traits, there are 48 loadings and trait variances, 15 trait covariances, 1 method variance, 1 acquiescence factor variance, and 1 social desirability factor variance, leading to $48 + 15 + 3 = 66$ parameters.

We conducted a simulation study using the SEM software Mplus 6.12 to investigate whether, with a 50% response rate, the precision, coverage, and power of the variance parameters of interest would still be adequate. We used the PATMISS option in Mplus to simulate the planned missingness pattern. Assuming 750 responses, two traits,[2] reliability coefficients around 0.7, and method, social desirability, and acquiescence standardized effects around 0.3 (10% of total variance), the power to detect these factor variances is well over 0.9. The power for the social desirability factor is lowest and drops to 0.77 when all factor variances are set to 5% instead of 10%. Using 250 replications, the estimates are unbiased, and the standard errors are acceptable. The Mplus syntax for the simulation is provided in the online appendix.

24.5.1.5 How can data collection minimize memory effects?

In order to minimize possible memory effects, the routing of the questionnaire avoided asking the second form if fewer than five minutes passed since the last question of the first form. Additionally, time stamps/latencies and cognitive ability were measured and will be used in the future for sensitivity analyses. This procedure would enable us to investigate some of the assumptions of the model that cannot be controlled for using the experimental design.

One possible way to use this information for sensitivity analysis after data collection could be to run the model separately for those that had the smallest difference between answering the two forms, which was five minutes, and those that had a longer period between them, for example, around 20 minutes. If memory effects are an issue, the correlation between the forms should be higher for the first group compared to the second one. Of course, this is not a perfect assessment of memory effects due to possible confounding factors. For example, some people might answer fewer questions due to routing in the questionnaire and their characteristics might be related to sources of error. Cognitive ability could provide a powerful control variable in this context.

24.5.2 Estimating MTME with four stochastic errors

As presented previously, the design matrix (Table 24.8) can be used to understand what latent variables must be estimated and what constraints have to be employed. In this case, we have six

Table 24.8 Design matrix for MTME model measuring attitudes toward immigrants in UKHLS-IP.

Form	Method	Acquiescence	Social desirability
F_1	0	+1	+1
F_2	0	+1	−1
F_3	1	+1	+1
F_4	1	+1	−1
F_5	0	−1	+1
F_6	0	−1	−1
F_7	1	−1	+1
F_8	1	−1	−1

2 We use just two traits to make the simulation easier. In principle, the addition of traits should add more information to the model, making it easier to estimate.

trait measures of attitudes toward immigrants. Additionally, we have three types of systematic errors estimated as latent variables: method, acquiescence, and social desirability. The effect of the second method (11-point scale) is estimated by constraining all the items from forms 3, 4, 7, and 8 to +1. Acquiescence is measured as in the previous example with forms 1 to 6 having the loadings constrained to +1 (agree/disagree formats), while questions measured using forms 5 to 8 have the loadings constrained to −1. Using a similar approach, the social desirability latent variable model is estimated by constraining all the items in forms 1, 3, 5, and 7 to +1 and as −1 for the rest of them.

The relationships can also be written in equation form as shown in the formula below. Here we add the effect of social desirability, S_m, as measured by the λ_{Sjklm} loadings.

$$y_{jklm} + \lambda_{Tjklm} T_j + \lambda_{Mjklm} M_k + \lambda_{Ajklm} A_l + \lambda_{Sjklm} S_m + E_{klm}$$

The formula can also be illustrated in a SEM graphic (Figure 24.2). To keep things manageable, we have specified the model just for the first question/trait, but this can be easily extended to the other five. This time we have four sources of variation. In addition to acquiescence (A in the figure) and method (M) now, we also estimate social desirability (S), which we manipulated using the positive and negative wording of the questions. Once again we can use the design matrix in Table 24.8 to understand the restrictions in the coefficients. For example, questions asked in forms 1, 2, 5, and 6 were all measured using method 1 (2-point response scale). As such, they all have arrows restricted to 1 coming form M (which will represent method 1). Acquiescence and social desirability follow a similar pattern, having +1 for the forms where we expect a stronger effect and −1 when we expect a smaller one. We also see that T_1, the estimated "true" or "trait" score for question 1, has loadings fixed to 1 for all 8 questions (i.e. same question measured in eight different forms). In the case of our data, we would have five more questions, each measured using the eight forms. So, the model would be extended to include T2 to T6. The patterns of loadings would be similar to those shown in Figure 24.2.

24.6 Results

The design was implemented in the 7th wave of the UKHLS-IP (University of Essex. Institute for Social and Economic Research 2016). This is a longitudinal household survey in the United Kingdom used for methodological research. Wave 7 achieved a 54% household response rate (1505 households) and a 67% (2337 respondents) individual response rate. For more details regarding the data collection, see Al-Baghal et al. (2015).

Figure 24.3 presents the initial results from this MTME experiment. The analysis decomposes the total variance of the observed items into different components: trait, random error, method, acquiescence, and social desirability. This is done for each of the eight forms for five of the traits in the stacked bars as shown in Figure 24.3.

The total "quality" of the items can be described as those parts of the stacked bars that are trait variance (Saris and Gallhofer 2007). This quality varies considerably, from approximately 0.4 for form 8 and when asking "Allow different race" to approximately 0.9 for form 2.

With the current analysis, the highest quality was observed for question forms 1, 2, 5, and 6, which use the two-point response scale. It appears that their variance is less biased by the systematic errors included in this MTME design as compared to an 11-point answer scale.

These findings correspond to those of Revilla et al. (2014), who observed greater method variance for agree–disagree scales with 11 points than with fewer scale points, whereas they observed the converse with item-specific scales. Whether this finding can be generalized or is a consequence of our model formulation remains a topic for further investigation.

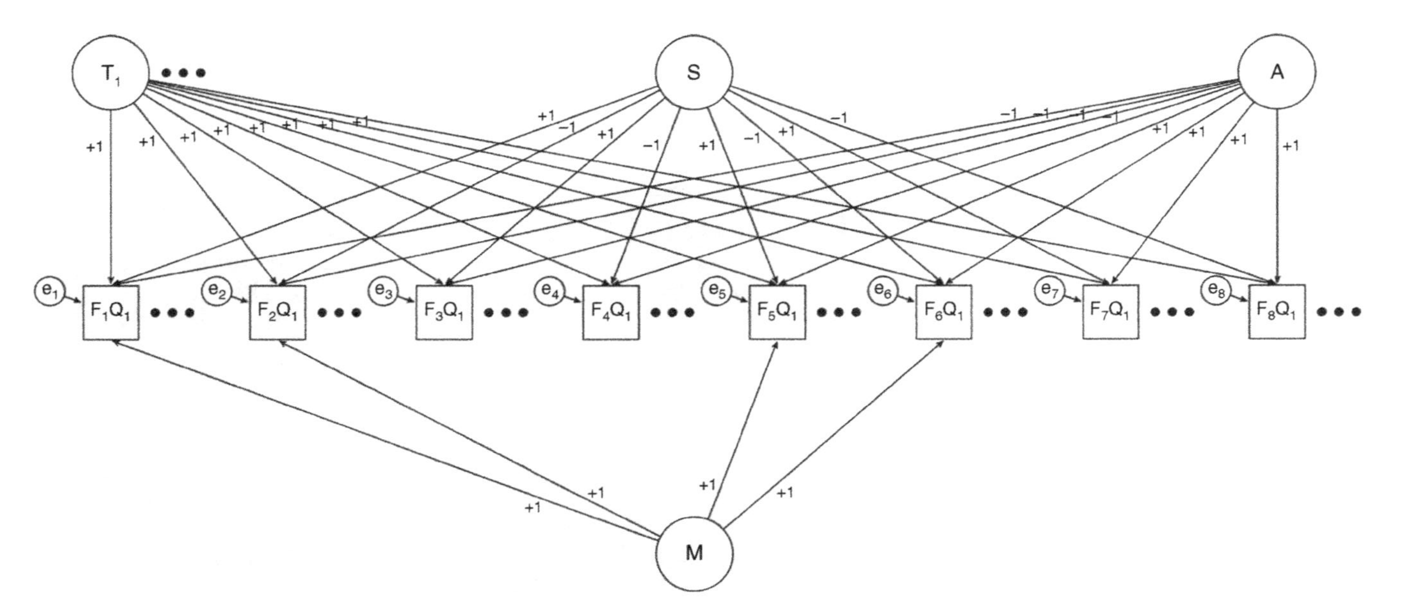

Figure 24.2 Visual representation of the Structural Equation Model for MTME with three stochastic errors. The analysis included six questions (Q_1–Q_6) regarding attitudes toward immigrants. Here only the model for Q_1 is shown for brevity (ellipses show how the model would expand for Q_2–Q_5).

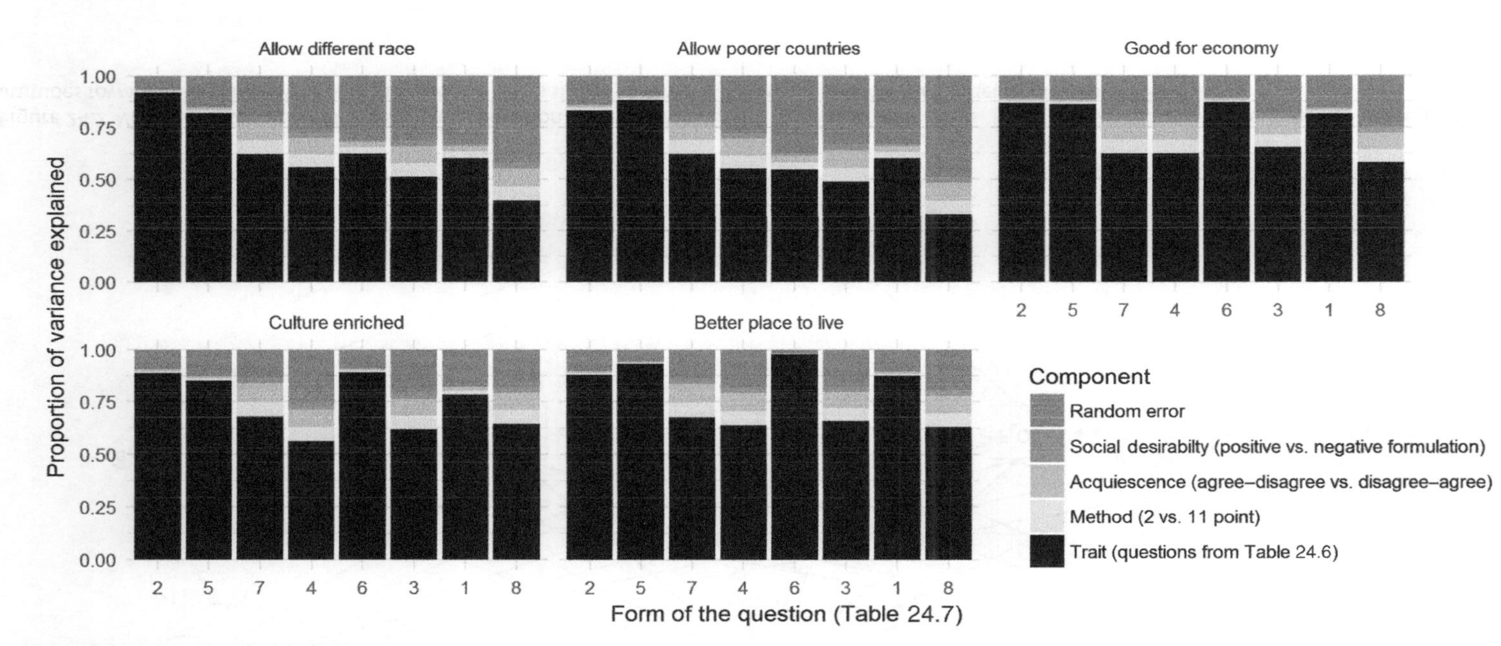

Figure 24.3 Variance decomposition in measures of attitudes toward immigrants.

Interestingly, the largest amount of nontrait variance is explained by random error in all the different questions and forms. This is followed by method and acquiescence, whereas social desirability does not explain much of the variance in the responses in our model.

24.7 Conclusions and Future Research Directions

In this chapter, we have argued that within-persons experimental designs are essential in survey research as they enable us to estimate and correct for stochastic errors that can bias substantive results. We have proposed a new design, the MTME design that directly tackles two of the problems with previous approaches. First, it enables researchers to concurrently estimate multiple types of systematic errors. Second, by randomizing the order of question forms, it makes it possible to control for some of the carryover or test effects.

We have encouraged the reader to answer five essential questions in order to design an MTME experiment:

1. *What are the main types of measurement errors that should be estimated?*
2. *How can the questions be manipulated in order to estimate these types of error?*
3. *Is it possible to manipulate the form pair order?*
4. *Is there enough power to estimate the models?*
5. *How can data collection minimize memory effects?*

We have also shown an application of the MTME in the UKHLS-IP, which is publicly available.[3] In this design, we have used six traits measuring attitudes toward immigrants and estimated concurrently four types of stochastic errors: method effects, acquiescence, social desirability, and random error. We have seen that around 70% of the variation comes from "trait," or is valid variation, while the rest is explained by the other components. While this is in line with some other research (e.g. Alwin 2007; Saris and Gallhofer 2007), there are very different estimates of data quality depending on the wording of the questions. Again, this is in line with the large body of research in survey methodology that has tried to develop a set of best practices for wording and formatting questions. Interestingly, our results indicate that random error represents the biggest proportion of nontrait variance. This is unexpected given the sensitive nature of the questions, in which we expect higher social desirability bias.

This application is just an example of the possible ways in which survey questions can be manipulated and stochastic error variances estimated using MTME. An area ripe for future development is how to extend this approach and think of new and creative ways to manipulate questions in order to estimate stochastic variance.

There is one important assumption of within-persons experiments that the MTME does not tackle: the memory effect. In design terms, we are left with a potential confounding of Person × Repetition with Person × Topic interactions. If these are present, they could bias the results from MTME experiments. Creative thinking is also needed here. In this chapter, we have proposed using paradata to see how the amount of time between the two measures or individual memory capabilities influence the model coefficients. This information should come at relatively low cost to the data collection agency. If possible, design solutions should be implemented to solve this issue. For example, in a web survey environment, it would be relatively easy to have the second reinterview at a later date, thus minimizing memory effects. If such an approach is used, two new aspects must be considered. First, the second measurement must be chosen such as to minimize both memory effects and changes in the true score. Second, the additional wave of data collection might increase the chances of nonresponse.

3 https://www.understandingsociety.ac.uk/about/innovation-panel

In summary, since stochastic errors are an unavoidable part of surveys and we are often interested in studying relationships, within-persons designs are indispensible. The MTME approach goes some of the way toward clarifying how such designs help and how various sources of methodological bias can be mitigated while studying several sources of survey error simultaneously. At the same time, we recognize the potential problem of memory effects and suggest that future research programs should focus on reducing this concern with within-persons designs.

Acknowledgments

This paper makes use of data from the Understanding Society Innovation Panel. Understanding Society is an initiative funded by the Economic and Social Research Council and various Government Departments, with scientific leadership by the Institute for Social and Economic Research, University of Essex, and survey delivery by NatCen Social Research and Kantar Public. The research data are distributed by the UK Data Service.

References

Al-Baghal, T., Bloom, A., Burton, J. et al. (2015). Understanding society innovation panel wave 7: results from methodological experiments. *Understanding Society Working Paper Series* 2015–3: 1–62.

Alwin, D. (2007). *Margins of Error: A Study of Reliability in Survey Measurement*. New York: Wiley-Interscience.

Alwin, D. (2011). Evaluating the reliability and validity of survey interview data using the MTMM approach. In: *Question Evaluation Methods: Contributing to the Science of Data Quality*, 1e (ed. J. Madans, K. Miller, A. Maitland and G. Willis), 265–294. Hoboken, NJ: Wiley.

Andrews, F.M. (1984). Construct validity and error components of survey measures: a structural modeling approach. *Public Opinion Quarterly* 48 (2): 409–442. https://doi.org/10.1086/268840.

Ansolabehere, S. and Hersh, E. (2010). *The Quality of Voter Registration Records: A State-by-State Analysis*. Cambridge, MA: Department of Government, Harvard University https://elections.wi.gov/node/1234.

Battese, G., Fuller, W., and Hickman, R. (1976). Estimation of response variances from interview reinterview surveys. *Journal of the Indian Society of Agricultural Statistics* 28: 1–14.

Beullens, K. and Loosveldt, G. (2014). Interviewer effects on latent constructs in survey research. *Journal of Survey Statistics and Methodology* 2 (4): 433–458. https://doi.org/10.1093/jssam/smu019.

Biemer, P. (2011). *Latent Class Analysis of Survey Error*. New York: Wiley.

Billiet, J. and Davidov, E. (2008). Testing the stability of an acquiescence style factor behind two interrelated substantive variables in a panel design. *Sociological Methods and Research* 36 (4): 542–562. https://doi.org/10.1177/0049124107313901.

Billiet, J. and McClendon, M. (2000). Modeling acquiescence in measurement models for two balanced sets of items. *Structural Equation Modeling: A Multidisciplinary Journal* 7 (4): 608–628. https://doi.org/10.1207/S15328007SEM0704_5.

Campbell, D.T. and Fiske, D.W. (1959). Convergent and discriminant validation by the multitrait-multimethod matrix. *Psychological Bulletin* 56 (2): 81–105. https://doi.org/10.1037/h0046016.

Campbell, D.T. and Stanley, J.C. (1963). *Experimental and Quasi-Experimental Research*. Houghton Mifflin Company.

Carroll, R.J., Ruppert, D., Stefanski, L.A., and Crainiceanu, C.M. (2006). *Measurement Error in Nonlinear Models: A Modern Perspective*. CRC Press.
Cox, D.R. and Reid, N. (2000). *The Theory of the Design of Experiments*. CRC Press.
DeMaio, T. (1984). Social desirability and survey measurement: a review. In: *Surveying Subjective Phenomena* (ed. C. Turner and E. Martin), 257–282. New York: Russell Sage Foundation.
Eid, M. (2000). A multitrait-multimethod model with minimal assumptions. *Psychometrika* 65 (2): 241–261. https://doi.org/10.1007/BF02294377.
Eid, M., Lischetzke, T., Nussbeck, F.W., and Trierweiler, L.I. (2003). Separating trait effects from trait-specific method effects in multitrait-multimethod models: a multiple-indicator CT-C(M-1) model. *Psychological Methods* 8 (1): 38–60. https://doi.org/10.1037/1082-989X.8.1.38.
Enders, C.K. (2010). *Applied Missing Data Analysis*, 1e. New York: The Guilford Press.
Fuller, W.A. (1987). *Measurement Error Models*. New York: Wiley.
Groen, J.A. (2012). Sources of error in survey and administrative data: the importance of reporting procedures. *Journal of Official Statistics (JOS)* 28 (2): 173–198.
Groves, R.M. and Lyberg, L. (2010). Total survey error: past, present, and future. *Public Opinion Quarterly* 74 (5): 849–879. https://doi.org/10.1093/poq/nfq065.
Katosh, J.P. and Traugott, M.W. (1981). The consequences of validated and self-reported voting measures. *Public Opinion Quarterly* 45 (4): 519–535.
Krosnick, J.A. (1999). Survey research. *Annual Review of Psychology* 50 (1): 537–567. https://doi.org/10.1146/annurev.psych.50.1.537.
Krosnick, J.A. (2011). Experiments for evaluating survey questions. In: *Question Evaluation Methods: Contributing to the Science of Data Quality*, 1e (ed. J. Madans, K. Miller, A. Maitland and G. Willis), 265–294. Hoboken, NJ: Wiley.
Moors, G. (2003). Diagnosing response style behavior by means of a latent-class factor approach. Socio-demographic correlates of gender role attitudes and perceptions of ethnic discrimination. *Quality and Quantity* 37: 277–302.
Moors, G., Kieruj, N.D., and Vermunt, J.K. (2014). The effect of labeling and numbering of response scales on the likelihood of response bias. *Sociological Methodology* 44 (1): 369–399. https://doi.org/10.1177/0081175013516114.
Myers, J.L. (1972). *Fundamentals of Experimental Design*, 2e. Boston: Allyn and Bacon.
Oberski, D., Weber, W., and Révilla, M. (2012). The effect of individual characteristics on reports of social desirable attitudes towards immigration. In: *Methods, Theories, and Empirical Applications in the Social Sciences: Festschrift for Peter Schmidt*, 2012e (ed. S. Salzborn, E. Davidov and J. Reinecke), 151–158. VS Verlag für Sozialwissenschaften.
Oberski, D.L. (2014). lavaan.survey: an R package for complex survey analysis of structural equation models. *Journal of Statistical Software* 57 (1): 1–27. https://doi.org/10.18637/jss.v057.i01.
Revilla, M.A., Saris, W.E., and Krosnick, J.A. (2014). Choosing the number of categories in agree–disagree scales. *Sociological Methods and Research* 43 (1): 73–97. https://doi.org/10.1177/0049124113509605.
Saris, W. and Andrews, F. (1991). Evaluation of measurement instruments using a structural modeling approach. In: *Measurement Errors in Surveys* (ed. P. Biemer, R. Groves, L. Lyberg, et al.), 575–597. New York: Wiley-Interscience Publication.
Saris, W. and Gallhofer, I. (2007). Estimation of the effects of measurement characteristics on the quality of survey questions. *Survey Research Methods* 1 (1): 29–43.
Saris, W., Révilla, M., Krosnick, J.A., and Shaeffer, E.M. (2010). Comparing questions with agree/disagree response options to questions with item-specific response options. *Survey Research Methods* 4 (1): 61–79.

Saris, W., Satorra, A., and Coenders, G. (2004). A new approach to evaluating the quality of measurement instruments: the split-ballot MTMM design. *Sociological Methodology* 34 (1): 311–347.

Shadish, W., Cook, T., and Campbell, D. (2002). *Experimental and Quasi-Experimental Designs for Generalized Causal Inference*. Belmont, CA: Wadsworth Cengage Learning.

Tourangeau, R., Rips, L.J., and Rasinski, K. (2000). *The Psychology of Survey Response*, 1e. Cambridge University Press.

University of Essex. Institute for Social and Economic Research (2016). *Understanding Society: Innovation Panel, Waves 1–8, 2008–2015*, 7e. UK Data Service. SN: 6849.

Index

Experimental Methods in Survey Research: Techniques that Combine Random Sampling with Random Assignment, First Edition.
Edited by Paul J. Lavrakas, Michael W. Traugott, Courtney Kennedy, Allyson L. Holbrook, Edith D. de Leeuw, and Brady T. West.

Companion Website: www.wiley.com/go/Lavrakas/survey-research

j

k

l

WILEY SERIES IN SURVEY METHODOLOGY

Established in Part by WALTER A. SHEWHART and SAMUEL S. WILKS

Editors: MICK P. COUPER, GRAHAM KALTON, LARS LYBERG, J. N. K. RAO, NORBERT SCHWARZ, CHRISTOPHER SKINNER

Editor Emeritus: ROBERT M. GROVES

The Wiley Series in Survey Methodology covers topics of current research and practical interests in survey methodology and sampling. While the emphasis is on application, theoretical discussion is encouraged when it supports a broader understanding of the subject matter.

The authors are leading academics and researchers in survey methodology and sampling. The readership includes professionals in, and students of, the fields of applied statistics, biostatistics, public policy, and government and corporate enterprises.

ALWIN · Margins of Error: A Study of Reliability in Survey Measurement

BETHLEHEM · Applied Survey Methods: A Statistical Perspective

BETHLEHEM, COBBEN, and SCHOUTEN · Handbook of Nonresponse in Household Surveys

BIEMER · Latent Class Analysis of Survey Error

*BIEMER, GROVES, LYBERG, MATHIOWETZ, and SUDMAN · Measurement Errors in Surveys

BIEMER and LYBERG · Introduction to Survey Quality

BIEMER · Latent Class Analysis of Survey Error

BIEMER, DE LEEUW, ECKMAN, EDWARDS, KREUTER, LYBERG, TUCKER, and WEST · Total Survey Error in Practice

BRADBURN, SUDMAN, and WANSINK · Asking Questions: The Definitive Guide to Questionnaire Design—For Market Research, Political Polls, and Social Health Questionnaires, Revised Edition

BRAVERMAN and SLATER · Advances in Survey Research: New Directions for Evaluation

CALLEGARO, BAKER, BETHLEHEM, GÖRITZ, KROSNICK, and LAVRAKAS (editors) · Online Panel Research: A Data Quality Perspective

CHAMBERS and SKINNER (editors) · Analysis of Survey Data

COCHRAN · Sampling Techniques, Third Edition

CONRAD and SCHOBER · Envisioning the Survey Interview of the Future

COUPER, BAKER, BETHLEHEM, CLARK, MARTIN, NICHOLLS, and O'REILLY (editors) · Computer Assisted Survey Information Collection

COX, BINDER, CHINNAPPA, CHRISTIANSON, COLLEDGE, and KOTT (editors) · Business Survey Methods

*DEMING · Sample Design in Business Research

DILLMAN · Mail and Internet Surveys: The Tailored Design Method

FULLER · Sampling Statistics

GROVES and COUPER · Nonresponse in Household Interview Surveys

GROVES · Survey Errors and Survey Costs

GROVES, DILLMAN, ELTINGE, and LITTLE · Survey Nonresponse
GROVES, BIEMER, LYBERG, MASSEY, NICHOLLS, and WAKSBERG · Telephone Survey Methodology
GROVES, FOWLER, COUPER, LEPKOWSKI, SINGER, and TOURANGEAU · Survey Methodology, Second Edition
*HANSEN, HURWITZ, and MADOW · Sample Survey Methods and Theory, Volume 1: Methods and Applications
*HANSEN, HURWITZ, and MADOW · Sample Survey Methods and Theory, Volume II: Theory
HARKNESS, BRAUN, EDWARDS, JOHNSON, LYBERG, MOHLER, PENNELL, and SMITH (editors) · Survey Methods in Multinational, Multiregional, and Multicultural Contexts
HARKNESS, van de VIJVER, and MOHLER (editors) · Cross-Cultural Survey Methods
HUNDEPOOL, DOMINGO-FERRER, FRANCONI, GIESSING, NORDHOLT, SPICER, and DE WOLF · Statistical Disclosure Control
KALTON and HEERINGA · Leslie Kish Selected Papers
KISH · Statistical Design for Research
*KISH · Survey Sampling
KORN and GRAUBARD · Analysis of Health Surveys
KREUTER (editor) · Improving Surveys with Paradata: Analytic Uses of Process Information
LAVRAKAS, TRAUGOTT, KENNEDY, HOLBROOK, DE LEEUW, and WEST (editors) · Experimental Methods in Survey Research: Techniques that Combine Random Sampling with Random Assignment
LEPKOWSKI, TUCKER, BRICK, DE LEEUW, JAPEC, LAVRAKAS, LINK, and SANGSTER (editors) · Advances in Telephone Survey Methodology
LESSLER and KALSBEEK Nonsampling Error in Surveys
LEVY and LEMESHOW · Sampling of Populations: Methods and Applications, Fourth Edition
LUMLEY · Complex Surveys: A Guide to Analysis Using R
LYBERG, BIEMER, COLLINS, de LEEUW, DIPPO, SCHWARZ, TREWIN (editors) · Survey Measurement and Process Quality
LYNN · Methodology of Longitudinal Surveys
MADANS, MILLER, and MAITLAND (editors) · Question Evaluation Methods: Contributing to the Science of Data Quality
MAYNARD, HOUTKOOP-STEENSTRA, SCHAEFFER, and VAN DER ZOUWEN · Standardization

and Tacit Knowledge: Interaction and Practice in the Survey Interview
MILLER, WILLSON, CHEPP, and PADILLA (editors) · Cognitive Interviewing Methodology
PORTER (editor) · Overcoming Survey Research Problems: New Directions for Institutional Research
PRESSER, ROTHGEB, COUPER, LESSLER, MARTIN, MARTIN, and SINGER (editors) · Methods for Testing and Evaluating Survey Questionnaires

RAO · Small Area Estimation
REA and PARKER · Designing and Conducting Survey Research: A Comprehensive Guide, Third Edition
SARIS and GALLHOFER · Design, Evaluation, and Analysis of Questionnaires for Survey Research, Second Edition
SÄRNDAL and LUNDSTRÖM · Estimation in Surveys with Nonresponse
SCHWARZ and SUDMAN (editors) · Answering Questions: Methodology for Determining Cognitive and Communicative Processes in Survey Research
SIRKEN, HERRMANN, SCHECHTER, SCHWARZ, TANUR, and TOURANGEAU (editors) · Cognition and Survey Research
SNIJKERS, HARALDSEN, JONES, and WILLIMACK · Designing and Conducting Business Surveys
STOOP, BILLIET, KOCH and FITZGERALD · Improving Survey Response: Lessons Learned from the European Social Survey
SUDMAN, BRADBURN, and SCHWARZ · Thinking about Answers: The Application of Cognitive Processes to Survey Methodology
UMBACH (editor) · Survey Research Emerging Issues: New Directions for Institutional Research
VALLIANT, DORFMAN, and ROYALL · Finite Population Sampling and Inference: A Prediction Approach
WALLGREN and WALLGREN · Register-based Statistics: Administrative Data for Statistical Purposes, Second Edition

*Paperbacks are available.

Printed and bound by CPI Group (UK) Ltd, Croydon, CR0 4YY

16/04/2025

14658372-0002